Biomechanics

Motion, Flow, Stress, and Growth

Other titles by the same author:

Biomechanics: Mechanical Properties of Living Tissues (1981)
Biodynamics: Circulation (1984)

Springer

New York
Berlin
Heidelberg
Barcelona
Hong Kong
London
Milan
Paris
Singapore
Tokyo

Y.C. Fung

Biomechanics

Motion, Flow, Stress, and Growth

With 254 Illustrations

 Springer

Y.C. Fung
Department of Applied Mechanics
    and Engineering Science/Bioengineering
University of California, San Diego
La Jolla, CA 92093
USA

Library of Congress Cataloging-in-Publication Data
Fung, Y.C. (Yuan-cheng)
    Biomechanics : motion, flow, stress, and growth / Y.C. Fung.
        p.   cm.
    Includes bibliographical references.
    ISBN 0-387-97124-6 (alk. paper)
        1. Biomechanics.   2. Human mechanics.   3. Hemodynamics.
    4. Biophysics.   I. Title.
    QP303.F86   1990
    591.19′1—dc20   89-22017

Printed on acid-free paper.

Typeset by Asco Trade Typesetting Ltd., Hong Kong.
Printed and bound by Edwards Brothers, Inc., Ann Arbor, MI.
Printed in the United States of America.

9 8 7 6 5 4 3

ISBN 0-387-97124-6
ISBN 3-540-97124-6   SPIN 10833887

Springer-Verlag   New York Berlin Heidelberg
A member of BertelsmannSpringer Science+Business Media GmbH

Dedicated to the Memory of My Father 馮曜字重光 *Fung Chung-Kwang (1896–1952), Educator, Painter.*

He painted this scroll in 1939. He loved monkeys because "they are like children."

Preface

Biomechanics aims to explain the mechanics of life and living. From molecules to organisms, everything must obey the laws of mechanics. Clarification of mechanics clarifies many things. Biomechanics helps us to appreciate life. It sensitizes us to observe nature. It is a tool for design and invention of devices to improve the quality of life. It is a useful tool, a simple tool, a valuable tool, an unavoidable tool. It is a necessary part of biology and engineering.

The method of biomechanics is the method of engineering, which consists of observation, experimentation, theorization, validation, and application. To understand any object, we must know its geometry and materials of construction, the mechanical properties of the materials involved, the governing natural laws, the mathematical formulation of specific problems and their solutions, and the results of validation. Once understood, one goes on to develop applications. In my plan to present an outline of biomechanics, I followed the engineering approach and used three volumes. In the first volume, *Biomechanics: Mechanical Properties of Living Tissues*, the geometrical structure and the rheological properties of various materials, tissues, and organs are presented. In the second volume, *Biodynamics: Circulation*, the physiology of blood circulation is analyzed by the engineering method. In the third volume, the present one, the methods of problem formulation, solution, and validation are further illustrated by studying the motion of man and animals, the internal and external fluid flow, the stress distribution in the bodies, strength of tissues and organs, and the relationship between stress and growth. Thus the three volumes form a unit, although each retains a degree of independence.

The plan of this book is as follows. In Chapter 1, Newton's laws of motion and the basic equations of solid and fluid mechanics are presented and illustrated by biological examples. In Chapter 2, the motion of a system of connected elastic or rigid bodies is considered. Generalized coordinates, generalized forces, and Lagrange's equations are used. This method is especially useful in the analysis of the musculoskeletal system because it offers a systematic way of dealing with the many muscles involved in locomotion.

Then we consider external flow in Chapters 3 and 4, and internal flow in Chapters 5 through 9. External flow is the flow around bodies moving in wind and water, in locomotion, flying, and swimming. Internal flow is the flow of blood in blood vessels, gas in airways, water and other body fluid in interstitial space, urine in kidney, ureter and urethra. Although blood flow is treated in *Biodynamics* (Fung, 1984), a sketch of its salient features is included in Chapters 5 and 6 to make this volume somewhat self-contained, while some new topics are added. Chapter 7 treats the flow of gas in the lung. Chapter 8 derives the basic equations of fluid movement in the interstitial space between blood vessels and cells, as well as the transport mechanisms in cell membranes. Multiphasic mixtures are also discussed. Chapter 9 presents the analysis of fluid movement from capillaries to the interstitial space and lymphatics, the indicator dilution method of measurement, and transport by peristalsis.

The rest of the book is devoted to the biological effects of stress and strain. In Chapter 10 we set down the basic concepts of stress and strain in bodies subjected to large deformation. A simple introduction to the finite deformation theory is offered. I hope to acquaint the reader with the physical meaning of the most important formulas of the theory, and with the distinctions between Green's and Almansi's strains, and Cauchy's, Kirchhoff's, and Lagrange's stresses. With this easy introduction, I trust that the reader will be confident to use these quantities and formulas when dealing with soft tissues.

In Chapter 11, stress and strain in organs is studied. In Chapter 12, the strength of tissues and organs is considered. Trauma due to blunt impacts in airplane or automobile crashes is discussed.

In the last chapter, we consider the phenomena of growth and resorption of cells, tissues, and organs in relation to stress: phenomena such as hypertrophy due to high blood pressure, resorption due to too high or too low a stress, and normal growth with exercise in proper range. These phenomena are controlled by biological, chemical, and physical stimuli, including physical stress and strain. Furthermore, growth or resorption of tissues changes the zero-stress configurations of organs. Conversely, changes of zero-stress configurations are convenient indications of tissue remodeling.

The rapid progress of bioengineering makes a serious attempt at *tissue engineering* possible. A person's own cells may be grown rapidly in a polymer scaffold for use as tissue substitute in surgical repair of serious disease or trauma. Close attention to the study of growth in the future can be predicted.

Completeness is not claimed. Our subject is young, and progress is rapid. Seeking greater permanency, I have limited the scope of this book to the more fundamental aspects of biomechanics. For other aspects and handbook material, the reader must turn to the library. Liberal but selected lists of references are given in each chapter. Problems are proposed for solution, some of them with references to articles in the literature. These problems will serve to broaden the scope of our discussion or point to further applications of the principles, or new directions of research.

When I began writing these books, I made a 10-year plan. I did not anticipate that it would take much longer. The first draft of the manuscript of this volume was completed in 1980. Successive revisions, simplifications, and amplifications were made to reflect the advancing frontiers of our subject, and to gain a greater clarity.

There are many people whose help I must acknowledge. First, on classical subjects I followed classical books, especially Sir James Lighthill's book on Biofluiddynamics, and my friend Chia Shun Yih's book on Fluid Mechanics. Next, on biomechanics research I am indebted to many friends and colleagues who have influenced me greatly: Sidney Sobin, Michael Yen, Benjamin Zweifach, Geert Schmid-Schönbein, Hans Krumhaar, Dick Skalak, Savio Woo, Ted Yao-Tsu Wu, Maw Chang Lee, Zulai Tao, Charles C.J. Chuong, Evan Evans, Frank Yin, Paul Zupkas, and others. To Sid Sobin and Mike Yen, I must record my pleasure of having collaborated with them for more than 20 years. My students Jack Debes, Julius Guccione, Ghassan Kassab, Shu Qian Liu, Bruce Bedford, Carmela Rider, and Nina Tang read the final manuscript and gave me errata and other valuable suggestions. I also wish to express my thanks to the many authors and publishers who permitted me to quote from their publications and reproduce their figures and data in this book, especially to Professors Arnold Kuethe on flying, and James Hay on sports techniques. To my students at Caltech, UCSD, and several universities in China, I want to say thank you. You shaped this book by your questions, your enthusiasm, and your dismay. I am grateful to Perne Whaley for typing the manuscript.

La Jolla, California   Yuan-Cheng Fung

Caligraphy by
鄭燮字板橋
Cheng Sih, alias
Banchiao (1693–1765)
"On Literary Style."
Inscription says:
"Simplified as trees
in late Fall,
New as flowers
before Spring."

Contents

Chapter 3
External Flow: Fluid Dynamic Forces Acting on Moving Bodies   62

Chapter 4
Flying and Swimming   106

Chapter 5
Blood Flow in Heart, Lung, Arteries, and Veins   155

CHAPTER 1

Motion

1.1 Introduction

To live is to move. In this book we consider locomotion, motion of organs, motion of fluids around the body as in swimming and flying, and motion of fluids inside the body, such as gas in respiration, blood in circulation, body fluids in tissues. We then go on to consider stresses in the body, the effect of stresses on the physiological function of the organs, the pathological development when stresses are either too large or too small, and biological reaction to stresses in the phenomenon of growth and change.

To analyze motion we accept the following axioms. (1) The space is Euclidean, and time is an independent variable. (2) Each material particle has a mass m, which is a constant, invariable with respect to time. (3) The location of a particle can be described relative to an *inertial* rectangular Cartesian frame of reference, with respect to which the Newtonian equations of motion listed below are valid. Let the position of the particle be denoted by a vector* $\mathbf{x}$. Let t be the time. Let the velocity and acceleration vectors be defined by the equations:

$$\mathbf{v} = \frac{d\mathbf{x}}{dt}, \qquad \mathbf{a} = \frac{d\mathbf{v}}{dt}. \tag{1}$$

The concept of *force* is introduced through Newton's law. Let $\mathbf{F}$ be a *force* acting on the particle. Then *Newton's first law* states that if $\mathbf{F} = 0$, then

$$\mathbf{v} = \text{constant.} \tag{2}$$

---

* Vectors are represented by boldface characters in this book.

If $F \neq 0$, then *Newton's second law* states that

$$m\frac{d\mathbf{v}}{dt} = \mathbf{F} \quad \text{or} \quad \mathbf{F} = m\mathbf{a}. \tag{3}$$

When Eq. (3) is rewritten as

$$\mathbf{F} + (-m\mathbf{a}) = 0, \tag{4}$$

it appears as an equation of equilibrium of two forces. The term $-m\mathbf{a}$ is called the *inertial force*. Equation (4) states that the inertial force balances the external force. Stated in this way, Newton's law is called *D'Alembert's principle*.

Now consider a body that is composed of a system of particles that interact with each other. Every particle is influenced by all the other particles in the system. Let an index I denote the Ith particle. Let $\mathbf{F}_{IJ}$ denote the force of interaction exerted by the Jth particle on the Ith particle, and $\mathbf{F}_{JI}$, that of the Ith particle on the Jth particle, $I \neq J$. Then Newton's third law states that

$$\mathbf{F}_{IJ} = -\mathbf{F}_{JI} \quad \text{or} \quad \mathbf{F}_{IJ} + \mathbf{F}_{JI} = 0. \tag{5}$$

If $I = J$ we set $\mathbf{F}_{II} = 0$. Let K be the total number of particles in the system. The force $\mathbf{F}_I$ that acts on the Ith particle consists of an external force $\mathbf{F}_I^{(e)}$, such as gravity, and an internal force that is the resultant of the mutual interactions between the particles of the system. Thus

$$\mathbf{F}_I = \mathbf{F}_I^{(e)} + \sum_{J=1}^{K} \mathbf{F}_{IJ}. \tag{6}$$

The equation of motion of the Ith particle is, therefore,

$$m_I \frac{d\mathbf{v}_I}{dt} = \mathbf{F}_I^{(e)} + \sum_{J=1}^{K} \mathbf{F}_{IJ}, \qquad (I = 1, 2, \ldots, K). \tag{7}$$

Each particle is governed by such an equation. The totality of the K equations describes the motion of the system.

To make further progress we must specify how the forces of interaction $\mathbf{F}_{IJ}$ can be computed. Such a specification is a statement of the property of the system of particles, and is referred to as a *constitutive equation of the system*.

If the number of particles is very large, then it is awkward to deal with the large number of equations of type (7). One then asks whether it is possible to consider the system as a continuum, whether there exists a simplified way to write down the constitutive equation, and whether one can replace the system of ordinary differential equations (7) with a partial differential equation. The answer was provided by Euler, see Sec. 1.7.

1.2 Equilibrium

A special type of motion is equilibrium, i.e., one in which there is no acceleration for any particle in the body. At equilibrium, Eq. (7) of Sec. 1.1 (henceforth designated as Eq. 1.1:7) becomes

$$F_I^{(e)} + \sum_{J=1}^{K} F_{IJ} = 0, \qquad (I = 1, 2, \ldots, K). \tag{1}$$

Summing over I from 1 to K yields

$$\sum_{I=1}^{K} F_I^{(e)} + \sum_{I=1}^{K} \sum_{J=I}^{K} F_{IJ} = 0. \tag{2}$$

In the last sum, whenever F_{IJ} appears, F_{JI} also appears; according to Eq. (1.1:5) they add up to zero. Therefore, Eq. (2) is reduced to

$$\sum_{I=1}^{K} F_I^{(e)} = 0. \tag{3}$$

Thus, *for a body in equilibrium, the sum of all external forces acting on the body is zero.*

Next let us consider the tendency of a body to rotate. A body rotates if there is a *couple* acting on it. A couple is a pair of forces that are equal in magnitude, opposite in direction, and separated by a certain distance. In Fig. 1.2:1(a), the couple is formed by the forces F and $-F$, at a distance D apart. The product FD is the *moment* of the couple, where F is the magnitude of F.

If the couple in Fig. 1.2:1(a) is applied to a free body as in Fig. 1.2:1(b) then the body will rotate. Evidently we would have to distinguish the direction of rotation. We therefore define a moment as a vector, as in Fig. 1.2:1(c), whose magnitude is FD, whose direction is perpendicular to the plane containing the forces F and $-F$, and whose sense is determined by the right-handed screw rule. Finally, consider a body pivoted at a point P as in Fig. 1.2:1(d). When a force F acts on this body, the body tends to rotate about the pivot P. If we add a pair of equal and opposite forces at P, as in Fig. 1.2:1(c), we see that the

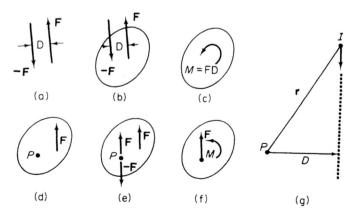

FIGURE 1.2:1 Couples and moments.

action of **F** is equivalent to a force **F** and a moment FD (Fig. 1.2:1(f)). Thus, the moment FD is the cause for rotation about P.

The moment of a force **F** about a point P is easily expressed in vector notation. Let **F** be the force vector and I be a point on the force vector (see, e.g., Fig. 1.2:1(g)). Let **r** be the position vector from P to I. Then the moment of **F** about P is

$$\mathbf{M} = \mathbf{r} \times \mathbf{F}, \tag{4}$$

where the symbol $\times$ denotes a vector product.

Now consider a body that is acted on by a system of external and internal forces. The moment of all the forces about a point P is the sum of the vector products of r_I and the individual forces, $\mathbf{F}^{(e)}$ and $\mathbf{F}_{IJ}$ for I and J varying from 1 to K. But the internal forces $\mathbf{F}_{IJ}$ and $\mathbf{F}_{JI}$ occur in equal and opposite pairs, so that the sum of their moments about P vanishes. Hence the total moment is

$$\mathbf{M} = \sum_{I=1}^{K} \mathbf{r}_I \times \mathbf{F}_I^{(e)}. \tag{5}$$

If the body is in equilibrium then there is no tendency to rotate, and $\mathbf{M} = 0$. Hence we obtain the second necessary condition of equilibrium: *The sum of the moments of all the external forces acting on the body about an arbitrary point is zero.*

We may rewrite the conditions of equilibrium in terms of the components of the forces and moments as follows: Let the components of **F** in the x, y, z directions of an inertial frame of reference be written as F_x, F_y, F_z and those of the moment **M** as M_x, M_y, M_z; then the conditions of equilibrium are

$$\sum F_x = 0, \qquad \sum F_y = 0, \qquad \sum F_z = 0, \tag{6}$$

$$\sum M_x = 0, \qquad \sum M_y = 0, \qquad \sum M_z = 0. \tag{7}$$

The moments are obtained by a vector multiplication of the radius vector **r** and the force **F**, Eq. (4). If the components of the radius vector **r** are written as x, y, z, then

$$M_x = yF_z - zF_y, \qquad M_y = zF_x - xF_z, \qquad M_z = xF_y - yF_x. \tag{8}$$

1.3 Dynamics

The motion of a particle in a body is governed by Eq. 1.1:7. By summing these equations for all particles of the body, and noting that the sum of the last terms vanishes as in Eqs. (1.2:2) and (1.2:3), one obtains

$$\sum_{I=1}^{k} m_I \mathbf{a}_I = \sum_{I=1}^{k} \mathbf{F}_I^{(e)}, \tag{1}$$

where $\mathbf{a}_I$ is the acceleration of the particle I, i.e., dv_I/dt, and k is the total number of the particles.

This equation is simplified by introducing the concept of *center of mass* (i.e., *center of gravity*). If we have a set of masses $m_I (I = 1, 2, \ldots, k)$ located at places $\mathbf{r}_I$ with coordinates (x_I, y_I, z_I) in reference to a rectangular Cartesian frame of reference, then the radius vector of the center of mass $\mathbf{r}_{C.G.}$ with components $x_{C.G.}, y_{C.G.}, z_{C.G.}$, is defined by the equations:

$$\sum_I m_I x_I = m x_{C.G.}, \qquad \sum_I m_I y_I = m y_{C.G.}, \qquad \sum_I m_I z_I = m z_{C.G.}, \qquad (2)$$

where

$$m = \sum_I m_I \qquad (3)$$

is the *total mass* of the body. If we differentiate both sides of Eqs. (2) with respect to time, we obtain equations which say that the sum of the momentum of a set of particles is equal to the product of the total mass of the particles and the velocity of the center of mass. If we differentiate the resulting equations again with respect to time, we obtain the result:

$$\sum_I m_I \mathbf{a}_I = m \mathbf{a}_{C.G.}. \qquad (4)$$

Here m is the total mass of the body, and $\mathbf{a}_{C.G.}$ is the acceleration of the *center of mass* of the body. Then Eq. (1) becomes

$$m \mathbf{a}_{C.G.} = \sum_I \mathbf{F}_I^{(e)}. \qquad (5)$$

This states that the *total mass multiplied by the acceleration of the center of mass is equal to the sum of all the external forces acting on the body.*

To obtain an equation that describes the tendency of the body to rotate about the center of mass, we use the same method that was used in deriving Eq. (1.2:5). Let x, y, z be an inertial frame of reference in which Newton's laws are valid, Fig. 1.3:1. Let 0 be a *fixed* origin. Let $\mathbf{r}_{C.G.}$ and $\mathbf{r}_I$ be the radius

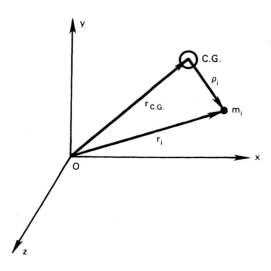

FIGURE 1.3:1 Position vectors referred to a fixed frame of reference x, y, z, and the center of mass (C.G.).

vectors from 0 to the center of mass (C.G.) and the Ith particle, respectively. Let $\boldsymbol{\rho}_I$ be the position vector from the C.G. to the Ith particle. Then

$$\mathbf{r}_I = \mathbf{r}_{\text{C.G.}} + \boldsymbol{\rho}_I. \tag{6}$$

By differentiating with respect to time and denoting the derivative by a dot, we obtain

$$\mathbf{v}_I = \dot{\mathbf{r}}_I = \dot{\mathbf{r}}_{\text{C.G.}} + \dot{\boldsymbol{\rho}}_I. \tag{7}$$

Equation (1.1:7) then reads

$$\frac{d}{dt}[m_I(\dot{\mathbf{r}}_{\text{C.G.}} + \dot{\boldsymbol{\rho}}_I)] = \mathbf{F}_I^{(e)} + \sum_{J=1}^{K} \mathbf{F}_{IJ}. \tag{8}$$

Forming a vector product with $\boldsymbol{\rho}_I$ and then summing over I, we obtain

$$\sum_I \boldsymbol{\rho}_I \times \frac{d}{dt}[m_I(\dot{\mathbf{r}}_{\text{C.G.}} + \dot{\boldsymbol{\rho}}_I)] = \sum_I \boldsymbol{\rho}_I \times \mathbf{F}_I^{(e)}, \tag{9}$$

because the summation over the internal forces $\mathbf{F}_{IJ}$ again vanishes in pairs. The term on the left-hand side is

$$\sum_I m_I \boldsymbol{\rho}_I \times \frac{d}{dt}\dot{\mathbf{r}}_{\text{C.G.}} + \sum_I m_I \boldsymbol{\rho}_I \times \frac{d}{dt}\dot{\boldsymbol{\rho}}_I.$$

The first term vanishes by the definition of the C.G. The second term is equal to

$$\frac{d}{dt}\sum_I m_I \boldsymbol{\rho}_I \times \dot{\boldsymbol{\rho}}_I - \sum_I m_I \dot{\boldsymbol{\rho}}_I \times \dot{\boldsymbol{\rho}}_I \tag{10}$$

as can be seen by carrying out the differentiation in Eq. (10). But the last term in Eq. (10) is identically zero. Hence Eq. (9) becomes

$$\frac{d}{dt}\left(\sum_I m_I \boldsymbol{\rho}_I \times \dot{\boldsymbol{\rho}}_I\right) = \sum \boldsymbol{\rho}_I \times \mathbf{F}_I^{(e)}. \tag{11}$$

The quantity:

$$\sum_I m_I \boldsymbol{\rho}_I \times \dot{\boldsymbol{\rho}}_I \equiv \mathbf{L}_c \tag{12}$$

is called the *angular momentum* (or *moment of momentum*) of the body about the center of mass. Thus Eq. (11) states that the *moment of the external forces acting on the body about the center of mass is equal to the time rate of change of the angular momentum of the body about the center of mass.*

As a special case, we have the *conservation theorem of angular momentum: With the absence of external moments about the center of mass the total angular momentum about the C.G. is conserved.*

Equations (5) and (11) are useful in biomechanical studies of crawling, walking, running, jumping, swimming, gait, and sports. They can be applied to the whole animal, but can be used equally well for an arm, a leg, a bone. Some examples will be discussed in the following sections.

An entirely similar derivation can be made to prove an angular momentum theorem in which the words "the center of mass" are replaced by "an arbitrary fixed point." Thus, with $\mathbf{r}$ denoting a radius vector from a fixed point 0 (Fig. 1.3:1), we have

$$\frac{d}{dt}\left(\sum_I m_I \mathbf{r}_I \times \dot{\mathbf{r}}_I\right) = \sum_I \mathbf{r}_I \times \mathbf{F}_I^{(e)}. \tag{13}$$

If the system of particles forms a *rigid body*, then the angular momentum can be expressed as the product of the *moment of inertia* and the *angular velocity*. In this case, we use a system of coordinates whose origin coincides with the center of mass and whose axes are parallel to a set of fixed inertial coordinates x, y, z. Let the angular velocity of the body be denoted by $\boldsymbol{\omega}$. Then the velocity of a point of the body relative to the C.G. is

$$\dot{\boldsymbol{\rho}}_I = \dot{\mathbf{r}}_I - \dot{\mathbf{r}}_{\text{C.G.}} = \boldsymbol{\omega} \times \boldsymbol{\rho}_I. \tag{14}$$

Now, by vector analysis,

$$(\boldsymbol{\rho} \times \dot{\boldsymbol{\rho}}) = \boldsymbol{\rho} \times (\boldsymbol{\omega} \times \boldsymbol{\rho}) = (\boldsymbol{\rho} \cdot \boldsymbol{\rho})\boldsymbol{\omega} - (\boldsymbol{\rho} \cdot \boldsymbol{\omega})\boldsymbol{\rho}. \tag{15}$$

Substituting (15) into (12), we obtain the angular momentum:

$$\mathbf{L}_c = \sum_I m_I \boldsymbol{\rho}_I \times \dot{\boldsymbol{\rho}}_I = \sum_I m_I(\boldsymbol{\rho}_I \cdot \boldsymbol{\rho}_I)\boldsymbol{\omega} - \sum_I m_I(\boldsymbol{\rho}_I \cdot \boldsymbol{\omega})\boldsymbol{\rho}_I. \tag{16}$$

This vector equation can be rewritten in terms of components relative to a set of rectangular Cartesian coordinates. With unit vectors $\mathbf{i}, \mathbf{j}, \mathbf{k}$ in the direction of the x, y, z axes, respectively, we have

$$\boldsymbol{\rho} = x\mathbf{i} + y\mathbf{j} + z\mathbf{k}, \qquad \boldsymbol{\omega} = \omega_x\mathbf{i} + \omega_y\mathbf{j} + \omega_z\mathbf{k}. \tag{17}$$

Then, from Eq. (15),

$$\begin{aligned}
(\boldsymbol{\rho} \times \dot{\boldsymbol{\rho}}) &= (x^2 + y^2 + z^2)\boldsymbol{\omega} - (x\omega_x + y\omega_y + z\omega_z)\boldsymbol{\rho} \\
&= [(y^2 + z^2)\omega_x - xy\omega_y - xz\omega_z]\mathbf{i} \\
&\quad + [-yx\omega_x + (x^2 + z^2)\omega_y - yz\omega_z]\mathbf{j} \\
&\quad + [-zx\omega_x - zy\omega_y + (x^2 + y^2)\omega_z]\mathbf{k},
\end{aligned} \tag{18}$$

and it can be shown that Eq. (16) may be written as a matrix equation:

$$\begin{bmatrix} (L_c)_x \\ (L_c)_y \\ (L_c)_z \end{bmatrix} = \begin{bmatrix} I_{xx} & -I_{xy} & -I_{xz} \\ -I_{yx} & I_{yy} & -I_{yz} \\ -I_{zx} & -I_{zy} & I_{zz} \end{bmatrix} \begin{bmatrix} \omega_x \\ \omega_y \\ \omega_z \end{bmatrix}, \tag{19}$$

where

$$I_{xx} = \sum (y_I^2 + z_I^2)m_I, \qquad I_{yy} = \sum (z_I^2 + x_I^2)m_I, \qquad I_{zz} = \sum (x_I^2 + y_I^2)m_I \tag{20}$$

are called the *moments of inertia*, and

$$I_{xy} = \sum x_I y_I m_I, \qquad I_{yz} = \sum y_I z_I m_I, \qquad I_{zx} = \sum z_I x_I m_I \tag{21}$$

are called the *products of inertia*. Together they are the *moment of inertia tensor*. If the mass in the rigid body is continuously distributed, then the summation in the equations above can be replaced by integrals. In particular, we may write

$$I_{xx} = \int_V (y^2 + z^2)\,dm, \qquad I_{xy} = \int_V xy\,dm, \tag{22}$$

etc., the integration being taken over the entire volume of the body, V, and x, y, z being the location of the mass element dm.

Many of the equations above can be written in a simpler form if we introduce the *index notation* and *summation convention*. A set of variables x_1, x_2, x_3 is denoted as x_i, $i = 1, 2, 3$. When written singly, the symbol x_i stands for anyone of the variables x_i, x_2, x_3. The symbol i is an index. In a product such as $a_i x_i$, the summation convention means that the repetition of an index in a term denotes a summation with respect to that index over its range. Thus

$$a_i x_i = a_1 x_1 + a_2 x_2 + a_3 x_3 = \sum_{i=1}^{3} a_i x_i. \tag{23}$$

Introducing the index notation by writing $I_{11} = I_{xx}$, $I_{12} = -I_{xy}$, etc., and using the summation convention, we may write Eq. (19) as

$$(L_c)_i = I_{ij}\omega_j, \tag{24}$$

and Eq. (11) becomes

$$\frac{d}{dt}(I_{ij}\omega_j) = \left(\sum_I \rho_I \times F_I^{(e)}\right)_i = e_{ijk}\rho_{Ij}F_{Ik}^{(e)}, \tag{25}$$

where e_{ijk} is the permutation symbol, which is defined by equations such as

$$e_{111} = e_{222} = e_{333} = e_{112} = e_{113} = e_{221} = e_{223} = e_{331} = e_{332} = 0$$
$$e_{123} = e_{231} = e_{312} = 1 \tag{26}$$
$$e_{213} = e_{321} = e_{132} = -1$$

i.e., e_{ijk} vanishes when the values of any two indices coincide; $e_{ijk} = 1$ when the subscripts permute like 1, 2, 3; and $e_{ijk} = -1$ otherwise.

If the moment of inertia tensor I_{ij} does not change with time, then we have

$$I_{ij}\dot{\omega}_j = \left(\sum_I \rho_I \times \mathbf{F}_I^{(e)}\right)_i = e_{ijk}\rho_{Ij}F_{Ik}^{(e)}. \tag{27}$$

Equations (5) and (27) describe the mechanics of rigid bodies completely. For rigid bodies there is no need to know the forces of interaction, F_{IJ}, between the particles of the body. If the body is not rigid, and if one wishes to analyze the motion of a particle in the body relative to others, then we must return to Eq. (1.1:7). In that case, we must know the forces of interaction, F_{IJ}.

Many machines (e.g., an automobile engine, an electric motor, a bicycle) can be considered an assemblage of rigid bodies connected by joints, springs, and dashpots. The forces of interaction between the individual rigid parts are often the unknown variables in the analysis of the motion of the machine. In analyzing the locomotion of an animal, the body of the animal may be treated in a similar manner.

1.4 Modeling

Sometimes it is easy to make a mathematical model of a natural system, e.g., an arm holding a load shown in Fig. 1.4:1. The flexor muscle force that balances the load and the weight of the forearm are computed easily. However, modeling the chest for the analysis of breathing is not so easy, because the large number of muscles that control the chest motion is difficult to idealize. Therefore creating a mathematical model that simplifies the real thing while retaining all the essential factors is considered an art that must be based on experience and careful thought.

An example of difficult modeling is that of the human lumbar spine to which many muscles and ligaments are attached. Nachemson and Elfström (1970) measured the pressure in the *nucleus pulposus*, the gelatinous center of the lumbar disk (L3), with a miniature pressure transducer, and from the data calculated (with the help of in vitro disk experiments) the loads on the lumbar spine. Their results are shown in Table 1.4:1. Note how large the load in the lumbar spine is when one lifts weight the 'wrong way'. Data of this kind

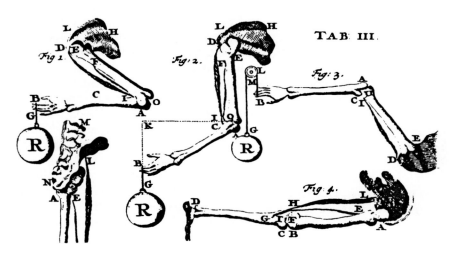

FIGURE 1.4:1 A leaf from Giovanni Alfonso Borelli's book, *On the Movement of Animals*, published in 1680/1681, illustrating his research on biomechanics. Borelli (1608–1679) was professor of mathematics at Pisa. This book was translated recently by Paul Maquet, and published in 1989 by Springer-Verlag.

TABLE 1.4:1 Approximate load on the lumbar disk L3 in a 70 kg individual in different positions, movements and maneuvers. Data from Nachemeson and Elfström (1970)

Position/movement/maneuver	Load, Newton (kgf)
Supine	490 (50)
Supine in traction (30 kg)	343 (35)
Standing	980 (100)
Upright sitting, no support	1373 (140)
Walking	1128 (115)
Twisting	1177 (120)
Bending sideways	1225 (125)
Coughing	1373 (140)
Jumping	1373 (140)
Isometric abdominal muscle exercise	1373 (140)
Laughing	1471 (150)
Bending forward 20°	1471 (150)
Bending forward 20° with 10 kg in each hand	2108 (215)
Lifting 20 kg, back straight, knees bent	1814 (185)
Lifting 20 kg, back bent, knees straight	3825 (390)
Sit-up exercise, knees bent	2059 (210)
Sit-up exercise, knees extended	2010 (205)
Active back hyperextension, prone	1765 (180)

are significant when one tries to understand the etiology of the low back pain which afflicts many people.

Schultz (1986) approached the same problem by measuring the myoelectric activity, the electric signals sent through nerves that cause muscles to contract. There are evidences that under the right circumstances the myoelectric activity is linearly related to muscle contraction forces. His results on the forces in the spine and muscles when a man is attempting maximum backward bend are shown in Table 1.4:2. It is impressive how large these forces can be! Merely bending the trunk forward 30 degrees triples the muscle force that prevails at relaxed-standing condition. Attempting maximum backward bending caused compression in the spine ten times the relaxed standing value; and very large tension in the major trunk muscles.

As a further illustration of the complexity of modeling, consider *referred pains* which are aches and pains originating in a distant organ. Familiar examples are pain in the shoulder accompanying diaphragmatic disorders, pain in the knee and arthritis of the hip, the sacral (lower back, just above buttocks) pains of childbirth, pain in the left shoulder and left arm in angina pectoris, and the testicular pain referred from lower back. One of the first crucial experiments on this subject was done by Thomas Lewis in 1936. Wishing to investigate muscular pain, he injected an irritant deep into the lower lumbar region. He found a diffuse pain running down the lower limb but little

TABLE 1.4:2 Typical magnitude of model-estimated lumbar trunk internal forces imposed by physical exertion when an adult man stands and attempts maximum backward bend. (From Schultz, 1986)

Forces in the spine	Newton
Compression	5,050
Forward/back shear	600
Side/side shear	370
Contraction forces per side in major muscles	
Rectus abdominus	340
Oblique abdominals (sum of 4)	1,700
Erector spinae	
multifidus	320
longissimus	630
iliocostalis	630
Lumbar latissimus dorsi	150
Quadratus lumborum	230
Psoas	440

discomfort at the site of the injection (Lewis, 1942). In 1938, Kellgren published the results of a systematic examination of the phenomenon of referred pain, showing them to radiate segmentally and not to cross the midline (Kellgren, 1938). Cyriax (1978) presents considerable details of this subject.

Referred pain is a subject for biomechanics because pressure on nerves is often the cause. When pressure is applied close to a nerve's distal extremity, the sensation of pins and needles is felt, but the main symptom is numbness. When pressure is applied on the trunk of a nerve and then released, the sensation is pins and needles rather than numbness. But if pressure is applied on a nerve root in the spine, the sensation of pins and needles is felt only as long as the pressure is sustained, but disappears as soon as the pressure is relieved. Stretching a nerve root is painful, and a common cause is a disk protrusion.

Figure 1.4:2 illustrates the concept of a cartilaginous disk lesion. An annular crack has led to a posterior displacement by hinging. The posterior longitudinal ligament is bulged out backward and pressure is exerted on the dura mater, causing a backache. No nuclear material has extruded, hence reduction by manipulation is simple.

Figure 1.4:3 illustrates the concept that a disk lesion may push the posterior ligament until it presses on the dura mater, causing backache. If the protrusion proceeds posterolaterally and impinges on the nerve root, then pain will be felt in the limb. In these cases the treatment of choice is a sustained traction. Spontaneous recovery is possible if the posteroinferior aspect of the vertebral bone is resorbed and the protrusion is accommodated so that the dura mater and nerve roots are no longer subjected to pressure.

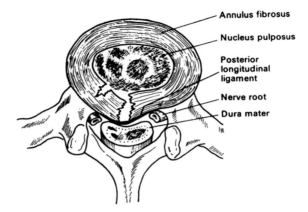

FIGURE 1.4:2 A cartilaginous disk lesion. From Cyriax (1978). Reproduced by permission.

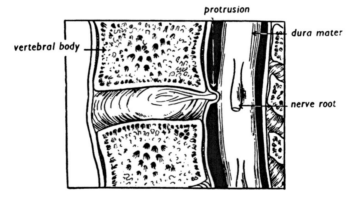

FIGURE 1.4:3 Illustration of a pulp and disk lesion. From Cyriax (1978). Reproduced by permission.

To model referred pain, we must consider the whole system: the organ in pain and the nervous lesion. The analysis of a good model may lead to designs to relieve such pains.

Further discussion of modeling is presented in Secs. 2.9 and 12.11.

1.5 Sports Techniques

In jumping, there are phases when the body is in free flight and the trajectory of the body's center of gravity is a parabola which is determined solely by the initial conditions. The trajectory cannot be changed by the athlete's action while in flight, but the body position can. Controlling the body position is of course important. Let us illustrate:

Long Jump

In the long jump, the distance covered consists of three parts (Fig. 1.5:1): L_1, the horizontal distance between the front edge of the takeoff board and the athlete's center of gravity at the instant of takeoff; L_2, the flight distance; and L_3, the distance between his C.G. at the instant his heels hit the sand and the marks in the sand. The records of good athletes show that L_1 is usually about 3.5% of the total, L_2 is about 88.5%, and L_3 is about 8% (see Hay, 1978, p. 409).

The takeoff distance L_1 is a function of the accuracy with which the athlete places his foot on the takeoff board, his physique, and his body position. The flight distance L_2 depends on his speed, angle of takeoff, height of takeoff, and the air resistance. The landing distance L_3 depends on his body position at touchdown and the actions he takes to avoid falling backward.

The most important of these variables is the athlete's speed at the instant of takeoff. This speed depends on his running and on the losses associated with the adjustments made when preparing for takeoff. The ratio of the vertical speed (or lift) acquired at takeoff to the horizontal speed is the tangent of the angle of takeoff. It is well known that, for a given initial speed, the largest distance is achieved by any projectile when the angle of takeoff is 45°. However, the lift that the athlete develops at takeoff is influenced by the speed of his run-up. The faster his run, the less time his foot spends on the ground at takeoff and the less vertical speed he can develop. Thus, the angle of takeoff used by top-class jumpers lies in the range of 19–22°.

Once the athlete is in the air, he should try to obtain the optimum body position for landing. Usually the jumper acquires a forward rotation of the body at takeoff, which tends to bring his feet beneath his center of gravity at the very instant of landing, while he actually wants them to be well forward. Thus, the athlete's principal problem is to overcome this forward rotation.

FIGURE 1.5:1 Contributions to the length of a hang-style long jump. From J. G. Hay (1978), p. 409. Reproduced by permission.

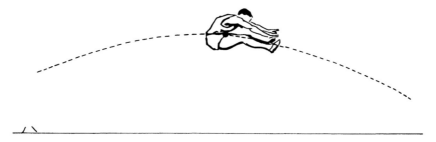

FIGURE 1.5:2 The in-the-air position adopted in the sail technique. From J. G. Hay (1978), p. 415. Reproduced by permission.

Compare the two techniques used in the long jump as illustrated in Figs. 1.5:1 and 1.5:2. The *sail* technique (Fig. 1.5:2) is used by most jumpers. Its weakness is that it places the athlete's mass close to his transverse axis and thus obtains a small moment of inertia, facilitating the forward rotation instead of retarding it. In the *hang* technique (Fig. 1.5:1) the athlete reaches forward with his leading leg and then sweeps it downward and backward until he has both legs together and somewhat behind the line of his body. He swings his arms backward, too. This results in an increase in the moment of inertia and a slowing of forward rotation. He continues to swing his arms upward, then bends his knees and begins the forward movement of his legs in preparation for landing. At the time of landing, the *extended* position, (trunk inclined slightly backward and hands beside hips), as shown in Fig. 1.5:1, is preferable to the *jackknife* position (trunk inclined forward and arms extended toward the feet), as shown in Fig. 1.5:2. Studies show that the extended position gives the advantage of approximately 12 more inches of jumping distance than the jackknife position.

There is a third in-the-air technique called the *hitch-kick*, or the *run-in-the-air*. Its first part is similar to that of the hang technique. Coordinated with this movement is a pulling through of the takeoff leg. Because of the difference between the moments of inertia of the extended and flexed legs, the angular momentum of the leading leg as it swings downward and backward far exceeds that of the takeoff leg that is moving in the opposite direction. As a result of the preservation of the total angular momentum, the trunk rotates backward, which is, of course, the objective to be achieved. Now, the athlete's legs are in a position that is essentially the reverse of that at takeoff. At this point, the leg that is to the rear is brought forward, with the knee fully-flexed, to join the other in preparation for landing. For a longer jump an additional full stride may be attempted. The hitch-kick method is used by high performers.

High Jump

In the high jump the height that an athlete clears is the sum of three heights: the height of the athlete's center of gravity (C.G.) at the instant of takeoff, the

TABLE 1.5:1 Relative contributions to height in the high jump

The three parts of performance	Height (cm)	Percentage
Height of C.G. at takeoff	144.0	67.5%
Height C.G. is lifted	78.2	36.6%
Height cleared — max C.G. height	−8.6	−4.1%

height of his C.G. raised during the flight, and the difference between the maximum height reached by his C.G. and the height of the crossbar. Hay (1978) gives the data for a jump of 7 feet by then-world-record-holder Pat Matzdorf in Table 1.5:1. According to the table the height of C.G. at takeoff is very important. This explains why high jumpers are tall men with long legs. The height of his flight is governed by his vertical velocity at takeoff, which depends on his actions during the last one or two strides of his run-up. If at the end of his penultimate stride he has sunk low over his supporting leg and then takes a low, fast step with his takeoff foot, his C.G. is likely to have no downward vertical velocity at the instant this foot touches down. Then all his effort will be used to raise his C.G. This is far better than if he sinks down with the penultimate stride.

The vertical force involved at takeoff results from the swing of his arms and leading leg and from the extension of the hip, knee, and ankle joints of his jump leg. The period of time during which the jump foot is in contact with the ground depends on the style of jumping. Athletes who use the straddle

FIGURE 1.5:3 A frontal approach and a front-piked position over the bar suggested by J. G. Hay (1978), pg. 433. Reproduced by permission.

style have takeoff times in the 0.17–0.23 sec range, whereas those who use the flop style, popularized by Fosbury, tend to have takeoff times in the 0.12–0.17 sec range. It also depends on the free limbs. Athletes who use a straight leading leg action generally have a longer time of takeoff than do those who use a bent leading leg. But in spite of the fact that impulse = force × time, it has been found that, within limits that are specific to the individual athlete, the shorter the time of takeoff the greater is the vertical lift (impulse) that the athlete obtains. This fact is hard to explain unless it is a feature of the muscle.

With the common styles: *scissors, straddle, cutoff*, and *roll*, the maximum C.G. height is higher than the height cleared, (Table 1.5:1). In 1970 Dick Fosbury originated the *Fosbury flop*, in which the athlete arches backward over the bar in such a way that his C.G. lies outside his body. Hay (1978) then proposed a similar method shown in Fig. 1.5:3, bending the upper body *forward* toward the feet. Note that the trajectory of the C.G. does not clear the crossbar while the athlete's body does.

Use of Friction

If two bodies are in contact along dry surfaces, the limiting friction is equal to the normal reaction (force of interaction between the two bodies in the direction normal to the surface of contact) multiplied by a constant whose value depends only on the nature of the surfaces. Thus,

$$F = \mu N, \tag{1}$$

where F = the limiting friction, N = the normal reaction, and μ = a constant called the *coefficient of limiting friction*. If the tangential force between the surfaces is less than the limiting friction, F, the bodies will remain stationary relative to each other. If F is exceeded, relative motion ensues. When a body is actually sliding on the surface of another body, the magnitude of the friction is given by the equation

$$F_s = \mu_s N, \tag{2}$$

where F_s = sliding friction, N = the normal reaction, and μ_s = a constant called the *coefficient of sliding friction*. μ_s is smaller than μ. Our shoes on solid ground have a coefficient of limiting friction of the order of one. The surfaces with the smallest coefficient of sliding friction are probably those of the articular cartilage surfaces of our joints. In *Biomechanics* (Fung, 1981, p. 410) we have shown that in cyclic loading a whole bovine synovial joint has a coefficient of sliding friction of 0.0026 at a normal stress of 500 kPa (~ 5 kgf/cm^2).

All athletes pay attention to ground friction by choosing suitable shoes according to the nature of sports and the condition of ground surface. Shoes affect the coefficient of friction. The other factor that can be controlled to a certain extent is the normal force N. N is affected by the athlete's posture and body dynamics. For example, a rock climber can increase the limiting friction between his feet and the rock by leaning well away from the rock face (see Prob. 1.11), because N depends on the location of the center of gravity of his body and the direction of the rope used for the climb.

The dynamics of the athlete also affects the limiting friction because in arresting a downward movement toward the ground, an upward inertial force must be created by the ground reaction. Thus the ground reaction acting on the feet of a man running exceeds that when he is standing, and the technique of running does have an influence on ground friction.

Use of Hands in Swimming

The forces exerted on the swimmer in reaction to the movements of his arms is a prime source of propulsion. Normally, this propulsive force is equal to the drag force acting on the swimmer's arms. However, Counsilman (1971) and Brown and Counsilman (1971) have suggested that lift may be used. With a small angle of attack, the hand can function as a hydrofoil. As a hydrofoil, the lift is larger than the drag, and can be used advantageously with a proper arm cycle. In a breaststroke cycle, the hands move forward, outward, inward, and forward again. Now, with an appropriate angle of attack, the lift force acting on the hands can be directed forward and thus become propulsive. Similar circumstances can be shown to exist in other competitive strokes: front crawl, butterfly stroke, and back crawl. The same principle can be applied to the propulsion derived from the legs in applicable circumstances.

Birds, insects, and fish use lift force for their locomotion. More details about the fluid mechanical principles of lift generation are given in Chapters 3 and 4.

1.6 Prosthesis

A prosthesis, by definition, replaces a missing body part for structural, functional, or cosmetic reasons. Thus, dentures, artificial heart valves, artificial limbs are prostheses. Most of them require extensive biomechanical research before an optimal design can be obtained.

Take an artificial leg or a joint replacement as an example. To make such a prosthesis, we must know the force that acts in it as the person moves. This force depends on the adjacent musculature. Records of the variation with time of the forces in hip and knee joints during level walking show two and three maximum values respectively in the stance phase of each cycle. An indication of the variation in joint forces with body weight W and the ratio of stride length L to height H is given in Figs. 1.6:1 and 1.6:2. Here the average of the maxima is plotted to a base of WL/H (Paul, 1970). Note how large the joint forces are. They are much larger than the body weight. The increase is due to the action of the muscles, which must keep the basically unstable structure in balance.

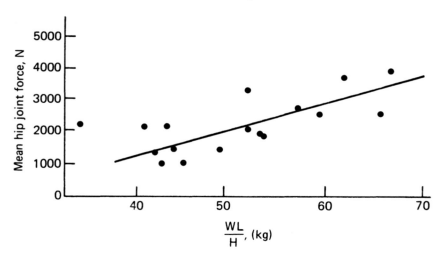

FIGURE 1.6:1 Variation of mean hip joint force in level walking with body weight W, stride length L, and height H. From J. P. Paul (1970). Reproduced by permission.

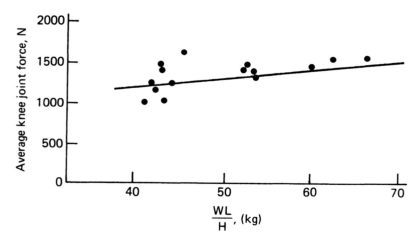

FIGURE 1.6:2 Variation of mean knee force in level walking with body weight W, stride length L, and height H. From J. P. Paul (1970). Reproduced with permission.

1.7 Continuum Approach

In the preceding sections, we presented Newton's equations of motion for a system of particles in general, and rigid bodies in particular. In biomechanics, there are many occasions (e.g., in the study of locomotion) in which we can regard an animal as an assemblage of rigid bodies connected by joints and muscles. If the mechanical properties of the connectors are known, then a system of equations of the type (1.3:5) and (1.3:25) will suffice to describe the motion. There are other occasions, however, in which this is not efficient. For example, we may wish to know the stresses in the bones, muscles, spine, heart, or brain in a man when walking. Then it is no longer convenient to write the equations of motion in the form of (1.1:7), because the number of particles to be considered is excessively large, and the way to specify the forces of interaction between the particles, F_{IJ} of Eq. (1.1:7), becomes very complex. In these occasions it is simpler to consider the body as a continuum, and search for a simplified way to specify the forces of interaction between particles by means of constitutive equations.

The equation of motion of a continuum was derived by Euler, by applying Newton's laws and several additional axioms. These axioms are: (1) The material particles form a one-to-one isomorphism with real numbers in a three-dimensional Euclidean space. (2) The mass distribution is characterized by the density, ρ, the mass per unit volume, defined as a piecewise continuous function over the volume of the continuum. (3) A particle can interact only with its immediate neighbors. "Interaction at a distance" is ruled out. The force of interaction of particles on the two sides of an arbitrary infinitesimal surface in a continuum can be expressed as a surface traction (force per unit area). The surface traction acting on an arbitrarily oriented infinitesimal surface can be computed from a well defined stress tensor.

Let x_1, x_2, x_3 be an inertial rectangular Cartesian frame of reference. Consider an infinitesimal volume $dx_1 dx_2 dx_3$. Let σ_{ij}, e_{ij} be the stress and strain tensors, respectively. Let v_i be the velocity vector, Dv_i/Dt be the acceleration vector, and X_i be the body force per unit volume. All Latin indicies range from 1–3. Then Newton's law takes on the form of *Euler's equation*:

$$\rho \frac{Dv_i}{Dt} = \frac{\partial \sigma_{ij}}{\partial x_j} + X_i, \qquad (i = 1, 2, 3). \tag{1}$$

See a text on continuum mechanics for details, e.g., Fung (1977). In Eq. (1) the summation convention of tensor analysis is used. A repetition of an index in any given term means a summation over the range of that index. Thus,

$$\frac{\partial \sigma_{ij}}{\partial x_j} = \frac{\partial \sigma_{i1}}{\partial x_1} + \frac{\partial \sigma_{i2}}{\partial x_2} + \frac{\partial \sigma_{i3}}{\partial x_3}. \tag{2}$$

$\frac{Dv_i}{\partial t}$ is the *material derivative* of v_i:

$$\frac{Dv_i}{Dt} = \frac{\partial v_i}{\partial t} + v_j \frac{\partial v_i}{\partial x_j} = \frac{\partial v_i}{\partial t} + v_1 \frac{\partial v_i}{\partial x_1} + v_2 \frac{\partial v_i}{\partial x_2} + v_3 \frac{\partial v_i}{\partial x_3}. \tag{3}$$

The conservation of mass is expressed by the *equation of continuity*:

$$\frac{\partial \rho}{\partial t} + \frac{\partial \rho v_j}{\partial x_j} = 0. \tag{4}$$

If the material is *incompressible*, then Eq. (4) is reduced to the form

$$\frac{\partial v_j}{\partial x_j} = 0. \tag{5}$$

Further development requires specification of the properties of the material in the form of a *constitutive equation* which relates stress with strain or strain rate, or strain history. For the convenience of the rest of the book let us collect a few important examples. A material is called a *Newtonian fluid* if it obeys the following stress–strain–rate relationship:

$$\sigma_{ij} = -p\delta_{ij} + \lambda\left(\frac{\partial v_k}{\partial x_k}\right)\delta_{ij} + \mu\left(\frac{\partial v_i}{\partial x_j} + \frac{\partial v_j}{\partial x_i}\right), \tag{6}$$

where p is the pressure, and λ and μ are two constants called the *coefficents of viscosity*. If the fluid is *incompressible*, then according to Eq. (5) the constitutive equation of a Newtonian fluid becomes

$$\sigma_{ij} = -p\delta_{ij} + \mu\left(\frac{\partial v_i}{\partial x_j} + \frac{\partial v_i}{\partial x_i}\right). \tag{7}$$

Substituting Eq. (7) into Eq. (1), one obtains the *Navier–Stokes equation*

$$\frac{\partial u_i}{\partial t} + u_j\frac{\partial u_i}{\partial x_j} = -\frac{1}{\rho}\frac{\partial p}{\partial x_i} + \frac{\mu}{\rho}\frac{\partial^2 u_i}{\partial x_j\partial x_j} + X_i \tag{8}$$

in which the constant μ/ρ is the *kinematic viscosity* of the fluid. To solve Eq. (8), appropriate boundary conditions must be used. If a Newtonian fluid comes in contact with a solid body, then the appropriate boundary condition is that there be no relative motion between the solid and fluid. This is the *no-slip condition*. The fluid adheres to the solid whether the surface is wettable or not. This condition is not so intuitively obvious, but has been found to be valid in all cases examined in the past, except for rarefied gases in which the mean free path of the molecules between collisions is comparable to the size of the body.

Fluids whose constitutive equation does not obey Eq. (6) are said to be *non-Newtonian*. Blood is non-Newtonian. Most body fluids are non-Newtonian. At shear strain rate above $100 \sec^{-1}$ blood is almost Newtonian. Air and water are Newtonian.

A material is called an *isotropic Hookean solid* if it obeys the following stress–strain relationship:

$$\sigma_{ij} = \lambda e_{kk}\delta_{ij} + 2Ge_{ij} \tag{9}$$

or its inverse:

$$e_{ij} = \frac{1+v}{E}\sigma_{ij} - \frac{v}{E}\sigma_{kk}\delta_{ij}. \tag{10}$$

Here e_{ij} is the strain tensor, λ, G, E, v are constants. λ and G are the *Lamé* constants, G is the *shear modulus*, E is the *Young's modulus*, v is the *Poisson's ratio*. δ_{ij} is the *Kronecker delta*, which is equal to 1 when $i = j$, and zero when $i \neq j$. Equations (9) and (10) are called *Hooke's law*.

On substituting Eq. (9) into Eq. (1) we obtain

$$\rho \frac{Dv_i}{Dt} = \lambda \frac{\partial}{\partial x_i} e_{\alpha\alpha} + 2G \frac{\partial e_{ij}}{\partial x_j} + X_i. \tag{11}$$

Here Dv_i/Dt is given by Eq. (3). The strain must be referred to a configuration of the body in which the stress is zero, because according to Eqs. (9) and (10), zero stress implies zero strain and vice versa. Let the displacement vector of a point in the body be measured with respect to an inertial rectangular Cartesian frame of reference and be denoted by u_i. If u_i is finite, the strain-displacement relationship is nonlinear, see Chap. 9. However, if $u_i(x_1, x_2, x_3, t)$ is infinitesimal, then

$$e_{ij} = \frac{1}{2} \left(\frac{\partial u_i}{\partial x_j} + \frac{\partial u_j}{\partial x_i} \right), \tag{12}$$

$$v_i = \frac{\partial u_i}{\partial t}, \qquad \frac{Dv_i}{Dt} = \frac{\partial^2 u_i}{\partial t^2}. \tag{13}$$

To the same order of approximation, the material density is a constant:

$$\rho = \text{const.} \tag{14}$$

On substituting Eqs. (12)–(14) into (11), one obtains the well-known *Navier's equation*:

$$G \frac{\partial^2 u_i}{\partial x_j \partial x_j} + (\lambda + G) \frac{\partial}{\partial x_i} \left(\frac{\partial u_j}{\partial x_j} \right) + X_i = \rho \frac{\partial^2 u_i}{\partial t^2}. \tag{15}$$

If we introduce the Poisson's ratio as in Eq. (10),

$$v = \frac{\lambda}{2(\lambda + G)}, \tag{16}$$

then Navier's equation can be written as

$$G \frac{\partial^2 u_i}{\partial x_j \partial x_j} + \frac{1}{1 - 2v} \frac{\partial}{\partial x_i} \frac{\partial u_j}{\partial x_j} + X_i = \rho \frac{\partial^2 u_i}{\partial t^2}. \tag{17}$$

Navier's equation is the basic field equation of the linearized theory of elasticity. It must be solved with appropriate initial and boundary conditions.

Living bodies, however, often take on finite deformations and obey more complex constitutive equations than the above-mentioned. The field equations are therefore more complex. Examples of formulation and solutions of organs subjected to finite deformations and nonlinear constitutive equations are discussed in Chapters 10 and 11.

Problems

1.1 Hold your arm in a horizontal position while lifting a weight in the hand. Name the major muscles that must provide the tension. How can you determine the tension in these muscles? Invent a theoretical and/or an experimental way to determine the tension in individual muscles.

1.2 What factors determine the pressure in the abdomen when one (a) lifts a heavy weight, (b) swims, (c) relaxes in bed? What muscles are involved?

1.3 When a giraffe moves its head up and down 6 m, how much does the hydrostatic pressure in its cerebral blood vessels change? In which way would this change in pressure affect the blood circulation in the giraffe's head? If such large change in blood pressure occurs in human brain, what ill effect may be expected?

1.4 Locusts (Schistocerca) can jump a distance up to about 80 cm on level ground. If a locust takes off at an angle of 45°, show that the initial velocity should be 280 cm/sec in order to reach 80 cm.

   Figure P1.4(a) shows the hind leg of a locust. The solid outline shows the leg before it starts to jump. The dotted outline shows its configuration at takeoff. Figure P1.4(b) shows the skeleton at some intermediate stage of jumping (AB = tarsus, BC = tibia, CD = femur, G = center of gravity of locust, C' = the point of attachment of muscle). R. H. J. Brown has taken cinephotographs of locust jumps (*Times Sci. Rev.* 6–7, 1963) and observed that locusts actually take off at at least 55°, because they jump to get airborne, and then start flying. He reports that the takeoff speed is about 340 cm/sec for an 80 cm jump. He shows that the muscle is roughly parallel to the femur, and that the distance CC' in Fig. P1.4(b) is about

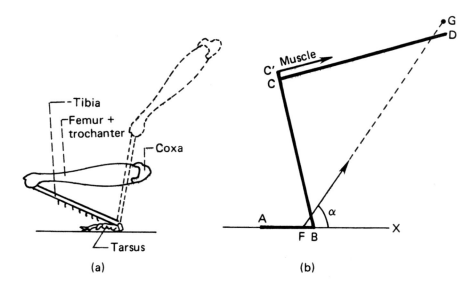

FIGURE P1.4 The hind leg of a locust: (a) Before it starts to jump (solid outline), and at the moment of lifting off (dotted lines); (b) The diagram of forces and angles.

BC/35. The locust accelerates its body by extending the hind leg rapidly. This moves the locust's coxa through a distance of about 4 cm before the tarsus leaves the ground.

Using Brown's data, compute the force exerted by the muscle during the jump by first deriving the formulas for the velocity v and the distance s at time t for the free fall of a particle subjected to a constant gravitational acceleration g. If $v = u$, $s = 0$ when $t = 0$, we have

$$v = u + gt, \qquad s = ut + \tfrac{1}{2}gt^2,$$
$$v^2 = u^2 + 2as, \qquad s = (u + v)t/2.$$

Show that a projectile fired from level ground with initial velocity u and an elevation angle of α will reach a height $u^2 \sin^2 \alpha/2g$ and a range $u^2 \sin 2\alpha/g$ if air resistance is ignored. Then show the following results by Brown successively:

(a) With $\alpha = 55°$ and a jump of distance $s = 80$ cm, the initial velocity u must be 290 cm/sec. The actual take-off speed of 340 cm/sec is greater than 290 cm/sec in order to compensate for air resistance.

(b) If $u = 0$, $v = 340$ cm/sec, and $s = 4$ cm, the acceleration is $a = 14{,}500$ cm/sec^2 if a can be assumed to be a constant.

(c) Find the resultant of the inertial force and the weight of the locust. This force (shown as *FG* in Fig. P1.4b must pass through the center of gravity of the locust if it does not send the animal spinning.

(d) Assuming an intermediate configuration with the angles $\angle CBX = 100°$, $\angle GFX = 57°$, $\angle BCD = 90°$, and $FB = BC/20$, calculate the force in the muscle. Show that for a 3g locust this is about 5 Newton. Quite a force!

Discussion of Problems 1.1–1.4

Questions like those in Problems 1.1 and 1.2 would provide motivation to study anatomy. Problem 1.2 suggests that we can control the pressure in our abdomen by voluntary muscle action. In *Biomechanics* (Fung, 1981, p. 14), it is shown that in order to relieve excessive loading on the spine it is necessary to tense up the abdominal muscles and increase the internal pressure. Weight lifters should have strong abdominal muscles.

In a relaxed state the human abdomen is soft. An effective way to make it more rigid is to increase the internal pressure. This is analogous to a soft garden hose which can be stiffened by turning on the water pressure. An erectile organ becomes erect by squeezing blood with a sphincter muscle. For example, Vathon reports (*Annual Conference of Engineering in Medicine and Biology*, Nov. 19, 1968, Houston, Texas) that the blood pressure in the corpus cavernosum artery of a bull's penis in erection is as high as 1,750 mm Hg.

Problem 1.3 takes on a physiological significance if you know that the diastolic pressure in the left ventricle is normally not much higher than the atmospheric pressure; nor is the pressure in the vena cava at the level of the heart. (What are the basic reasons for this?) Hence as the giraffe lifts its head, the pressure in its cerebral veins will be lower than atmospheric. Hence there is a tendency for the veins to collapse. If the vessels were collapsed the resistance to blood flow would increase. Then, how does the giraffe's heart supply the needed blood flow to the brain?

On the other hand, when its head is lowered, wouldn't the blood vessels expand and be gorged with blood? Would the brain tissue be squeezed? With what consequences?

Similarly in a standing man the hydrostatic pressure in the leg veins will cause the veins to bulge and become a reservoir of blood. If he lies down, the hydrostatic pressure in the leg decreases and the volume of the circulating blood will be increased. Thus, consider the following question: If a person who is walking suddenly felt a chest pain and thought a heart attack imminent, should he lie down immediately in order to minimize the work of the heart?

The large tension revealed in Problem 1.4 is significant when you realize that our own joints are similarly constructed. The muscles that move our legs and arms are almost parallel to the long bones, and the distances between the muscles and the fulcrum are small compared with the length of the limb. Thus the forces in our leg muscles can be large compared with the weight of our bodies. If a standing man sways a little his leg muscle must provide tension to control his posture. Hence, even when standing still the tensions in the muscle are significant.

Compare your elbow with your knee joint. Your have the patella at the knee but not at the elbow. What is the function of the patella? What is the consequence of this difference?

Problems

1.5 Design a human-powered hydrofoil for sports.

1.6 What polymer is pressure sensitive? Design a thin-film pressure transducer which can be used to measure pressure in narrow spaces such as intrapleural space, space between vertebra and disk, and in joints. Describe possible applications to sports, sports injury and repair research, gait, diagnosis, mechanical cardio-pulmonary resuscitation, etc.

1.7 Design recreational sports for the handicapped. Design helpful devices for the handicapped in their daily life.

1.8 Certain problems in industry are biomechanically oriented. For example, an airline stewardess often has a sore neck and spine strain due to leaning over to passengers. Coal miners often have loss of hearing or black lung disease. On the other hand, assembly line workers may reduce fatigue and improve efficiency by scheduling frequent short digression and exercise in the course of the day. Elaborate on these "work physiology" or "industrial biomechanics" by developing a program for an industry. Discuss the more fundamental, scientific part of the program.

1.9 Figure P.1.9 shows the type of exercise used effectively by patients with back pain at the Tientzin Hospital in China. Explain the rationale.

FIGURE P1.9 Hyper-extension exercises to strengthen the back of lumbago patients. Keep your back hollow.

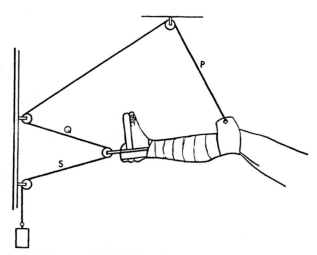

FIGURE P1.10 A Russell traction for the leg.

1.10 A Russell traction for immobilizing femoral fractures is shown in Fig. P.1.10. Determine the resultant force applied to the femur.

1.11 Consider a rock climber. Show that the normal reaction acting on the climber's feet is equal to the product of the weight of the climber and the perpendicular distance from the anchor of the rope to the vertical vector passing through the center of gravity of the climber, divided by the distance between the anchor of the rope and the climber's foot. How can the climber increase the limiting friction between his feet and the rock?

1.12 A patient suffering from low back pain often needs to stretch his spine. Design a stretcher that can be used when the patient lies in a hospital bed (e.g., a pulley system).

1.13 Design a stretcher that can be used by the patient at home. An example is a tilt table into which the patient can strap his feet and then tilt to an angle so that he is hung upside down.

1.14 Design an artificial aortic valve. Discuss the pros and cons of your design.

1.15 Since veins have valves and can serve as a one-way tunnel, design a mechanism which can apply external pressure on the legs in a certain way so that the mechanism can serve as a heart-assist device. Discuss possible uses and the pros and cons of your design.

1.16 Acceleration and deceleration can create pressure gradient which may help blood circulation. Design a machine which can shake a bed on which a patient lies in such a way as to serve as a heart-assist device. Discuss the pros and cons of your design.

1.17 Consider a very simple model of a scoliotic spine and three methods of correcting the deformities, as shown in Fig. P.1.17. Calculate the corrective bending moment obtained at the apex of the curve, C, due to the three types of loading. Show that for severely deformed spines ($\theta > 53°$) method (a) leads to a larger corrective moment than (b); for milder curves ($\theta < 53°$) method (b) is preferred, whereas (c) is better for all degrees of deformity. Cf. Panjabi and White (1980).

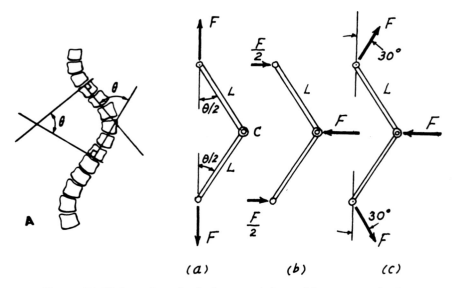

FIGURE P1.17 Several mechanical ways to help straighten a curved spine.

1.18 Model a subsystem of two vertebrae and a disk by a mass–spring–damper system, and write down the governing differential equation. For simplicity, consider only one degree of freedom, the vertical compression. Let the lower vertebrae be fixed, and denote the displacement of of the upper vertebrae by x.

1.19 Soldiers with chest wounds (perforated parietal pleura) often have collapsed lungs because of pleural space exposure to atmospheric pressure. The local increase of pleural pressure to atmospheric value reduces the effective pressure that inflates the lung.

Automatic collapse of lung (pneumothorax) sometimes occurs to healthy athletes while playing in the field. Explain how this could happen?

1.20 An accident broke a person's neck. Design a mechanism that will hold the head in place. (Such a device is sometimes called a "halo" device by its shape.)

1.21 A person has one of his facial bones knocked slightly inward during an automobile crash. A surgeon wants to pull that bone out to its original position. Design a mechanical device to do it.

1.22 A patient has a broken femur due to a fall. The most common treatment is to put the leg in a plaster of Paris cast while it heals slowly. The Chinese school (Chinese Academy of Traditional Medicine, Institute of Bone and Trauma, Beijing and Tientsin) has evidence that quicker healing can be achieved by a less rigid approach. They advocate the use of small boards (slats) of soft wood (e.g. balsa, with a width of about an inch or so) bandaged reasonably tightly (but not too tightly) around the broken bone. The patient is then taught a proper course of exercises which puts some compressive stress in the bone but minimizes shear stress or torsion. Design such a slat-bandage. Analyze its effect on the broken bone. Compare the biological process of healing that occurs in the Chinese way with the conventional way of plaster cast by taking the muscle forces into consideration.

1.23 Design a saw that can cut a plaster of Paris cast over a broken bone without hurting the soft tissues. (Note: How about using high frequency small amplitude oscillations?)

1.24 Scotch adhesive tape is used for bandage. The mechanics of tearing off a adhesive tape is very important to the surgeon and the patient. Develop a theory of peeling off adhesive tape.

1.25 A proper strategy of surgery and placement of bandages afterwards can be of great importance to healing. Consider a patient with skin cancer. A piece of skin, the size of a dollar, has to be removed. Design a strategy to excise the diseased skin and suture the healthy skin to cover the wound. Design a scheme to put on the bandage after surgery.

1.26 If a beam is simply supported and is loaded by a weight W at its center, the reaction at each of the supports is $W/2$. Consider a man standing. His two legs support his upper body with ball and socket joints (i.e., simply supported). Yet the reactions at the joints are far greater than $W/2$, and are very sensitive to his posture. Explain why with a good sketch of the anatomy and free-body diagrams for the forces, and analytical calculations.

References

This bibliography lists books and papers referred to in the text and problems, and a few selected entries that are not specifically discussed in the text, but are of importance to the topics. Since this volume is a sequel to the author's books on mechanics, the following references are quoted throughout this book:

Fung, Y. C. (1955; Revised 1969). *Theory of Aeroelasticity*. Wiley, New York. Revised, Dover, New York.

Fung, Y. C. (1965). *Foundations of Solid Mechanics*. Prentice-Hall, New Jersey.

Fung, Y. C. (1969; 2nd ed. 1977). *A First Course in Continuum Mechanics*. Prentice-Hall, New Jersey.

Fung, Y. C. (1981). *Biomechanics: Mechanical Properties of Living Tissues*. Springer-Verlag, New York.

Fung, Y. C. (1984). *Biodynamics: Circulation*. Springer-Verlag, New York.

For the classical mechanics used in Secs. 1.2, 1.3, and 1.7, see Fung (1965, 1977), Greenwood (1965), and Yih (1969, 1989). For anatomy, see Gray (1973). For modeling, see Sec. 12.11 and References of Ch. 12.

Brown, R. M. and Counsilman, J. E. (1971). The role of lift in propelling the swimmer. In *Selected Topics on Biomechanics: Proc. CIC Symp. on Biomechanics*, (J. M. Cooper, ed.), The Athletic Institute, Chicago, pp. 179–188.

Burns, B. H. and Young, R. H. (1951). Results of surgery in sciatica and low back pain. *Lancet* **260**(1):245–249, (Correction) 358.

Counsilman, J. E. (1971). The application of Bernoulli's principle to human propulsion in water. In *Proc. First Intern. Symp. on Biomechanics in Swimming, Waterpolo, and Diving*, (L. Lewille and J. P. Clarys, eds.), Univ. Libre de Bruxelles Lab. de L'effort, Brussels.

Cyriax, J. (1978). *Orthopedic Medicine, Vol. 1, Diagnosis of Soft Tissue Lesions*, 7th Ed. Bailliere Tindall, London.

Gray, H. (1973). *Anatomy*, A classic which has been revised and expanded by many authors. 35th British ed. (R. Warwick and P. L. Williams, eds.), W. B. Saunders, Philadelphia, PA.

Greenwood, D. T. (1965). *Principles of Dynamics*, Prentice-Hall, Englewood Cliffs, N. J.

Hay, J. (1978). *The Biomechanics of Sports Techniques*, Prentice-Hall, Englewood Cliffs, N. J.

Kellgren, J. H. (1938). Observations on referred pain arising from muscle. *Clin. Sci.* **3**:175–190. See also *Clin. Sci.* **4**:35–46.

Lewis, T. (1942). *Pain*, MacMillan, N. Y.

Nachemson, A. and Elfström, G. (1970). Intravital dynamic pressure measurements in lumbar discs. A study of common movements, maneuvers, and exercises, *Scand. J. Rehab. Med. Suppl.* **1**, 1–40. See also author's article in *Perspectives in Biomedical Engineering*, (R. M. Kenedi, ed.), Proc. of a Symposium. University Park Press, Baltimore, 1973, pp. 111–119.

Panjabi, M. M. and White, A. A. III. (1980). Spinal mechanics. In *Perspectives in Biomechanics*, Vol. 1, Part B, (H. Reul, D. N. Ghista and G. Rau, eds.), Harwood Academic Pub., New York, pp. 617–682.

Paul, J. P. (1970). The effect of walking speed on the force actions transmitted at the hip and knee joints. *Proc. Roy. Soc. Med.* **63**(2):200–202.

Schultz, A. B. (1986). Loads on the human lumbar spine. *Mech. Eng.* **108**:36–41.

Williams, M. and Lissner, H. R. (1977). *Biomechanics of Human Motion*. 2nd Ed. (B. LeVeau, ed.), Saunders, Philadelphia, PA.

Yih, C. S. (1969, 1989). *Fluid Mechanics*, McGraw-Hill, New York (1969). West River Press, Ann Arbor, Michigan (1989).

CHAPTER 2

Segmental Movement and Vibrations

2.1 Introduction

In the preceding chapter we described motion in the Newtonian form. An animal is considered to be a collection of particles, and particle movement is expressed in terms of displacement, velocity, acceleration, external forces, and forces of interaction between particles. In application to biomechanics, one finds that the terms F_{IJ} in Eq. (1.1:7), that describe the mutual interaction between particles, are the most troublesome. For example, in the analysis of human locomotion, we must know the forces in all the muscles of the legs. But the human musculoskeletal system is highly redundant and the determination of the forces in the muscles is one of the most difficult problems in biomechanics (see Sec. 2.8 infra). Hence there is a need for a method that does not require such detailed information. The method of Joseph Louis Lagrange (1736–1813) offers such an alternative in terms of work and energy. If the kinetic and potential energies are known as functions of the generalized coordinates and their derivatives with respect to time, and if the work done by the external forces can be computed when a generalized coordinate changes, then the equations of motion can be written down.

This beautiful method is very useful in biomechanics, especially in dealing with problems of locomotion and vibrations of internal and external organs. We will introduce this method as an alternative to the Newtonian approach (Secs. 2.2–2.5). Then we will discuss the normal modes of vibration. The normal modes of vibration are the most useful generalized coordinates, with which the equations of motion can be decoupled in general (Secs. 2.6–2.7).

Equations describing systems with damping and fluid dynamic forces and the possibility of decoupling these equations are discussed in Secs. 2.10 and 2.11.

2.2 Examples of Simple Vibration Systems

Example 1. Figure 2.2:1. A mass connected to the ground by a spring and a damper. The mass has one degree of freedom: vertical displacement x from a position of equilibrium $x = 0$. The tension in the spring is linearly proportional to x. The resistance of the damper is linearly proportional to the velocity $\dot{x}$. Thus the spring and damper offer a restoring force $Kx + \beta\dot{x}$, where K is the *spring constant*, β is the *damping constant*, and dot over x signifies a differentiation of x with respect to time. Following Eq. (1.1:7) the equation of motion of the particle, with an external force $F^{(e)}$ acting on it, is

$$m\ddot{x} + Kx + \beta\dot{x} = F^{(e)}. \tag{1}$$

Example 2. Figure 2.2:2. Several masses connected by linear springs and dampers, all confined to move in one direction. An inertial frame of reference is erected on the ground which is stationary. The mass m_i is located at x_i. The force acting on the mass m_1 due to the motion of the masses m_1, m_2 is

$$F_{12} = -K_1(x_1 - x_2) - \beta_1(\dot{x}_1 - \dot{x}_2).$$

That acting on m_2 due to the motion of m_1, m_2, m_3 is

$$F_{21} + F_{23} = -K_1(x_2 - x_1) - \beta_1(\dot{x}_2 - \dot{x}_1) - K_2(x_2 - x_3) - \beta_2(\dot{x}_2 - \dot{x}_3).$$

That acting on m_3 due to the motion of m_2, m_3 and the stationary ground is

$$F_{32} + F_{30} = -K_2(x_3 - x_2) - \beta_2(\dot{x}_3 - \dot{x}_2) - K_3x_3 - \beta\dot{x}_3.$$

These forces can be represented in the form

$$-\sum_{j=1}^{3} k_{ij}x_j - \sum_{j=1}^{3} \beta_{ij}\dot{x}_j, \qquad (i = 1, 2, 3)$$

with the appropriate choice of the constants k_{ij}, β_{ij}. Hence, on substituting

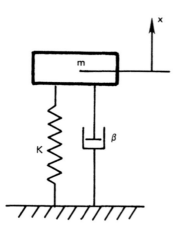

FIGURE 2.2:1 A mass-spring-damper system. The mass is supported on ground by a spring on the left, and by a damper on the right.

FIGURE 2.2:2 Several masses connected by springs and dampers.

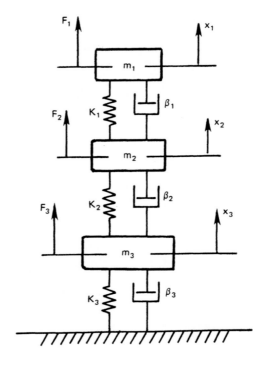

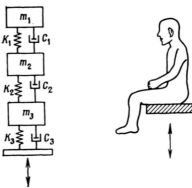

FIGURE 2.2:3 A freely moving body.

into Eq. (1.1:7), we obtain the equations of motion

$$m_i \ddot{x}_i + \sum_{j=1}^{3} k_{ij} x_j + \sum_{j=1}^{3} \beta_{ij} \dot{x}_j = F_i^{(e)}, \qquad (i = 1, 2, 3). \qquad (2)$$

Example 3. *Figure* 2.2:3, *representing a free-flying astronaut*. In this case, the inertial force depends on the acceleration of the body, whereas the elastic and damping forces depend only on the deformation of the body. Hence it is useful to introduce a local frame of reference attached to the body, with origin located at the center of mass, to describe the elastic deformation. Let us consider again a unidirectional motion without rotation. Let X_i be the coordinate of particle m_i referred to an inertial frame of reference, x_i that referred to the center of mass, and X_0 be the location of the origin of the moving frame of reference (center of mass). Then

$$X_i = X_0 + x_i \tag{3}$$

and the acceleration in the inertial frame of reference is

$$\ddot{X}_i = \ddot{X}_0 + \ddot{x}_i. \tag{4}$$

On the other hand, the spring and damper forces depend only on x_i. Hence the equations of motion are

$$\left(\sum_{i=1}^{3} m_i \right) \ddot{X}_0 = \sum_{i=1}^{3} F_i^{(e)} \tag{5a}$$

$$m_i(\ddot{X}_0 + \ddot{x}_i) + \sum_j k_{ij} x_j + \sum_j \beta_{ij} \dot{x}_j = F_i^{(e)}, \qquad (i = 1, 2, 3). \tag{5b}$$

The solution of these equations will be discussed in the following sections. An application is suggested in Problem 2.1. Literature on modeling human bodies is cited in Chapter 11.

2.3 Strain Energy and the Properties of the Influence Coefficients

Consider an elastic body obeying Hooke's law. Let it be rigidly supported in a fixed space and subjected to a set of forces $Q_1, Q_2, \ldots, Q_n$ acting at points $1, 2, \ldots, n$, respectively. According to Hooke's law, the *deflection* q_i at point i *in the direction of the force* Q_i is linearly proportional to the forces $Q_1, \ldots, Q_n$, and vice versa:

$$q_i = c_{i1} Q_1 + c_{i2} Q_2 + \cdots + c_{in} Q_n, \qquad (i = 1, 2, \ldots, n) \tag{1}$$

$$Q_i = k_{i1} q_1 + k_{i2} q_2 + \cdots + k_{in} q_n, \qquad (i = 1, 2, \ldots, n). \tag{2}$$

The constants of proportionality c_{ij}, k_{ij} are independent of the forces and displacements. c_{ij} are the *flexibility influence coefficients*. k_{ij} are the *stiffness influence coefficients*. The physical meaning of k_{ij} is the force that is required to act at the point i due to a unit deflection at the point j while all points $1, \ldots, n$ other than j are held fixed. In the case of a single degree of freedom, the stiffness influence coefficient is the familiar spring constant. The physical meaning of the flexibility constant c_{ij} is the deflection at i due to a unit force acting at j.

In the following, we shall show that the constants k_{ij} and c_{ij} have very special properties.

The linear relationships (1) or (2) imply that there exists a unique unstressed state of the body to which the body returns whenever all the external forces are removed. They imply that the principle of *superposition* applies, and that the total work done by a set of forces is independent of the order in which the forces are applied. In particular, if we slowly apply all the forces $Q_1, \ldots, Q_n$ together, beginning with zero and ending with full values, always keeping their ratios constant, then the displacements $q_1, \ldots, q_n$ will also increase slowly in constant ratios. The work done by each force is $\frac{1}{2}Q_i q_i$, and the total work done by the system of forces is

$$W = \frac{1}{2}\sum_{i=1}^{n} Q_i q_i = \frac{1}{2}\sum_{i=1}^{n}\sum_{j=1}^{n} k_{ij} q_i q_j = \frac{1}{2}\sum_{i=1}^{n}\sum_{j=1}^{n} c_{ij} Q_i Q_j. \tag{3}$$

This work is stored as *strain energy* in the elastic body. Now, there is a well-known thermodynamic argument (see Fung, 1965, pp. 351–352) that states that if the material is *stable* (i.e., if the body will return to its natural, unstressed state when all the loads are removed), then the strain energy must be *positive definite*, i.e., W must be positive, and can be zero only if all the q's and Q's vanish. The conditions of positive definiteness of the quadratic forms in Eq. (3) are well known (see Fung, 1965, pp. 29, 30) and are

$$k_{ii} > 0, \qquad (i \text{ not summed}), \tag{4a}$$

$$\begin{vmatrix} k_{ii} & k_{ij} \\ k_{ji} & k_{jj} \end{vmatrix} > 0, \qquad (i, j = 1, 2, \ldots, n), \tag{4b}$$

$$\begin{vmatrix} k_{ii} & k_{ij} & k_{im} \\ k_{ji} & k_{jj} & k_{jm} \\ k_{mi} & k_{mj} & k_{mm} \end{vmatrix} > 0, \qquad (i, j, m = 1, 2, \ldots, n), \tag{4c}$$

$$\cdots$$

That is, all the *principal minors*, including the determinant of the full k_{ij} matrix, must be greater than zero. Similar conditions hold for c_{ij}'s.

One of the most important properties of the influence coefficients is that they are symmetric:

$$c_{ij} = c_{ji}, \qquad k_{ij} = k_{ji}. \tag{5}$$

In other words, the displacement at point i due to a unit force acting at another point j is equal to the displacement at j due to a unit force acting at i, provided that the displacement and force *correspond*, i.e., they are positive in the same direction at each point. The proof is simple: Consider two forces Q_1 and Q_2. When the forces are applied in the order Q_1, Q_2 the work done by the forces is

$$W = \frac{1}{2}(c_{11}Q_1^2 + c_{22}Q_2^2) + c_{12}Q_1 Q_2.$$

When the order of application of the forces is reversed, the work done is

$$W' = \frac{1}{2}(c_{22}Q_2^2 + c_{11}Q_1^2) + c_{21}Q_1 Q_2.$$

But $W = W'$ for arbitrary Q_1, Q_2. Hence $c_{12} = c_{21}$, and the theorem is proved.

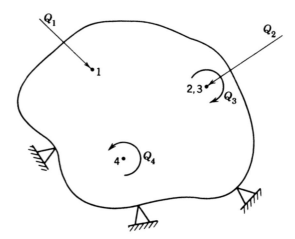

FIGURE 2.3:1 Generalized forces.

The concept of *corresponding* forces and displacements can be generalized to include torques and angles (Fig. 2.3:1). If M_j is a couple acting at a point j and θ_j is the rotation of the material at point j in the direction of the couple, then $M_j\theta_j$ is the work done by the couple M_j, and M_j and θ_j are said to be corresponding to each other. The couple M_j may be called a *generalized force* at point j, and θ_j is the *corresponding generalized displacement*. We can then denote M_j, θ_j by Q_j, q_j and show that Eqs. (1)–(5) hold for generalized forces and displacements.

With these concepts and notations, we can write the equations of motion of a set of masses embedded in an elastic body. The simplest way is to apply D'Alembert's principle, which states that the particles can be considered to be in a state of equilibrium if the inertial forces (mass × acceleration in an inertial frame of reference) are applied in reversed direction on the particles. Thus, if the body is supported by stationary supports, then the force Q_i applied on the mass m_i is $-m_i\ddot{u}_i + F_i^{(e)}$. Here u_i is the displacement of the particle and $F_i^{(e)}$ is the external force acting on it. Both u_i and $F_i^{(e)}$ are vectors. We may treat each component of u_i and $F_i^{(e)}$ as a separate variable. This set of forces Q_i will keep the body in equilibrium with displacements u_i. Thus, using a dot over a variable to indicate the differention of that variable with respect to time, Eq. (2) becomes

$$m_i\ddot{u}_i + \sum_{j=1}^{n} k_{ij}u_j = F_i^{(e)}, \qquad (i = 1, 2, \ldots, n). \qquad (6)$$

This is the generalization of Eq. (2.2:2) to a three-dimensional elastic body which is supported at fixed points in an inertial frame of reference.

If the body is free to translate and rotate in space while it deforms elastically, then we can use a moving frame of reference x, y, z which is attached to the body with the origin located at the center of mass. The force-deformation relationships remain linear as in Eqs. (1) and (2). To apply D'Alembert's principle, we compute the acceleration that is referred to an inertial frame of reference. As we have discussed in Sec. 1.3, let $\mathbf{r}$ be the position vector of a point in an inertial frame of reference, $\mathbf{R}_o$ be the position vector of the origin of the moving coordinates, and $\boldsymbol{\rho}$ be the position vector of the point relative to the moving coordinates. See Fig. 1.3:1. Then

$$\mathbf{r} = \mathbf{R}_o + \boldsymbol{\rho} = \mathbf{R}_o + x\mathbf{i}_x + y\mathbf{i}_y + z\mathbf{i}_z \tag{7}$$

$$\dot{\mathbf{r}} = \dot{\mathbf{R}}_o + \dot{\boldsymbol{\rho}} = \dot{\mathbf{R}}_o + \dot{x}\mathbf{i}_x + \dot{y}\mathbf{i}_y + \dot{z}\mathbf{i}_z + x(\mathbf{i}_x)^{\boldsymbol{\cdot}} + y(\mathbf{i}_y)^{\boldsymbol{\cdot}} + z(\mathbf{i}_z)^{\boldsymbol{\cdot}}, \tag{8}$$

where $\mathbf{i}_x$, $\mathbf{i}_y$, $\mathbf{i}_z$ are unit vectors, and $(\mathbf{i})^{\boldsymbol{\cdot}} = \boldsymbol{\omega} \times \mathbf{i}$, $\boldsymbol{\omega}$ being the angular velocity of the moving frame. Equation (8) may be written as

$$\dot{\mathbf{r}} = \dot{\mathbf{R}}_o + \dot{\boldsymbol{\rho}}_r + \boldsymbol{\omega} \times \boldsymbol{\rho}. \tag{9}$$

This equation defines $\dot{\boldsymbol{\rho}}_r = \dot{x}\mathbf{i}_x + \dot{y}\mathbf{i}_y + \dot{z}\mathbf{i}_z$ as the *relative velocity* of the point. A further differentiation yields, after simplification,

$$\ddot{\mathbf{r}} = \ddot{\mathbf{R}}_o + \boldsymbol{\omega} \times \boldsymbol{\omega} \times \boldsymbol{\rho} + \dot{\boldsymbol{\omega}} \times \boldsymbol{\rho} + \ddot{\boldsymbol{\rho}}_r + 2\boldsymbol{\omega} \times \dot{\boldsymbol{\rho}}_r. \tag{10}$$

Treating the three components as the inertial force $-m\ddot{\mathbf{r}}$ resolved along the moving coordinates x, y, z as three forces, using D'Alembert's principle, and denoting the three components of the vector ( ) by ( )$_i$, we obtain the equations of motion

$$m_i(\ddot{\mathbf{R}}_o + \boldsymbol{\omega} \times \boldsymbol{\omega} \times \boldsymbol{\rho} + \dot{\boldsymbol{\omega}} \times \boldsymbol{\rho} + \ddot{\boldsymbol{\rho}}_r + 2\boldsymbol{\omega} \times \dot{\boldsymbol{\rho}}_r)_i + \sum_{j=1}^{n} k_{ij}u_j = F_i^{(e)}, \tag{11}$$

$(i = 1, \ldots, n)$. Equation (11) describes the deformation of the body. The translation and rotation of the body, $R_o(t)$ and $\omega(t)$, respectively, are described by Eqs. (1.3:5) and (1.3:25).

In Sec. 2.2 we considered dampers. If the body is linearly viscoelastic we can add damping terms so that Eq. (6) is generalized to

$$m_i\ddot{u}_i + \sum_{j=1}^{n} k_{ij}u_j + \sum_{j=1}^{n} \beta_{ij}\dot{u}_j = F_i^{(e)}. \tag{12}$$

Damping, however, can be much more complex than this. It can be aerodynamic in origin (cf. Chs. 3, 4), or nonlinearly viscoelastic, and not necessarily stabilizing. See Sec. 2.10.

2.4 Generalized Coordinates

So far we have written the equations of motion in terms of the Cartesian coordinates of particles. We note, however, that there are occasions in which the movement of the particles in a body can be described by quantities other than the Cartesian coordinates. For example, the movement of all the points on a rigid rod can be described by the displacement of one point on the rod and the rotation of the rod about that point. In general, a set of quantities q_1, $q_2, q_3, \ldots, q_n$ are called the *generalized coordinates* of a system if they have the following properties: (1) The displacement of the system is described completely by the q's. (2) They are independent, so that one q_i can be varied while the remaining q_j's are held constant. Thus each q_i describes a degree of freedom of the system.

Let x_i, y_i, z_i, the Cartesian coordinates of a particle of number i in an inertial frame of reference, be related to the q's by relations of the form:

$$x_i = \Phi_i(q_1, q_2, \ldots), \qquad (i = 1, \ldots, N) \tag{1}$$

which do not contain time t explicitly. By differentation, we have the velocities

$$\dot{x}_k(q, \dot{q}) = \sum_{i=1}^{n} \frac{\partial x_k}{\partial q_i} \dot{q}_i. \tag{2}$$

Now, let us express the total kinetic energy of the system

$$K = \frac{1}{2} \sum_{k=1}^{N} m_k(\dot{x}_k^2 + \dot{y}_k^2 + \dot{z}_k^2) \tag{3}$$

in terms of the generalized coordinates. Here m_k is the mass of the kth particle. On substituting Eq. (2) into Eq. (3), we obtain

$$K = \frac{1}{2} \sum_{k=1}^{N} m_k \left[\left(\sum_{j=1}^{n} \frac{\partial x_k}{\partial q_j} \dot{q}_j \right)^2 + \left(\sum_{j=1}^{n} \frac{\partial y_k}{\partial q_j} \dot{q}_j \right)^2 + \left(\sum_{j=1}^{n} \frac{\partial z_k}{\partial q_j} \dot{q}_j \right)^2 \right]. \tag{4}$$

This can be written as

$$K = \frac{1}{2} \sum_{i=1}^{n} \sum_{j=1}^{n} m_{ij} \dot{q}_i \dot{q}_j, \tag{5}$$

in which the coefficients m_{ij} are functions of the q's:

$$m_{ij} = m_{ji} = \sum_{k=1}^{N} m_k \left(\frac{\partial x_k}{\partial q_i} \frac{\partial x_k}{\partial q_j} + \frac{\partial y_k}{\partial q_i} \frac{\partial y_k}{\partial q_j} + \frac{\partial z_k}{\partial q_i} \frac{\partial z_k}{\partial q_j} \right). \tag{6}$$

As the system moves through small displacements dq_i, an increment of work will be done by the forces acting on the system. This work can be written in the form

$$dW = \sum_{j=1}^{n} Q_j dq_j. \tag{7}$$

The Q_j's are called the *generalized forces* corresponding to the generalized coordinates q_i.

Example 1. A lower leg is hinged at the knee ($x = 0$) and acted on by forces perpendicular to its longitudinal axis at an intensity of $p(x)$ per unit length (as in swimming), see Fig. 2.4:1. With the angle θ chosen as a generalized coordi-

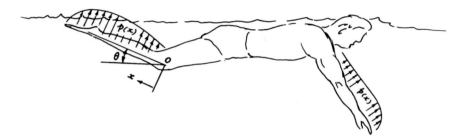

FIGURE 2.4:1 Forces acting on the arms and legs in swimming.

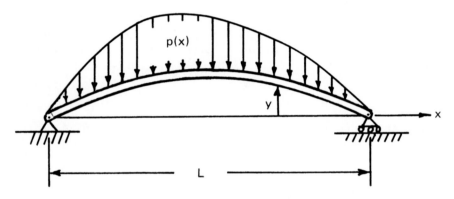

FIGURE 2.4:2 A simply-supported beam.

nate, the work done on the leg due to a small change of angle θ is

$$dW = \left(\int_0^L p(x)x\, dx \right) d\theta = Q\, d\theta. \qquad (8)$$

The quantity enclosed in the parenthesis is the generalized force corresponding to θ.

Example 2. A simply supported beam vibrates in its first mode. The displacement is a half sine wave as shown in Fig. 2.4:2,

$$y = a\left(\sin \frac{\pi x}{L} \right) \sin \omega t. \qquad (9)$$

A generalized coordinate for this mode is a quantity that specifies the displacement, e.g., the displacement at $x = L/2$. Thus

$$q_1 = a \sin \omega t \qquad y = q_1 \sin(\pi x/L). \qquad (10)$$

If the beam is acted on by a distributed lateral load of magnitude $p(x)$ per unit length, then the work done during displacement dq_1 is

$$dW = \int_0^L p(x)\,dx\,dy = \left(\int_0^L p(x)\sin\frac{\pi x}{L}\,dx\right)dq_1 = Q_1\,dq_1. \tag{11}$$

Hence the generalized force corresponding to the generalized coordinate q_1 is the quantity enclosed in the parentheses:

$$Q_1 = \int_0^L p(x)\sin\frac{\pi x}{L}\,dx. \tag{12}$$

2.5 Lagrange's Equations

The kinetic energy K is, according to Eq. (2.4:5), a homogeneous quadratic function of the $\dot{q}$'s. According to Euler's theorem for homogeneous functions, K must be of the form:

$$\sum_{j=1}^n \frac{\partial K}{\partial \dot{q}_j}\dot{q}_j = 2K. \tag{1}$$

On the other hand, the coefficients m_{ij} in Eq. (2.4:6) are functions of the q's. Hence the rate of change of K, as a function of q_j and $\dot{q}_j$, can be obtained by following the general rules of differentiation:

$$\dot{K} = \sum_j \left(\frac{\partial K}{\partial \dot{q}_j}\ddot{q}_j + \frac{\partial K}{\partial q_j}\dot{q}_j\right)$$
$$= \sum_j \left[\frac{d}{dt}\left(\frac{\partial K}{\partial \dot{q}_j}\dot{q}_j\right) - \frac{d}{dt}\left(\frac{\partial K}{\partial \dot{q}_j}\right)\dot{q}_j + \frac{\partial K}{\partial q_j}\dot{q}_j\right]. \tag{2}$$

On substituting Eq. (1) into the first term on the right-hand side of (2), we obtain

$$\dot{K} = 2\dot{K} - \sum_j \left[\frac{d}{dt}\left(\frac{\partial K}{\partial \dot{q}_j}\right)\dot{q}_j - \frac{\partial K}{\partial q_j}\dot{q}_j\right]$$

i.e.,

$$\dot{K} = \sum_j \left(\frac{d}{dt}\frac{\partial K}{\partial \dot{q}_j} - \frac{\partial K}{\partial q_j}\right)\dot{q}_j. \tag{3}$$

Now, the kinetic energy is one form of energy. For a biological system, other forms of energy are involved, such as the gravitational potential G, the internal energy U, and the chemical energy C. According to the *first law of thermodynamics*, the energy of a system can be changed by absorption of heat, H, and by doing work on the system. The rate of change of total energy must be equal to the sum of the rates of heat input, $\dot{H}$, and work done, $\dot{W}$:

$$\dot{K} + \dot{G} + \dot{U} + \dot{C} = \dot{H} + \dot{W}. \tag{4}$$

A part of the internal energy U is the strain energy discussed in Sec. 2.3. It is a quadratic function of the generalized coordinates $q_1, q_2, \ldots, q_n$, and is, in general, a function of temperature. If the temperature remains constant then

$$\dot{U} = \sum_{j=1}^{n} \frac{\partial U}{\partial q_j} \dot{q}_j. \tag{5}$$

The rate at which work is done by the forces acting on the system is, according to Eq. (2.4:7),

$$\dot{W} = \sum_{j=1}^{n} Q_j \dot{q}_j. \tag{6}$$

Substituting Eqs. (3), (5), and (6) into Eq. (4), we obtain

$$\sum_{j=1}^{n} \left(\frac{d}{dt} \frac{\partial K}{\partial \dot{q}_j} - \frac{\partial K}{\partial q_j} + \frac{\partial U}{\partial q_j} - Q_j \right) \dot{q}_j = \dot{H} - \dot{G} - \dot{C}. \tag{7}$$

In the special case when

$$\dot{H} - \dot{G} - \dot{C} = 0, \tag{8}$$

then

$$\sum_{j=1}^{n} \left(\frac{d}{dt} \frac{\partial K}{\partial \dot{q}_j} - \frac{\partial K}{\partial q_j} + \frac{\partial U}{\partial q_j} - Q_j \right) \dot{q}_j = 0. \tag{9}$$

But the q_j's are independent variables that can assume arbitrary values. In particular, we may set $\dot{q}_1 \neq 0$ while all other q_j's vanish. Then the coefficient of $\dot{q}_1$ must vanish. Similarly we can show that the quantity in the parentheses of every term of Eq. (9) must vanish. Hence

$$\frac{d}{dt} \frac{\partial K}{\partial \dot{q}_j} - \frac{\partial K}{\partial q_j} + \frac{\partial U}{\partial q_j} = Q_j, \qquad (j = 1, 2, \ldots, n). \tag{10}$$

These are *Lagrange's equations*.

If the generalized forces are partial derivatives of a function V of $q_1, q_2, \ldots, q_n$,

$$Q_j = -\frac{\partial V}{\partial q_j}, \tag{11}$$

then the forces are said to be *conservative* and V is called the *potential energy*. Since U and V are independent of $\dot{q}_j$, $\partial U/\partial \dot{q}_j = \partial V/\partial \dot{q}_j = 0$. Then Lagrange's equations may be written

$$\frac{d}{dt} \frac{\partial L}{\partial \dot{q}_j} - \frac{\partial L}{\partial q_j} = 0, \tag{12}$$

in which

$$L = K + U - V \tag{13}$$

is called the *Lagrangian function* of the system.

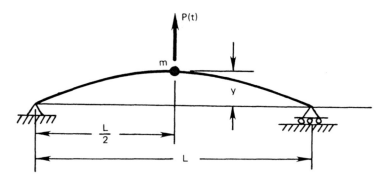

FIGURE 2.5:1 A simple example.

The aerodynamic or hydrodynamic forces acting on an animal moving in a fluid are nonconservative. They cannot be derived from a potential by differentiation. Hence, for problems involving these forces, it is better to use Eq. (10). (For simple problems, there is no particular advantage in using Lagrange's equations; advantages often show up in complex problems.)

Bioengineers are constantly faced with the question of how to approximate real systems. Lagrange's equations tell us that in mathematical modeling one should concentrate on approximating the energies. The merit of a scheme may be judged by how well the energies are approximated.

Example 1. Consider a weightless beam with a point mass m at its midspan (Fig. 2.5:1). A force $P(t)$ acts on the mass. The stiffness influence coefficient for the load at midspan due to unit deflection is k. Then, if y is the deflection at midspan, we have

$$K = \tfrac{1}{2}m\dot{y}^2, \qquad U = \tfrac{1}{2}ky^2, \qquad Q = P(t),$$

$$m\ddot{y} + ky = P(t).$$

Example 2. Consider a simply-supported beam of uniform material and uniform cross section (Fig. 2.5:1). Since the deflection curve $w(x, t)$ of the beam is continuous and satisfies the boundary conditions $w = \partial^2 w/\partial x^2 = 0$ at $x = 0$, L, it can be developed into a Fourier sine series:

$$w = \sum_{n=1}^{\infty} a_n \sin \frac{n\pi x}{L}.$$

The coefficients a_n may be considered the generalized coordinates $q_i = a_i$, $(i = 1, 2, 3, \ldots)$. When the beam vibrates, w and q_i are functions of time. Let m be the mass per unit length of the beam. Show that

$$K = \frac{1}{2} \int_0^L m\dot{w}^2 \, dx = \frac{mL}{4} \sum_{n=1}^{\infty} \dot{a}_n^2(t),$$

$$U = \frac{1}{2} \int_0^L EI \left(\frac{\partial^2 w}{\partial x^2}\right)^2 dx \qquad (EI = \text{rigidity} = \text{const})$$

$$= \frac{\pi^4}{4} \frac{EI}{L^3} \sum_{n=1}^{\infty} n^4 a_n^2(t).$$

If a lateral load of magnitude $P(x, t)$ per unit length acts on the beam, the generalized forces can be calculated as follows. Let the generalized coordinate a_n be given a virtual displacement δa_n. The beam configuration undergoes a virtual displacement $\delta w(x) = \delta a_n \sin(n\pi x/L)$. The work done by the external load is

$$Q_n \delta a_n = \int_0^L P(x, t) \delta w(x) \, dx = \int_0^L P(x, t) \delta a_n \sin \frac{n\pi x}{L} \, dx.$$

But δa_n is arbitrary, hence,

$$Q_n = \int_0^L P(x, t) \sin \frac{n\pi x}{L} \, dx.$$

The equations of motion are

$$\frac{mL}{2} \ddot{a}_n + \frac{\pi^4 EI}{2L^3} n^4 a_n = Q_n, \qquad (n = 1, 2, 3, \ldots).$$

Example 3. Consider a cantilever beam of uniform cross section (Fig. 2.5:2). The deflection of the beam can be expanded into a series

$$w(x, t) = \sum_{n=1}^{\infty} q_n(t) f_n(x)$$

where $f_n(x)$ is the nth mode of undamped free vibration of the beam. The functions $f_n(x)$ are orthogonal and can be normalized so that

$$\int_0^L f_m(x) f_n(x) \, dx = \begin{cases} 1, & \text{when } m = n, \\ 0, & \text{when } m \neq n. \end{cases}$$

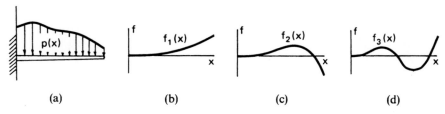

(a)   (b)   (c)   (d)

FIGURE 2.5:2 A cantilever beam of uniform cross section and its first three modes of natural vibration.

Show that

$$K = \frac{mL}{2} \sum_{n=0}^{\infty} \dot{q}_n^2, \qquad U = \frac{EIL}{2} \sum_{n=0}^{\infty} \kappa_n^4 q_n^2,$$

$$Q_n = \int_0^L P(x,t) f_n(x)\, dx,$$

in which κ_n are the solutions of the equation $\cos \kappa_n \cosh \kappa_n + 1 = 0$. Therefore, under a lateral load $P(x,t)$ the equations of motion are

$$mL\ddot{q}_n + EIL\kappa_n^4 q_n = \int_0^L P(x,t) f_n(x)\, dx \qquad (n = 1, 2, \dots).$$

Example 4. *A simple model of posture control.* A man is represented by a mass m, located at the center of gravity at height L, and a centroidal moment of inertia I (Fig. 2.5:3). The moment produced at the ankles is M, the ground reaction forces are F_x, F_y. Then, the motion of the center of mass (located at x, y) and the rotation about it are governed by the equations:

$$m\ddot{y} = F_y - mg, \qquad I\ddot{\theta} = F_x L \sin\theta - F_y L \cos\theta + M. \qquad (14)$$

Since, according to Fig. 2.5:3,

$$x = L \cos\theta, \qquad y = L \sin\theta, \qquad (15)$$

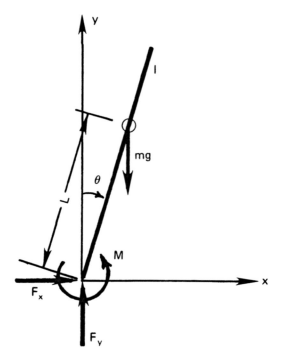

FIGURE 2.5:3 A simple inverted-pendulum model of posture control.

we obtain by differentiation

$$\ddot{x} = -\ddot{\theta}L\sin\theta - \dot{\theta}^2 L\cos\theta,$$
$$\ddot{y} = \ddot{\theta}L\cos\theta - \dot{\theta}^2 L\sin\theta. \tag{16}$$

These are five equations for the five unknowns $\ddot{x}$, $\ddot{y}$, $\ddot{\theta}$, F_x, F_y.

Generalized Coordinate, θ. The kinetic energy is

$$K = \tfrac{1}{2}(m\dot{x}^2 + m\dot{y}^2 + I\dot{\theta}^2) = \tfrac{1}{2}(mL^2 + I)\dot{\theta}^2. \tag{17}$$

The potential energy is

$$V = mgL\sin\theta. \tag{18}$$

The virtual work relation is

$$\delta W = Q\delta\theta = M\delta\theta, \qquad Q = M. \tag{19}$$

Hence the Lagrangian equation is a single equation

$$\ddot{\theta}(mL^2 + I) + mgL\cos\theta = M. \tag{20}$$

Note that in Lagrangian approach, the ground reactions F_x, F_y do not appear in the final equation.

Feedback Control. Camana et al. (1977) considered four sensing modalities for maintaining an erect posture:

$$M = k_1\theta + k_2\dot{\theta} + k_3\dot{\theta}_{sc} + k_4\theta_0, \tag{21}$$

where θ and $\dot{\theta}$ are ankle angle and ankle angle rates, as shown in Fig. 2.5:3, θ_{sc} is an approximation to the angular rate information sensed by the semicircular canals, and θ_0 is a similar approximation to angular position as sensed by the otolith system. For a normal person, one may also wish to add *visual* sensing.

Example 5. A simple model of gait (Fig. 2.5:4). The entire mass of a simulated biped is concentrated in a single rigid body. Let r, ϕ_1, ϕ_2 be the three generalized coordinates, and let J, F, M be the moment of inertia, tangential leg force, and hip moment, respectively, all normalized to the system mass, m. For simplicity, take the ankle moment as zero. Then the Lagrangian equations of motion are

$$\ddot{r} + L\ddot{\phi}_2\sin(\phi_1 - \phi_2) - r\dot{\phi}_1^2 - L\dot{\phi}_2^2\cos(\phi_1 - \phi_2) + g\cos\phi_1 = F$$
$$r^2\ddot{\phi}_1 + rL\ddot{\phi}_2\cos(\phi_1 - \phi_2) + 2r\dot{r}\dot{\phi}_1 + rL\dot{\phi}_2^2\sin(\phi_1 - \phi_2) - gr\sin\phi_1 = -M$$
$$L\ddot{r}\sin(\phi - \phi_2) + Lr\ddot{\phi}_1\cos(\phi_1 - \phi_2) + (J + l^2)\ddot{\phi}_2 + 2Lr\dot{\phi}_1\cos(\phi - \phi_2)$$
$$- Lr\dot{\phi}_1^2\sin(\phi_1 - \phi_2) - Lg\sin\phi_2 = M.$$

McGhee (1980) discussed gait and posture on the basis of these equations, and a comparison of these equations with those obtained by free-body approach. The latter requires more equations and has to bring the constraint

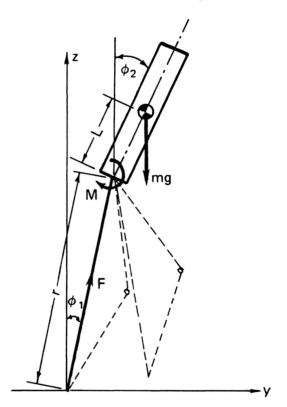

FIGURE 2.5:4 A simplified model of biped locomotion. From McGhee (1980), reproduced by permission.

forces such as the ground reactions into account; but each individual equation appears simpler.

2.6 Normal Modes of Vibration

The most commonly used generalized coordinates for an elastic body are the vibration modes of the body. By using the vibration modes, the equations of motion can often be decoupled into a set of independent equations, (see Sec. 2.7). Then the motion in each degree of freedom can be solved separately. This efficient procedure is the principal reason why people are so interested in normal modes. The basic theory is presented below.

The equations of free vibration of an elastic body (without damping) is given by Eq. (2.3:6) with $F_i^{(e)} = 0$ (without external forces). For simplicity of notation and manipulation, we shall consider it as a matrix equation. Define the column and square matrices

$$\mathbf{X} = \begin{Bmatrix} u_1 \\ u_2 \\ \vdots \\ u_n \end{Bmatrix}, \quad \mathbf{K} = \begin{bmatrix} K_{11} & K_{12} & \cdots & K_{1n} \\ K_{21} & K_{22} & \cdots & \\ & & \ddots & \\ K_{n1} & K_{n2} & \cdots & K_{nn} \end{bmatrix}, \quad \mathbf{M} = \begin{bmatrix} m_1 & 0 & \cdots & 0 \\ 0 & m_2 & \cdots & 0 \\ & & \ddots & \\ 0 & 0 & & m_n \end{bmatrix}. \tag{1}$$

Then the equation of free vibration is

$$\mathbf{M}\ddot{\mathbf{X}} + \mathbf{K}\mathbf{X} = 0. \tag{2}$$

The kinetic and potential energies, given by Eqs. (2.2:1) and (2.3:3), respectively, may be expressed in the following form:

$$K = \tfrac{1}{2}\dot{\mathbf{X}}^{\mathrm{T}}\mathbf{M}\dot{\mathbf{X}}, \qquad U = \tfrac{1}{2}\mathbf{X}^{\mathrm{T}}\mathbf{K}\mathbf{X}. \tag{3}$$

The superscript T means the *transpose* of the matrix. The kinetic energy K is positive definite (Sec. 2.3). The strain energy U is also positive definite if the body cannot move as a rigid body; whereas $U = 0$ in the rigid-body mode. Hence $\mathbf{M}$ and $\mathbf{K}$ are subjected to conditions listed in Eq. (2.3:4). In particular, the determinant of $\mathbf{M}$ does not vanish.

Since Eq. (2) is a set of linear differential equations with constant coefficients, it is expected that the solution will be an exponential function of time, in the form of

$$\mathbf{X}(t) = \mathbf{X}_0 e^{i\omega t}, \tag{4}$$

where $\mathbf{X}_0$ is a column of real or complex constants. The real and imaginary parts of (4) both represent simple harmonic oscillations. The problem is to determine the *circular frequency* ω and the corresponding amplitude function $\mathbf{X}_0$.

On substituting Eq. (4) into Eq. (2), dropping the subscript 0, and writing λ for ω^2, we obtain

$$\mathbf{K}\mathbf{X} - \lambda\mathbf{M}\mathbf{X} = 0, \qquad \lambda = \omega^2. \tag{5}$$

Now, the set of linear simultaneous equations (5) can have a nontrivial solution $\mathbf{X}_v \neq 0$ if and only if the determinant of the coefficients vanishes:

$$\det|\mathbf{K} - \lambda\mathbf{M}| = 0. \tag{6}$$

Since the degree of the polynomial

$$\det|\mathbf{K} - \lambda\mathbf{M}| = (-1)^n \lambda_n \det|\mathbf{M}| + \cdots + \det|\mathbf{K}|$$

is exactly n, $\mathbf{M}$ being nonsingular, there exist exactly n roots. The roots are called *eigenvalues*, and the corresponding solutions $\mathbf{X}$ are called *eigenvectors*. Let λ_1 be an eigenvalue and let $\mathbf{X}_1 \neq 0$ be a corresponding eigenvector, so that

$$\mathbf{K}\mathbf{X}_1 = \lambda_1 \mathbf{M}\mathbf{X}_1. \tag{7}$$

X_1 might be complex valued. Let $\bar{X}_1$ be the complex conjugate of X_1. Pre-multiply Eq. (7) with $\bar{X}_1^T$, we obtain

$$\bar{X}_1^T(KX_1) = \lambda_1\bar{X}_1^T(MX_1). \tag{8}$$

But M and K are real, symmetric matrices. Hence the products $\bar{X}_1^T(MX_1)$ and $\bar{X}_1^T(KX_1)$ are real numbers; the first is positive, the second is nonnegative. It follows that λ_1 is a real number and is nonnegative, and $\omega_1 = \sqrt{\lambda_1}$ is a real number, called an *eigenfrequency*. $\omega_1 = 0$ if the body moves as a rigid body, $\omega_1 \neq 0$ if the body deforms during vibration without rigid-body motion.

Hence we obtain the important result: *all the eigenfrequencies of free vibration of a linear elastic solid are real numbers.*

When λ_1 is real valued, it is clear that the solution X_1 of Eq. (7) can be normalized to be a real-valued column matrix. Hence we conclude that all the eigenvectors can be normalized to be real valued. In other words, all free vibration modes of a linear elastic body are real, i.e., all masses move in phase in each mode.

Now we can prove the following:

If the equation of free vibration of a linear elastic solid, Eq. (5), has two unequal eigenvalues λ_1 and λ_2, then any two eigenvectors X_1 and X_2, corresponding to λ_1 and λ_2, respectively, are orthogonal with respect to M and K, i.e.,

$$(X_1, MX_2) = 0, \tag{9}$$

$$(X_1, KX_2) = 0. \tag{10}$$

Here the notation (A, B) denotes the scalar product of two vectors A, B; i.e., $A^T B$ when A, B are represented by matrices.

Proof. By assumption, $KX_1 = \lambda_1 MX_1$, $KX_2 = \lambda_2 MX_2$. According to what was said earlier, λ_1 and λ_2 are real numbers. Hence,

$$(KX_1, X_2) = (\lambda_1 MX_1, X_2), \tag{11a}$$

$$(X_1, KX_2) = (X_1, \lambda_2 MX_2) = \lambda_2(X_1, MX_2). \tag{11b}$$

But $(KX_1, X_2) = (X_1, KX_2)$, $(MX_1, X_2) = (X_1, MX_2)$ because K and M are symmetric. Therefore, using these equations and subtracting (11b) from (11a), one obtains $(\lambda_1 - \lambda_2)(X_1, MX_2) = 0$, whence Eq. (9) follows, because, by assumption, $\lambda_1 - \lambda_2 \neq 0$. A glance at Eq. (11) then proves Eq. (10). Q.E.D.

Combining the results stated above, one sees that if the determinantal equation (6) has single roots only, then we can assert that:

The free vibration equation of a linear elastic solid, Eq. (5), has n real eigenvalues $\lambda_1, \lambda_2, \ldots, \lambda_n$ and n corresponding eigenvectors $X_1, X_2, \ldots, X_n$ called the normal modes, satisfying the relations

$$KX_\nu = \lambda_\nu MX_\nu, \qquad (\nu = 1, \ldots, n),$$

which are orthogonal by pairs with respect to M and can be normalized so that

$$(\mathbf{X}_\mu, \mathbf{M}\mathbf{X}_\nu) = \delta_{\mu\nu}, \qquad (\mu, \nu = 1, \dots, n), \tag{12}$$

where $\delta_{\mu\nu} = 1$ if $\mu = \nu$; $\delta_{\mu\nu} = 0$ if $\mu \neq \nu$.

Extending this result, it can be shown that *if λ_ν is a k-fold multiple root of the characteristic Eq. (6), then to λ_ν there belongs exactly k of the eigenvectors* $\mathbf{X}_1, \dots, \mathbf{X}_n$. The proof can be found in Bellman (1970).

The *orthonormal* character of the normal modes of vibration exhibited by Eq. (12) is the basis on which the importance of normal modes rests. With Eq. (12), we see that an arbitrary motion of the body, represented by a vector $\mathbf{X}$ in an *n*-dimensional space, can be represented in the form

$$\mathbf{X} = \sum_{\nu=1}^{n} q_\nu \mathbf{X}_\nu, \tag{13}$$

where q_ν are constants:

$$q_\nu = \frac{(\mathbf{X}, \mathbf{M}\mathbf{X}_\nu)}{(\mathbf{X}_\nu, \mathbf{M}\mathbf{X}_\nu)} = (\mathbf{X}, \mathbf{M}\mathbf{X}_\nu). \tag{14}$$

This is easily proven by premultiplying Eq. (13) by $\mathbf{X}_\mu^T\mathbf{M}$ and using Eq. (12). Because of Eq. (13), we see that $\mathbf{X}_\nu$ can serve as the basis and q_ν the *generalized coordinates*. Since $\mathbf{X}_\nu$ are normal modes, q_ν are called *normal coordinates*. Physically, we say that any displaced configuration of a body can be represented by a linear combination of the normal modes of free vibration of a linear elastic body of the same geometry and dimensions.

2.7 Decoupling of Equations of Motion

Now we shall show that the equations of motion become particularly simple when expressed in terms of normal coordinates. Let $\mathbf{X}_1, \dots, \mathbf{X}_n$ be the normal modes and $\lambda_1, \dots, \lambda_n$ be the corresponding eigenvalues of the free vibration equation:

$$\mathbf{M}\ddot{\mathbf{X}} + \mathbf{K}\mathbf{X} = 0 \quad \text{or} \quad \mathbf{K}\mathbf{X} = \lambda\mathbf{M}\mathbf{X}. \tag{1}$$

We have

$$\mathbf{K}\mathbf{X}_\nu = \lambda_\nu\mathbf{M}\mathbf{X}_\nu, \qquad (\nu = 1, \dots, n), \tag{2}$$

$$(\mathbf{X}_\nu, \mathbf{M}\mathbf{X}_\mu) = \delta_{\mu\nu}, \qquad (\mu, \nu = 1, \dots, n), \tag{3}$$

$$(\mathbf{X}_\nu, \mathbf{K}\mathbf{X}_\mu) = \lambda_\mu\delta_{\mu\nu}, \qquad (\mu, \nu = 1, \dots, n). \tag{4}$$

With these normal modes we can build a square matrix $\boldsymbol{\Phi}$:

$$\boldsymbol{\Phi} = (\mathbf{X}_1, \mathbf{X}_2, \dots, \mathbf{X}_n). \tag{5}$$

The first column of $\boldsymbol{\Phi}$ is the column matrix $\mathbf{X}_1$, the second column of $\boldsymbol{\Phi}$ is the column matrix $\mathbf{X}_2$, etc. It can be easily verified, on account of Eqs. (3) and (4), that if all the eigenvalues $\lambda_1, \dots, \lambda_n$ are different, i.e., none of them is a multiple root of the characteristic equation (2.6:6), then

$$\mathbf{\Phi}^T \mathbf{M} \mathbf{\Phi} = \mathbf{I} = \text{unit matrix}, \tag{6}$$

$$\mathbf{\Phi}^T \mathbf{K} \mathbf{\Phi} = \begin{bmatrix} \lambda_1 & 0 & 0 & \cdots & 0 \\ 0 & \lambda_2 & 0 & \cdots & 0 \\ \vdots & \vdots & \vdots & & \vdots \\ 0 & 0 & 0 & \cdots & \lambda_n \end{bmatrix} \equiv \mathbf{\Lambda}. \tag{7}$$

$\mathbf{\Phi}^T$ is the transpose of $\mathbf{\Phi}$. Since the vectors $\mathbf{X}_v$ are linearly independent, the matrix $\mathbf{\Phi}$ is nonsingular. Hence $\mathbf{\Phi}^{-1}$ exists, and according to Eq. (6) this is equal to

$$\mathbf{\Phi}^{-1} = \mathbf{\Phi}^T \mathbf{M}. \tag{8}$$

The normal coordinates for an arbitrary vector $\mathbf{X}$, given by Eqs. (2.6:13) and (2.6:14), can thus be written in matrix form

$$\mathbf{X} = \mathbf{q}^T \mathbf{\Phi}, \tag{9}$$

$$\mathbf{q} = \mathbf{\Phi}^T \mathbf{M} \mathbf{X} = \mathbf{\Phi}^{-1} \mathbf{X}, \tag{10}$$

where $\mathbf{q}$ is a column matrix whose components are the normal coordinates $q_1, q_2, \ldots, q_n$.

With these ingredients we can now simplify the equations of motion of a linear elastic system subjected to a set of forces $\mathbf{F} = (F_1, F_2, \ldots, F_n)$:

$$\mathbf{M} \ddot{\mathbf{X}} + \mathbf{K} \mathbf{X} = \mathbf{F}. \tag{11}$$

Premultiplying both sides with $\mathbf{\Phi}^T$ and inserting $\mathbf{\Phi} \mathbf{\Phi}^{-1} = \mathbf{I}$ we obtain

$$\mathbf{\Phi}^T \mathbf{M} \mathbf{\Phi} \mathbf{\Phi}^{-1} \ddot{\mathbf{X}} + \mathbf{\Phi}^T \mathbf{K} \mathbf{\Phi} \mathbf{\Phi}^{-1} \mathbf{X} = \mathbf{\Phi}^T \mathbf{F}. \tag{12}$$

Using Eqs. (6), (7) and (10), and defining the column matrix of the generalized forces $\mathbf{Q}$ as

$$\mathbf{Q} = \mathbf{\Phi}^T \mathbf{F} \tag{13}$$

we obtain from Eq. (12)

$$\ddot{\mathbf{q}} + (\mathbf{\Phi}^T \mathbf{K} \mathbf{\Phi}) \mathbf{q} = \mathbf{Q} \tag{14}$$

or, using (7),

$$\ddot{\mathbf{q}} + \mathbf{\Lambda} \mathbf{q} = \mathbf{Q}. \tag{15}$$

This is the equation of motion in its simplest form.

To emphasize the difference between Eqs. (11) and (15), we write them in extenso. Equations (11) are

$$\begin{aligned} m_1 \ddot{u}_1 + k_{11} u_1 + k_{12} u_2 + \cdots + k_{1n} u_n &= F_1(t), \\ m_2 \ddot{u}_2 + k_{21} u_1 + k_{22} u_2 + \cdots + k_{2n} u_n &= F_2(t), \\ &\vdots \\ m_n \ddot{u}_n + k_{n1} u_1 + k_{n2} u_2 + \cdots + k_{nn} u_n &= F_n(t). \end{aligned} \tag{16}$$

They are simplified into

$$\ddot{q}_1 + \omega_1^2 q_1 = Q_1,$$
$$\ddot{q}_2 + \omega_2^2 q_2 = Q_2,$$
$$\cdots \tag{17}$$
$$\ddot{q}_n + \omega_n^2 q_n = Q_n.$$

In these equations, we have written ω_ν^2 for λ_ν to emphasize the physical meaning of the constant ω_ν, which is the circular frequency of free vibration of the νth normal mode.

It can be shown that each generalized force Q_ν, defined formally in Eq. (13), has a simple physical interpretation. It is numerically equal to the work done by the external forces acting on the body when the body deforms in the νth mode, which is so normalized in amplitude that

$$m_1 u_1^2 + m_2 u_2^2 + \cdots + m_n u_n^2 = 1, \tag{18}$$

where $u_1, u_2, \ldots, u_n$ are the displacements of the particles $m_1, m_2, \ldots, m_n$ in the νth normal mode.

We see that if the Q_ν's are functions of time only (i.e., if they are independent of the elastic displacements), Eq. (15) is completely decoupled. Each normal coordinate varies with time as a single mass–spring oscillator of one degree of freedom, independent of all other modes. Such is the simplicity introduced by the use of normal coordinates.

Use of normal modes also relieves the need to account for any *constraints* imposed on the body. For example, if we want to study the response of the spine to shock loading, the end conditions at the neck and the sacrum must be satisfied. If we study the stresses in the leg in certain maneuver, the constraint of the hip joint must be imposed. If normal modes are used as generalized coordinates, the boundary conditions are satisfied by every mode and no further attention to the constraints is needed.

Fortunately, in most practical problems the deformation of a continuous body can usually be described to a sufficient degree of accuracy by using only a few normal modes associated with the lower frequencies of free vibration. In such cases the continuous body is approximated by one possessing only a few degrees of freedom.

Unfortunately, if the system were damped, complete decoupling is in general not possible. See Secs. 2.10 and 2.11.

2.8 Muscle Forces

The high degree of redundancy of animal body structure is the main difficulty for a detailed analysis of the musculoskeletal system. By *degree of redundancy* is meant the excess of the number of unknowns over the total number of

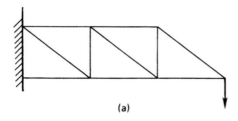

(a)

FIGURE 2.8:1 Simple trusses illustrating the concept of degree of redundancy.

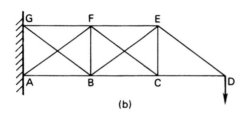

(b)

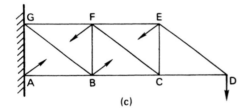

(c)

available independent equations of equilibrium (or motion) and constraints. The idea can be illustrated by a simple example. Fig. 2.8:1(a) shows a simple truss made of steel members that can transmit tension and compression. If we wish to know the force in each member when a weight is hung on the truss, we have to determine eight unknown forces. If we make various free-body diagrams and write down the equations of equilibrium, we find that there are eight linearly independent equations available. This is exactly the right number of equations to determine the eight unknowns, so the structure is said to be *statically determinate*.

In the truss shown in Fig. 2.8:1(b), the engineer has added two more members. Under an external load, there are now ten unknown forces while the total number of linearly independent equations of equilibrium remains at eight. Thus the forces cannot be determined by the equations of equilibrium alone, and the structure is said to be *statically indeterminate*; and the degree of redundancy is $10 - 8 = 2$.

Now, practically everything in our body is statically indeterminate. You may compare our finger with the truss shown in Fig. 2.8:1(b); and you will find that our finger has a much higher degree of redundancy.

A statically determinate structure is usually very efficient in transmitting load from one place to another, and economical in the use of materials, (e.g., of low weight, or low cost of construction). But every part has to function. Breaking one member of the truss shown in Fig. 2.8:1(a) and the whole structure will collapse.

A statically indeterminate structure may achieve some degree of *safety against failure*. For example, two members of the truss shown in Fig. 2.8:1(b) may be broken without causing the truss to collapse. Engineering structures which must avoid catastrophic failures are often built with high degrees of redundancy. An example is the airplane. In this sense, animal bodies are beautiful *fail-safe* designs.

How can the forces in an statically indeterminate structure be determined? In the case of the truss shown in Fig. 2.8:1(b), the solution calls for an application of the theory of elasticity. The principle is that the structure must remain an integral one, so the deformation of every member of the structure can be consistent with each other as a whole. The deformation of each member is related to the force in the member according to the law of elasticity. One analytical procedure introduces imagined cuts in a number of members to make the structure statically determinate, (e.g., cut the members AF, BE, in Fig. 2.8:1(b)), then applies forces on the cut surfaces as in Fig. 2.8:1(c), computes the displacements of the joints $B\ C\ D\ E\ F$ as a functon of these forces, and finally determines these forces by the fact that the lengths of the members AF and BE must be changed by exactly the same amount as those caused by the displacements of the joints B, E, F, and A that were computed in the preceding step. This method is called the method of *consistent deformation*.

Similarly, in the musculoskeletal system of an animal the forces in the muscles, tendons, ligaments, capsules, and joint contact surfaces can be determined according to the principle of consistent deformation. Each component obeys its constitutive equation. The analysis is obviously more complex than that for an inanimate structure because muscles can contract actively, and the feedback control of the neuromuscular system must be considered.

Few rigorous analysis of muscle forces in the musculoskeletal system exists. Most publications adopt some gross simplifying assumptions. Examples are:

(a) *Functional grouping*. Muscles with similar functions or common anatomical insertions and orientations are grouped together and regarded as one.

(b) *Optimization*. Based on heuristic reasoning, it is hypothesized that nature works in such a way that some quantities are minimized. Famous successful examples are the Hamilton principle in dynamics, the minimum strain energy principle in elasticity, Fermat's minimum time principle in optics, Maupertius's principle of least action. So people have proposed the minimum principle for muscle forces at joints. The mathematical problem takes the form of minimizing

$$J = f(x_1, x_2, \ldots, x_n) \tag{1}$$

subjected to the conditions

$$g_j(x_1, x_2, \ldots, x_n) = 0, \qquad (j = 1, 2, \ldots, m) \tag{2}$$

and

$$0 \leqslant b_i \leqslant x_i \leqslant a_i, \qquad (i = 1, 2, \ldots, n). \tag{3}$$

Here x_i stands for unknown muscle and joint forces, g_i stands for equations of motion or constraints. J is debated. The following have been proposed for J, but none has been validated.

$$\Sigma \text{ (muscle force)} \qquad \Sigma \text{ (muscle force)}^n$$

$$\Sigma \text{ (joint force)} \qquad \Sigma \text{ (joint force)}^2$$

$$\Sigma \text{ (joint moment)} \qquad \Sigma \text{ (muscle stress)}^n$$

$$\Sigma \text{ (muscle stress)} \qquad \Sigma \text{ (muscle energy)}$$

Experimental approach using myoelectricity is discussed by Schultz et al. (1982) and Bean et al. (1988).

2.9 Segmental Movement and Vibrations

In the above, we presented several ways by which the equations of motion of the musculoskeletal system can be written down, and some of the basic properties of these equations. We showed the existence of vibration modes in certain instances and the opportunity of greatly simplifying the equations of motion by using the normal modes as generalized coordinates.

In the musculoskeletal system, segmental movement and vibrations, of course, occur together. The major bones move essentially like rigid bodies, but they respond to dynamic loads by propagating elastic waves and vibrations. When one is interested in stresses in the bones, in their strength and failure, in their repair and interface with prosthesis, the vibrational features will come to the foreground.

For internal organs, the segmental movement is less evident. Yet the response of every organ to dynamic loads can be examined by first looking at the movement of the center of mass of the organ, and then analyzing the deformation of the organ relative to the center of mass; e.g. Ex. 3 of Sec. 2.2. For example, in considering possible injuries to the heart in a car crash, one may first regard the heart as a rigid body in the chest, then analyze its stresses and strains under the dynamic loading.

2.10 Systems with Damping and Fluid Dynamic Loads

If the material is viscoelastic or some of the forces are hydrodynamic or aerodynamic in origin, the system of equations are more complex than those discussed above. We question whether eigenvalues and normal modes exist, whether decoupling of the equations of motion is possible or not. For simplicity, we limit our discussion to linear systems.

The most general constitutive equation of a linear viscoelastic solid is due to Ludwig Boltzmann (1844–1906). Applying Boltzmann's formulation to the system considered in Sec. 1.3, we may write the internal force acting on a particle i located at point i by a displacement u_j located at point j as (see, e.g. Fung, 1965, 1981):

$$F_{ij}(t) = \int_{-\infty}^{t} G_{ij}(t - \tau) \frac{du_j(\tau)}{dt} d\tau \tag{1}$$

where $G_{ij}(t)$ are the *relaxation functions*. For solids with *fading memory*, which include Maxwell, Voigt, and Kelvin models,

$$G_{ij}(t) = K_{ij} + \beta_{ij} e^{-tv_1} + \cdots + \cdots + \gamma_{ij} e^{-tv_n} \tag{2}$$

where $v_1, v_2 \ldots v_n$ are the *relaxation frequencies*, $K_{ij}, \beta_{ij} \ldots \gamma_{ij}$ are spectral constants.

If the system is subjected to fluid dynamic forces, (e.g. in swimming, in wind, or with internal flow), the external force at the point i due to motion at point j may be written as (Chapter 3, and Fung, 1969)

$$F_{ij}^{(e)}(t) = \int_{-\infty}^{t} A_{ij}(t - \tau) \frac{du_j(\tau)}{d\tau} d\tau \tag{3}$$

where $A_{ij}(t)$ are *aerodynamic influence functions*. It is a special property of aerodynamics that A_{ij} is, in general, unequal to A_{ji}; i.e. the aerodynamic influence functions are unsymmetric.

When these expressions are inserted into Eq. (1.1:7), we obtain the equation of motion of a mass number i in an n-mass system embedded in a viscoelastic medium:

$$m_i \frac{d^2 u_i(t)}{dt^2} + \sum_{j=1}^{n} \int_{-\infty}^{t} G_{ij}(t - \tau) \frac{du_j(\tau)}{d\tau} d\tau = \sum_{j=1}^{n} \int_{-\infty}^{t} A_{ij}(t - \tau) \frac{du_j(\tau)}{d\tau} d\tau + F_i^{(e)} \tag{4}$$

where $F_i^{(e)}$ is the external force other than the fluid dynamic forces. To avoid writing integrals, let us use Laplace transformation. The Laplace transformation of a function $u(t)$ is defined by multiplying $u(t)$ by e^{-st}, integrating the product from $t = 0$ to $t = \infty$, and denoting the product by $\bar{u}(s)$. Thus,

$$\bar{u}(s) = \int_{0}^{\infty} e^{-st} u(t) \, dt. \tag{5}$$

Taking Laplace transformation of Eq. (4), with the initial values of the displacement $u_i(\tau)$ and velocity $\dot{u}_i(\tau)$ equal to u_{i0} and $\dot{u}_{i0}$ respectively at $t = 0$, we obtain

$$m_i s^2 \bar{u}_i(s) + \sum_{j=1}^{n} \left[-\bar{A}_{ij}(s) + \beta_{ij} \frac{1}{s + v_1} + \cdots + \gamma_{ij} \frac{1}{s + v_n} \right] [s\bar{u}_j(s) - \bar{u}_{j0}]$$

$$+ K_{ij} \bar{u}_j(s) = \bar{F}_i^{(e)} + m_i s u_{i0} + m_i \dot{u}_{i0}, \qquad (i = 1, \ldots, n). \tag{6}$$

These equations can be written as a matrix equation

$$s^2 M\bar{X} + K\bar{X} - sA\bar{X} + \frac{s}{s+v_1} B\bar{X} + \cdots + \frac{s}{s+v_n} \Gamma\bar{X} = \bar{F} \qquad (7)$$

in which M is a diagonal matrix (m_i), A, K, $B\ldots\Gamma$ are $(n \times n)$ square matrices (A_{ij}), (K_{ij}), (β_{ij}), $\ldots$, (γ_{ij}) respectively, X is a column matrix $(\bar{u}_i)$, F is a column matrix $\left(\bar{F}_i^{(e)} + m_i s u_{i0} + m_i \dot{u}_{i0} + \sum_{j=1}^{n} \left[-\bar{A}_{ij} + \beta_{ij} \frac{1}{s+v_1} + \cdots \right] u_{j0} \right)$. Compared with the equation treated in Secs. 2.6 and 2.7, our new equation involves additional matrices A, B, $\ldots$, Γ.

Our question is whether there exists a transformation from $(\bar{u}_1, \bar{u}_2, \ldots, \bar{u}_n)$ to generalized coordinates $(\bar{q}_1, \ldots, \bar{q}_n)$ so that Eq. (7) is decoupled into a set of independent equations of the form

$$s^2 \bar{q}_i + h_i(s)\bar{q}_i = \bar{Q}_i(s), \qquad (i = 1, \ldots, n) \qquad (8)$$

where $h_i(s)$, $\bar{Q}_i(s)$ are functions of s, independent of the $\bar{q}$'s. If this reduction is possible, then each individual coordinate q_i can be solved separately, and the features of normal coordinates are obtained.

Suppose that this reduction of Eq. (7) to Eq. (8) is possible and that the transformation from $(\bar{u}_1, \ldots, \bar{u}_n)$ to $(\bar{q}_1, \ldots, \bar{q}_n)$ is accomplished by a nonsingular real-valued matrix Φ so that

$$\begin{pmatrix} \bar{u}_1 \\ \bar{u}_2 \\ \vdots \\ \bar{u}_n \end{pmatrix} = \begin{pmatrix} \varphi_{11} & \varphi_{12} & \cdots & \varphi_{1n} \\ \varphi_{21} & \varphi_{22} & \cdots & \varphi_{2n} \\ \vdots & & & \vdots \\ \varphi_{n1} & \varphi_{n2} & \cdots & \varphi_{nn} \end{pmatrix} \begin{pmatrix} \bar{q}_1 \\ \bar{q}_2 \\ \vdots \\ \bar{q}_n \end{pmatrix} \equiv \Phi\bar{q}. \qquad (9)$$

Then, each column of Φ represents a deformation mode. These modes can be orthonormalized by the Schmidt process (Sec. 2.6) so that

$$\Phi^T M \Phi = I = \text{identity matrix.} \qquad (10)$$

When such a normalization is done, it is justified to call each column of Φ a *normal mode*.

We shall now discuss the conditions under which normal modes exist for a system described by Eq. (7).

On assuming that the reduction of Eq. (7) to Eq. (8) is possible and writing Eq. (9) and its inverse as

$$\bar{X} = \Phi\bar{q}, \qquad \bar{q} = \Phi^{-1}\bar{X} \qquad (11)$$

we premultiply Eq. (7) by Φ^T, and using (10) and (11) to obtain

$$s^2 I\bar{q} + (\Phi^T K \Phi)\bar{q} - s((\Phi^T A \Phi)\bar{q} = \cdots = \Phi^T G. \qquad (12)$$

In order that Eq. (12) be of the form of Eq. (8), it is necessary that

$$\Phi^T K \Phi = \text{diagonal}, \qquad \Phi^T A \Phi = \text{diagonal}, \qquad \Phi^T B \Phi = \text{diagonal}. \qquad (13)$$

Conversely, if such a Φ is available so that Eqs. (13) are satisfied, then (7) is diagonalized into (8). Hence, *the necessary and sufficient condition for the decoupling of the equations of motion by a transformation of the type* (11) *is the existence of a transformation matrix* Φ *which diagonalizes all the matrices* $\mathbf{M}$, $\mathbf{A}$, $\mathbf{B}$, ..., $\mathbf{K}$ *simultaneously.*

Now, in Sec. 2.6, we have shown that the columns of Φ that satisfies Eq. (10) and diagonalizes the matrix $\mathbf{K}$ are the eigenvectors of the equation

$$\mathbf{KX} = \lambda \mathbf{MX}. \tag{14}$$

Extending the reasoning to other matrices $\mathbf{A}$, $\mathbf{B}$, ..., we can state the result as follows: *The necessary and sufficient condition for Eq.* (7) *to be decoupled into Eq.* (8) *is that the eigenvalue problems*

$$\mathbf{KX} = \lambda \mathbf{MX}, \ \mathbf{AX} = \lambda \mathbf{MX}, \ \mathbf{BX} = \lambda \mathbf{MX}, \ ..., \ \mathbf{\Gamma X} = \lambda \mathbf{MX} \tag{15}$$

have a common set of linearly independent eigenvectors $\mathbf{X}_1, ..., \mathbf{X}_n$.

2.11 Sufficient Conditions for Decoupling Equations of System with Damping

For arbitrary matrices $\mathbf{K}$, $\mathbf{A}$, $\mathbf{B}$, ..., $\mathbf{\Gamma}$, decoupling of Eq. (2.10:7) is, in general, not possible. But several sufficient conditions for the possibility are known. We shall quote them without proof.

1) The case in which $\mathbf{A}$, $\mathbf{B}$, ..., $\mathbf{K}$ are real and symmetric.

If the matrices $\mathbf{A}$, $\mathbf{B}$, ..., $\mathbf{K}$ *are real and symmetric, and can be diagonalized simultaneously, then among these matrices there is one, say* $\mathbf{K}$, *which is commutable with all the rest; i.e.,*

$$\mathbf{AK} = \mathbf{KA}, \ \mathbf{BK} = \mathbf{KB}, \ ..., \ etc. \tag{1}$$

If one of the matrices, say $\mathbf{K}$, *is commutable with all other matrices* $\mathbf{A}$, $\mathbf{B}$, ..., *and if all the eigenvalues of* $\mathbf{K}$ *are distinct* (*no multiple eigenvalues*), *then* $\mathbf{A}$, $\mathbf{B}$, ..., $\mathbf{K}$ *can be simultaneously diagonalized.*

This result is due to Bellman (1970). The condition that $\mathbf{K}$ should have no multiple eigenvalues weakens the theorem, but it cannot be removed. However, if only two matrices $\mathbf{A}$, $\mathbf{K}$ are involved, then multiple eigenvalue can be permitted. In this case one must show that if λ_v is an r-fold multiple root, then among the linear combinations of the r mutually orthogonal eigenvectors of $\mathbf{K}$ belonging to λ_v, there are r eigenvectors of $\mathbf{A}$.

2) The case in which $\mathbf{A}$, $\mathbf{B}$, ..., $\mathbf{K}$ are normal matrices.

If we do not wish to restrict ourselves to real and symmetric matrices, we should demand at least the existence of n linearly independent eigenvectors for the eigenvalue problems (Eq. (2.10:15)). Now, there is a well-known theorem which states that *if* $AX = \lambda X$ *possesses a system of n linearly independent eigenvectors, then* A *is a normal matrix*; i.e., it is commutable with its adjoint:

$$A \cdot (\text{adj } A) = (\text{adj } A) \cdot A. \tag{2}$$

[The adjoint of a square matrix $A = (a_{ij})$ is the matrix adj $A = (\bar{a}_{ji})$, i.e., the transpose of a matrix whose elements are the complex conjugate of those of (a_{ij}). If A is real, then adj A is just the transpose of A.]

We quote without proof Bellman's (1970) results:

If M *is real, symmetric, and positive definite, and* $A, B, \ldots, K$ *are all normal matrices commutable with* M,

$$AM = MA, BM = MB, \ldots, KM = MK, \tag{3}$$

then the matrices $M, A, B, \ldots, K$ *can be simultaneously diagonalized if one of the matrices,* $A, \ldots, K$, *say* K, *is commutable with all the rest*:

$$AK = KA, BK = KB, \text{ etc.}, \tag{4}$$

and if all the eigenvalues of K *are single.*

Examples. Rayleigh (1894) considered the vibration of a Voigt solid described by the matrix equation

$$M\ddot{X} + B\dot{X} + AX = F, \tag{5}$$

where M, B, A are real, symmetric, and positive definite. This equation can be decoupled; if B is a linear combination of M and A:

$$B = \alpha M + \beta A, \tag{6}$$

where α, β are numerical constants.

Caughey (1960) considered Eq. (5) under the assumption that A is real, symmetric, and positive definite, and showed that B and A can be simultaneously diagonalized if B can be represented as a power series of A,

$$B = \sum_{m=0}^{n=1} \alpha_m A^m. \tag{7}$$

Caughey (1960) showed further that if $A = B^\kappa$, $\kappa = $ an integer, then A and B are commutable. Define $A^{1/\kappa}$ as a matrix B which satisfies the equation $A = B^\kappa$. Then A and B can be simultaneously diagonalized if B can be expressed in a series

$$B = \sum_{m=0}^{n=1} \beta_m (A^{1/\kappa})^m. \tag{8}$$

Problems

2.1 Solve Eq. (2.2:2) when $F^{(e)}$ is a sinusoidal excitation:

$$F^{(e)} = F_0 \sin \omega t.$$

Sketch the relationship between the amplitude of motion $|x|$ and the frequency of excitation ω. Discuss the three cases:

$$\beta^2 < 4Km, \qquad \beta^2 > 4Km, \qquad \beta^2 = 4Km.$$

Determine also the response of this system to an impulsive impact load which is a delta function of time. Assume that the initial conditions are $x = \dot{x} = 0$ at $t = 0$. Propose a possible application of the solution to a clinical problem.

2.2 Write down the Lagrangian equation of motion for a weightless cantilever beam with a mass attached to the free end. The mass has a finite moment of inertia, I, and is rigidly attached to the beam.

2.3 Derive the Lagrangian equation for the vertebrae–disks system (Sec. 1.4, Figs. 1.4:3, 1.4:4) by allowing six degrees of freedom: three relative translations of two adjacent vertebrae, two bending and one twisting. Consider the lower vertebrae as fixed while the upper vertebra and the disk in between can be displaced.

2.4 Formulate a model of the head and spine of a man to be used in the investigation of head injury to drivers in automobile crashes, or neck injury of an aircraft pilot in an ejection seat. Derive the equations of motion.

2.5 Show that in normal coordinates, the kinetic and strain energies are

$$K = \tfrac{1}{2}(\dot{\mathbf{q}}, \dot{\mathbf{q}}) = \tfrac{1}{2}(\dot{q}_1^2 + \cdots + \dot{q}_n^2),$$
$$U = \tfrac{1}{2}(\mathbf{q}, \mathbf{\Phi}^{\mathsf{T}} \mathbf{K} \mathbf{\Phi} \mathbf{q}) = \tfrac{1}{2}(\omega_1^2 q_1^2 + \cdots + \omega_n^2 q_n^2).$$

2.6 In Fig. 2.5:3, a man standing on ground is modeled as an inverted pendulum. Consider this as a dynamic model of posture control. Assume a feedback control system of the body which yields the following moment at the ankle: $M = k_1 \theta + k_2 \dot{\theta}$, where k_1 and k_2 are constants. Considering only very small angle θ, what are the restrictions on the values of k_1, k_2 so that the body remains *stable*, so that he will not fall down if slightly disturbed. What are the restrictions on the values of k_1, k_2 so that he is *dynamically stable* (so that a small disturbance will not result in an oscillation whose amplitude will increase with increasing time)? (Cf. Nashner (1971) and McGhee (1980).)

2.7 By increasing the number of joints and degrees of freedom in the inverted pendulum model, present a better model for posture control. (Cf. McGhee (1980) and Pedotti (1980). Data on muscles of the leg, their mean length, fiber length, cross-sectional area, maximum velocity, maximum isometric force, and constants in Hill's equation are presented in Pedotti's paper (p. 113).)

2.8 Discuss the aerodynamic forces acting on an athlete doing one of the following sports: parachuting, free fall from an airplane, ski jumping or ski flight, jumping from a diving board into water, pole vaulting, motor cycling. (Cf. J. Maryniak (1977). Static and dynamic investigations of human motion. In *Euromech Colloquium*, Varna, 1975: *Mechanics of Biological Solids*, (edited by G. Brankov). Bulgarian Academy of Science, Sofia, pp. 151–174.)

2.9 To investigate possible injuries to the knee in playing soccer football, make a model of the leg as a double pendulum, and derive the Lagrangian equation of motion using the angles of the thigh and the lower leg from the verticle as

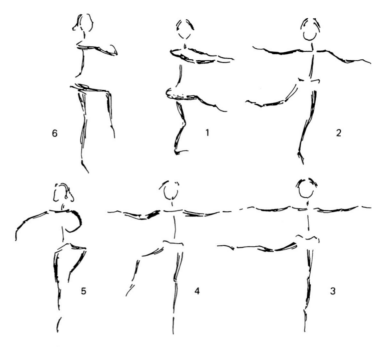

FIGURE P2.10 A fouetté turn. The performer's body rotates very little in views 2–4 while her extended leg retains most of the angular momentum. In views 5–6 she turns rapidly while her leg is held close to her body, where it has a smaller moment of inertia.

generalized coordinates. (Cf. L. Lindbeck (1977). Theoretical analysis of reactions in the knee caused by impact. In *Euromech Coll.*, Varna, 1975, loc cit. pp. 142–150.)

2.10 Analyze the dance steps of pirouettes (a rotation of the body around a vertical axis over one supporting foot on the floor) in terms of the principle of conservation of angular momentum. There are many types and styles of pirouettes, from a low turn on bent legs in modern dance to attitude or arabesque turns, and to the spectacular multiple fouetté turns often seen in classical ballet. See sketches in Fig. P2.10. To commence a pirouette, the dancer must exert opposite horizontal forces with the two feet in order to apply the necessary torque. How is this done? Can the dancer apply some additional torque in the course of the turn to maintain the angular momentum against necessary losses? (See Laws, K. (1984). *The Physics of Dance*. Schirmer Books.)

2.11 Two dancers, one 5 ft. tall, the other 6 ft tall, are similarly proportioned, and their muscles are similar. Compare their body masses, cross sectional areas of their legs, the forces their muscles can exert, the heights they can jump, and the lengths of time they can remain in the air. Show that in order to allow the larger dancer to jump to the same height as the smaller dancer a 73% force larger is necessary. If they want to jump to the heights in same proportion to their body length, the larger dancer would have to exert twice the muscular force. If they perform "beats" with legs at the same rate and angular amplitude, the taller dancer must exert 2.5 times the torque at the hip. (A beating motion is executed during a vertical jump, in which the legs, while straight are kicked out to each side, then crossed, then kicked out, often several times in one leap.) (See Laws, K. (1984). loc cit.)

2.12 T. Kenner designed a very sensitive instrument to measure the density of blood, plasma, and other body fluid (Paar Inc., Graz, Austria). The principle is to set the fluid flowing through a small cylindrical tube which is bent into a U shape and excite the tube to measure its frequency response. The resonance frequencies can be measured very accurately with an electronic counter. These frequencies are related to the density of the fluid. Knowing this much about the principle, reinvent such an instrument with your own design. (Cf. Applications to pulmonary problem by J.S. Lee (1988). *Microvas. Res.* **35**: 48–62.)

2.13 Formulate a model of the arm to study the tennis elbow, pains that often afflict tennis players.

2.14 Formulate a model of the human leg with the objective to measure the load on the articular cartilage of the knee. Discuss the possible relevance of such studies to the problem of cartilage breakdown and joint degeneration (cf. Mizrahi and Susak (1982)).

2.15 Review the meaning of Eulerian angles which describe the rotation of a rigid body in space; cf. e.g. Greenwood (1965, 1977), Kane (1968), Huston and Perrone (1980). In the modeling of arms, legs in locomotion, the choice of parameters to measure angular position is very important. There are special choices of the rotation axes with which the final angular position of the rigid body is specified by three angles whose values do not depend on the order of the rotations. A general theory for such choices has been developed by Roth (1967). A *gyroscopic* system has been used by Chao (1980), Chao and Morrey (1978), Grood and Suntay (1983), in which the angles are the *clinical* measures used by orthopedic surgeons: the flexion-extension, the internal-external rotation, and the adduction-abduction.

References

For a thorough discussion of Lagrangian equations, see Greenwood (1977). For influence coefficients (Sec. 2.3), see Fung (1955), and Fung (1965). For aerodynamic loading, see Fung (1955). For muscle mechanics, see Fung (1981).

Bean, J.C., Chaffin, D.B., and Schultz, A.B. (1988). Biomechanical model calculation of muscle contraction forces: a double linear programming method. *J. Biomech.* **21**: 59–66.

Bellman, R.E. (1970). *Introduction to Matrix Analysis*, 2nd ed. McGraw-Hill, New York.

Camana, P.C., Hemami, H., and Stockwell, C.W. (1977). Determination of feedback for human posture control without physical intervention. *J. Cybernet.* **7**: 199.

Caughey, T.K. (1960). Classical normal modes in damped linear dynamic systems. *J. Appl. Mech.* **27**: 269–271.

Chao, E.Y.S. (1980). Justification of triaxial goniometer for the measurement of joint motion. *J. Biomech.* **13**: 989–1006.

Chao, E.Y.S. and Morrey, B.F. (1978). Three-dimensional rotation of the elbow. *J. Biomech.* **11**: 57–74.

Fung, Y.C. (1955, 1969). *The Theory of Aeroelasticity*. Wiley, New York. Revised, Dover Publications, New York.

Fung, Y.C. (1965). *Foundations of Solid Mechanics*, Prentice-Hall, Englewood Cliffs, N.J.

Ghista, C.N. (1982). *Osteoarthromechanics*. McGraw-Hill, New York.

Greenwood, D.T. (1965). *Principles of Dynamics*. Prentice-Hall, Englewood Cliffs, N.J.

Greenwood, D.T. (1977). *Classical Dynamics*. Prentice-Hall, Englewood Cliffs, N.J.

Grood, E.S. and Suntay, W.J. (1983). A joint coordinate system for the clinical description of three-dimensional motions: Application to the knee. *J. Biomech. Eng.* **105**: 136–144.

Huston, R.L. and Perrone, N. (1980). Dynamic response and protection of the human body and skull in impact simulation. In *Perspective in Biomechanics*. (H. Reul, D.N. Ghista and G. Rau, eds.), Vol. 1, Part B, pp. 531–571, Harwood Academic Publishers, New York.

Kane, T.R. (1968). *Dynamics*. Holt, Rinehart, and Winston, New York.

McGhee, R.B. (1980). Computer simulation of human movements. In *Biomechanics of Motion* (A. Morecki, ed.), Springer-Verlag, Wien and New York, pp. 41–78.

Mizrahi, J. and Susak, Z. (1982). In vivo elastic damping response of the human leg to impact forces. *J. Biomech. Eng.* **104**: 63–66.

Nashner, L.M. (1971). A model describing vestibular detection of body sway motion. *Acta Otolaryng.* **72**: 429–436.

Pauwels, F. (1980). *Biomechanics of the Locomotor Apparatus*. (Trans. from the German by P. Maquet and R. Furlong) Springer-Verlag, New York.

Pedotti, A. (1980). Motor coordination and neuromuscular activities in human locomotion. In *Biomechanics of Motion* (A. Morecki, ed., Springer-Verlag, New York, pp. 79–129.

Rayleigh, Baron, John William Strutt (1894). *The Theory of Sound*. Republished by Dover, New York, Vol. 1, p. 131.

Roth, B. (1967). Finite position theory applied to mechanism synthesis. *J. Appl. Mech.* **34**: 599–606.

Schultz, A.B., Anderson, G.B.J., Haderspeck, K., Örtengren, R., Nordin, M., and Björk, R. (1982). Analysis and measurement of lumbar trunk loads in tasks involving bends and twists. *J. Biomech.* **15**: 669–675.

HAWK IN THE AUTUMN by Lin Liang (林良). Light color and ink on silk, 136.8 × 74.8 cm, wall scroll, in National Palace Museum, Taipei. Lin was a native of Kwantung. He lived in the latter half of the 15th century, in Ming Dynasty.

CHAPTER 3

External Flow: Fluid Dynamic Forces Acting on Moving Bodies

3.1 Introduction

When humans exercise, birds fly, fish swim, animals run, and trees sway, we want to know the forces they experience. The calculation of the forces they experience from the surrounding fluid is the realm of fluid dynamics. In this chapter we present the classical theory. In the next chapter we discuss flying and swimming in nature.

3.2 Flow Around an Airfoil

A real fluid is viscous and compressible. But, if the speed of flow is much less than the speed of propagation of sound, then the variation of density caused by the motion of a body in the fluid is so small that the fluid may be regarded as *incompressible*. Furthermore, for birds and fish moving in air and water at Reynolds number much greater than one, the effect of the viscosity of the fluid is felt only in a thin layer (the *boundary layer*) next to solid wall of the body. Outside the boundary layer the fluid may be regarded as *nonviscous*. A nonviscous and incompressible fluid is a *perfect fluid*. In many problems of locomotion, it is sufficient to consider the fluid as a perfect fluid. Yet the viscosity, however small, has profound effects, for it controls the boundary layer which may become detached from part of the solid body, and thus affects the macroscopic picture of the flow.

The force exerted by the fluid on a solid body depends on the *relative velocity* between them. The fluid dynamic force consists of two components: the *pressure force* normal to the surface of the body, and the *skin friction*, or *shearing force*, tangential to the surface of the body.

The parameters affecting the force acting on a body in a flow can be determined by a *dimensional analysis*. Obviously the force depends on the geometry of the body and its attitude relative to the flow; these can be characterized by a *typical length* L and a *typical angle* α. The force depends also on the *density* of the fluid ρ, the *viscosity* of the fluid μ, the *speed* of flow U, the *compressibility* of the fluid, and the nonstationary characteristics of the flow, e.g., the *frequency* ω if the motion is periodic. The compressibility of the fluid may be expressed in various ways. A simple index of the compressibility is the *speed of propagation of sound* in the fluid, because sound is propagated as longitudinal elastic waves.

Let the speed of sound propagation be denoted by c. Then, in an oscillating flow of a compressible fluid, the force experienced by a solid body will depend on the following variables:

$$L, \alpha, \rho, U, \mu, \omega, c. \tag{1a}$$

A dimensional analysis shows that, for geometrically similar bodies, the force F acting on the body can be expressed as

$$F = f\left(\alpha, \frac{UL\rho}{\mu}, \frac{\omega L}{U}, \frac{U}{c}\right)\frac{1}{2}\rho U^2 L^2 \tag{1b}$$

where f is a function of α, $UL\rho/\mu$, $\omega L/U$, and U/c. It is easy to verify that the parameters $UL\rho/\mu$, $\omega L/U$, and U/c are *dimensionless*. They are known as:

$$R = \frac{UL\rho}{\mu} = \frac{UL}{\nu} = \text{Reynolds number}$$

$$k = \frac{\omega L}{U} = \text{reduced frequency or Strouhal number} \tag{2}$$

$$M = \frac{U}{c} = \text{Mach number.}$$

R and k are also written as N_R and N_S, resp., elsewhere in this book. The quantity $q = \frac{1}{2}\rho U^2$ is known as the *dynamic pressure*. The factor $\nu = \mu/\rho$ is the *kinematic viscosity*.

The *speed of sound* propagation in a gas is given by the equation

$$c = \sqrt{\frac{\gamma p}{\rho}}, \tag{3}$$

where p is the static pressure, ρ is the density, and γ is the ratio of the specific heat at constant pressure to the specific heat at constant volume. For dry air at 15°C and 1 atm, $\nu = 0.145$ cm^2/sec, $\gamma = 1.401$, $\rho = 1.225 \times 10^{-3}$ gm/cm^3, $c = 340.6$ m/sec.

For water under the same conditions, $\nu = 1.138 \times 10^{-2}$ cm^2/sec, $\rho = 0.9991$ gm/cm^3, $c = 1{,}445$ m/sec.

In flying and swimming, we are concerned primarily with bodies like an airfoil. The wing geometry and the conventional terminology are illustrated in Fig. 3.2:1. A *chord* is a line defined in a cross section, passing through the trailing edge, pointing in the direction of relative wind of no-lift. The angle between the chord and the direction of flight is the *angle of attack*. Two components of force and one component of moment that act on the body are

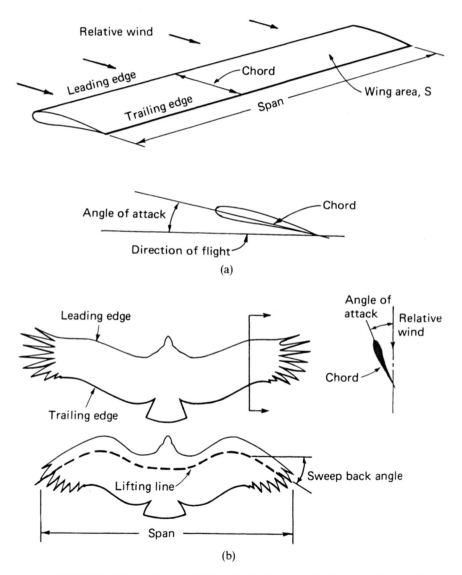

FIGURE 3.2:1 Definitions of terms. (a) An idealized wing. (b) Upper: A black vulture with wings outstretched in soaring flight. Lower: Wings flexed in fast gliding flight.

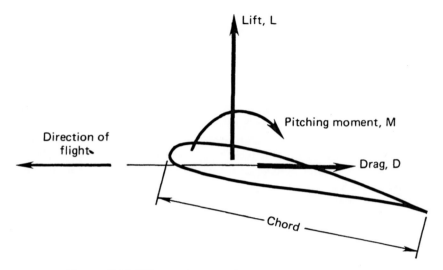

FIGURE 3.2:2 Wing geometry and conventional terminology.

of interest (Fig. 3.2:2):

Lift $= L =$ force perpendicular to the direction of motion,

Drag $= D =$ force in the direction of motion, positive when the force acts
in the downstream direction.

Pitching moment $= M =$ moment about an axis perpendicular to both the
direction of motion and the lift vector, positive
when it tends to raise the leading edge of the
body.

For the wings of birds or airplane, the chords of all cross sections lie
approximately in a plane defined as the *mean chord plane*. The mean chord
length of the wing is usually taken as the characteristic length of the wing. The
area of the wing projected on the mean chord plane is defined as the *wing area*.
Let c be the mean chord length, S be the wing area, q be the dynamic pressure.
Then the three primary dimensionless coefficients of interest are:

$$C_L = \text{lift}/(qS) = \textit{lift coefficient}$$

$$C_D = \text{drag}/(qS) = \textit{drag coefficient} \qquad (4)$$

$$C_M = \text{pitching moment}/qSc = \textit{pitching-moment coefficient}.$$

C_L, C_D, and C_M are functions of the Reynolds number, Mach number, Strouhal
number, and the body's shape and attitude with respect to the flow.

In a steady flow of an incompressible fluid, the Strouhal number and the
Mach number both vanish, and C_L, C_D, C_M depend on the Reynolds number

and angle of attack. The Reynolds number characterizes the effect of fluid viscosity. For large birds, fish and animals, the Reynolds number is much larger than one. In the remainder of this section, unless mentioned otherwise, we shall consider only the steady flow of an incompressible fluid at large Reynolds number. Flow at small Reynolds number, relevant to microbes and cells, is discussed in Chapter 4.

The lift increases linearly with increasing angle of attack until a certain value around 10–20° is exceeded. Beyond that value, the lift levels off, eventually reaches a maximum, and then drops off rapidly. The maximum lift is reached at a critical angle of attack, $\alpha_{L\,max}$, which is the *stalling angle*. For angle of attack α greater than $\alpha_{L\,max}$ the wing is *stalled*. The stalling angle and the maximum lift coefficient are characteristic numbers for each wing design.

The lift of a wing is zero at some angle of attack. It is convenient to define that angle of attack as zero, and measure the angle of attack from the *zero-lift line*. Such an angle of attack is denoted by α. The *chord* of an airfoil is defined along the zero-lift line.

For an unstalled wing, the lift coefficient can be expressed as

$$C_L = a\alpha \tag{5}$$

in which α is the angle of attack and a is the *lift curve slope*. If α is measured in radians, hydrodynamic theory (see Sec. 3.11) gives the lift-curve slope for thin airfoils of infinite span in a two-dimensional flow as

$$a_0 = 2\pi \quad \text{(theory, incompressible fluid).} \tag{6}$$

Experimental values of the lift-curve slope are somewhat smaller than this for most wings, but a/a_0 is greater than 1 for the so-called NACA low-drag sections designed and tested by NACA (US National Advisory Committee of Aeronautics, predecessor of NASA).

According to the theory of thin airfoil (Sec. 3.11), the center of pressure of the lift is located at $\frac{1}{4}$-chord aft of the leading edge. This point is the *aerodynamic center*. If the moment coefficient is computed about the aerodynamic center, it does not vary with C_L. The symbol $C_{Mc/4}$ is used to denote the moment coefficient that refers to an axis located at the $\frac{1}{4}$-chord point. The aerodynamic center remains close to the $\frac{1}{4}$-chord point in a compressible fluid as long as the flow is *subsonic*; but it moves close to the midchord point if the flow becomes *supersonic*.

The subscript "0" of the lift-curve slope a_0 in Eq. (6) signifies that a_0 is the value pertaining to an airfoil of infinitely long span. (The *span* is the distance from wing tip to wing tip.) For wings of *finite span*, the lift-curve slope is smaller. In Prandtl's finite-wing theory (see Sec. 3.10), a wing is replaced by a *vortex line*. Since a vortex line cannot end at the wing tip, it must continue laterally out of the wing and become a *free trailing vortex* in the fluid. This will be discussed later in Secs. 3.9 and 3.10. The vertical velocity induced by the vortex line and trailing vortices is called the *induced velocity*, or *downwash*,

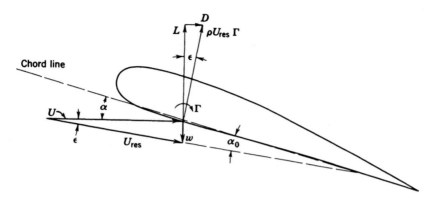

FIGURE 3.2:3 Downwash and induced angle.

and is denoted by w in Fig. 3.2:3. Because of the induced velocity, the direction of flow at the airfoil is changed by an amount ε indicated in the figure. If U is the flight speed, then

$$\tan \varepsilon = w/U. \tag{7}$$

From Fig. 3.2:3 it is seen that the *effective angle of attack* α_0 is smaller than the *geometric angle of attack* α according to the relation

$$\alpha_0 = \alpha - \varepsilon \doteq \alpha - \frac{w}{U} \tag{8}$$

when ε is so small that $\tan \varepsilon \doteq \varepsilon$. The force induced by the effective angle of attack is proportional to the velocity resultant U_{res}, and acts in a direction normal to the velocity vector U_{res}. See Sec. 3.7 infra and vector $\rho U_{res} \Gamma$ in Fig. 3.2:3. It can be resolved into a *lift component L* perpendicular to the velocity of flow U and a *drag component (induced drag) D* in the direction of U. The resultant $\rho U_{res} \Gamma$ is proportional to the circulation Γ (Sec. 3.7) and angle of attack α_0. By using Eqs. (5), (6), and (8), we obtain

$$C_L = a\alpha = a_0\alpha_0 = a_0\left(\alpha - \frac{w}{U}\right). \tag{9}$$

The downwash w is uniform over the entire wing if the wing planform is an elongated *ellipse* and is *untwisted* (having a constant angle of attack across the span) (see Sec. 3.10). In this case

$$\frac{w}{U} = \frac{C_L}{\pi AR}, \tag{10}$$

where

$$AR = aspect\ ratio = \frac{(\text{span})^2}{\text{wing area}} = \frac{\text{wing span}}{\text{chord length}} \tag{11}$$

Substituting (10) into (9) yields a, the *lift-curve slope* of a wing of *finite span*:

$$a = \frac{a_0}{1 + (a_0/\pi AR)} \quad \text{(elliptic wing).} \tag{12}$$

Nonelliptic wings would have nonuniform downwash across the span and a somewhat smaller lift curve slope than that given by Eq. (12). See Glauert (1947).

These results are applicable to birds. To improve the stalling characteristics of the wing and to obtain a high maximum lift coefficient, birds spread their feathers to produce a more curved wing cross-section, and use many other features that are copied by aeronautical enginers. See Sec. 4.3.

3.3 Flow Around Bluff Bodies

For a blunt body (such as a man) moving in a fluid, usually the drag force predominates but an oscillatory lift may exist. We shall discuss drag in Sec. 3.15 and lift in this section.

Whoever has rowed a boat must have observed the trail of vortices leaving the oar. Figure 3.3:1 shows the wake behind a circular cylinder. Vortices are "shed" alternately from the sides of the cylinder. This shedding of vortices induces a periodic force in the direction perpendicular to the line of motion, i.e., an oscillatory lift.

The vortex shedding phenomenon is relevant to the swaying of trees, rustling of leaves, and bending of the blades of grass. We feel it on our legs

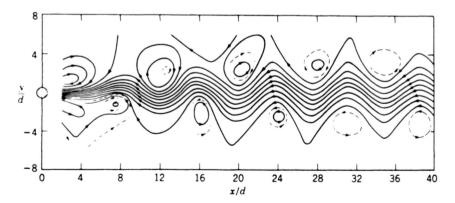

FIGURE 3.3:1 The wake behind a circular cylinder. Reynolds number 56. Measurements by Kovasznay (1949). Figure shows the streamline pattern viewed relative to the undisturbed flow at infinity. The development and decay of the vortices can be seen. The lines correspond to differences in the stream function $\Delta \psi = 0.1\ Ud$; the dotted lines are half-values between two full lines.

when we wade in running water in a creek. Cranes and ducks must know it well. In the manmade world, telephone wires "sing," and smokestacks, submarine periscopes, oil pipe lines, and television antennas vibrate for the same reason. These vibrations can be controlled either by stiffening the structures so that the natural frequency is higher than the frequency of the vortex shedding in wind, or by introducing vibration dampers into the system to absorb the energy.

The flow around a long circular cylinder will be explained in greater detail. The flow changes with the Reynolds number, R, defined as vd/v, d being the diameter of the cylinder. The variations of the drag coefficient (Eq. 3.2:4) and the Strouhal number (Eq. 3.2:2) of the flow with Reynolds number are shown in Fig. 3.3:2. At low Reynolds number, the flow is smooth and unseparated,

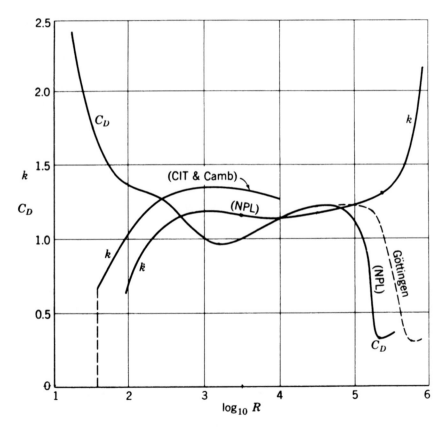

FIGURE 3.3:2 Variation of the Strouhal number and drag coefficient against Reynolds number for a circular cylinder. C_D and R are based on the diameter of the cylinder. Sources of data are: NPL; Relf and Simmons, Aeronaut. Research Com. R. and M. 917 (1924). Cambridge; Kovasznay, Proc. Roy. Soc. A. **198** (1949). CIT; Roshko, NACA Tech. Note 2913 (1953). Göttingen; Ergebnisse AVA Göttingen, 2 (1923).

but the fluid at the back of the cylinder is appreciably retarded. At higher values of R, two symmetrical standing vortices are formed at the back. When R reaches about 40, the vortices become asymmetrical, detach from the obstacle, and move downstream as if they were discharged alternately from the two sides of the cylinder. An eddying motion in the wake is set up. As the flow moves downstream the eddying motion is gradually diffused and "decays" into a general turbulence. For R in the range of 40 to 150, the "shedding" of vortices is regular. The range of R between 150 and 200 is a transition range, in which the vortex shedding is less regular and its frequency appears to be somewhat erratic. For $R \sim 300$, the vortex shedding is irregular, for although a predominant frequency exists, the amplitude appears to be random. Finally, at R of the order 2×10^5, the separation point of the boundary layer moves rearward on the cylinder. Consequently, the drag coefficient of the cylinder decreases appreciably, as shown in Fig. 3.3:2.

The geometry of the wake, when the Reynolds number is in a range in which vortices are regarded as shedding, is as follows: The frequency at which the vortices are shed, expressed nondimensionally as the *Strouhal number k*, is a function of the Reynolds number, as shown in Fig. 3.3:2. Here the Strouhal number k is defined as $\omega d/U$, where ω is the frequency in radians per second and d is the diameter of the cylinder. The number of vortices shed from each side of the cylinder every second is $n = \omega/2\pi$:

$$n = \frac{kU}{2\pi d} \text{ per second.} \tag{1}$$

The distance A between two consecutive vortices in a row is

$$A = \frac{U - v}{n}, \qquad \left(\frac{v}{U} \sim 0.25 \text{ to } 0\right), \tag{2}$$

where v is the relative velocity of the vortices with respect to the free stream. The flow pattern is approximated by a regular array of vortices shown in Fig. 3.3:3. This flow pattern was analyzed by theodore von Kármán and is known as the von Kármán vortex street. The theoretical ratio of the distance H between the rows of vortices to the distance A is given by von Kármán to be

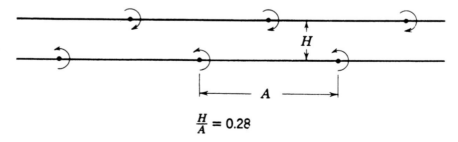

$$\frac{H}{A} = 0.28$$

FIGURE 3.3:3 A Kármán vortex street.

$H = 0.281A$ at small distance from the cylinder, but H/A increases as the distance from the cylinder increases. At large distance, H/A is of the order 0.9. The maximum intensity of the velocity fluctuations occurs in the vicinity of 7 diameters downstream. Thus it appears that the vortices are not really shed from the cylinder, but are developed gradually.

The shedding of vortices creates an oscillatory *lift*. In the Reynolds number range 100–1,000, the periodic lift coefficient has an amplitude of 0.45, while the drag coefficient is about 1.09. Some experiments indicate that the lift coefficient can be as large as 1.

3.4 Steady-State Aeroelastic Problems

Let us illustrate the application of the information presented in Secs. 3.2 and 3.3, by considering some aeroelastic instabilities.

Divergence

If a wing in steady flight is accidentally deformed, an aerodynamic moment will be induced which tends to twist the wing. This twisting is resisted by elastic moment. However, since the elastic stiffness is independent of the speed of flight, whereas the aerodynamic moment is proportional to the square of the flight speed, there may exist a *critical speed* of flight, at which the elastic stiffness is barely sufficient to hold the wing in a stable position. Above the critical speed, an infinitesimal accidental deformation of the wing will lead to a large angle of twist. This critical speed is called the *divergence speed*, and the wing is then said to be *torsionally divergent*.

As a *two-dimensional* example, let us consider *a strip of unit span* of an infinitely long wing of uniform cross-section. As shown in Fig. 3.4:1 let the elastic restraint imposed on this strip be regarded as a torsional spring with an axis at a point G. If the spring is linear then the torque is directly proportional to the angle of twist.

The action of the aerodynamic force on the airfoil can be represented by a lift force acting through the aerodynamic center, and a moment about the same point. Let us write the distance from the aerodynamic center to the axis

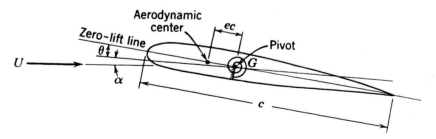

FIGURE 3.4:1 A two-dimensional airfoil.

of the torsional spring as ec, c being the chord length and e being a ratio expressing the *eccentricity* of the aerodynamic center (positive if the spring lies behind the aerodynamic center). The lift coefficient C_l (we use lower case l to indicate lift per unit span) is proportional to the angle of attack, whereas the coefficient of moment about the aerodynamic center, C_{m0}, is practically independent of the angle of attack. Hence, the *lift* and *moment per unit span* acting on the airfoil are

$$L' = qcC_l = qca(\theta + \alpha),$$
$$M'_0 = qC_{m0}c^2 \quad \text{(about aerodynamic center)},$$
(1)

respectively, where a is the slope of the lift-curve (C_l vs. α), q is the dynamic pressure ($\frac{1}{2}\rho U^2$), α is the initial angle of attack, and θ is the angle of twist. The prime on L' denotes force per unit length in the spanwise direction.

Using Eqs. (1), the aerodynamic moment *per unit span* about the torsional spring is

$$M'_a = M'_0 + L'ec = qc^2 C_{m0} + qec^2 a(\theta + \alpha)$$
$$= qec^2 a\theta + qec^2 a(\alpha + C_{m0}/ea).$$
(2)

By redefining the angle of attack by absorbing C_{m0}/ea into α we may be rewrite Eq. (2):

$$M'_a = qec^2 a(\theta + \alpha).$$
(3)

When equilibrium prevails, the aerodynamic moment is balanced by the elastic restoring moment. Let K_α be the spring constant, then the elastic restoring moment per unit span is $K_\alpha \theta$. On equating this with the aerodynamic moment given by Eq. (3), and solving for θ, we obtain

$$\theta = \frac{qec^2 a\alpha}{K_\alpha - qec^2 a}.$$
(4)

For a given nonvanishing α, the angle θ will increase when the dynamic pressure q increases. When q is so large that the denominator vanishes, the angle θ tends to infinity and the airfoil becomes divergent. Hence, the condition of divergence is

$$K_\alpha - qec^2 a = 0.$$
(5)

The dynamic pressure at divergence, q_{div}, and the divergence speed of flight, U_{div} are given by the equations

$$q_{\text{div}} = \frac{K_\alpha}{ec^2 a} \quad \text{and} \quad U_{\text{div}} = \sqrt{\frac{2K_\alpha}{\rho ec^2 a}}.$$
(6)

Thus the critical divergence speed increases with increasing rigidity of the wing and decreasing chord length and eccentricity. The ratio of the actual angle of twist of an elastic wing to that of a rigid wing varies with the ratio of the dynamic pressure to the critical divergence pressure, as shown in Fig. 3.4:2.

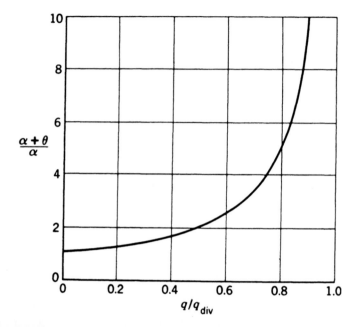

FIGURE 3.4:2 Ratio of angles of twist of an elastic wing to that of a rigid wing.

It is seen that the twist becomes very large when the divergence speed is approached.

The analysis presented above can be extended to three-dimensional wings. For details see Fung (1955, 1969), Bisplinghoff and Ashley (1962), Dowell, et al. (1978). Among other things it is shown that *sweeping back* the wing (wing sheared backward with wing tip pointing towards the tail) *increases* the critical divergence speed; whereas *sweeping forward* (wing so sheared that the tip points to the head) *decreases* it. Most bird's wings are sweptback. In the example shown in Fig. 3.2:1(b), the inner span is swept forward, the outer span is swept back.

Loss of Control

Airplanes pitch, roll, and yaw by deflecting their ailerons, flaps, elevators and rudders. Birds do these maneuvers by feathering their wings and deflecting their tails. The effectiveness of the deflection of control surface, however, depends on the speed of flight. The loss of effective control as flight speed increases puts an upper limit on the speed of flight.

The phenomenon can be explained by considering a simplified model as shown in Fig. 3.4:3. Consider a wing of chord length c that is held by a torsional spring located at a distance ec from the aerodynamic center. The wing has an aileron of width Ec. e and E are constant fractions. The relative wind comes

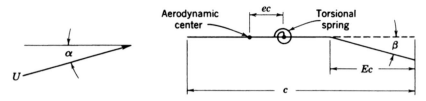

FIGURE 3.4:3 A two-dimensional wing with aileron.

at an angle of attack α against the zero-lift line. The angle of deflection of the aileron (positive downward) with respect to the main airfoil is denoted by β.

The lift coefficient and the coefficient of moment about the aerodynamic center can be written in the form

$$C_l = a\alpha + \beta \frac{\partial C_l}{\partial \beta}, \tag{7}$$

$$C_m = \beta \frac{\partial C_m}{\partial \beta} + C_{m0}, \tag{8}$$

where a is the lift-curve slope and C_{m0} is the coefficient of moment about the aerodynamic center of the airfoil with undeflected aileron. According to Glauert (1947), the coefficients $\partial C_l/\partial \beta$ and $\partial C_m/\partial \beta$ for a two-dimensional airfoil in an incompressible fluid are

$$\frac{1}{a}\frac{\partial C_l}{\partial \beta} = \frac{1}{\pi} \left[\arccos(1 - 2E) + 2\sqrt{E(1 - E)} \right] \tag{9}$$

$$\frac{\partial C_m}{\partial \beta} = \frac{a}{\pi}(1 - E)\sqrt{E(1 - E)} \quad (\beta \text{ in radians}), \tag{10}$$

where E is the ratio of the flap chord to the total chord (Fig. 3.4:3). According to Eq. (7), the lift per unit length of this airfoil is

$$L' = qc \left(a\alpha + \frac{\partial C_l}{\partial \beta} \beta \right). \tag{11}$$

If no lift can be produced when the aileron is deflected, then the control is lost. The critical reversal condition is given by the vanishing of the derivative $dL'/d\beta$. This happens, according to Eq. (11) and noting that $\partial C_l/\partial \beta$ is a constant, Eq. (9), when

$$\frac{dL'}{d\beta} = qc \left(a \frac{\partial \alpha}{\partial \beta} + \frac{\partial C_l}{\partial \beta} \right) = 0. \tag{12}$$

Solving this equation for $\partial \alpha/\partial \beta$, we have

$$\frac{\partial \alpha}{\partial \beta} = -\frac{1}{a}\frac{\partial C_l}{\partial \beta}. \tag{13}$$

To evaluate $\partial\alpha/\partial\beta$ we must consider the elastic constraint. The aerodynamic pitching moment per unit length of span about the axis of rotation is

$$M' = qc^2\left(eC_l + \beta\frac{\partial C_m}{\partial\beta} + C_{m0}\right). \tag{14}$$

In this equation, the first term $qc^2 eC_l$ is the product of the lift qcC_l acting at the aerodynamic center times the distance from the aerodynamic center to the axis of rotation. The last two terms are the moment about the aerodynamic center according to Eq. (8). This pitching moment is balanced by the elastic restoring moment αK, K being the stiffness of the torsional restraint per unit span of the airfoil. Hence

$$\alpha K = qc^2\left(eC_l + \beta\frac{\partial C_m}{\partial\beta} + C_{m0}\right) = qc^2\left(ea\alpha + e\beta\frac{\partial C_l}{\partial\beta} + \beta\frac{\partial C_m}{\partial\beta} + C_{m0}\right). \tag{15}$$

Differentiating with respect to β, we obtain (C_{m0} being a constant),

$$K\frac{\partial\alpha}{\partial\beta} = qc^2\left(ea\frac{\partial\alpha}{\partial\beta} + e\frac{\partial C_l}{\partial\beta} + \frac{\partial C_m}{\partial\beta}\right). \tag{16}$$

Substituting $\partial\alpha/\partial\beta$ from Eq. (13), we obtain the critical dynamic pressure for control reversal:

$$q_{\text{rev}} = \frac{1}{a}\frac{\partial C_l}{\partial\beta}\left(-\frac{\partial C_m}{\partial\beta}\right)^{-1}\frac{K}{c^2}. \tag{17}$$

Hence, the *critical reversal speed* is given by

$$U_{\text{rev}} = \left(\frac{1}{a}\frac{\partial C_l}{\partial\beta}\right)^{1/2}\left(-\frac{\partial C_m}{\partial\beta}\right)^{-1/2}\left(\frac{2K}{\rho c^2}\right)^{1/2}, \tag{18}$$

which shows that the critical reversal speed increases with increasing stiffness K of the airfoil. Note that the quantity e is absent from this formula. Hence the position of the torsional axis does not affect the reversal speed. This is because the net change of lift due to aileron displacement is zero at the reversal speed.

If we define the *elastic efficiency* of the control surface, or *control surface efficiency*, as the ratio of the lift force produced by a unit deflection of the aileron on an elastic wing to that produced by the same aileron deflection on a fictitious rigid wing of the same chord length, then it can be shown that (Fung, 1969)

$$\text{Elastic efficiency} = \frac{1 - q/q_{\text{rev}}}{1 - q/q_{\text{div}}}. \tag{19}$$

This result is shown in Fig. 3.4:4. It is seen that the control surface efficiency drops to zero very rapidly when the flight speed approaches the critical reversal speed.

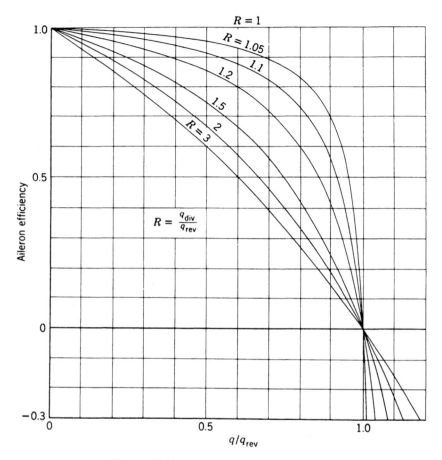

FIGURE 3.4:4 Aileron efficiency versus dynamic pressure when $q_{\text{div}} > q_{\text{rev}}$.

3.5 Transient Fluid Dynamic Forces Due to Unsteady Motion

When bodies make unsteady motion, the surrounding fluid will of course respond in a transient manner. To understand flying and swimming we must study the fluid dynamics of unsteady motion.

The effect of unsteadiness is to delay lift generation and stall. For example, in steady state a two-dimensional wing flying at a velocity U at an angle of attack α will generate a lift force per unit span equal to

$$L = 2\pi c \tfrac{1}{2}\rho U^2 \alpha, \tag{1}$$

according to Eqs. (3.2:4)–(3.2:6), c being the chord length, and ρ being the density of the fluid. But if the same wing starts motion impulsively from rest to a uniform velocity U at the same angle of attack, the lift force per unit span

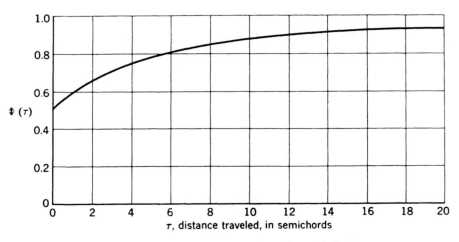

FIGURE 3.5:1 Impulsive motion of an airfoil.

would be a function of time:

$$L' = 2\pi c \tfrac{1}{2}\rho U^2 \alpha \Phi(\tau), \qquad \Phi(\tau) = 0 \quad \text{if } \tau < 0, \tag{2}$$

where

$$\tau = \frac{2U}{c}t. \tag{3}$$

Thus τ is time measured by the distance traveled in units of semichord length. The function $\Phi(\tau)$ was first determined by H. Wagner (1925) and is known as *Wagner's function*. It is illustrated in Fig. 3.5:1. An approximate expression which agrees within 2 percent of the exact value in the entire range $0 < \tau < \infty$ is given by Garrick (1938):

$$\Phi(\tau) \doteq 1 - \frac{2}{4 + \tau}, \qquad (\tau > 0). \tag{4}$$

Another approximate expression is given by R.T. Jones (1940):

$$\Phi(\tau) = 1 - 0.165e^{-0.0455\tau} - 0.335e^{-0.300\tau}. \tag{5}$$

From these formulas, and also from Fig. 3.5:2, we see that only half of the lift is generated immediately following the start of motion. The lift then grows with time. When 20 semichord lengths are traveled, the lift reaches about 94% of the steady-state value.

By the same token, if the wing oscillates, then the unsteady effect will cause the lift to have a phase lag behind the angle of attack. The theories of Theodorsen (1935), Küssner (1936), and Biot (1942) refer to oscillating wings of infinite span in a two-dimensional flow at speeds much lower than the speed of sound. For simplicity, the fluid is assumed incompressible and inviscid. The wing is assumed to be planar and thin and the amplitude of oscillation is so small that the velocity of the wing due to oscillation is small compared with a constant flight velocity of the wing relative to the fluid at infinity. The Reynolds number of flow is assumed to be much larger than 1, and the effect of fluid viscosity is taken into account by the Kutta condition (Sec. 3.8).

Vertical Translational Oscillations

Using complex representation of harmonic motion, the wing surface is described by

$$y = y_0 b e^{i\omega t} = y_0 b e^{iUkt}, \tag{6}$$

where y_0 is a dimensionless amplitude representing the ratio of the amplitude of vertical motion to the semichord b. U is the speed of flight, ω is the circular frequency of oscillation in rad./sec. The reduced frequency k is equal to $\omega b/U$.

The solution involves a function $C(k)$ introduced by Theodorsen (1935):

$$C(k) = \frac{K_1(ik)}{K_1(ik) + K_0(ik)} = F(k) + iG(k). \tag{7}$$

where K_0, K_1 are modified Bessel functions of the second kind of orders zero and one, respectively, with arguments ik. The function $C(k)$ is often referred to as *Theodorsen's function*. Its numerical value is given in Fig. 3.5:2. The complex amplitude of the total lift per unit span, $Le^{i\omega t}$, is given by

$$L = \pi \rho U^2 y_0 k^2 b \left[1 - \frac{2i}{k} C(k) \right]. \tag{8}$$

The moment about the midchord point is (positive in the nose-up sense)

$$M_{1/2} = -\pi \rho U^2 i y_0 k b^2 C(k). \tag{9}$$

A comparison between Eqs. (8) and (9) shows that part of the lift that is proportional to $C(k)$ has a resultant acting at the $\frac{1}{4}$-chord point. This part of the lift can be identified as that caused by the bound vorticity over the airfoil. The other part of the lift has a resultant that acts through the midchord point. This latter term arises from a noncirculatory origin, and is equal to the product of the apparent mass and the vertical acceleration. For a flat plate the mass of the fluid enclosed in a circumscribing cylinder having the airfoil chord as a diameter is the theoretical apparent mass associated with the vertical motion.

Rotational Oscillations

The skeleton airfoil, which executes rotational oscillation with a small amplitude about the origin (the midchord point), is represented by the equation

$$y = -\alpha_0 x e^{i\omega t} = -\alpha_0 x e^{iUkt}. \tag{10}$$

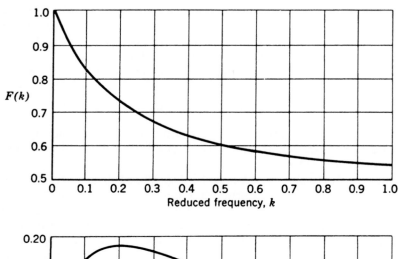

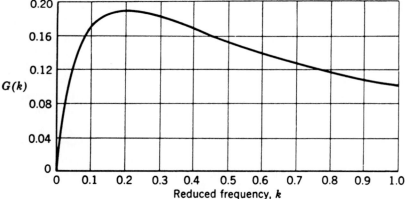

FIGURE 3.5:2 The real and imaginary parts of Theodorsen's function $F(k)$ and $G(k)$. Note the difference in vertical scale in these two figures. $F(k)$ tends to $\frac{1}{2}$ and $G(k)$ tends to zero as k tends to infinity.

Writing

$$\alpha = \alpha_0 e^{iUkt}, \qquad \dot\alpha = \frac{d\alpha}{dt}, \quad \text{etc.,} \tag{11}$$

then the theory yields

$$L = \pi\rho Ub\dot\alpha + 2\pi\rho U^2 b\left(1 + \frac{ik}{2}\right)C(k)\alpha, \tag{12}$$

$$M_{1/2} = -\frac{\pi\rho U}{2}b^2\dot\alpha - \frac{\pi\rho}{8}b^2\ddot\alpha + \pi\rho U^2 b^2\left(1 + \frac{ik}{2}\right)C(k)\alpha. \tag{13}$$

Comparing the expressions L and $M_{1/2}$, we see that the term $\pi\rho U\dot\alpha$ represents a lift that acts at the $\frac{3}{4}$-chord point, the term proportional to $C(k)$ represents a lift that acts at the $\frac{1}{4}$-chord point, and the term $(\pi\rho/8)\ddot{\alpha}$ is a pure couple. It can be shown that the term proportional to $C(k)$ represents the lift due to circulation. The other two terms are of noncirculatory origin.

In both translation and rotation cases, the lift due to circulation can be written as

$$L_1 = -2\pi\rho U C(k)w_{3/4} \tag{14}$$

where $w_{3/4}$ stands for *upwash* at the $\frac{3}{4}$-chord point. Thus the upwash at the $\frac{3}{4}$-chord point has a unique significance. For this reason, the $\frac{3}{4}$-chord point is called the *rear aerodynamic center*.

The theory of oscillating wings has been developed along several directions. Theodorsen (1935) gave the first rigorous solution to the problem of oscillating wings of infinite span in a two-dimensional flow of an incompressible inviscid fluid by conformal mapping. Theodorsen and Garrick (1942) extended the theory to a wing–aileron–tab combination. Küssner (1936) and Küssner and Schwarz (1940) solved the same problem by the method of superposition of singularities. Biot (1942) presented a simplified solution using acceleration potential.

Then the theories were extended to compressible fluid, and to arbitrary motion of the wing, while experimental results were accumulated.

The indicial response to gust and sudden motion was first solved by Wagner (1925) and developed by Küssner (1940), Sears (1941), and many others.

For literature review, we again refer to Fung (1955, 1969), Bisplinghoff and Ashley (1962), and Dowell et al. (1978). Experimental results are reviewed in these references also. New experimental results are surprisingly lacking. It is hoped that advances in laser velocimeter, ultrasound, and data handling techniques will soon yield important results.

Further theoretical analyses of more complex situations, especially in the nonlinear world, are needed. For example, to study fish propulsion, it is necessary to consider the interaction of the unsteady motion of neighboring fins. Since the vortex sheets shed by fins in the front interact with the motion of fins in the back, the effects could be quite complex. See Chapter 4. The nonlinear interaction of the vortex sheets has not been explored in detail. Great advancements in computational fluid mechanics are sure to bear fruits.

3.6 Flutter

One possible effect of the phase shift between motion and force is to cause an important phenomenon of *flutter*, which is defined as a self-excited oscillation of the body. Flutter of leaves of trees, flags, and tents are familiar to all of us. For birds, insects, fish, and aircraft, the flutter phenomenon imposes another limitation to their possible speed of motion, in addition to divergence and loss of control studied in the preceding sections. Power availability is not the only factor that determines speed.

To understand flutter, it may be useful to consider a wind tunnel experiment. Let a wing be mounted in a wind tunnel. When the wind speed is zero and the model is disturbed by a poke with a rod, oscillation may set in, which is gradually damped. When the speed of flow in the wind tunnel is increased continuously, the rate of damping of the oscillation will first increase, then decrease. Eventually, at a critical speed of flow the damping becomes zero. At the *critical flutter speed*, a disturbed wing will oscillate at a steady amplitude. At speeds above the critical, an accidental disturbance can trigger a violent oscillation which is flutter.

Thus it is important to understand flutter. As a first step, let us use dimensional analysis to identify the relevant parameters, as was done in Sec. 3.2. In addition to the variables listed in Eq. (3.2:1a) let us also consider σ, a characteristic material density of the wing structure of dimensions $[ML^{-3}]$, and a characteristic torsional stiffness constant of the wing, K, of dimension $[ML^2T^{-2}]$. Out of the five variables L, U, ρ, σ, and K, two independent nondimensional parameters can be formed, e.g.,

$$\frac{\rho}{\sigma}, \quad \frac{K}{\sigma L^3 U^2}. \tag{1}$$

Any nondimensional quantity relating to the motion can be expressed as a function of these parameters. Thus, if, in a free oscillation, the deflection at a point on the wing is described by an expression $e^{-\varepsilon t}\cos \omega t$, the damping factor ε, of dimension $[T^{-1}]$, can be combined with U and L to form a nondimensional parameter $\varepsilon L/U$, which then must be a function of the parameters listed in Eq. (1):

$$\varepsilon L/U = F[\rho/\sigma, K/(\sigma L^3 U^2)]. \tag{2}$$

At the critical flutter condition, $\varepsilon = 0$, the right-hand side of Eq. (2) vanishes. Hence a relation exist between the parameters ρ/σ and $K/(\sigma L^3 U^2)$. This relation may be written as

$$U_{flutter}^2 = \frac{K}{\sigma L^3} f\left(\frac{\rho}{\sigma}\right) \tag{3}$$

which says that the square of the critical speed of flight is directly proportional to the torsional stiffness of the wing, inversely proportional to the cube of the wing dimension, and inversely proportional to the wing material density. The constant of proportionality is a function of the ratio of the densities of the wing material and the fluid.

The frequency of flutter oscillation ω (radians per second at flutter), with dimension $[T^{-1}]$, can be expressed nondimensionally in the parameter

$$k = \frac{\omega L}{U} \tag{4}$$

which is the *reduced frequency* or *Strouhal number* (Sec. 3.2). Hence the Strouhal number of flutter is a function of ρ/σ, and $K/(\sigma L^3 U_{flt}^3)$.

A physical interpretation of the Strouhal number is as follows. If a periodic deflection occurs at a point on a body while the fluid moves downstream with a velocity U, then the spacing, or the "wave length" of the disturbance in the fluid, is $2\pi U/\omega$. The ratio of the characteristic length L of the body to this wave length is the Strouhal number. Thus the Strouhal number characterizes the way a disturbance at one point is felt at other points in the flow field.

We can show that the phase shift between force and motion is the cause of energy exchange between a wing and the surrounding fluid. Consider a wing in horizontal flight performing a vertical translational oscillation, whose downward velocity $\dot{h}(t)$ is

$$\dot{h} = \dot{h}_0 e^{i\omega t}. \tag{5}$$

The lift force created by this motion is oscillatory and out of phase with $\dot{h}$. We may write the lift as

$$L = L_0 e^{i(\omega t + \psi)}, \tag{6}$$

where ψ is the phase angle by which the lift leads the deflection.

When the airfoil moves through a distance dh, the work done by the lift is, in real variables,

$$dW = -L\,dh = -L\dot{h}\,dt. \tag{7}$$

It must be recognized that, when L and $\dot{h}$ are expressed in the complex form as in Eqs. (5) and (6), the physical quantities are represented by the *real parts* of the complex representations. Thus, in complex representation,

$$dW = -R1[L] \cdot R1[\dot{h}]\,dt. \tag{8}$$

Integrating through a cycle of oscillation, we obtain the work done *by* the air *on* the airfoil per cycle:

$$W = -\int_0^{2\pi/\omega} L_0 \cos(\omega t + \psi)\dot{h}_0 \cos \omega t\,dt$$

$$= -\frac{\pi}{\omega} L_0 \dot{h}_0 \cos \psi. \tag{9}$$

Hence, the gain of energy W by the airfoil from the airstream is proportional to $(-\cos \psi)$. If $-\pi/2 < \psi < \pi/2$, W is negative; i.e., the oscillating airfoil will lose energy to the airstream. ψ can be evaluated by comparing Eqs. (5), (6) with Eqs. (3.5:7), (3.5:8). It is seen that free vertical translational oscillation will be damped.

It can be shown (Fung, 1955) that free pitching oscillation of the wing will also be damped. However, a wing moving with a combination of translation and rotation can, under certain conditions of frequency and amplitudes, gain energy from the surrounding fluid stream which sustains flutter. In aircraft, flutter is dreaded, and it is the designer's duty to know the critical flutter speed accurately, and the pilot's duty never to exceed it. In nature, it is not known whether it is used to advantage by some animals.

3.7 Kutta–Joukowski Theorem

The rest of this chapter is devoted to the wing theory. In the idealized case, we assume that the wing has an infinite span and a cylindrical body; the fluid is incompressible; and the Reynolds number is so large that the boundary layer is very thin and the fluid can be considered as nonviscous outside the boundary layer. The flow field is two-dimensional; the velocity component in the spanwise direction is zero. A rectangular Cartesian system of coordinates xyz, with origin fixed in the wing, and the z-axis in the direction of the span, perpendicular to the direction of flight, will be used.

Under these assumptions, the Eulerian equations of motion (see Sec. 1.7) are simplified to

$$\frac{\partial u}{\partial t} + u\frac{\partial u}{\partial x} + v\frac{\partial u}{\partial y} + \frac{1}{\rho}\frac{\partial p}{\partial x} = 0, \tag{1a}$$

$$\frac{\partial v}{\partial t} + u\frac{\partial v}{\partial x} + v\frac{\partial v}{\partial y} + \frac{1}{\rho}\frac{\partial p}{\partial y} = 0, \tag{1b}$$

in which u, v denote the components of the velocity vector of the fluid in the x, y direction, respectively; ρ is the density of the fluid, and p is the pressure. The equation of continuity for an incompressible fluid (see Sec. 1.7), describing the law of conservation of mass, is

$$\frac{\partial u}{\partial x} + \frac{\partial v}{\partial y} = 0. \tag{2}$$

The curl of the velocity field $\mathbf{v}$ is the *vorticity* of the field. If the vorticity vanishes in the whole field, then the flow is said to be *irrotational*. If the flow field is irrotational we have

$$\frac{\partial u}{\partial y} - \frac{\partial v}{\partial x} = 0. \tag{3}$$

Equations (2) and (3) can be solved for specified boundary conditions. Then Eq. (1) can be used to compute the pressure distribution. It can be shown that the solution for a flow field satisfying a suitably specified boundary condition is unique. Hence any method that yields a solution provides the right solution.

By direct substitution, we see that Eq. (2) can be satisfied by an arbitrary function $\psi(x, y)$ if the velocity components are calculated according to

$$u = \frac{\partial \psi}{\partial y} \quad \text{and} \quad v = -\frac{\partial \psi}{\partial x}. \tag{4}$$

Equation (3) can be satisfied by an arbitrary function $\phi(x, y)$ if

$$u = \frac{\partial \phi}{\partial x} \quad \text{and} \quad v = \frac{\partial \phi}{\partial y}. \tag{5}$$

ψ is the *stream function*. ϕ is the *velocity potential*. On substituting (5) into (2), and (4) into (3), we see that

$$\frac{\partial^2 \phi}{\partial x^2} + \frac{\partial^2 \phi}{\partial y^2} = 0 \quad \text{and} \quad \frac{\partial^2 \psi}{\partial x^2} + \frac{\partial^2 \psi}{\partial y^2} = 0. \tag{6}$$

Equations (6) are *Laplace equations* which are also known as *harmonic equations*; their solutions are *harmonic functions*. Thus the stream and potential functions are both harmonic functions. One method of solving the flow problem is to look for a harmonic function that satisfies the boundary conditions. It so happens that the real and imaginary parts of any analytic function of a complex variable $z = x + iy$ are harmonic. Thus, a famous method of solving the flow problem is to look for an analytic function of a complex variable, $w(z) = \phi(x, y) + i\psi(x, y)$ that satisfies the boundary conditions.

The following examples are given in every textbook:

$$w(z) = (U - iV)z, \qquad \phi = Ux + Vy, \qquad \psi = -Vx + Uy, \tag{7}$$

$$w(z) = m \ln z, \qquad \phi = m \ln r, \qquad \psi = m\theta, \tag{8}$$

$$w(z) = \frac{i\Gamma}{2\pi} \ln z, \qquad \phi = -\frac{\Gamma}{2\pi} \theta, \qquad \psi = \frac{\Gamma}{2\pi} \ln r, \tag{9}$$

$$w(z) = -\frac{\mu}{z}, \qquad \phi = -\frac{\mu x}{x^2 + y^2}, \qquad \psi = \frac{\mu y}{x^2 + y^2}. \tag{10}$$

Equation (7) represents a uniform flow with velocity components U and V. Equation (8) represents a source of strength $2\pi m$ per unit length. Equation (9) represents a vortex with circulation Γ. Equation (10) represents a doublet of strength $|\mu|$. The streamlines of these flows are illustrated in Fig. 3.7:1.

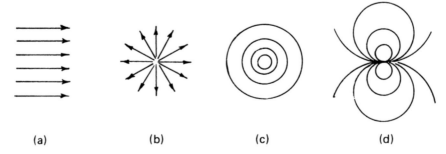

(a)   (b)   (c)   (d)

FIGURE 3.7:1 Streamlines of (a) a uniform flow, (b) a source, (c) a vortex, (d) a source-sink doublet.

Since the governing equations (2), (3), and (6) are linear, a solution can be superposed to obtain new solutions. Thus the flow past a circular cylinder of radius a without circulation can be obtained by superposition of a uniform flow and a doublet:

$$w(z) = Uz + U\frac{a^2}{z}. \tag{11}$$

The flow about a noncircular cylinder can be obtained by a superposition of sources and sinks and a uniform flow. The flow past a circular cylinder of radius a with a clockwise circulation of Γ can be obtained by adding a vortex to Eq. (11)

$$w(z) = U\left(z + \frac{a^2}{z}\right) + \frac{i\Gamma}{2\pi}\ln z. \tag{12}$$

The streamlines in the last case are illustrated in Fig. 3.7:2.

We now integrate the equations of motion to obtain the pressure p and then calculate the total force acting on a solid body moving in a fluid. For an irrotational flow of an incompressible fluid, this is quite easy to do. For the derivatives $\partial u/\partial t$, $\partial v/\partial t$ we use the velocity potential defined in Eq. (5). We then multiply Eq. (1a) by dx and Eq. (1b) by dy, then add and integrate the result along an arbitrary curve C in the fluid. The result is zero because the right-hand sides of Eqs. (1a), (1b) are zero. Thus

$$\int_C \left[\left(\frac{\partial}{\partial x}\frac{\partial\phi}{\partial t} + u\frac{\partial u}{\partial x} + v\frac{\partial u}{\partial y} + \frac{1}{\rho}\frac{\partial p}{\partial x}\right)dx \right.$$
$$\left. + \left(\frac{\partial}{\partial y}\frac{\partial\phi}{\partial t} + u\frac{\partial v}{\partial x} + v\frac{\partial v}{\partial y} + \frac{1}{\rho}\frac{\partial p}{\partial y}\right)dy\right] = 0.$$

This can be reduced to

$$\int_C d\left(\frac{\partial\phi}{\partial t} + \frac{u^2 + v^2}{2} + \frac{p}{\rho}\right) = 0.$$

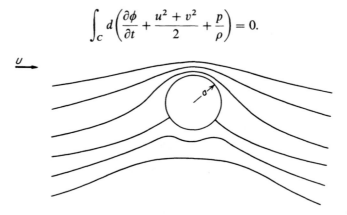

FIGURE 3.7:2 Streamlines of a flow around a circular cylinder with circulation. This flow can be obtained by a proper superposition of a vortex, a doublet, and a uniform flow of velocity u. If the circulation is Γ, a lift equal to $\rho U\Gamma$ is created.

Hence

$$\frac{\partial \phi}{\partial t} + \frac{u^2 + v^2}{2} + \frac{p}{\rho} = \text{const.} \tag{13}$$

Since the integration is taken with respect to x and y and not to t, the constant could be a function of time, that can be absorbed in $\partial \phi / \partial t$. In the case of steady flow, the time-dependent term vanishes, and the last equation, after multiplying through by ρ, and writing V^2 for $u^2 + v^2$, becomes the well-known *Bernoulli equation*

$$\rho \frac{V^2}{2} + p = \text{const.} \tag{14}$$

Thus the pressure at any point in the field of flow is given by

$$p = \text{const} - \tfrac{1}{2}\rho V^2. \tag{15}$$

Apply this result to the flow about a circular cylinder specified by Eq. (12). Consider the forces acting on the wall of the cylinder. On an element of length ds located at polar coordinates (a, θ), as shown in Fig. 3.7:2, the pressure force $p\,ds$ is directed radially; its components in the x and y directions are

$$-p \cos \theta \, ds \quad \text{and} \quad -p \sin \theta \, ds,$$

respectively. By an integration over the whole circle, we obtain the x and y components of the resulting force acting on the cylinder per unit length in the spanwise direction:

$$F_x = -\int_0^{2\pi} pa \cos \theta \, d\theta, \qquad F_y = -\int_0^{2\pi} pa \sin \theta \, d\theta. \tag{16}$$

To calculate F_x, F_y, we substitute p from Eq. (15), and compute V^2 from Eq. (12). Now, since

$$w(z) = \phi + i\psi, \tag{17}$$

$$\frac{dw}{dz} = \frac{\partial \phi}{\partial x} + i \frac{\partial \psi}{\partial x} = u - iv, \tag{18}$$

we have

$$\left| \frac{dw}{dz} \right|^2 = u^2 + v^2 = V^2. \tag{19}$$

From Eq. (12) we obtain

$$\frac{dw}{dz} = U \left(1 - \frac{a^2}{z^2} \right) + \frac{i\Gamma}{2\pi} \frac{1}{z}. \tag{20}$$

On the circle $z = ae^{i\theta} = a(\cos \theta + i \sin \theta)$, we have

$$\left| \frac{dw}{dz} \right|^2 = \left[U(1 - \cos 2\theta) + \frac{\Gamma \sin \theta}{2\pi a} \right]^2 + \left[U \sin 2\theta + \frac{\Gamma}{2\pi a} \cos \theta \right]^2$$

$$= 4U^2 \sin^2 \theta + \left(\frac{\Gamma}{2\pi a} \right)^2 + \frac{2U\Gamma}{\pi a} \sin \theta. \tag{21}$$

Clearly $|dw/dz|^2$ or V^2 at (a, θ) is equal to V^2 at $(a, \pi - \theta)$, hence p is symmetric with respect to the y-axis. But $\cos \theta$ is antisymmetric with respect to the y-axis. Hence $F_x = 0$. On the other hand, $\sin \theta$ is symmetric with respect to the y-axis; hence the contributions of the right and left half of the cylinder to F_y are equal, and

$$F_y = -2 \int_{-\pi/2}^{\pi/2} pa \sin \theta \, d\theta. \tag{22}$$

On substituting (15), (19), and (21) into (22), and noting that the integrals of $\sin^3 \theta$, $\sin \theta$ from $-\pi/2$ to $\pi/2$ are zero, whereas that of $\sin^2 \theta$ is $\pi/2$, we obtain

$$F_y = \rho U \Gamma. \tag{23}$$

This shows that the cylinder experiences a lift force equal to the product of the velocity of flight U, the circulation Γ, and the density of the fluid ρ. There is no drag. The doublet makes no contribution to the lift force. Similarly, by integration around the wall of a cylinder it can be shown that any enclosed sources and sinks make no contribution to the lift and drag. Although these sources and sinks define the shape of the wing, they do not affect the lift. Thus we obtain the famous *Kutta–Joukowski theorem*:

The force per unit length acting on a cylindrical wing of any cross-section whatever is equal to $\rho U \Gamma$ and acts in a direction perpendicular to U.

Thus, the lift is proportional to the strength of circulation of the vortex line Γ, and to the relative velocity U with which the vortex line moves with respect to the free stream. If the vortex moves with the free stream, then $U = 0$ and there will be no lift.

3.8 The Creation of Circulation Around a Wing

The Kutta–Joukowski theorem tells us that the lift per unit span acting on the wing is equal to $\rho U \Gamma$. But what determines Γ, the circulation around the wing, and how is it created?

Insects, birds, and fish solved the problem of creating circulation around the wing by providing the wing with a sharp trailing edge. The inventors of the airplane copied the insects and birds and made aeronautics a success.

The importance of a sharp trailing edge can be seen in the illustrations of Fig. 3.8:1. The airfoil is stationary and the flow comes from the left. Figure 3.8:1(a) shows streamlines of a potential flow around the airfoil. Among all the streamlines there is one that ends on the body, and defines a *front stagnation point* and a *rear stagnation point* on the body. If the rear stagnation point does not coincide with the trailing edge, then the flow must turn around that very sharp corner. This will create a very high velocity gradient and hence a large shear stress, which will tend to retard the flow around the corner, causing the rear stagnation point to move toward the trailing edge. The process will continue until the rear stagnation point coincides with the trailing edge, as shown in Fig. 3.8:1(b). Such a flow requires a certain amount of

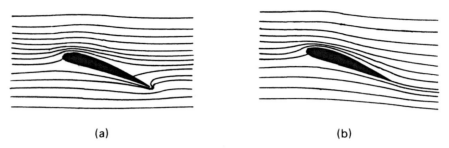

(a)   (b)

FIGURE 3.8:1 Flow around an airfoil with a sharp trailing edge. (a) Streamlines about
an airfoil starting to move. The circulation has not been established yet. (b) Streamlines
about an airfoil in steady motion.

circulation Γ. Thus Γ is fixed by the condition that the rear stagnation point
coincides with the trailing edge. A mathematical statement of this fact is
known as the *Kutta condition*, which says that *a body with a sharp trailing edge
moving through a fluid will create about itself a circulation of sufficient strength
to hold the rear stagnation point at the trailing edge.*

3.9 Circulation and Vorticity in the Wake

So far we have considered two-dimensional airfoils in steady flow. To understand finite wings in oscillatory motion, we must go deeper into fluid
dynamics. The key to this matter is Kelvin's theorem concerning circulation
in the fluid. The *circulation* $\Gamma(C)$ associated with any closed circuit C in the
fluid is defined by the line integral:

$$\Gamma(C) = \int_C \mathbf{v} \cdot \mathbf{dl} = \int_C v_i \, dx_i, \tag{1}$$

where the integrand is the scalar product of the velocity vector $\mathbf{v}$ (with
components v_1, v_2, v_3) and the vector $\mathbf{dl}$ (with components dx_1, dx_2, dx_3),
which is tangent to the curve C and of length dl (Fig. 3.9:1). The rate of change
of $\Gamma(C)$ with respect to time, when C is a fluid line (i.e., a curve formed by the
same set of fluid particles at all times), and the fluid is nonviscous and
barotropic (fluid density is a unique function of pressure) is given by the Kelvin
theorem, which states that

$$\frac{D\Gamma}{Dt} = 0 \tag{2}$$

if the external force field is conservative.

  To prove this theorem, we note that since C is a fluid line composed always
of the same particles, the order of differentiation and integration may be

FIGURE 3.9:1 Quantities defining circulation.

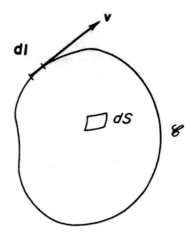

interchanged in the following integral:

$$\frac{D}{Dt}\int_c v_i\,dx_i = \int_c \frac{D}{Dt}(v_i\,dx_i) = \int_c\left(\frac{Dv_i}{Dt}dx_i + v_i\frac{Ddx_i}{Dt}\right). \tag{3}$$

But $D(dx_i)/Dt$ is the rate at which dx_i is changing as a consequence of the motion of the fluid; hence it is equal to the difference of the velocities parallel to x_i at the ends of the element, i.e., dv_i. Hence the last term in Eq. (3) is $v_i\,dv_i$. Further, Dv_i/Dt is the fluid particle acceleration, which, according to the Eulerian equation of motion, is equal to the sum of the negative pressure gradient and the body force per unit volume, ρX_i, divided by the density, ρ. Thus Eq. (3) becomes

$$\frac{D\Gamma}{Dt} = \int_c\left[\left(-\frac{1}{\rho}\frac{\partial p}{\partial x_i} + X_i\right)dx_i + v_i\,dv_i\right]$$

$$= -\int_c\frac{dp}{\rho} + \int_c X_i\,dx_i + \int_c dv^2. \tag{4}$$

Of the terms on the right-hand side, the first vanishes because the fluid is barotropic (ρ is a unique function of p) and C is a closed curve; the second term vanishes because the external force field X_i is assumed to be conservative; the third term vanishes because the value of the line integral is equal to v^2 at a point on C minus v^2 at the same point after one goes around C in a closed circuit. Thus we get Eq. (2). Q.E.D.

Applying Eq. (2) to the flight of birds and insects, we note that since the Reynolds number is much larger than 1, the boundary layer is thin, and outside the boundary layer the fluid may be considered to be nonviscous. The Mach number of bird's flight is small compared with 1; hence the air may be considered incompressible (thus barotropic). Gravity is the only external force and is conservative. Hence all the assumptions for the Kelvin theorem are valid, and Eq. (2) applies. Now, consider a wing starting to move from a stationary position (Fig. 3.9:2). The circulation Γ about any fluid line outside the boundary layer is zero because it is zero before the motion takes place, however, the volume occupied by the airfoil and the boundary layer must be excluded. A fluid line C enclosing the boundary of the wing becomes elongated when the wing moves forward as shown in Fig. 3.9:2. According to Kelvin's theorem, the circulation about C is zero. But one cannot conclude that the vorticity actually vanishes everywhere inside C. *In the region occupied by the wing, and in the wake behind the wing, vorticity does exist if there is a lift force acting on the wing.* Although Fig. 3.9:2 shows $I(C) = 0$, Fig. 3.9:3 shows that $I(C)$ may be regarded as the sum of the circulations $I(C')$ and $I(C'')$; where $C' + C'' = C$. If $\Gamma(C') = \Gamma$, then $\Gamma(C'') = -\Gamma$, so the circulation about the entire wake is equal and opposite to that around the wing.

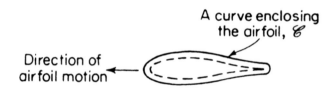

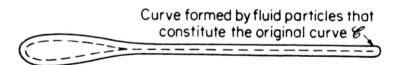

FIGURE 3.9:2 Fluid line enclosing an airfoil and its wake. *Top*: A stationary airfoil starting to move. *Bottom*: The airfoil has moved forward for a distance, and the fluid line is elongated.

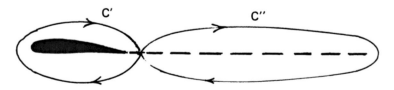

FIGURE 3.9:3 The circulation about the curve c in Fig. 3.9:2 is the same as that about the curve $c' + c''$ in this figure. The circulation about the wing (c') is equal and opposite to that about the wake (c'').

3.10 Vortex System Associated with a Finite Wing in Nonstationary Motion

An important result deducible from Kelvin's theorem is the *Helmholtz theorem* which states that *the vortex lines in an inviscid fluid move with the fluid*, meaning that the fluid particles constituting a vortex line will continue to constitute a vortex line as the fluid flows. We present a proof by Yih (1969).

The proof is facilitated by noting that according to Gauss' theorem, the circulation defined by Eq. (3.9:1) can be expressed as a surface integral:

$$\Gamma(C) = \int_S \operatorname{curl} \mathbf{v} \cdot \mathbf{v} \, dS, \tag{1}$$

where S is a surface bounded by C, and $\mathbf{v}$ is the unit normal vector of the surface. **Curl v** is the vorticity of the flow field. Thus, circulation is the sum of the normal component of vorticity over a surface.

Now let us consider a continuous field of vorticity, Fig. 3.10:1(a). A vortex line is a line tangent to the vorticity vectors. On a point A on a vortex line, draw two lines BAC and DAE intersecting the vortex line, and draw vortex vectors passing through BAC and DAE. These vectors form two surfaces, S_1 and S_2, that intersect in the vortex line under consideration. According to Eq. (1), the circulation along any closed circuit on S_1 or S_2 is zero. According to Kelvin's theorem, it will continue to be zero as the fluid moves. Thus the circulation along any closed circuit situated on the surface S_1 or S_2 remains

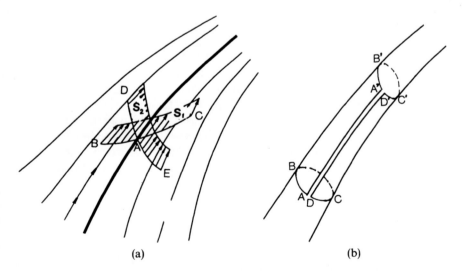

(a)   (b)

FIGURE 3.10:1 (a) A vector field of vorticity. The arrows represent the vorticity at the points where the arrows start. (b) A circuit on a vortex tube. When D and A and D' and A' coalesce, there results two circuits on a vortex tube.

zero. Hence S_1 and S_2 remain vorticity surfaces consisting of vortex vectors. Their intersection will continue to be a vortex line, for otherwise at some point on it a vortex vector would intersect at least one of the surfaces S_1 or S_2, contradicting the fact that the circulation along any circuit in S_1 or S_2 is zero for all times.   Q.E.D.

A vortex tube is formed by vortex lines, Fig. 3.10:1(b). If a circuit $ABCD-D'C'B'A'$ is drawn on the vortex tube, the circulation along this circuit is zero. If the points D and A as well as D' and A' coalesce, we obtain two circuits $ABCD$ and $D'C'B'A'$, of which the sum of the circulation is zero. Thus Γ of $ABCD$ and Γ of $D'C'B'A'$ are equal and opposite; Γ of $ABCD$ and Γ of $A'B'C'D'$ are exactly equal. Therefore the circulation around any vortex tube does not change.

It follows that a vortex tube cannot end in a fluid, because the circulation is zero at the end, which, according to the result just derived, requires the circulation to be zero along the entire vortex tube; a trivial case. Thus, a vortex line can end only on a solid surface, or on the boundary of the fluid. It can end on itself, forming a ring, or extend to infinity if the fluid field is infinite.

With the Helmotz theorem, we can now understand the vortex system associated with a real wing. Let us approach the real wing step by step.

Horseshoe Vortex

Consider a rectangular wing in a steady flow with free stream velocity U. Attach a frame of reference that moves with the wing so that the wing appears stationary while the free stream comes from the left. In Fig. 3.10:2(a) the wing is represented by a *bound vortex*. Since the vortex line cannot end at the wing tips, it has to turn around and move with the free stream, forming the *trailing vortices*. Together, the bound and traling vortices look like a *horseshoe*. The circulation is constant along the entire vortex line.

If the wing planform is elliptical and untwisted (having a constant angle of attack across the span), it can be shown that the lift distribution is also elliptical across the span (see Fig. 3.10:2(b)). Then the wing can be represented by a series of bound vortices, the total strength of which varies elliptically across the span. Since each bound vortex has two trailing vortices, the wing and its wake can be represented by a series of horseshoe vortices. The trailing vortices interact with each other, and have a tendency to roll up, as indicated in Fig. 3.10:2(b).

Downwash

Each vortex line induces a velocity field (see Fig. 3.10:3). For simplicity we show only the uniform free-stream velocity U, and the velocity on the vortex lines induced by themselves. Since the trailing vortices move with the free stream in the direction of U, the induced velocity is perpendicular to U, and

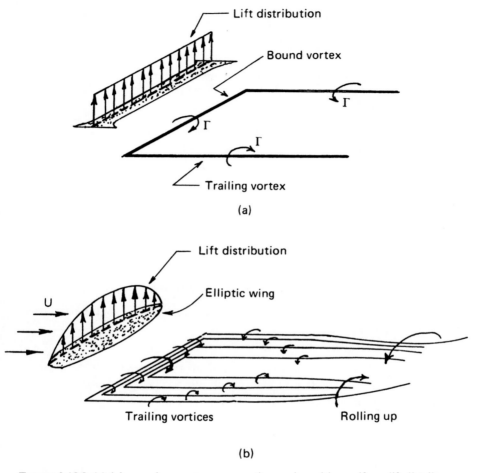

FIGURE 3.10:2 (a) A horse-shoe vortex representing a wing with a uniform lift distribution. (b) Lift distribution on an elliptic wing.

is called the *downwash*. In Fig. 3.10:3(a), the downwash over a rectangular wing is seen to be nonuniform across the span. In Fig. 3.10:3(b), the downwash over an untwisted elliptical wing is seen to be uniform across the span.

Induced drag: The Price you Pay for the Lift

In Sec. 3.2 we discussed the effect of the downwash on the wing due to trailing vortices on the slope of the lift coefficient versus the angle of attack. Here we shall consider its effect on the drag. As shown in Fig. 3.2:3, the downwash velocity w deflects the free-stream velocity vector U by an angle ε, with

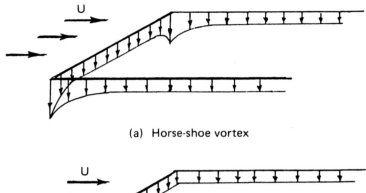

(a) Horse-shoe vortex

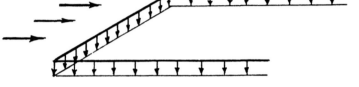

(b) Elliptic wing, uniform downwash

FIGURE 3.10:3 (a) Nonuniform downwash on a horse-shoe vortex. (b) Uniform downwash on an elliptic wing.

$\tan \varepsilon = w/U$. Relative to the wing, the free stream comes in the direction of the velocity $\mathbf{U}_{res}$. When the circulation is Γ, the resulting force $\rho U_{res}\Gamma$ acts in a direction normal to the velocity vector $\mathbf{U}_{res}$. It can be resolved into a *lift* component L perpendicular to the velocity of flow $\mathbf{U}$, and a *drag* component D in the direction of $\mathbf{U}$. From Fig. 3.2:3, we see that

$$D = L \tan \varepsilon = \frac{w}{U} L. \tag{2}$$

This drag is induced by the downwash as a consequence of creating a lift. Known as the *induced drag*, it accompanies the lift, and is a price a flying animal must pay for the lift.

In Fig. 3.10:3 it is shown that the downwash w is uniform across the span of an untwisted elliptical wing, whereas it is nonuniform if the wing planform is not elliptical. This is a result derived by method described in Sec. 3.12. Glauert showed that for a given total lift the wing with a uniform downwash yields the minimum induced drag.

For an elliptic lift distribution, it has been shown theoretically that the downwash is given by

$$\frac{w}{U} = \frac{C_L}{\pi AR}, \tag{3}$$

where C_L is the lift coefficient and AR is the *aspect ratio* (see Sec. 3.2). Hence the induced drag given by Eq. (1) is equal to

$$D_i = \frac{C_L^2}{\pi} \frac{1}{2}\rho U^2 S, \quad \text{or} \quad D_i = \frac{C_L}{\pi}(\text{weight}). \tag{3}$$

S is the wing area and W is the weight of the flying object. For a lift distribution other than elliptical, the induced drag is greater than this. Equation (3) shows that the induced drag is reduced if the aspect ratio is increased. This is undoubtedly why a soaring bird spreads out the wing to make its aspect ratio as large as possible (see Fig. 3.2:1(b)).

Oscillating Wings

If the wing oscillates, either by flapping up and down, or by changing its angle of attack periodically, then the strength of the bound vortex varies with time, and there will be vortices shed from the trailing edge into the wake. A crude sketch is shown in Fig. 3.10:4. Every time an increment of bound vortex is created, an equal and opposite vortex is left to the wake. Thus the vortex structure in the wake becomes quite complex. The complexity is increased if the amplitude of the oscillation is large. We can then appreciate how difficult it is to analyze the flight of birds and insects (see Fig. 3.10:5). Fortunately, the

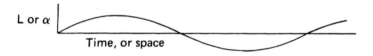

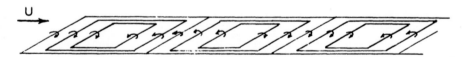

FIGURE 3.10:4 The vortices in the wake of an oscillating wing, idealized under the assumption that the lift fluctuation is very small so that the distortion of the wake due to the vortices in it is also very small. L is the lift, α is the angle of attack.

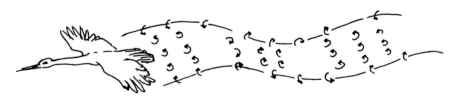

FIGURE 3.10:5 The vortex wake behind a stork in level flight.

reduced frequency (or Strouhal number based on radian frequency and chord length) for birds and insects is relatively small (in the order of 0.2 to 0.6), and it is often permissible to use the quasi-steady approximation to obtain rough estimates. The same quasi-steady approximation would be less valid for the lunate tails of fish such as a shark or whale. These tails function as wings, but operate at Strouhal numbers of the order of 1.

3.11 Thin Wing in Steady Flow

Birds can change their wing planform and airfoil cross-section. The concepts developed above can be used to estimate the effects of these changes.

Consider first the effect of airfoil cross-section. Let the wing be very thin as shown in Fig. 3.11:1; it has an infinite span, and the flow is two-dimensional. Let us use a rectangular cartesian frame of reference with the origin located at the leading edge, x-axis along the chord line, and y-axis perpendicular to it. The airfoil camber line is described by the equation

$$y = Y(x), \qquad (0 \leqslant x \leqslant c).$$ (1)

If a steady two-dimensional flow with undisturbed speed U and angle of attack α streams past the airfoil, disturbances are introduced into the flow by the airfoil in such a manner that the resulting flow is tangent to the airfoil. The thin airfoil can be replaced by a continuous distribution of vortices. Let the strength of the vorticity over an element of unit length in the spanwise direction and of length dx in the chordwise direction be $\gamma(x)\,dx$. According to the Kutta–Joukowski theorem, the lift force contributed by the element dx is

$$dL = \rho U \gamma(x)\,dx.$$ (2)

The total lift per unit span is therefore

$$L = \rho U \int_0^c \gamma(x)\,dx,$$ (3)

where c is the chord length of the airfoil.

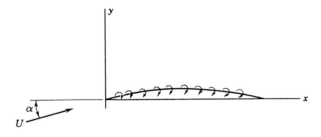

FIGURE 3.11:1 Steady flow over a two-dimensional airfoil.

For a thin airfoil of small camber, $Y(x) \ll c$, the surface of the airfoil differs only infinitesimally from a flat plate. The induced velocity over the airfoil surface, to the first order of approximation, can be calculated by assuming that the vortices are situated on the x-axis. Since the velocity at x induced by a vortex of strength $\gamma(\xi)\,d\xi$ located at ξ is $\gamma(\xi)\,d\xi/[2\pi(\xi - x)]$, the y component of the induced velocity at a point x on the x-axis is

$$v_i(x) = \int_0^c \frac{\gamma(\xi)\,d\xi}{2\pi(\xi - x)}, \tag{4}$$

which, to the first order of approximation, is the same as the component of velocity normal to the airfoil surface at the chordwise location x. The slope of the fluid stream on the airfoil is then $\alpha + v_i/U$. This must be equal to the slope of the airfoil surface dY/dx. Hence, the boundary condition on the airfoil is

$$\alpha + \frac{v_i}{U} = \frac{dY}{dx}. \tag{5a}$$

The vorticity distribution $\gamma(x)$ must be determined from Eqs. (4) and (5a). Thus we obtain the integral equation for the vorticity distribution $\gamma(x)$:

$$\int_0^c \frac{\gamma(\xi)\,d\xi}{\xi - x} = 2\pi U\left(\frac{dY}{dx} - \alpha\right). \tag{5b}$$

In addition, the Kutta condition $\gamma(c) = 0$ must be satisfied, i.e., the fluid must leave the trailing edge smoothly.

Equation (5a) is a singular integral equation of the Cauchy type which has been treated most thoroughly by Muskhelishvili (1953a,b). The methods of Glauert, Lotz, Hilderbrand, Multhopp, and Sears are also well known. In Glauert's method, we introduce a new independent variable ψ so that

$$x = \frac{C}{2}(1 - \cos\psi). \tag{6}$$

When x varies from 0 to c along the chord, ψ varies from 0 to π. The vorticity distribution can be written as

$$\gamma = 2U\left(A_0 \cot\frac{\psi}{2} + \sum_1^\infty A_n \sin n\psi\right) \tag{7}$$

with unknown coefficients, $A_0, A_1 \ldots$. Substituting (7) into (4), we obtain (see Glauert, 1947)

$$v_i = U\left(-A_0 + \sum_1^\infty A_n \cos n\psi\right). \tag{8}$$

Equation (5) then implies

$$\alpha - A_0 + \sum_1^\infty A_n \cos n\psi = \frac{dY}{dx}. \tag{9}$$

The left-hand side is a Fourier series. The coefficients can therefore be determined by the usual method. Multiplying Eq. (9) by $\cos n\psi$ ($n = 0, 1, 2, \ldots$), and integrating from 0 to π, we obtain

$$\alpha - A_0 = \frac{1}{\pi} \int_0^\pi \frac{dY}{dx} d\psi, \qquad A_n = \frac{2}{\pi} \int_0^\pi \frac{dY}{dx} \cos n\psi \, d\psi \tag{10}$$

and the problem is solved.

From Eqs. (3) and (7), the total lift can be obtained. The result, expressed as the lift coefficient, is

$$C_L = \pi(2A_0 + A_1). \tag{11}$$

Similarly, the moment about the leading edge, expressed as the moment coefficient, is

$$(C_M)_{\text{l.e.}} = -\tfrac{\pi}{2}(A_0 + A_1 - \tfrac{1}{2}A_2) = -\tfrac{\pi}{4}(A_1 - A_2) - \tfrac{1}{4}C_L. \tag{12}$$

If the airfoil is a flat plate, $Y = 0$, $A_0 = \alpha$, the lift curve slope is seen to be 2π. From Eq. (12), the lift force is seen to act through the $\frac{1}{4}$-chord point.

3.12 Lift Distribution on a Finite Wing

To analyze the stress and strain in a swimming fish or a flying bird, one of the most important problems is to know the spanwise distribution of lift and moment, which must be resisted by their bones, muscles, and tendons. We shall present a classical approximate theory credited to Prandtl. The solution also yields information on downwash and induced drag that is needed to evaluate the energy cost of locomotion.

Theoretically, a wing can be replaced by a system of vortices as discussed in Sec. 3.11. For example, if a wing is straight, without significant sweepback or sweepforward, then in first approximation it can be replaced by a straight line vortex. Consider such a wing in a symmetric flight. Let a rectangular cartesian frame of reference be used, with origin located on the vortex line at midspan, x-axis in the direction of undisturbed flow, y-axis in the direction of span. The bound vortex lies on the y-axis from $-s$ to s, s being the length of semispan. The circulation around the vortex line, $\Gamma(y)$, is a function of y. In a steady flight of velocity U, the lift force acting on a segment of the wing of width dy at a station y is $\rho U \Gamma(y) \, dy$ (see Sec. 3.7), where ρ is the density of the fluid. If the chord length is c and the slope of the curve of lift coefficient versus angle attack is a_0 (see Sec. 3.2), then the angle of attack needed to generate this lift is

$$\alpha_0(y) = \frac{\rho U \Gamma(y) \, dy}{a_0 \frac{1}{2} \rho V^2 c \, dy} = \frac{2\Gamma(y)}{a_0 V c}. \tag{1}$$

The actual angle of attack is equal to $\alpha_0(y)$ plus the induced angle of attack due to downwash, w/U (see Fig. 3.2:2). The downwash caused by the bound vortex on itself is zero when the vortex line is straight. Downwash caused by the trailing vortices must be calculated. Now on the segment of the wing of width dy at station y, the strength of the bound vortex is changed by an amount $(d\Gamma/dy)\,dy$ in the spanwise direction. This change must become the trailing vortex moving with the fluid stream in the direction of the x-axis. Thus the trailing vortex attached to this segment of the wing at y is of strength $(d\Gamma/dy)\,dy$, and is a semi-infinite straight line extending from $x = 0$ to $x = \infty$ if the rolling up of the trailing vortex sheet (due to interaction of the vortices) is neglected. The downwash at the point $(x = 0, y = \xi)$ caused by this trailing vortex is equal to

$$dw(\xi) = \frac{1}{4\pi(\xi - y)}\frac{d\Gamma}{dy}\,dy, \tag{2}$$

which is half of what what would have been induced by a vortex line extending from $-\infty$ to ∞ (see Sec. 3.7). Hence the downwash at a point $(x = 0, y = y)$ due to the whole system of trailing vortices can be obtained by an integration of $dw(y)$ over the entire span from $-s$ to s:

$$w(y) = \frac{1}{4\pi}\int_{-s}^{s}\frac{1}{y - \eta}\frac{d\Gamma(\eta)}{d\eta}\,d\eta. \tag{3}$$

Note the change of symbols from ξ to y and y to η in using Eq. (2) to arrive at Eq. (3). The absolute angle of attack required to produce the lift distribution is, therefore, the sum of $\alpha_0(y)$ given by Eq. (1) plus $w(y)/U$, i.e.,

$$\alpha_a(y) = \frac{2\Gamma(y)}{a_0 Uc} + \frac{1}{4\pi U}\int_{-s}^{s}\frac{1}{y - \eta}\frac{d\Gamma(\eta)}{d\eta}\,d\eta. \tag{4}$$

This is the integral equation used to solve for $\Gamma(y)$, subject to the boundary condition that $\Gamma(y)$ must vanish at the wing tips, $y = -s$ and s.

Equation (4) is again a singular integral equation of the Cauchy type, and can be solved by the same methods mentioned in Sec. 3.11. A solution in the form of a Fourier sine series is most convenient:

$$\Gamma = \frac{a_0 c_0 U}{2}\sum_{n=1}^{\infty} A_n \sin n\theta, \tag{5}$$

with $y = s\cos\theta$. Details can be found in Kuethe and Chow (1986).

Applying this analysis to birds and fish, we must make at least two more extensions: First, real wings have sweepback and sweepforward (see Fig. 3.2:1) so the lifting line is not a straight line. Sometimes the aspect ratio is not much larger than 1, making it necessary to use a lifting surface theory in order to achieve the desired accuracy. For fish, the lifting surface is usually not symmetric. Second, the elastic deflection of the structure under load is usually significant, and the reaction and control by muscle must be taken into account.

Hence the angle of attack, $\alpha_a(y)$ of Eq. (4), (and similarly, the terms dy/dx and α in Eq. (3.11:5a) is a function of the muscle function, as well as of bones, tendons, tissues, feathers, or fins. The method of approach used in the theory of *aeroelasticity* can be used to account for the elastic deflection and active control (see Fung, 1969, Bisplinghoff and Ashley, 1967, Dowell et al., 1978), however, a thorough work on animals remains to be done.

3.13 Drag

So far we have considered lift, moment, and control which makes flight and swimming possible. Now let us turn to drag which determines the price an animal must pay to move in a fluid. Drag means force acting on a body in the direction of motion relative to the fluid at infinity (if the domain is unbounded). For an airfoil, a dimensionless coefficient, C_D, called the *drag coefficient*, is defined by Eq. (3.2:4).

For an airfoil in an unbounded fluid, drag can arise from skin friction, from wake due to boundary layer separation, and from downwash due to vortices associated with lift force generation. Skin friction is shear stress associated

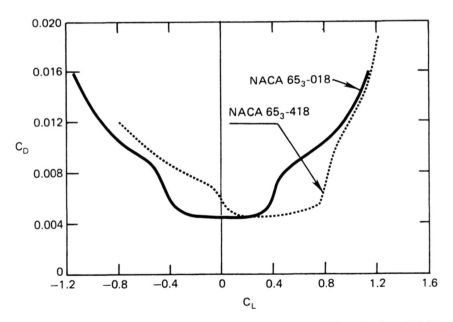

FIGURE 3.13:1 The drag coefficient plotted against the lift coefficient C_L of two NACA airfoils. Experimental data measured at Reynolds number 6×10^6 for smooth airfoil. Data from Abbott and von Doenhoff (1949). Courtesy of NASA.

with fluid viscosity and shear strain rate in the boundary layer attached to the solid surface of the airfoil. Wake is a front-back asymmetric flow pattern caused by detachment of the boundary layer from the solid wall. It occurs on blunt bodies (Sec. 3.3). It occurs also on airfoils when the angle of attack is too large. When a wake exists, the resulting force due to normal stress acting on the surface of the body will not be zero in the direction of motion. This drag is the *form drag*.

The third source of drag is associated with downwash which is velocity induced by the vortex system around the airfoil and in the wake. It affects the *effective angle of attack* of the airfoil. Lift associated with the change of effective angle of attack due to downwash is the *induced drag* (Sec. 3.2). Skin friction,

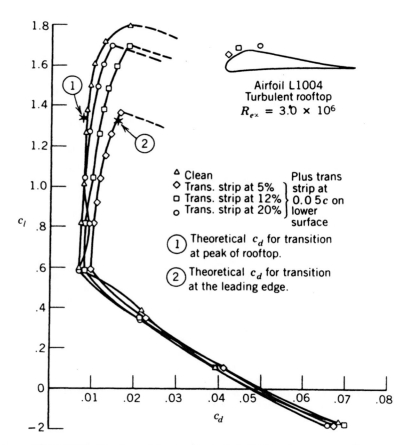

FIGURE 3.13:2 C_L vs C_D plots of insect wings and airfoils. Reprinted with permission from A.M. Kuethe and C.Y. Chow, *Foundations of Aerodynamics*, 4th Ed., copyright © 1986, John Wiley & Sons, Inc. Data adapted from Nachtigall (1974) and Thom and Swart (1940).

form drag, and induced drag are the three principal components of the total drag.

It is convenient to consider skin friction and form drag separately from induced drag. If a wing is cylindrical (does not vary in the spanwise direction) and is tested in a wind tunnel, the flow around the wing can approach a two-dimensional condition if large flat plates are installed at the tips of the wing. Then the aspect ratio of the wing is effectively infinity and the induced drag tends to zero (see Eq. (3.2:10)). The form drag and skin friction can then be measured by measuring the velocity distribution around a control volume. The drag coefficient so measured varies with the angle of attack, hence with the lift coefficient. Figure 3.13:1 shows the measured drag coefficient-lift coefficient relationship of two smooth NACA airfoils at a Reynolds number of 6×10^6. It is seen that C_D rises drastically when certain C_L value is exceeded, but it is possible to design the airfoil so as to obtain the "low-drag bucket" at a designed range. Both of the airfoils shown in this figure are so-called "laminar flow" airfoils. Figure 3.13:2 shows the C_L vs C_D relationship of a number of insect wings and airfoils.

Methods of wind tunnel measurement of drag, and means of achieving high maximum lift coefficient and low minimum drag coefficient are discussed in Kuethe and Chow (1986), Liebeck (1978), Smith (1975), Stratford (1959), Walsh (1980).

The airfoil theory is applicable to the fins and tails of fish. But fish swimming near the surface of the water must bear additional drag arising from making waves on the free surface of the water (Sec. 4.11).

Problems

3.1 Consider the aerodynamics of racing bicycles and cyclists. In order to strive for the best, what can you do to reduce the drag of the man-machine combination? (Cf. Kyle, C. and Burke, E.: *Mechanical Engineering*, 106: 35–45, Sept. 1984. For an interesting history of human-powered vehicles, see Wilson, D.G.: *American Scientist*, 74: 350–357, 1986.)

3.2 Human-powered flight excites people's imagination. How much has man succeeded? Looking over the horizon, can you suggest some new improvements? (Cf. "Human-Powered Flight." *Mechanical Engineering*, 106: 46–55, 1984.)

3.3 There have been attempts to fly nonstop around the world on a single tank of fuel. What kind of features must such an airplane have? What kind of wing span to chord length ratio? What kind of engine? What kind of control system? What kind of safety devices must the plane and flyers have? For a two-seater, how large and how heavy is the plane likely to be? (Cf. "Another World Aviation Record for Voyager?" *Mechanical Engineering*, 108: 41–44, 1986.)

3.4 Mechanical and geometric properties of the upper and central airways between the mouth and carina can be inferred noninvasively in individual subjects from high frequency acoustic reflection data measured at the month. Write down the equations of motion, continuity, and constitutive equations, and formulate the inverse problem of determining mechanical and geometric properties from acoustic waves. (Cf. Fredberg, J.J.: Acoustic Determinants of Respiratory System Properties. *Ann. Biomed. Eng.*, 9: 463–473, 1981. The inverse problem is similar to one of geophysics, see Ware, J.A. and Aki, K.: Continuous and Discrete Inverse Scattering Problem in a Stratified Elastic Medium. I. Plane Waves at Normal Incidence. *J. Acoust. Soc. Am.*, 54: 911–921, 1969.)

3.5 To prevent aeroelastic instabilities such as flutter, stall, and divergence, modern airplane designers use electronic feedback control to *actively* control the structural parameters such as wing stiffness and wing shape (aspect ratio, sweepback or sweepforward angle) as well as the deflections of the control surfaces. Survey the parameters and discuss the possibilities of active control for a high performance airplane.

3.6 Following the thin-wing theory outlined in Sec. 3.11, find the lift distribution on an airfoil whose cross-section is (1) a straight line, (2) a shallow sine curve, $Y = a\sin(\pi x/c)$, and (3) an S-shaped curve given by $Y = a\sin(2\pi x/c)$. Is there any difference in the lift coefficient and the moment coefficient about the leading edge in these three cases?

3.7 Following the lifting-line theory of a finite wing outlined in Sec. 3.12, determine wing planforms that will produce (1) an elliptic distribution and (2) a uniform distribution.

3.8 If the lifting line of Sec. 3.12 is not a straightline, but consists of two straightline segments meeting at an angle at the middle, representing a wing swept back or forward, what major change in the theory has to be introduced?

3.9 The vortex system of fish's fins has important effect on the dynamics of force generation in fish swimming. Make a sketch of the vortices of a fish of your choice, and discuss the dynamics of the system during fish swimming. What is the difference in the pattern between normal swimming and a condition for hunting or fighting-for-life situation? Formulate a rigorous mathematical theory to study the fluid dynamics of this problem. As a biologist, think of the biological side of the problem also: the muscle system, the nervous control, the neuronetwork. Formulate an interesting problem to study.

3.10 Describe the vortex system (bound and free) acting on a flapping wing of finite span.

3.11 Describe the changes occurring in the vortex system when a bird maneuvers to pitch, yaw, and roll. Discuss the time lag between a movement of a control surface and the forces and moments generated on the wing.

3.12 Based on the free and bound vortex system, formulate a mathematical approach to analyze the two-dimensional lift distribution on a bird's wing (not as a single straight line) under the assumption that the wing is rigid.

3.13 Under the assumption of a rigid wing (as in Prob. 3.12), determine the forces and moments acting on the bird.

3.14 Bird's wings are generally swept forward near the body and swept back in outer span. With the anatomical structure and muscle system taken into consideration, discuss qualitatively the effect of these sweep angles on the stability and control of the bird.

3.15 An eagle dives from a high altitude to catch a squirrel. What determines the diving speed? If it dives too fast, would the speed interfere with the ability to catch the object? How does the eagle pull out of the dive?

3.16 At high speed of diving, what effect does the Corioli's acceleration due to earth's rotation have on the flight of the bird?

References

Material of this chapter is taken mostly from the author's book *An Introduction to the Theory of Aeroelasticity*. Additional references can be obtained from the following:

Abbott, I.H. and von Doenhoff, A.E. (1949). *Theory of Wing Sections, Including a Summary of Airfoil Data*, McGraw-Hill, New York. (Paperback edition, Dover, New York, 1959.)

Batchelor, G.K. (1967). *An Introduction to Fluid Dynamics*. Cambridge University Press, Cambridge, United Kingdom.

Biot, M.A. (1942). Some simplified methods in airfoil theory. *J. Aeronaut. Sci.* **9**: 186–190.

Bisplinghoff, R.L. and Ashley, H. (1962). *Principles of Aeroelasticity*. Wiley, New York.

Bradshaw, P. (1964). *Experimental Fluid Mechanics*, MacMillan, New York. Pergamon Press, (p. 235). *Introduction to Turbulence and Its Measurement* (1975), Pergamon Press, New York.

Cebeci, T. and Bradshaw, P. (1978). *Momentum Transfer in Boundary Layers*, McGraw-Hill, New York.

Chow, C.-Y., Huang, M.-K., and Yan, C.-Z. (1985). Unsteady Flow About a Joukowski Airfoil in the Presence of Moving Vortices. *AIAA J.* **23**(5): 657–658.

Dowell, E.H., Curtiss, H.C., Scanlan, R.H., and Sisto, F. (1978). *A Modern Course in Aeroelasticity*. Sijthoff and Noordhoff, Alphen aan den Rijn, The Netherlands.

Fung, Y.C. (1955, 1969). *An Introduction to the Theory of Aeroelasticity*. John Wiley, New York (1955). Paperback, expanded, Dover Publications, New York (1969).

Glauert, H. (1947). *The Elements of Aerofoil and Airscrew Theory*. 2nd Ed. Cambridge Univ. Press, London.

Kovasznay, L.S.G. (1949). Hot wire investigation of the wake behind cylinders at low Reynolds numbers. *Proc. Roy. Soc. London*, A, **198**, 174.

Kuethe, A.M. and Chow, C.Y. (1986). *Foundations of Aerodynamics*, 4th Edn. Wiley, N.Y.

Küssner, H.G. (1936). Zusammen fassender Bericht der instationären Auftrieb von Flügeln. *Luftfahrt-Forsch.* **13**: 410–424.

Küssner, H.G. (1940). Das zweidimensionale Problem der beliebig bewegten Tragfläche unter Berücksichtigung von Partialbewegungen der Flüssegkeit. *Luftfahrt-Forsch.* **17**: 355–361.

Küssner, H.G. and Schwarz, L. (1940). The oscillating wing with aerodynamically balanced elevator. *Luftfahrt-Forsch.* **17**: 337–354. English translation: *NACA Tech. Memo* 991 (1941).

Liebeck, R.H. (1978). Design of Subsonic Airfoils for High Lift. *J. Aircraft* **15**(9): 547–561. Design of Airfoils on High Lift, Proc. AIAA Symposium on Aircraft Design, 1980.

Nachtigall, W. (1974). *Insects in Flight*. H. Oldroyd et al., trans. McGraw-Hill, New York.

Pai, S.I. (1956). *Viscous Flow Theory*. Van Nostrand, New York.

Reynolds, O. (1883). An Experimental Investigation of the Circumstances Which Determine Whether the Motion of Water Shall be Direct or Sinuous, and of the Law of Resistance in Parallel Channels, *Phil. Trans. R. Soc.* London, **174**: 935–982.

Sears, W.R. (1941). Some aspects of non-stationary airfoil theory and its practical application. *J. Aeronaut. Sci.* **8**: 104–108.

Smith, A.M.O. (1975). High-Lift Aerodynamics, *J. Aircraft* **12**(6): 501–531.

Stratford, B.S. (1959). The Prediction of Separation of the Turbulent Boundary Layer, *J. Fluid Mech.* **5**: 1–16.

Theodorsen, T. (1935). General theory of aerodynamic instability and the mechanism of flutter. *NACA Rept.* 496.

Theodorsen, T. and Garrick, I.E. (1942). Flutter calculations in three degrees of freedom. *NACA Rept.* 741.

Thom, A. and Swart, P. (1940). Forces on an airfoil at very low speeds. *J. Roy. Aero. Soc.* **44**: 761–769.

Van Dyke, M. (1982). *An Album of Fluid Motion*, Parabolic Press, Stanford, California.

Wagner, H. (1925). Über die Entstehung des dynamischer Auftriebes von Tragflügeln. *Z. angew. Math. u. Mech.* **5**: 17–35.

Walsh, M.J. (1980). *Viscous Drag Reduction. Progress in Astronautics and Aeronautics*, Vol. 72, (Gary R. Hough, ed.).

Yih, C.-S. (1969, 1989). *Fluid Mechanics*. McGraw-Hill, New York (1969). West River Press, 3530 West Huron River Dr., Ann Arbor, MI 48103 (1989).

CHAPTER 4

Flying and Swimming

4.1 Introduction

Locomotion is, of course, an extremely interesting subject. People are forever fascinated by sports. We cheer gold medal winners. How athletes are trained is certainly a legitimate question for biomechanics. There are people who suffer impairments in locomotion and others who try to help them recover or overcome their handicaps. These people, the sports lovers, educators, patients, orthopedic surgeons, engineers, physical therapists, nurses, prosthesis manufacturers, and hospital managers, will benefit from a good understanding of the biomechanics of locomotion. Then there is the world of animals around us. We see animals walking and crawling on land, flying in air, and swimming in fluid. From man and mice to birds, fishes, and sperms, there is a tremendous variety of questions one may wish to ask about locomotion.

The objective of the present chapter is to present a brief discussion of the mechanics of flying and swimming. We select a few topics from the point of view of fluid and solid mechanics in order to gain some insight to the problems of locomotion. We shall pass over walking, running, jumping, and crawling because they are familiar to us, and because their mathematical analysis is formidably complex.

4.2 The Conquest of the Air

About 400 million years ago, the earth was populated by plants and tall trees; the air was moist, and the ground was covered with decaying leaves; and the insects began to appear (Smart and Hughes, 1972). The insects achieved powered flight in the middle of the Carboniferous period, about 300 million years ago. Birds came on the scene later, about 150 million years ago. Reptiles, which flourished 100 to 200 million years ago, included an extensive group of flying animals, the pterosaurs. Certain flying mammals, including bats, appeared about 50 million years ago.

All the animals we can see are winners of evolution. They represent successful designs that have met all the constraints imposed by their environment. It is not useful to cover too large a territory in this chapter. To be brief, let us limit ourselves to two aspects: (1) identifying those features in nature that can be understood in terms of the fluid mechanics discussed in the previous chapter, and (2) recording observations of flying and swimming in nature that involve principles beyond the scope of the preceding chapter. In the latter category are items such as the hovering of hummingbirds, and the flying of tiny insects.

The wings of birds and insects not only differ in size and structure, but also in basic articulation mechanism. The bone structure of a bird does not differ too much from that of our arms (see Fig. 4.2:1). The bird moves its wings by

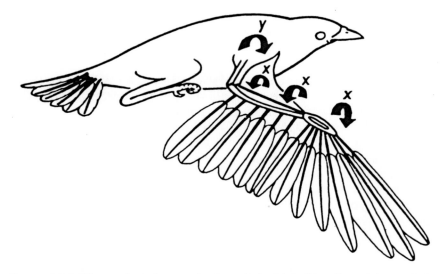

FIGURE 4.2:1 Wing articulation mechanism of the bird. The elbow joint, and the humerus, ulna, radius, and carpals bones are seen. The arrows x and y show the moments created by the muscles about the bone. The aerodynamic force is transmitted through the bases of the feathers to the wing bones. The resulting tendency to twist the manus and ulna in the nose-down direction is resisted by the pitch-up moments (x) supplied by the muscles. Owing to the sharp angle at which the elbow joint is held in flight, the center of the lift lies ahead of the axis of the humerus and the resulting nose-up moment is counteracted by the moment y supplied by the pectoralis muscle, which pulls downward on the deltoid crest, ahead of the axis of the bone. From Pennycuik (1972, p. 16), by permission.

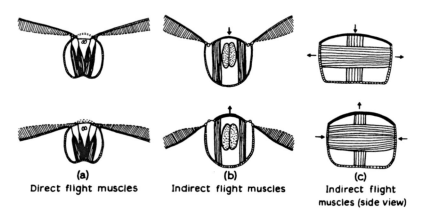

(a)	(b)	(c)
Direct flight muscles	Indirect flight muscles	Indirect flight
		muscles (side view)

FIGURE 4.2:2 Wing articulation mechanisms of insects. (a) In insects with *direct* flight
muscles, the muscles are connected to the wings. (b) In insects with *indirect* flight
muscles, the downstroke isproduced by the raising of the roof of the thorax which is
brought about by the contraction of the dorsal longitudinal muscles as shown in (c).
Upstroke is obtained by the contraction of the vertical muscles. In (b), a *double hinge
system* that connects the wings to the roof of the thorax is seen. The operation of the
hinges involves a skeletal click mechanism. From G. Goldspink (1977, p. 13) which is
modified from Pringle (1975). By permission.

muscle, so its flying muscles are big and strong. Insects use different arrangements (see Fig. 4.2:2). The insect thorax is shielded by cuticles of thin-walled
chitinous shells with good elasticity and rigidity, joined with an elastic material, *resilin*. The wings and the thorax shell form a vibration system. In some
insects the wing movements are produced by wing muscles directly inserted
into the base of the wing. In others, such as *Diptera*, the movements are
produced by muscles that pull on the thorax shell, while the shell deformation
moves the wings. In the latter cases [Fig. 4.2:2(b)] the hinges are so arranged
that the wing position is stable either fully raised or fully lowered, whereas in
moving from one stable position to the other it goes through an unstable
position very quickly. A twisting movement can be superposed on the up and
down movement by a proper arrangement of the hinges. Resilin is an almost
perfect rubber. It is capable of large deformation with a nearly linear force–
deflection relationship with very little internal friction. Its mechanical properties are similar to those of *elastin* which is discussed in *Biomechanics* (Fung,
1981, Chap. 7, pp. 197–201), except that the hysteresis loop is even smaller.
Elastin is found in the mammalian blood vessel wall. In the neck of the cow
or horse the large ligament nuchae is almost all elastin, and it holds the head
up without need of muscle action. In an analogous way the resilin can store
elastic energy and can release it very quickly when unloaded. The wing-thorax
resilin-and-muscle system is a mass-spring system which has a characteristic
frequency of vibration. Thus, some insect wings can beat automatically if

excited, and the muscle only needs to supply enough energy to overcome the aerodynamic drag, and to provide its own excitation.

Resilin exists also in the leg joints of those insects that can jump, e.g., fleas and click-beetles. To jump, these insects use the muscle to bend to joints and store energy in the resilin by elastic deformation. Upon release of the muscle tension the stored energy is released very quickly, causing the insect to be catapulted.

The insect's flight muscles are rather different from mammalian skeletal muscles in that they have a very short I-band in the sarcomere (Goldspink, 1977). As a consequence they operate in a very short range of length. This is consistent with the mechanism exhibited in Fig. 4.2:2.

Be it a bird or an insect, a flying animal must be able to generate thrust and lift, which is accomplished by superimposing an adequate positive angle of attack onto a thrust-generating oscillatory motion. For birds and larger insects in *fast forward flight*, the lift principle is similar to that of the airplane. See Sec. 4.3. However, the majority of insects are small, and make use of *hovering and slow flight*. To do this without a rotor, unlike a helicopter, most hovering animals use a mode in which the body stays head-up (almost erect) and the wings beat back and forth in a horizontal plane, preceding each beat with a wing rotation that always allows the same leading edge to move forward at an appropriate angle of attack. Some insects, such as the chalcid wasp (*Encarsia formosa*), clap their wings together dorsally once per beat and achieve a remarkably effective hovering (Weis-Fogh, 1975). The butterflies (*Papilionoidea*) can clap their wings ventrally as well as dorsally. The dragonfly can hover with an almost horizontal body axis by making a complicated coordinated motion of its four wings. These are discussed in Sec. 4.5.

The story of evolution tells us that out of all known animal species, fossil and living, those using hovering or slow flight form of locomotion represent three quarters, or 750,000, of them (Weis-Fogh, 1975). The great success of using active flapping flight as the principal mode of locomotion on earth is evident.

But many birds can also utilize natural air currents to soar and to glide over long distances. The albatross can even take advantage of the wind *shear* over the ocean (i.e., the increase in wind speed with the altitude) to spiral and soar with little expenditure of energy (Sec. 4.4). Nature is full of wonders.

4.3 Comparing Birds and Insects with Aircraft

To understand the mechanism of flight, it is useful to compare birds and insects with airplanes and helicopters. Man copied nature and developed the aircraft industry; but the constraints imposed on man's flight are not entirely the same as those in nature, and the final products are different.

Birds and insects use flapping motion for propulsion, hence their aerodynamics is predominantly unsteady. To understand their aerodynamic lift, thrust, moment, and induced drag, we must think of the vortices shed into the airstream at the trailing edge of the wing, as well as the oscillatory horseshoe vortices which are bounded to the wing in the spanwise direction and shed into the wake in the streamwise direction. These are discussed in the preceding chapter, see Secs. 3.9 and 3.10, especially Fig. 3.10:5. The most important consequence of the vortex shedding mechanism is the delay in lift force generation that causes the aerodynamic lift and moment to get out of phase with the angle of attack. The system is, therefore, nonconservative, and there is an exchange of energy between the wing and the airstream in every cycle of flapping (see Sec. 3.6). This is very different from the fixed-wing aircraft for which, unless the wing vibrates, only drag consumes energy at a steady flight.

The unsteady condition affects the stalling characteristics also (see Fig. 4.3:1). If a sudden change of angle of attack is so large that in normal steady flight the wing would have stalled (a drop in lift coefficient occurs in the $C_{L \, max}$ region shown in Sec. 3.13, Fig. 3.13:2), in unsteady motion the inertia of the air may be able to prevent flow separation from occurring. Thus, a higher $C_{L \, max}$ may be obtained. This is important for birds when hovering, landing, and taking off.

A jetliner may extend its flaps to increase wing area when taking off and landing, deflect the flaps to increase the curvature of the wing cross section to increase the maximum lift coefficient ($C_{L \, max}$), deflect the aileron to roll, move the rudder to yaw, and lift the elevators to pitch. It may have a leading edge slot to delay stall and increase $C_{L \, max}$, as well as to increase the wing area. It may have vortex generators (a row of protruding small plates arranged on the upper surface of the wing in the tip region) to help prevent tip stall. Some high performance planes are designed to increase their sweep angle (sweep back or sweep forward) as the flight speed increases (to avoid shock waves). All these features have prototypes in nature.

Figure 4.3:2 shows six major devices that enable birds to achieve high lift.

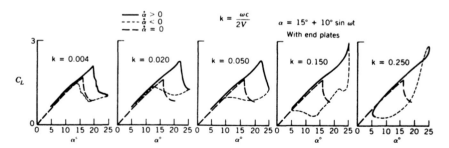

FIGURE 4.3:1 Unsteady stalling characteristics of wings. The maximum lift coefficient as a function of the rate of change of the angle of attack. Airfoil NACA 0012, 1.22 m chord, 1.98 m span with end plates. Oscillation about quarter-chord point at various reduced frequencies (Strouhal numbers k) in 29.5 m/sec airstream. From McAlister et al. (1978). Courtesy of U.S. Army.

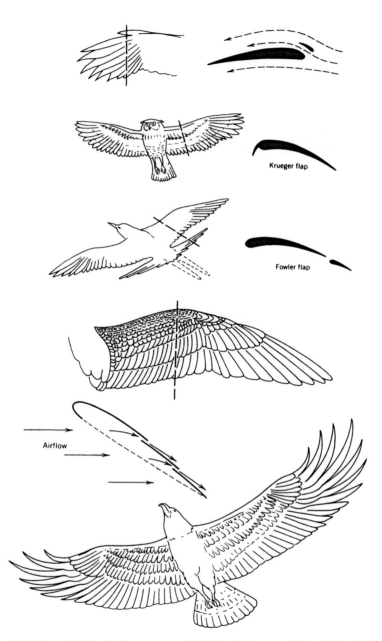

FIGURE 4.3:2 Arnold Kuethe's summary of bird's devices to achieve high lift. From top down. (1) The "thumb pinion" (alula) of a pheasant as a leading edge slot. The thumb pinion is particularly highly developed in woodland birds, presumably to better enable them to avoid obstacles. (2) The drooping leading edge clump of feathers at an owl as a Krueger flap. (3) The swept forward position of the tail of a split-tail falcon as a Fowler flap. (4) The "layered" wing feathers as a multi-surface air-foil. (5) The upward deflected flight feathers of the wing tips of a hawk as "winglets." Reproduced by courtesy of Professor Arnold Kuethe. From Kuethe (1975a, b). Kuethe and Chow (1986).

These pictures were drawn by Cyril H. Barnes to illustrate a paper by Arnold M. Kuethe (1975b). Some of them are redrawn from versions in von Holst and Küchemann (1974). In Fig. 4.3:2(a), the feathers are shown to twist in order to direct a component of the lift in the forward direction to provide thrust. The twisting is achieved automatically by a proper aeroelastic structural design of the feather by placing the elastic axis (a line of the centers of twist) and the line of the centers of aerodynamic pressure at a proper relative position (see Fung, 1969, p. 17).

The equivalent of airplane leading edge slot, split flap, Krueger flap, Fowler flap, and blown flap are shown in Fig. 4.3:2(b)–(f). The blown flap of advanced aircraft, using compressed air bypassed from the engine to blow out the stagnant air in the thickened boundary layer near the trailing edge of the wing, can raise the maximum lift coefficient by 400–500%. The birds, using the devices named above, probably can increase the lift coefficient by 50–100%.

Figure 4.3:3 shows the method used by a hummingbird to hover (Weis-Fogh, 1975). With its body nearly erect, a hummingbird swings its wings about a near-horizontal plane; each wing describes a "figure 8" through an arc of about 120°. During the interval when the wings are moving forward, the

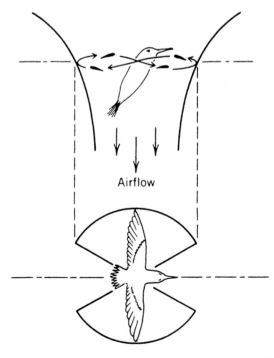

FIGURE 4.3:3 Hummingbird's hovering method as explained by Weis-Fogh (1973). Reproduced by permission.

"palms" are down; for the return stroke, they flip to palms-up. Both motions produce lift in roughly equal proportions.

The Chalcid wasp hovers by a clap-and-fling motion which is described in Sec. 4.5, Figs. 4.5:1 and 4.5:2. There is no man-made flying machine of this nature.

There are other similarities and differences. Both animals and aircraft are elastic, but the animals are more so. The aeroelastic phenomena, discussed in Sec. 3.4, especially divergence, loss of control, and flutter, will impose restrictions on animal flight as they do aircraft. In addition, animals have nerves and muscles, and exquisite sensory organs; and they achieve remarkable feedback and control which are not yet fully understood. Only the most advanced aircraft designs seek active electronic control of their basic structures (in addition to their control surfaces), whereas all birds do.

4.4 Forward Flight of Birds and Insects

Only by detailed observation can one gain a true understanding of animal locomotion. For insects, a monumental piece of work was done by Weis-Fogh and Jensen (1956), who analyzed the flight of the desert locust *Schistocerca gregaria* by observing them in a wind tunnel. They suspended the locust from a force balance, and used the force measured by the balance to control the speed of the wind-tunnel fan in order to simulate free sustained forward flight conditions. The wind speed, once the insect was beating its wings regularly, was automatically adjusted by feedback control so that the net horizontal force measured by the balance was zero. Thrust was then balancing drag as in free flight.

Typical values of the locust's weight, wing length, the speed of flight, U, and the wing-beat frequency, $\omega/2\pi$, are, respectively, 2 g, 4 cm, 4 m/sec, and 20 Hz. The Reynolds number Uc/v is approximately 200 based on the forewing chord c. The Strouhal number $k = \omega c/U$ (with ω in radians per sec) is about 0.25, low compared with $k = 2.2$ for most airplane wings in flutter condition, or roughly 1 for fish with lunate tails, but about the same as that for birds (Lighthill, 1975). Hence the effect of nonstationarity is not large.

Figure 4.4:1 shows the effective motion through the air of a forewing chord in the forward flight of a locust. In the downstroke the forewings have a positive angle of attack so that the lift on them is much larger than the drag and the resulting aerodynamic force supports both the animal's weight and the thrust needed for propulsion. This stroke is long and there is plenty of time for the airfoil to build up its steady–state force, nearly at a right angle to its instantaneous path through the air.

A rapid "supination" (in aeronautical terms, "pitch-up") marks the onset of an upstroke of the locust wing with the airfoil at a large positive angle of pitch relative to the direction of mean motion. At the same time, however, its angle of attack relative to its instantaneous direction of motion in the air is

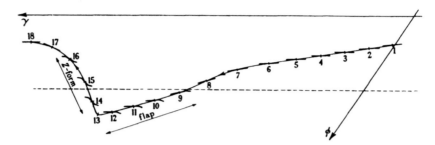

FIGURE 4.4:1 Tracings from a film of a tethered locust in a wind tunnel, showing the movement relative to the air of a chordwise section through the mid-point of a locust forewing in forward flight. From Weis-Fogh and Jensen (1956), by permission.

negative. At this stage of the stroke the wing is bent in the form of a Z-shaped cross-section, which has some rather peculiar aerodynamic properties. Wind tunnel tests show that in steady-state the lift is small and acts opposite to the normally expected direction, probably due to flow separation at the leading edge due to the bent head of the Z section. But the supination is rapid and the vortices that shed into the wake of the wing are important to the transient aerodynamic response. The actual nonstationary aerodynamic force acting during the upstroke is still unknown.

The motion of the hindwings is coordinated with that of the forewings. The forewing motion exhibits a phase lag behind that of the hindwings. This phase lag is characteristic of four-winged insects in general. Detailed estimations of the lift, drag, thrust, and the interference between the wings are given by Weis-Fogh and Jensen (1956).

For birds, corresponding pieces of beautiful work were done by Brown, Tucker, Pennycuick and others. Brown (1953) trained birds to fly along a 60 m long passage to a cage as he photographed them with a high-speed camera halfway along the passage. Tucker (1968) and Pennycuick (1968) trained birds to fly freely in wind tunnels. From these studies it was found, for example, that pigeons can fly at a speed of about 10 m/sec, wing beat frequency of about 5 Hz, corresponding to a Strouhal number of around 0.5, and Reynolds number of about 10^4 based on the wing chord. The downstroke, with the wing chord practically horizontal, is quite similar to the downstroke of the locust shown in Fig. 4.4:1. The rapid supination at the end of the downstroke is also just as marked. The upstroke is highly 'feathered': the bird's *wrist* is flexed in the upward movement, and then the primary feathers near the wing tips are swung violently backward to obtain a significant thrust in a manner depicted in Fig. 4.4:2.

The bird's lift is probably acquired mainly on the downstroke when the primary feathers are spread out, and the wing area is increased. In the meantime, the maximum lift coefficient is increased as we have discussed in

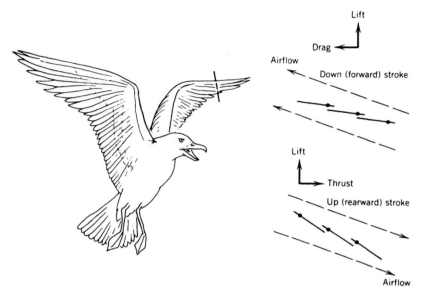

FIGURE 4.4:2 Hovering flight of a gull. The locations of the quills of the feathers are indicated. The force vectors are shown in the right. Courtesy of Dr. Arnold Kuethe.

Sec. 4.3, and stalling is avoided even when a large angle of attack is used at the end of the downstroke to initiate the supination and upstroke.

Compared with insects, birds have a higher power requirement per unit mass. Pennycuick (1968) estimated the specific power requirement for sustained pigeon flight to be around 20 W per kg, amounting to over 96% of the total metabolic rate. An oxygen consumption of over 130 ml/min is needed to sustain flight, in contrast to a resting metabolic requirement of only 5 ml/min. On the other hand, the sustained locust flight was estimated by Weis-Fogh and Jensen (1956) to require only about one-third as much muscular power output per unit mass of the animal.

Thus the bird needs a lung that can handle a large variation in ventilation to meet the enormous variation in its energy requirement in life. The bird's lung is encased in a rigid chest cage, and its ventilation is powered by a number of air sacs (Dunker, 1972). The lung maintains a constant volume, while the air sacs expand and contract to generate a flow. Figure 4.4:3 shows a schematic drawing of the avian respiratory system. The gas exchange apparatus consists of parallel bundles of thin-walled "air capillaries" of diameter 3 to 10 μm (Dunker, 1972) in close proximity to blood capillaries of about the same size. The construction is similar to some boilers in industrial power plants. Since the air and blood capillaries are about the same size, the birds are able to pack the blood–gas exchange surfaces into their lungs more compactly than mammals can in their reciprocating machinery. The exchange surface area per unit

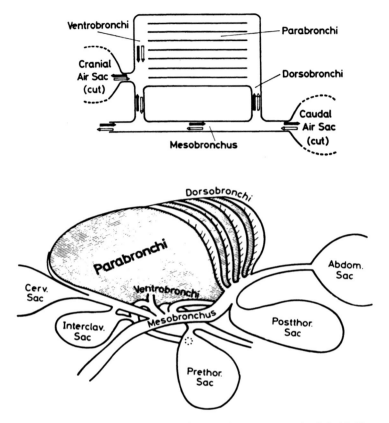

FIGURE 4.4:3 Schematic drawing of the avian respiratory system by Scheid, Slama, and Piiper (1972).

volume is about 250 mm^{-1} in the bird, as compared with 10–16 mm^{-1} for man. But the most remarkable fact about the bird's lung is that the flow in the air capillaries is unidirectional while the flow in the air sacs is reciprocating. This unidirectional flow is achieved without a valve anywhere! It is a valveless pump. One can understand such a pump only if the inertia of the fluid and the resistances to flow in the various bronchi and bronchial bifurcation points are considered.

Example. Take a surgeon's glove. Tie up the wrist and cut holes at the tips of the thumb and ring finger. Into the holes insert glass tubes of 3 to 5 mm diameter and tie up. Hold the tubes vertical and fill the glove with water to a level half way up the tubes. Now you have a valveless pump. Tap on the palm. With a little practice you can find the right frequency, amplitude and location of tapping that will create a substantial difference of the heights of the water columns in the tubes. Explain the mechanics!

Power Requirement for Flight

The thrust required for flight must balance the drag and the inertial force. The inertial force is equal to the product of the acceleration and the sum of the mass of the animal and virtual mass of the fluid associated with the mode of motion. The drag force consists of three parts: One part is due to skin friction in the boundary layer over the wing, body, and tail. Another part is the induced drag discussed in Secs. 3.2 and 3.10. The third part is due to the flapping motion and the transient vortex shedding into the wake.

The sum of the skin friction and induced drag is the drag at steady flight:

$$D = \tfrac{1}{2}\rho U^2 A C_{D_f} + K L^2/(\tfrac{1}{2}\rho U^2 b^2). \tag{1}$$

Here D is the drag force, ρ is air density, U is the speed of flight, A is the wing area, C_{D_f} is the coefficient of drag due to skin friction on the wings, body, and tail, K is a constant, L is the lift, b is the wing span. The first term on the right-hand side of (1) is Eq. (3.2:4), except that we now let C_{D_f} include the frictional drag of the body and tail. The second term in (1) is Eq. (3.10:4).

In steady flight the lift must be equal to the weight of the animal, W:

$$L = W = \tfrac{1}{2}\rho U^2 A C_L. \tag{2}$$

Hence the lift coefficients C_L, and the corresponding angle of attack (see Fig. 3.2:2) depend on the weight and the flight speed. The drag coefficient C_{D_f} varies with the angle of attack, and hence with C_L. When C_L is small, C_{D_f} is roughly constant, but C_{D_f} increases rapidly if the angle of attack is increased toward the stalling angle or beyond.

The power required for flight is equal to the product of the thrust T and the speed of flight U. If we consider steady flight, and very roughly assume the power required for the flapping motion of the wings to be a constant, P_0, then

$$\text{Power required} = TU = \text{const} + DU$$

$$= P_0 + \tfrac{1}{2}\rho U^3 A C_{D_f} + K W^2/(\tfrac{1}{2}\rho U b^2). \tag{3}$$

If the power required is plotted against the forward speed U we obtain the results shown in Fig. 4.4:4 Here the power required is seen to be a U-shaped curve. The minimum power point $P_{\min}$ corresponds with a velocity $U_{\min P}$. The point of tangency of the power curve with a straight line drawn from the origin defines the velocity of flight and the power required for the maximum range. This is because to maximize the range we must maximize distance travelled per unit work, and this is equivalent to finding the smallest ratio of power to speed. The maximum-range speed is at least 1.3 times the minimum-power speed, and usually more—in the pigeon it is around 1.8 times the minimum-power speed.

The power required to hover (at $U = 0$) is usually much larger than $P_{\min}$. Eq. (3) does not apply to the hovering mode.

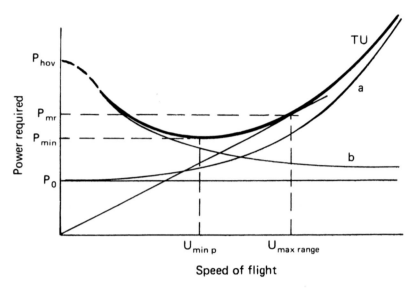

FIGURE 4.4:4 A plot of Eq. (3). P_0 represents the constant in Eq. (3). Curve (a) represents the second term on the right-hand side. Curve (b) represents the last term. Curve TU is the sum of these two, or the power required. Eq. (3) does not apply when $U \to 0$; that part of the curve is estimated separately.

Range

The work done by flying over a distance dX in still air is equal to the product of thrust times dX. The thrust is equal to the power P divided by the velocity U, see Eq. (3). Hence the work done dE is PdX/U. If we regard dE as the amount of fuel used in order to fly the distance dX, then we can write

$$dX = \frac{U \, dE}{P}. \tag{4}$$

We would like to express this in terms of the lift/drag ratio. For this purpose note that $P = TU$, and that at level flight, T is equal to the drag D, and the lift L is equal to the weight W. Hence

$$dX = \frac{U \, dE}{TU} = \frac{L}{W} \frac{dE}{D} = \frac{L}{D} \frac{dE}{W}. \tag{5}$$

The lift/drag ratio, L/D, depends on the angle of attack and the wing and body geometry. The energy dE is obtained by burning a small amount of fat dW. Writing $dE = JdW$ where J is the mechanical equivalent per unit weight of fuel (about 8×10^5 joules per kg weight), we obtain, on assuming L/D to be constant and integrating Eq. (5), the result

$$X = J\frac{L}{D}\ln\left(\frac{W_1}{W_2}\right), \tag{6}$$

where W_1, W_2 are the bird's take off and landing weights, respectively. Thus, to fly a long distance, as some migrating birds do, the lift/drag ratio should be large and the bird should be fattened as much as possible before taking off. A ratio of W_1/W_2 equal to 2 is not inconceivable.

Figure 4.4:4 tells us how to find the speed to fly for maximum range. In this figure the horizontal axis is the bird's airspeed. If there is a tail wind blowing, the ground speed is equal to the wind speed plus the bird's airspeed. If the bird's aim is to maximize distance travelled *over the ground* per unit work done, then the graph must be redrawn with groundspeed instead of airspeed as the abscissa. The effect of a tailwind is to shift the origin to the left, and the tangent now has to be drawn from the new origin, giving a lower air speed than before for maximum range. Similarly, airspeed must be increased to obtain maximum range against a headwind. Radar tracking of migrating birds flying high at night shows that they do adjust their speed of flight this way, although it is not clear how they find out the wind speed!

Stability and Control

Whereas an airplane needs ailerons to control roll, elevators to control pitch, rudder to control yaw, and fin and stabilizers to give stability, most insects and birds can achieve stability and control by their wings alone. The mechanisms used include: (1) variable sweepback of the wings and variable angle of attack across the span, (2) controlled variable camber, i.e., curvature of the wing in chordwise direction, and (3) variable wing tip control with feathers deflected downward. Some of the prehistoric birds (*Archaeopteryx*) had long tails like an airplane, but later members dispensed with the tail, suggesting that the tailless method of control is more efficient aerodynamically, but more difficult to operate.

Pennycuick (1972) suggests that the function of the bird's tail is like that of the flap of an airplane: to increase the maximum lift coefficient at low speed. The tail is typically spread and depressed at landing and take off. This adds lift surface area and sucks air downwards over the central portion of the wing, so stalling can be delayed. Birds with long forked tails, such as swifts (*Apodidae*) and swallows (*Hirundinidae*), are able to fly very slowly and hover, and use the spread tail for steep downward deflection of the airflow leaving the wings.

Gliding and Soaring

If a bird or insect of mass m flies without beating its wings, it can get thrust only from gravitational acceleration, i.e., by gliding downward. In a still air let the flight path be inclined at an angle θ to the horizon, then the lift L, drag D, and gliding angle θ are related by the following equations:

$$L = mg\cos\theta, \qquad D = mg\sin\theta, \qquad \theta = \tan^{-1}(D/L). \tag{7}$$

Since L is essentially the bird's weight, the last equation shows that the gliding angle θ will be a minimum if the drag is minimized. With drag given by Eq.(1), this occurs at a flight speed U which makes the two terms on the right-hand side of Eq. (1) equal:

$$U_{\text{min drag}} = \left(\frac{4K}{\rho^2 C_{D_f}} \frac{L^2}{Ab^2}\right)^{1/4}. \tag{8}$$

If U is smaller than that given in Eq. (8), an accidental rise in speed will reduce the drag and the speed will be increased. Hence gliding at a speed less than the minimum-drag speed is unstable. Gliding is stable only if U is greater than the minimum-drag speed, and, of course, also above the stalling speed.

If the bird catches a thermal, then while Eq. (7) remains true relative to the wind, the path relative to the earth may become inclined upward, and the bird soars. As this is well known, let us not elaborate any further but turn to the albatross, which "soars" according to a different principle.

"Soaring" of the Albatross

The famous "soaring" of the albatross *Diomedea*, by which it varies its height above the ocean in a periodic cycle with minimum expenditure of energy was analyzed by Rayleigh (1883) and Jameson (1958). It utilizes the wind *shear* over the sea, i.e., the variation of wind speed with height. Stanley Corrsin explained it as follows (see Lighthill, 1975, p. 169): Let the coordinates x, y, z be fixed on earth and let the mean natural wind velocity be in the x-direction, with magnitude $\bar{u}(z)$ varying with altitude z above the ocean. Let the three components of the velocity vector of the albatross relative to the earth be

$$(\bar{u}(z) + u', v', w'). \tag{9}$$

Then its velocity relative to the local wind is (u', v', w') with a resultant V. The dynamic pressure is $\frac{1}{2}\rho V^2$, ρ being the air density. At steady flight the lift must balance the bird's weight W, and the life coefficient is $C_L = W/(\frac{1}{2}\rho V^2 S)$, where S is the wing area (Eq. 3.2:4). The angle of attack is $\alpha = C_L/a$ if a denotes the lift curve slope (Eq. 3.2:5). To avoid stalling α must be smaller than the stalling angle α_{max}, and C_L must be smaller than the maximum possible, $C_{L\,\text{max}}$ (see Fig. 4.3:1). Since C_L varies inversely with V^2, to avoid stalling the square of the relative velocity must be so large that the lift coefficient remains below the maximum value $C_{L\,\text{max}}$. In other words the kinetic energy:

$$\tfrac{1}{2}m(u'^2 + v'^2 + w'^2) \tag{10}$$

must be steadily maintained above a minimum value against the dissipative action of aerodynamic resistance. Now, in the theory of turbulence we learn that any fluctuating motion with velocity (9) (whether of an eddy or of an albatross) extracts energy from the *mean* motion at a mean rate

$$-m\overline{u'w'}\frac{d\bar{u}}{dz}, \tag{11}$$

where the bar denotes a mean value. But $\overline{u'w'}$ can be large and negative, making (11) large and positive and thus maintaining the kinetic energy (10) above a minimum level. To do this u' and w' should have opposite signs: the albatross should move upward when upwind and downward when downwind.

Over the ocean the atmospheric boundary layer thickness can be of the order of 50 m. The wind shear is significant near the ocean surface. The albatross turns into the wind in order to gain height, and then turns downwind to gain speed. At sea level it soars along the slope of the windward face of a wave until it meets a suitable upward gust, when it turns into the wind and initiates another upwind climb. It zigzags in this way over the Antarctic Ocean, progressing on average downwind, and circulating in the prevailing westerlies 'round and 'round Antarctica.

An Impressive Record

The speed of most insect flight is not impressive, but the deer botfly has been clocked at 64 miles per hour (28 m/sec) or 300 body lengths per second (Nachtigall, 1974). By comparison, the Cheetah (fastest of the land animals) achieves 18 lengths per second, while man achieves 5, an automobile 5 to 15, a swift 60, a Starling 80, a jet fighter at Mach 3 reaches about 100.

4.5 Hovering and Other Modes of Motion

Most birds and insects hover like a hummingbird, with the body vertical and the wings flapping in a horizontal plane, see Fig. 4.3:3. A dragon fly hovers with body horizontal and two pairs of wings executing a coordinated motion. Weis-Fogh (1975) and Norberg (1975) analyzed this motion with quasi-steady strip theory (treating every cross section of the wing as if it were a two-dimensional airfoil reacting to a steady flow whose velocity is the same as that of the instantaneous field) and concluded that hovering is energy intensive. Indeed, large birds hover only a short time at take off or landing, and usually have to use anaerobic power to do so.

Weis-Fogh (1973) described another type of hovering. He studied a small wasp (Hymenoptera) *Encarsia formosa*, that is used in the biological control of greenhouse aphids. It has a wing span of 0.6 mm, an airborne weight of 25×10^{-8} N. Its fore- and hind-wings are hooked together and beat as one at a frequency of about 400 Hz. Like many insects *Encarsia* can jump but the wing movements seen in Fig. 4.5:1 relate to free hovering with a slow climb. Its wings move essentially in a horizontal plane both during the downstroke and upstroke. The sequence of its wing movements contains three phases: (a) the *clap* phase in which the two pairs of wings are brought together as a single vertical plate (b) a *fling* phase in which a very rapid pronation of the wings takes place and the wings are flung open in a manner reminiscent of the "flinging open" of a book, followed by a horizontal downstroke; (c) a *flip* phase during which a very rapid supination takes place before the upstroke. These

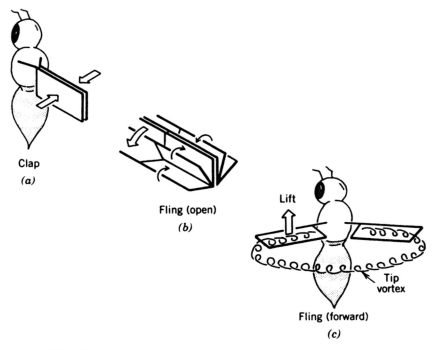

Clap
(a)

Fling (open)
(b)

Lift

Tip
vortex

Fling (forward)
(c)

FIGURE 4.5:1 The hovering motion of *Encarsia formosa* as sketched by Weis-Fogh (1973, Fig. 21). Reproduced by permission.

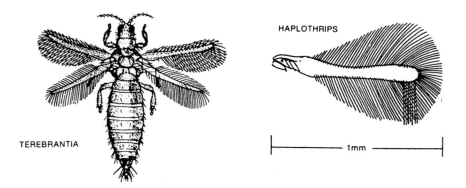

HAPLOTHRIPS

TEREBRANTIA

1mm

FIGURE 4.5:2 The bristles (setae) of the wings of *Terebrantia* and *haplothrips*, tiny insects commonly known as "thrips." Courtesy of Dr. Kuethe.

three phases are present in *Encarsia* during all kinds of flight. Moreover, the lift equalled the body weight long before the wings reached maximum angular velocity.

Weis-Fogh's explanation (1973) is supported by Lighthill's (1975) analysis. Lighthill explains that toward the end of the clap, the wings are essentially at

FIGURE 4.5:3 Scanning electron microscopic photographs of a Mulleinthrips *Haplothrips verbasci*. The setae (shown here in a stiffened condition) are 1×2 μm in cross section, and 200–300 μm long. × 130. Courtesy of Dr. Arnold Kuethe.

rest relative to the air. During the fling the wings open up, and a potential flow is induced so as to fill the triangular void created between the two upper wing surfaces, giving rise to two bound vortices of equal strength and opposite sign. When the wings break apart along the "hinge" and start the horizontal downstroke, they already have circulation and can produce lift in accordance with the Kutta–Joukowski Theorem, Eq. (3.7:23). Since the bound vortices are already there, the generation of this part of the lift has no delay and no Wagner effect (Sec. 3.5 and Sec. 3.14). In other words, the initial lift is generated in the fling stage.

The Reynolds number of this wasp's hovering motion based on the wing chord and wing velocity in downstroke is about 20. The Reynolds number based on the speed of the leading edge during its "fling" is about 30. These Reynolds numbers are sufficiently large so that the boundary layer theory and vortex wake concept discussed in Chapter 3 are presumably applicable.

For even smaller insects that operate at Reynolds number close to 1 or smaller than 1, boundary layer concept does not apply, and any analysis would have to be based on Navier–Stokes equations. In these cases definitive theory and experiments are both lacking.

Kuethe (1975) points out that many tiny insects have one striking feature in common. From the tips of their wings, a fringe of bristles (setae) projects forward, rearward, and outward. Figure 4.5:2 shows a *Terebrantia*, commonly known as "thrips", and a haplothrips. Figure 4.5:3 shows a magnified view of these bristles. Kuethe notes that the flow of air around each bristle has Reynolds number much smaller than 1, hence must be governed by Stokes equations. The analysis of flow around such moving bristles is similar to those around cilia and flagella (Secs. 4.8 and 4.9). It seems natural to suggest that by a proper wave motion of these bristles, lift can be generated in analogy with cilia propulsion. Kuethe's idea is that the wave motion of the sheet of bristles resembles that of the pectoral fin of a skate or ray. He estimated that the lift force generated by wave motion of the sheet of bristles is more than 100 times greater than that would be if the sheet were impervious to the flow.

4.6 Aquatic Animal Propulsion

From single cells to whales, most organisms live in water. The most elementary form of motion is chemotaxis. The protoplasm in cells is not stationary, it moves in response to chemical stimuli. Some cells, such as leucocytes and amoeba, move this way. Chemotaxis is one of the intensely studied subjects today; but the mechanism is more closely related to the molecular transformation of actin molecules and the sol-gel transformation of the protoplasm than hydrodynamics. See Sec. 4.12.

To survey the field of aquatic animal propulsion, we may begin with *Protozoa*. Among these single-celled creatures are those belonging to the class *Mastigophora*, which propel themselves by undulatory motions of a whiplike flagellum. The basic mode of motion is to pass a wave backward along the flagellum, either by a flexural motion, or by rotation in the manner of a screw. See Fig. 4.6:1. Fluid resistance to the motion provides the propulsive force to the organism. This method of propulsion has been most successful in the evolution process, and has been adopted and developed by practically all the aquatic animals that are successfully mobile.

Another class of Protozoa, the *Ciliophora*, includes organisms which propel themselves by movements of a large number of attached cilia. Each cilium behaves like a flagellum; but with many cilia there are certain organized

SWIMMING   ATTACHED

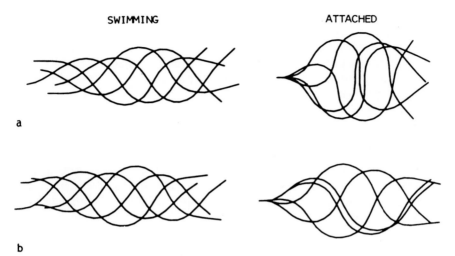

FIGURE 4.6:1 Flagella movement computed by Brokaw and Gibbons (1975) under the hypothesis that active moment is proportional to curvature, and viscous and elastic resistances are considered.

patterns. Ciliary propulsion is used by animals in the class *Ctenophora*, and most members of the phylum *Rotifera*. But this method has been less successful in the evolutionary process.

In man and animals, cilia also grow on the surfaces of respiratory airways, mesenteric membranes, uterian tubes, etc. They help move dust particles from the lung, mucus in the airway, ovum from the ovary, etc. Their function depends very much on the viscoelasticity of the fluid. Figure 4.6:2 shows the ciliary motion of the respiratory tract. Successive stages of the forward stroke are shown in (a). Successive stages of the retraction stroke are shown in (b). See Sec. 4.9.

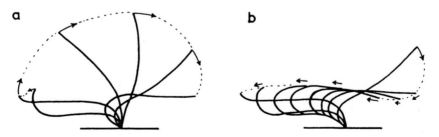

FIGURE 4.6:2 Successive stages of the movement of a cilium on the surface of a respiratory airway. (a) Forward stroke. (b) Return stroke. The length of arrows indicates the relative speeds of ciliary motion. From Blake and Sleigh (1975), by permission.

An alternative to undulatory propulsion is jet propulsion, which is used by animals in the classes *Hydrozoa* and *Scyphozoa* in the phylum *Coelenterata*. These animals have a characteristic jellyfish shape, and by contraction of the circular subumbral muscles can expel water from the subumbral cavity and thereby slowly propel themselves. In the phylum *Mollusca*, some animals in the class *Cephalopoda*, the squids, use a jet concentrated in a narrow funnel and can achieve quite high speeds of 2–4 m/sec, according to size. In the phylum *Chordata*, members of the class *Urochordata* (headless and accordingly invertebrate) characteristically draw in water and expel it for propulsion. But jet propulsion is rare among the vertebrates.

Thus undulation is the method of propulsion preferred by animals higher in the evolutionary ladder. It is remarkable that the same method can be used by both large and small animals. The forward velocity of the animal U is related to the speed of the transverse wave V passing down the animal's body. For a flagellate at a typical Reynolds number of 10^{-3}, U is of the order of $0.2\ V$. (Gray and Hancock, 1955). For a nematode with a typical Reynolds number of about 1 the forward speed U is of the order of $0.4\ V$ (Gray and Lissmann, 1964). For a leech with a Reynolds number of order 10^3 the forward speed is about $0.3\ V$ (Gray, 1939). For an eel U is typically of the order of $0.6\ V$. For goldfish U is about $0.7\ V$ (Bainbridge, 1963).

The most important modification of the undulatory method of propulsion introduced by fish is the transverse 'compression' of the body, i.e., flattening at the posterior end of the animal. The flattening makes the rear end more flexible transversely, and permits motions of larger amplitude at the tail region. The next important modification is the enlargement of the caudal fin to increase the reactive force of propulsion. It can be shown (Lighthill, 1975; see Sec. 4.10) that the motion at the tail region is really what counts in the propulsion of the fish, and the width of the caudal fin at the posterior end is the most important dimension of the fish as far as propulsion is concerned. Hence the evolution points in the direction of increasing the caudal fin dimension.

The most advanced development is the lunate tail on such fast swimmers as the shark, tuna, dolphin, and whale. The caudal fin is transformed into "wings" of advanced hydrodynamic design. See Figure 4.6:3.

Other fins of the fish, dorsal and anal, pectoral and pelvic, are important in stability and control, turning and stopping, as well as for auxiliary propulsion.

Breder (1926) classifies fish swimming into "anguilliform" and "carangiform" modes. The former is named after the common eel, of genus *Anguilla*. The latter is named after jacks and horse mackerels, of family *Carangidae*. In the former the whole body participates in an undulatory mode. In the latter the tail waves vigorously while the front body remains almost rigid.

This broad classification is widely used but is only indicative. There are many variations and fine details about each fish that make it unique. And often the situation is not so clear cut. For example, eels use strict anguilliform

FIGURE 4.6:3 Evolution of swimmers. The first row shows some early dwellers of the sea: protozoa, trilobite, sea scorpion. Rows 2–10 show the evolution of fish, beginning with Devonian period (405 ∼ 345 × 10^6 yrs ago) at left, to the present time at right. Row 2 shows Agnatha (Jawless fishes), Pteraspis and Lampetra (lamprey). Row 3 shows a Placodermi (Armored fish) Pterichthyodes. Row 4 shows Acanthodii (Spiny fishes) Climatius and Acanthodes. Rows 5 and 6 show Osteichthyes (Higher bony fishes) Cheirolepis, Scomber (mackerel), Oncorhynchus (salmon), Dorypterus, and Acipenser (sturgeon). Row 7 shows an Osteolepis. Row 8 shows a Dipnoi (Lungfish) Dipterus. Rows 9 and 10 show Chondrichthyes (Cartilaginous fishes) Cladoselache, Dogfish shark, Manta (devil ray), Helodus, and Chimaera (ratfish). The last row shows a mammal, a killer whale.

propulsion only in fast swimming; for slow motion they use dorsal and anal fins. They can swim backwards by putting the fin undulations to reverse. Cod, haddock, and whiting, having numerous separate dorsal fins and a substantial anal fin and a caudal fin, use these fins in something close to anguilliform. The vortex shed by the fins is essentially continuous in spite of the gaps between the fins, which function almost like a continuous one.

We present a mathematical framework for the analysis of flagella and cilia propulsion in Secs. 4.7–4.9. The theory discussed in Sec. 4.9 could probably be extended to the anguilliform swimming of some animals with 'poor' hydrodynamic form such as water snakes.

Lighthill's theory for carangiform propulsion is presented in Sec. 4.10. This theory reveals the predominant importance of the tail. In realizing that each stroke of the tail motion produces a large sidewise force, one must wonder how the fish can stay on an even course without hopelessly yawing from side to side. Lighthill (1969) suggests two answers. One is that by narrowing the depth of the fish cross-section before the caudal fin, the side force is reduced. The other is that by having a large depth of body and fins in the front part of the fish, the reaction to side force is reduced.

Bainbridge (1958, 1960,1963) has observed the details of carangiform propulsion of several species of fish and found that the maximum amplitude of the caudal fin movement (i.e., the lateral displacement from the unstretched position) to be about 0.2 times the length of the fish, L. The reduced frequency or Strouhal number $\omega L/U$ (where the radian frequency ω is 2π times the frequency in Hz) takes values clustering around 10. This is in sharp contrast to the reduced frequencies of insects' and birds' flight, which is less than 1, and usually of the order of 0.2 to 0.6. The influence of vortices in the transient wake (Sec. 3.10) is thus much more significant for the fish.

4.7 Stokeslet and Dipole in a Viscous Fluid

Microbes swimming with flagella (such as human sperm) or cilia (such as paramecium) are so small that the Reynolds number of their motion is much less than 1. In that case the inertial force is negligible compared with the viscous forces; the Navier–Stokes equation is reduced to the Stokes equation, and the flow is said to be in the Stokes flow regime. In constructing a hydrodynamic theory for the flagellar motion, use is made of singularities such as the Stokeslets and dipoles. Let us first explain what they are.

The Stokeslet

Consider an incompressible Newtonian fluid. Let p denote the pressure, $\mathbf{u}$ denote the velocity vector, ρ denote the fluid density, μ denote the fluid viscosity coefficient. Then, on neglecting the acceleration terms (because the Reynolds number $\to 0$), and assuming a body force field $\mathbf{X}$, the Navier–Stokes equation (see Sec. 1.7) becomes the Stokes equation

$$-\nabla p + \mu \nabla^2 \mathbf{u} + \mathbf{X} = 0. \tag{1}$$

The equation of continuity is

$$\nabla \cdot \mathbf{u} = 0. \tag{2}$$

Applying the ∇ operator on Eq. (1) and noting (2), we obtain

$$-\nabla^2 p + \nabla \cdot \mathbf{X} = 0. \tag{3}$$

Applying the Laplace operator ∇^2 on Eq. (1) and noting (3), we obtain

$$\mu \nabla^4 \mathbf{u} = \nabla \times \nabla \times \mathbf{X}. \tag{4}$$

If the body force vanishes, then $\mathbf{X} = 0$, and

$$\nabla^4 \mathbf{u} = 0. \tag{5}$$

Hence in a region in which body force vanishes the pressure is a *harmonic function* (because p satisfies the harmonic equation $\nabla^2 p = 0$), and the velocity components are *biharmonic functions* (because they satisfy the biharmonic equation (5)) in an inertialess flow.

Oberbeck (1876) obtained a general solution of Stokes' equation in terms of harmonic functions:

$$\mathbf{u} = 2\boldsymbol{\phi} - \nabla(\mathbf{x} \cdot \boldsymbol{\phi}) + \nabla \Phi_c \tag{6}$$

$$p = -2\mu \nabla \cdot \boldsymbol{\Phi}. \tag{7}$$

Here $\boldsymbol{\phi}$ is a vector with three components which are harmonic functions, $\nabla^2 \boldsymbol{\phi} = 0$, and Φ_c is a scalar harmonic function, $\nabla^2 \Phi_c = 0$. By direct substitution, it is easy to verify that $\mathbf{x} \cdot \boldsymbol{\phi} - \Phi_c$ is a biharmonic function. Hence $\mathbf{u}$ given by Eq. (6) is biharmonic, satisfying Eq. (5). The solution is *general* in the sense that any biharmonic function in a region whose boundary intersects any straight line in at most two points can be represented by a function of the form given in Eq. (6) (for a proof, see Fung (1965), p. 208).

On substituting $\boldsymbol{\phi} = \boldsymbol{\alpha}/r$ and $\Phi_c = 0$ into Eq. (6), one obtains a solution which is called the *Stokeslet* by Hancock (1953). The flow field of a Stokeslet is

$$\mathbf{u} = \frac{\boldsymbol{\alpha}}{r} + \frac{(\boldsymbol{\alpha} \cdot \mathbf{x})\mathbf{x}}{r^3} \tag{8a}$$

$$p = 2\mu \frac{(\boldsymbol{\alpha} \cdot \mathbf{x})}{r^3}. \tag{8b}$$

If $\boldsymbol{\alpha}$ is a unit vector in the direction in the x-axis, then $\boldsymbol{\alpha} = (\alpha, 0, 0)$, $\mathbf{x} = (x, y, z)$, $r^2 = x^2 + y^2 + z^2$; we have

$$\mathbf{u} = \alpha\left(\frac{x^2 + r^2}{r^3}, \frac{xy}{r^3}, \frac{xz}{r^3}\right), \tag{9a}$$

$$p = 2\mu \frac{\alpha x}{r^3} = -2\mu\alpha\partial(r^{-1})/\partial x. \tag{9b}$$

The solution is singular at the origin. To obtain a physical meaning of α, consider a flow generated by a concentrated force applied at the origin in the x-direction. Represent this force by the Dirac function $F_x\delta(\mathbf{r})$, where $\mathbf{r}$ is the radius vector and F_x can be a function of time. Equations (1) and (3) then become

$$-\nabla p + \mu\nabla^2\mathbf{u} + F_x\delta(\mathbf{r}) = 0 \tag{10}$$

$$\nabla^2 p = \nabla \cdot F_x\delta(\mathbf{r}) = F_x(\partial\delta(r)/\partial x). \tag{11}$$

At this point it is convenient to recall a familiar solution representing the source flow of an incompressible fluid (Sec. 3.7), which is described by the equation

$$\nabla^2\phi = \delta(r) \tag{12}$$

for the velocity potential ϕ. The source is located at the origin and its strength is a unit volume per second. The solution is

$$\phi = -\frac{1}{4\pi r}. \tag{13}$$

Comparing (13), (12) with (9b) and (11), we see that the solution

$$p = -\frac{F_x}{4\pi}\frac{\partial}{\partial x}\left(\frac{1}{r}\right) \tag{14}$$

represents the pressure field generated by a concentrated force F_x at the origin. Comparing (14) with (9b), we see that

$$F_x = 8\pi\mu\alpha. \tag{15}$$

Thus α is equal to the force at the origin divided by $8\pi\mu$.

The total force acting on a control volume V enclosed in a surface S is given by integrating the surface traction over the entire surface. If σ_{ij} denote the stress tensor and v_j denote the unit normal vector of the surface, then the force acting on S is, by Cauchy's formula (Sec. 1.7),

$$F_i = -\int_S (-p\delta_{ij} + \sigma_{ij})v_j\,dS. \tag{16}$$

By Gauss theorem, this can be transformed into a volume integral

$$F_i = -\int_V \left(-\frac{\partial p}{\partial x_i} + \frac{\partial\sigma_{ij}}{\partial x_j}\right)dV. \tag{17}$$

By Euler's equation (Sec. 1.7), the integrand is the body force. Hence

$$F_1 = \int_V F_x\delta(x_1)\,dV$$

$$= \int_V 8\pi\mu\alpha\delta(x_1)\,dV = 8\pi\mu\alpha = F_x, \tag{18}$$

$$F_2 = F_3 = 0.$$

The divergence of σ_{ij}, namely, $\partial\sigma_{ij}/\partial x_i$, is the term $\mu\nabla^2 u$ in Eq. (5). The force so computed by Eq. (18) is independent of the size and shape of the control volume V. Thus the force on any control surface is equal to the concentrated force acting at the origin.

At each point, the *radial* component of the velocity field of a Stokeslet is, according to Eq. (9a):

$$u_r = (\mathbf{r} \cdot \mathbf{u})/r = [F/(8\pi\mu)](2\cos\theta/r), \tag{19}$$

where θ is the spherical polar coordinate defined so that $\cos\theta = x/r$. The *transverse* component is

$$u_\theta = (u^2 - u_r^2)^{1/2} = [F/(8\pi\mu)](-\sin\theta/r). \tag{20}$$

Figure 4.7:1 shows the velocity components. Note that the velocity decreases rather slowly as r^{-1} as r increases, and that only the factor 2 in Eq. (19) prevents the Stokeslet from representing a unidirectional velocity field in the x-direction with a magnitude equal at all points on a spherical surface of radius r.

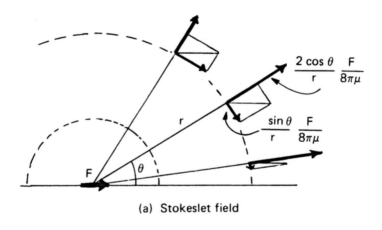

(a) Stokeslet field

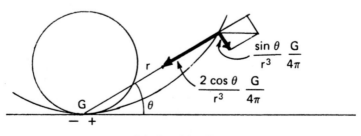

(b) Doublet field

FIGURE 4.7:1 The velocity components in spherical polar coordinates for (a) a stokeslet field of strength F, and (b) a dipole field of strength G.

Dipoles

The fow field $\nabla\phi$ derived from a *velocity potential*

$$\phi = Gx/(4\pi r^3) \tag{21}$$

is said to represent a *dipole* with strength $(G, 0, 0)$. A dipole is a source and a sink next to each other, and its potential is obtained by a differentiation of the potential of a source (Eq. (13). The gradient of the velocity potential is the velocity vector. The velocity and pressure fields for a dipole are:

$$\mathbf{u} = \frac{G}{4\pi}\left(\frac{1}{r^3} - \frac{3x^2}{r^5}, \; -\frac{3xy}{r^5}, \; -\frac{3xz}{r^5}\right), \qquad p = \text{const.} \tag{22}$$

The last equation follows from Eq. (1) since $u = \nabla\phi$ satisfied $\nabla^2 u = 0$ when $X = 0$. The velocity field has radial and transverse components

$$u_r = -\frac{G}{4\pi}\frac{2\cos\theta}{r^3}, \qquad u_\theta = -\frac{G}{4\pi}\frac{\sin\theta}{r^3}, \tag{23}$$

which decays like r^{-3}. This is illustrated in Fig. 4.7:1.

4.8 Motion of Sphere, Cylinder, and Flagella in Viscous Fluid

From Fig. 4.7:1 or Eqs. (19), (20), and (23) of Sec. 4.7, it can be seen that if one wants to represent the motion of a sphere of radius a, one should superpose a dipole of strength

$$G = Fa^2/(6\mu) \tag{1}$$

to a Stokeslet of strength F. Then the velocity on a sphere $r = a$ is a uniform velocity $U = F/(6\pi\mu a)$ parallel to the direction of the applied force. This yields the Stokes formula

$$F = 6\pi\mu aU \tag{2}$$

relating the external force required to move a sphere of radius a through a fluid of viscosity μ at a speed U when the Reynolds number approaches zero. For an alternative method of deriving this result, see Fung (1984), pp. 250–254.

Slender Cylinder

Since a Stokes flow around a sphere can be described by a superposition of a stokeslet and a dipole, it may be expected that a Stokes flow around a cylinder can be obtained by superposing an infinite number of stokeslets and dipoles on a straight line segment. Indeed, let a line segment stretch from $z = -b$ to $z = c$ on the z-axis, Fig. 4.8:1. Let the strength of the stokeslet on a line element of length dz located at z be $f\,dz$, whereas that of the dipole be $g\,dz$, f and g being constants. Then on moving the origin of the solutions given by Eqs. (9),

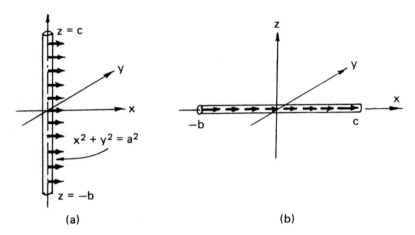

FIGURE 4.8:1 Motion of slender cylinders: (a) in a direction perpendicular to its axis, (b) in the direction of the axis.

(15), and (22) of Sec. 4.7 to Z (replacing z by $z - Z$), leaving the stokeslet and dipole pointing in the x-direction, changing F to $f dZ$, G to $g dZ$, and integrating the result from $Z = -b$ to $Z = c$, one obtains a solution which represents the results of the superposition. Lighthill (1975) shows that if one choses

$$g = \frac{f a^2}{4\mu} \tag{3}$$

(not, in fact, the same choice given by Eq. (1) that was needed for the sphere problem), then the velocity $\mathbf{u}_N$ on the circle

$$x^2 + y^2 = a^2, \qquad z = 0 \tag{4}$$

is uniform, with the components

$$\mathbf{u}_N = \frac{f}{8\pi\mu}\left(1 + \log\frac{4cb}{a^2}, 0, 0\right) \tag{5}$$

provided that $a \ll b$, $a \ll c$. This shows that the result represents approximate solution of the velocity field generated by a uniform translation of a cylinder of radius a in a direction normal to its axis. The force applied by the cylinder on the fluid is equal to f per unit length of the cylinder, and it acts in the direction of motion perpendicular to the axis of the cylinder. The force is equal to the strength of the Stokeslets because the dipoles contribute nothing to the forces.

This solution is approximate because it does not hold in the neighborhood of the ends of the cylinder where $z - b$ or $z - c$ is not large compared with a. Furthermore, a constant f produces a slowly varying velocity field along the length of the cylinder. The translational velocity is the maximum at the midpoint, but the percentage variation over most of the length of the cylinder is not large.

For the solution named above to be valid, the length of the cylinder $b + c$ must be larger compared with a, but it must remain finite, because the integrals do not converge if b, c tends to ∞. This is an illustration of the well-known paradox that the Stokes equations, Eqs. (1) and (2) of Sec. 4.7, possess no solution satisfying the boundary conditions of a uniform motion of an infinitely long circular cylinder.

The corresponding solution for a cylinder moving in the direction of its axis is obtained by superposing stokeslets of strength $(f\,dx, 0, 0)$ in each element of length dx in the segment $-b < x < c$ of the x-axis. See Fig. 4.8:2(b). The velocity on the cylinder $y^2 + z^2 = a^2$ at $x = 0$ is $\mathbf{u}_T$:

$$\mathbf{u}_T = \frac{f}{8\pi\mu}\left(-2 + 2\log\frac{4cb}{a^2}, 0, 0\right). \tag{6}$$

In this case, no dipole is needed.

These results are often expressed in terms of the *coefficients of resistances* to motions of the cylinder normal and tangential to itself, K_N and K_t, respectively, defined as the ratios of the force per unit length f to the velocities at the midsection of the cylinder:

$$K_N \equiv f/u_N, \qquad K_T \equiv f/u_T, \tag{7}$$

u_N and u_T being the magnitude of the vectors given by Eqs. (5) and (6), respectively. Hence

$$K_N = \frac{8\pi\mu}{1 + \log(4cb/a^2)}, \qquad K_T = \frac{4\pi\mu}{-1 + \log(4cb/a^2)}. \tag{8}$$

Note that the resistance decreases with increasing length. Note also that if $\log(4cb/a^2)$ is large compared with 1, the value of K_N is almost twice as large as K_T. Thus the resistance to normal motion is almost twice that to tangential motion.

This difference in resistance coefficients K_N and K_T offers the principal key to the understanding of the propulsion by undulatory motion of the flagellum or cilium. The transverse undulation in a flagellum exert a greater force on the fluid by its normal component than by its tangential component. The normal component can be made to propel while the tangential component drags.

Gray and Hancock (1955), Hancock (1953), Gray (1968), Lighthill (1969, 1975), Chwang and Wu (1971, 1974–1976), made extensive use of this kind of approach to analyze the propulsion of small organisms by flagella or cilia. An outline of their basic analysis is presented in the next section.

Parallel with the hydrodynamic study, there is another body of research looking into the biology of the flagella and cilia. The flagella of some organisms are spiral shaped and propel by rotation similar to a propeller. Other organisms have flagella that undulate in a plane. What are the intrinsic

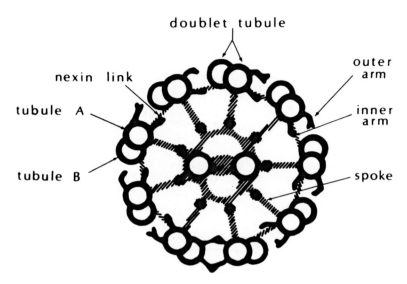

FIGURE 4.8:2 Interpretive diagram of the structures seen in electron micrographs of cross-sections of a flagellar or ciliary axoneme. (Modified from an original by K.E. Summers.) From C.J. Brokaw and I.R. Gibbons (1975), by permission.

structure and mechanism of these flagella that imparts the wave motion? Almost all flagella have a unique 9 + 2 structure illustrated in Fig. 4.8:2. The current concept of flagellar motion is built around the possibility of sliding of these internal filaments in analogy with the actin and myosin fibers in the muscle. The flagellar motion is modulated by ATP and Ca^{++} flux. Studies of these biochemical events yield valuable insights. A convenient reference to many investigations of this nature is the two-volume book edited by Wu et al. (1975).

4.9 Resistive-Force Theory of Flagellar Propulsion

Small animals which propel with flagella or cilia usually have lengths less than 1 mm. Some flagella perform undulatory motion in a plane. Some have a spiral form and rotate like a screw. Some combine undulation and rotation. The motion appears in general as a wave travelling along the body of the animal from head to tail. Let us consider a flagellum undulating in a plane. Figures 4.9:1–4.9:3 show several views of the flagellar motion. Figure 4.9:1 shows what appears to an observer who moves with the waveform. Figure 4.9:2 shows what appears to an observer who moves with the animal. Figure 4.9:3 shows the motion of the animal as it appears to an observer fixed in the laboratory. The animal moves with an average velocity U relative to the undisturbed fluid

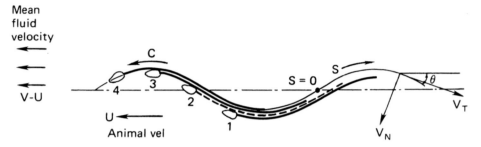

FIGURE 4.9:1 The appearance of a flagellum as seen by an observer who moves with the waveform. Animal appears to move with a velocity c along the wave. In projection on the x-axis the animal moves forward with an average speed $V = \alpha c$. The free stream appears to move at velocity $V-U$.

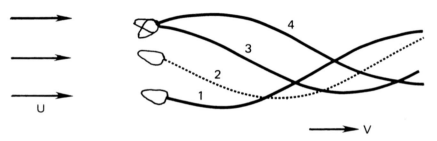

FIGURE 4.9:2 Observer moves with animal's head. The body wave appears to move backward with velocity V. The undisturbed free stream appears to flow at velocity U.

FIGURE 4.9:3 Animal swims in a pool in laboratory. Observer in laboratory frame of reference. The undisturbed fluid is stationary. Animal forward velocity is U.

(called 'free stream') The undulation wave of the flagellum moves backward (toward the tail) with a velocity **V** relative to the animal's body.

Lighthill (1975), following Gray and Hancock (1955), analyzed the motion shown in Fig. 4.9:1 as follows. A flagellum is assumed to have a length L when stretched straight. To describe the wave form he expresses the coordinate x of a point on the waveform by a function $X(s)$, s being the distance measured along the curve. Similarly, the coordinates y and z are expressed. Thus, the waveform of the flagellum is described by

$$(x, y, z) = (X(s), Y(s), Z(s)). \tag{1}$$

Since s is the arc length the derivatives of $X(s)$, $Y(s)$, $Z(s)$ satisfy the equation

$$X'^2(s) + Y'^2(s) + Z'^2(s) = 1. \tag{2}$$

Hence $X'(s)$, $Y'(s)$, $Z'(s)$ are the direction cosines of the tangent to the waveform curve. For an inextensible flagellum moving along the waveform curve with a constant speed c, the points on the flagellum can be described by

$$(x, y, z) = (X(s - ct), Y(s - ct), Z(s - ct)). \tag{3}$$

The point $s = 0$ is identified as the flagellum's head at time $t = 0$. The speed c differs from the horizontal velocity V only in that c is measured along the curved waveform whereas V is its horizontal projection. c is larger than V in the same proportion as the curved length of the flagellum L is larger than the horizontal projection of the flagellum.

If the forward velocity of the animal is $\mathbf{U}$ while the wave moves backward on the animal's body with a velocity $\mathbf{V}$, then the free stream would appear to move forward with a velocity $\mathbf{V} - \mathbf{U}$ relative to the waveform (see Fig. 4.9:1). The velocity relative to the fluid of any section of the flagellum is then the difference of the velocity $\mathbf{c}$ directed along the forward tangent of the waveform and the vector $\mathbf{V} - \mathbf{U}$. Figure 4.9:1 shows that its component along the backward tangent is

$$V_T = (V - U)\cos\theta - c, \tag{4}$$

where θ is the angle between the local tangent to the flagellum and the x-axis. Its component along the backward normal, V_N, is

$$V_N = (V - U)\sin\theta. \tag{5}$$

As we have noted earlier, following Eq. (2),

$$\cos\theta = X'(s - ct), \qquad \sin\theta = [1 - X'^2(s - ct)]^{1/2}. \tag{6}$$

Lighthill assumes the resistive force acting on any section of the flagellum to be linearly proportional to the local velocity of the section relative to the undistorted fluid. Thus the tangential and normal forces per unit length of the flagellum are

$$f_T = K_T V_T, \qquad f_N = K_N V_N, \tag{7}$$

respectively, where K_T, K_N are resistance coefficients as defined in Eq. (4.8:7). The sum of the horizontal components of these forces is the *thrust per unit length* of the flagellum, and an integration of this thrust over the length of the flagellum yields the *total thrust* T:

$$T = \text{total thrust} = \int_0^L (f_T\cos\theta + f_N\sin\theta)\, ds. \tag{8}$$

With Eqs. (4)–(7), this becomes

$$T = \int_0^L \{K_T[(V - U)X'(s - ct) - c]X'(s - ct)$$

$$+ K_N(V - U)[1 - X'^2(s - ct)]\} \, ds. \tag{9}$$

Let us write

$$\int_0^L X'(s - ct) \, ds = \int_0^L \cos\theta \, ds = \alpha L,$$

$$\int_0^L X'^2(s - ct) \, ds = \int_0^L \cos^2\theta \, ds = \beta L. \tag{10}$$

Then α, β are the average values of $\cos\theta$ and $\cos^2\theta$ over the length of the flagellum. α, β are functions of t. Hence, on noting that $V = \alpha c$, Lighthill obtains the total thrust T:

$$T = K_T L[(V - U)\beta - V] + K_N L(V - U)(1 - \beta). \tag{11}$$

For a free organism there is no net force acting on the body (inertial force being neglected since the Reynolds number is $\ll 1$). If the entire animal is like a flagellum (e.g., a nematode), then $T = 0$ and Eq. (11) can be used to solve for the speed of swimming U. If the flagellum propels an inert head (e.g. a spermatozoon), then T must balance the drag of the head. Lighthill (1975) writes the drag of the head as $K_N L U \delta$, where δ is a coefficient defined as

$$\delta = \frac{\text{resistance to forward motion of the head}}{\text{resistance to uniform normal motion of whole flagellum}}. \tag{12}$$

Then on equating (11) with $K_N L U \delta$, the ratio of the swimming speed U to the wave speed V is obtained:

$$\frac{U}{V} = \frac{(1 - \beta)(1 - r_K)}{1 - \beta + r_K\beta + \delta}, \tag{13}$$

where

$$r_K = K_T / K_N. \tag{14}$$

If $\beta = 1$ (the flagellum is flat and horizontal), then $U = 0$ (there is no forward velocity). As β decreases from 1 (amplitude of undulation increases) the ratio U/V increases but cannot exceed the limiting value (as $\beta \to 0$):

$$\left(\frac{U}{V}\right)_{\max} = \frac{1 - r_K}{1 + \delta}. \tag{15}$$

Thus the maximum achievable speed of swimming depends on the ratio r_K of the tangential and normal force coefficients.

Example 1. Consider a small organism whose length L is very short compared with the wave length of a planar undulation waveform, as illustrated in

Fig. 4.9:1. Then to an observer moving with the animal the organizm will be seen as shown in Fig. 4.9:2. Let the waveform in Fig. 4.9:1 be given by the equation

$$y = A \sin(2\pi s/\Lambda) \doteq A \sin(2\pi x/\lambda) \tag{16}$$

where Λ is the wavelength along the curve and λ is the wavelength in the x-direction. Then the configuration of the animal in Fig. 4.9:2 can be represented by

$$y = A \sin[2\pi(x - Vt)/\lambda] \tag{17}$$

in which x is measured from the head toward the tail. The slope of the animal is given by

$$\tan \theta = \frac{\partial y}{\partial x} = A \frac{2\pi}{\lambda} \cos[2\pi(x - Vt)/\lambda]. \tag{18}$$

Since the length L is assumed to be much shorter than λ, x/λ is very small, and we have the following approximate expression for the entire length of the organism

$$\tan \theta \doteq A \frac{2\pi}{\lambda} \cos \frac{2\pi Vt}{\lambda}, \qquad (0 \leqslant x \leqslant L). \tag{19}$$

Hence from (10),

$$\beta = \cos^2 \theta = (1 + \tan^2 \theta)^{-1} = \left[1 + A^2 \frac{4\pi^2}{\lambda^2} \cos^2 \frac{2\pi Vt}{\lambda} \right]^{-1}. \tag{20}$$

A substitution of Eq. (20) into Eq. (13) gives the ratio of the instantaneous speed of the organism's swimming, U, to the speed of undulation wave V.

Example 2. Ciliary Motion. The method developed above can be used to analyze the function of cilia. Cilia are cylindrical cellular projections having a uniform diameter of about $\frac{1}{4}$ μm and composed of a characteristic array of longitudinal fibrils. Oscillatory bending movements of cilia are responsible for the propulsion of fluids over the cell surfaces, resulting in the locomotion of small, unattached organisms such as *Opalina* and *Paramecium*, or in the maintenance of currents of water or mucus around the cells of larger or attached organisms. A Frog catches an insect by a flip of the tongue, which is coated with mucus that glues the insect to it, and then sends the insect down the throat by ciliary motion. Man clears the respiratory tract with cilia. Female mammals transport ova in the ovary duct with cilia.

Figure 4.6:2 shows a characteristic pattern of the ciliary beat cycle. Each cycle consists of two phases: an *effective* stroke during which the cilium remains fairly straight and moves through an arc around its basal attachment, and a recovery stroke in which a region of bending is propagated from base to top. Details can be found in Sleigh (1974). Typically, a cilium 12 μm long may beat at a frequency of 30 Hz, the effective stroke occupying about 2/5 of the cycle, the maximum tip speed is about 2.5 mm/sec. The cilia in the trachea are about 5 μm long. The typical Reynolds number of their motion is ≪ 1.

Further Development. The importance of the normal and tangential resistance coefficients K_N and K_N and their ratio r_k is shown in Eqs. (13) and (15). Equation 4.8:8 gives theoretical values of these constants; but it is based on an analysis which ignores the curvature of the flagellum, and the proximity of any other flagella or solid walls. In a flow that has a very low Reynolds number the mutual interference of solid bodies in a flow field is often surprisingly great. There is also a difficult question of how to assign values to the constants a, b, c in Eqs. (3) to (8) of Sec. 4.8: to a because sometimes the surface of the animal body is so soft (perhaps almost like water) that the effective radius is not what is seen in the microscope, to b and c, because they are somewhat arbitrary; especially for a curved flagellum. These questions, as well as the rotation of the animal, are discussed in Lighthill (1975), Chwang and Wu (1971, 1974–1976), and Wu (1976). The questions of swimming efficiency and scaling (size effect) are discussed by the same authors, especially Wu (1976).

4.10 Theories of Fish Swimming

There are three approximate theories of aquatic animal locomotion at higher Reynolds numbers ($\gg 1$). They are all developed for animals with elongated bodies, along the lines of the so-called "slender-body" theory:

(a) For aquatic animals such as the nematode and water snake, a "resistive force" theory similar to that presented in the preceding section is used. In this theory the main force of interaction between the animal and the surrounding fluid is considered to be the resistance caused by the viscosity of the fluid. The resistance law may be linear (proportional to the local tangential and normal velocity) or nonlinear (e.g., proportional to the square of velocity), depending on individual cases.

(b) For fish such as the carp and mackerel, a "reactive force" theory was developed by Lighthill (1975). In this theory, the force of interaction between the animal and the surrounding fluid is separated into two parts: the force tangential to the body of the fish is considered to be *resistive*. The force normal to the body of the fish is considered to be *reactive*, generated by the inertia of the fluid. The inertial force is equal to the rate of change of the momentum of the fluid. At large Reynolds numbers, the momentum of the fluid is essentially concentrated in a cylinder enveloping the animal with a radius not much greater than the lateral dimension of the animal, and can be calculated from the 'virtual mass' of the fluid associated with the moving body.

Lighthill's (1975) theory is based on the following three observations:

i) Water momentum near a section of fish is in a direction perpendicular to the backbone and has a magnitude equal to the virtual mass per unit length m times the component of the fish's velocity in that direction w.

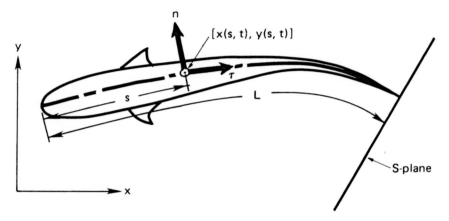

FIGURE 4.10:1 Coordinates and symbols used in the analysis of fish swimming. The view of a fast swimming fish as seen by one looking down the fish's back. The fish swims in the x, y-plane. Its caudal fin is parallel to the z-axis which is perpendicular to the x, y-plane. Hence the caudal fin appears only as a line. The heavy dash-dot line represents the fishes spine, which bends in the xy-plane as the fish swims.

ii) Thrust can be obtained by considering the rate of change of momentum within a control volume V which encloses the fish. For convenience, one surface of the control volume is selected to be a flat surface S perpendicular to the caudal fin through its posterior end at each instant of time, as shown in Fig. 4.10:1.

iii) In the momentum balance it is necessary to take into account transfer of momentum across the surface S named above, not only by convection but also by the action of the pressure generated by the motion within the plane S.

The fish's motion is then described by the motion of its spine, which is assumed to be inextensible, of total length L, and can bend only in the (x, y) plane. The analysis is quite sophisticated. The final result is as follows. The resulting force acting on the fish, F, is a vector in the (x, y) plane:

$$\mathbf{F} = \left[mw \left(\frac{\partial y}{\partial t}, -\frac{\partial x}{\partial t} \right) - \frac{1}{2} mw^2 \left(\frac{\partial x}{\partial s}, \frac{\partial y}{\partial s} \right) \right]_{s=L} - \frac{d}{dt} \int_0^L mw\mathbf{n}\, ds. \qquad (1)$$

In this formula, m, w is explained in i) above, $s = L$ is where caudal fin is located. The coordinates (x, y, z) of a point on the spine at time t are functions of s and t:

$$x = x(s, t), \qquad y = y(s, t), \qquad z = 0 \qquad (2)$$

from which the derivatives $\partial y/\partial t$, $\partial x/\partial t$, $\partial y/\partial s$, $\partial x/\partial s$ are computed. $\mathbf{n}$ is a unit vector perpendicular to the spine (Fig. 4.10:1) in the horizontal plane $z = 0$.

The unit vectors τ and $\mathbf{n}$ tangential and perpendicular to the spine are, respectively:

$$\tau = \left(\frac{\partial x}{\partial s}, \frac{\partial y}{\partial s}\right), \qquad \mathbf{n} = \left(-\frac{\partial y}{\partial s}, \frac{\partial x}{\partial s}\right). \tag{3}$$

The horizontal velocity vector $\mathbf{v}$ of a point on the spine is

$$\mathbf{v} = \left(\frac{\partial x}{\partial t}, \frac{\partial y}{\partial t}\right). \tag{4}$$

Its components tangential and normal to the spinal column are u and w, respectively:

$$u = \frac{\partial x}{\partial t}\frac{\partial x}{\partial s} + \frac{\partial y}{\partial t}\frac{\partial y}{\partial s}, \qquad w = \frac{\partial y}{\partial t}\frac{\partial x}{\partial s} - \frac{\partial x}{\partial t}\frac{\partial y}{\partial s}. \tag{5}$$

The resulting force $\mathbf{F}$ varies with time. If the swimming movement is periodic, then the mean value of the last term in Eq. (1), which is a time derivative, must be zero. Hence the mean force acting on the fish is:

$$\mathbf{F} = \overline{[-muw\mathbf{n} + \tfrac{1}{2}mw^2\tau]_{s=L}} = \overline{\left\{mw\left(\frac{\partial y}{\partial t}, \frac{\partial x}{\partial t}\right) - \frac{1}{2}mw^2\left(\frac{\partial x}{\partial s}, \frac{\partial y}{\partial s}\right)\right\}_{s=L}}, \tag{6}$$

where the bar denotes the mean. If the x-axis coincides with the mean direction of motion, then the mean thrust is the x component of $\bar{\mathbf{F}}$:

$$\bar{F}_x = \overline{\left[mw\frac{\partial y}{\partial t} - \frac{1}{2}mw^2\frac{\partial x}{\partial s}\right]_{s=L}}. \tag{7}$$

The same method gives an expression for the mean rate of energy dissipation

$$\bar{D} = \overline{[\tfrac{1}{2}mw^2 u]_{s=L}} \tag{8}$$

by shedding of kinetic energy $\tfrac{1}{2}mw^2$ per unit length at velocity u *across the plane S* into the vortex wake. The mean speed of swimming is

$$\bar{U} = \overline{\left[u\frac{\partial x}{\partial s}\right]_{s=L}}. \tag{9}$$

The efficiency of propulsion is

$$\eta = \frac{\bar{U}\bar{F}_x}{\bar{U}\bar{F}_x + \bar{D}}. \tag{10}$$

Thus it is seen that the mean terms $\bar{F}_x$, $\bar{D}$, $\bar{U}$, and η are all determined at the posterior end of the caudal fin. To obtain large thrust and high efficiency, the fish should have a deep caudal fin and execute a large amplitude movement. The shape and motion of the front portion of the fish really do not matter very much. In the evolution of the fish the tail end became "compressed," i.e., narrowed down so that the tail became more flexible and more able to execute large amplitude oscillations; whereas the caudal fin increased in depth (the virtual mass m is increased in proportion to the square of the depth of the caudal fin at the posterior end). These fish can achieve an efficiency in the 80% range, in sharp contrast to the values around 50% expected at a high Reynolds number from a resistive interaction with perpendicular motions.

Example. A fish swims in a pool. With x-, y- axes fixed in the pool and s measured along the fish's spine (Fig. 4.10:1), let the undulatory motion of the fish be described by functions specifying the coordinates of points on the spine as in Eq. (1). Assume that the fish swims in the way sketched in Figs. 4.9:1–4.9:3, so that

$$x = s - Ut, \qquad y = h(s - Vt), \tag{11}$$

where U is the forward speed of the fish, V is the speed of the undulatory wave passing backward along its spine, h is a function of the variable $\xi = s - Vt$. At $t = 0$, $y = h(s)$ describes the waveform of the fish. Then, according to Eqs. (5) and (11),

$$u = -U - V\left(\frac{\partial h}{\partial s}\right)^2, \qquad w = -(V - U)\frac{\partial h}{\partial s}. \tag{12}$$

Writing W as the lateral velocity of the fish relative to the pool

$$W = \frac{\partial y}{\partial t} = \frac{\partial h}{\partial t} = -V\frac{\partial h}{\partial s} \tag{13}$$

and substituting into (12), we obtain an important relation between the lateral velocity relative to water and that relative to the pool:

$$w = \left(1 - \frac{U}{V}\right)W. \tag{14}$$

(c) For fish such as the dolphin and whale, with well developed lunate tails which look like airplane wings, a "two-dimensional section" theory of the tail was developed by Wu (1971). These tails have high propulsive efficiency. The theory pays attention to the vortex wake due to the oscillatory motion of the tail, much like the wing theory for birds and airplanes described in Chapter 3.

In recent years, extensive computational aerodynamics was developed in the aeronautical field. Aeroelastic design of airplanes is based on computation of three-dimensional unsteady flow using concepts outlined in Chapter 3. A corresponding development for the analysis of flying of birds and swimming of fish has not taken place.

4.11 Energy Cost of Locomotion

To evaluate the energy cost of locomotion, we must evaluate drag; because the product of drag and velocity of motion is *power*. The integral of power over time is the *energy* or *work done* by the drag force. Thus, the energy cost in a time interval (t_1, t_2) in which the body moves from x_1 to x_2 is

$$\int_{t_1}^{t_2} DV \, dt = \int_{t_1}^{t_2} D \frac{dx}{dt} \, dt = \int_{x_1}^{x_2} D \, dx. \tag{1}$$

Here D is drag, V is velocity $= dx/dt$.

There are several sources of drag. For birds and insects, we have

a) the skin friction acting on the body surface;
b) the pressure drag produced by eddy or wake formation (see Sec. 3.15);
c) the induced drag, (see Secs. 3.2, 3.12, and 3.13)

For fish, we should add

d) the wave resistance,
e) the air resistance on the part above the water.

Of these, the skin friction is the hardest one to evaluate. Even today theoretical analysis based on Navier–Stokes equations has not been able to adequately predict the drag coefficient as a function of the Reynolds number for a flat plate in a uniform flow when turbulence is involved. Hence one must rely on experimental results.

Direct measurement of drag of animals is very difficult (cf. Sec. 3.15). The wide scatter of data in the literature is evidence of this difficulty. Hence animal physiologists often take an alternative approach: by measuring the animal's oxygen consumption, applying the standard energy conversion factors to obtain the metabolic power, and then using the energy conservation equation and flight efficiency to evaluate the drag. By the principle of conservation of energy, and writing D for drag, V for velocity, P for metabolic power, and η for efficiency, we have

$$DV = \eta P, \qquad \eta = \eta_m \eta_h. \tag{2}$$

The power P is appropriately identified if the basal metabolic rate is subtracted from the measured metabolic rate during active swimming. The overall efficiency η is equal to $\eta_m \eta_h$, where η_m is the "muscle efficiency" with which biochemical energy is converted to mechanical power of the muscle, and η_h is the fluid dynamic efficiency of propulsion.

Figure 4.11:1 shows the results presented by Wu (1984) on the basis of Brett's (1963) measurement of sockeye salmon (*Oncorhynchus nerka*) and the analysis of Wu and Yates (1978). Shown are the data points of the power coefficient C_p, and the dead-drag coefficient D_{Dd} versus the Reynolds number Re, where

$$C_p = \frac{P}{\rho U^3 S/2}, \qquad C_{Dd} = \frac{D}{\rho U^2 S/2}, \qquad Re = \frac{UL}{v}. \tag{3}$$

P is metabolic power of a swimming fish, ρ is the water density, v is the kinematic viscosity of water, L is the fish length (of a single-sized group with mean $L = 0.178$ m) and S is the wetted surface area of the fish. A least-square

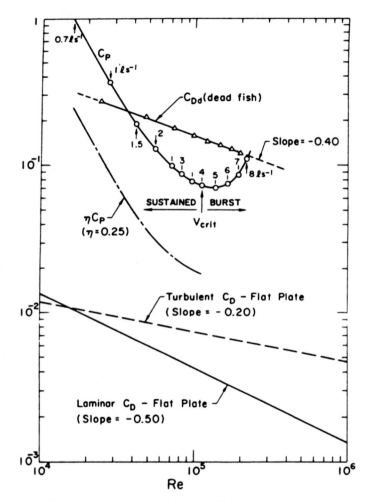

FIGURE 4.11:1 Data on the power and drag coefficients of sockeye salmon and comparison with the theoretical results of Wu and Yates (1978).

error fit to the drag data yields

$$C_{Dd} = 15.4 R_e^{-0.4}, \qquad (2.5 \times 10^4 < R_e < 2 \times 10^5). \qquad (4)$$

On the other hand, the metabolic power coefficient C_p has two branches meeting near the minimum value of 0.07: one branch for sustained swimming at lower speeds, another branch for a burst of energy. The efficiency η_m is estimated by Webb (1975) to be in the range 0.2–0.3. The efficiency η_h is about 0.9 (Wu, 1971), hence η is about 0.25. With $\eta = 0.25$, the values of ηC_p are calculated and plotted in Fig. 4.10:1. This curve lies considerably below the dead-drag coefficient C_{Dd}. It is suggested that the dead fish data are not useful;

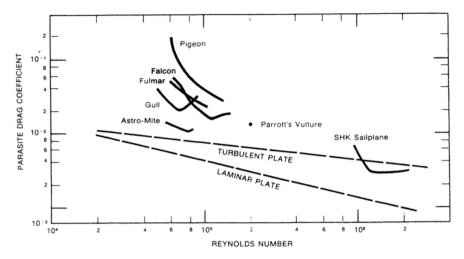

FIGURE 4.11:2 Data on drag coefficients of birds and sail planes. From Kuethe and Chow (1986), and from Tucker and Parrott (1970). Reproduced by permission.

and live fish has drag coefficients considerably lower than those of the dead fish.

A corresponding plot for the birds and sail planes is shown in Fig. 4.11:2. A comparison of Figs. 4.11:1 and 4.11:2 shows that birds and fish are in the same class as far as drag coefficients are concerned. They are not as efficient as contemporary sailplanes (Tucker and Parrott, 1970).

Knowing the drag as a function of speed of flight, one can compute the energy cost according to Eq. (1). It is interesting to examine some results in the following.

Cost of Distance

If distance covered is the major consideration, one should consider the energy cost of transport for moving a unit weight for unit distance. To maximize distance and minimize the cost, the cruising speed should be so chosen that dE/dx is a minimum (E being the energy, x being the distance). Since dE/dt = Power P, and dx/dt = speed V, we see that $dE/dx = P/V$. Hence we must determine V at which P/V is a minimum. For animals, this can often be done by measuring the metabolic energy (oxygen consumption, CO_2 release lactate) at various speeds of locomotion. With the speed at which P/V is a minimum so determined, a dimensionless parameter

$$\frac{P}{VW} = \text{specific resistance} = \text{cost of transport} \qquad (5)$$

can be plotted as a function of the weight of the animal as shown in Fig. 4.11:3.

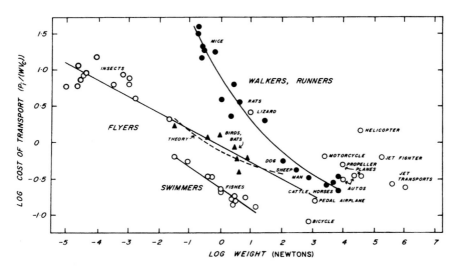

FIGURE 4.11:3 The minimum energetic cost of transport to move a unit of body weight for a unit of distance in swimming, flying, walking and running animals of various weights. Data also are included for various machines. The dashed line labeled "theory" is that predicted for flying birds by Eq. (26) in the text. Other curves are least squares regression lines. Data for swimmers from Schmidt-Nielsen (1972). Other data from Tucker (1970), (1973b). Reproduced by permission from Tucker (1975).

Here one sees that the fish are the best performers, followed by the flyers, runners, walkers, and manmade vehicles. The "theoretical" curve shown in Fig. 4.11:3 is Tucker's (1975) semiempirical formula for birds and bats in level flight:

$$P_i = (0.00723h + 105.9)m^{1.382}b^{-1.236} \qquad (6)$$

in which P_i is the power, h is the altitude, b is the wing span, in SI units of Watts and meters.

The specific resistance is a kind of global friction coefficient for an animal or a machine; it is a ratio of the total resistance to the gross weight of the body.

What Price Speed?

In the competition of the fittest, a great role is played by the speed of locomotion. A possible figure of merit is the "minimum" value of the "power per unit weight" as a function of the "maximum speed." Theodore von Kármán and Gabrielli (1950) chose to use the same dimensionless parameter, P/WV, defined in Eq. (5) but evaluated at the maximum speed V_{max}, to represent the cost of speed. Here P is the power, W is the weight, V is the speed, in consistent units such as kg-m/s, kg, m/s; or lbs-ft/sec, lbs, and ft/sec, respectively. They plotted available data of specific resistance versus V_{max} for various classes of

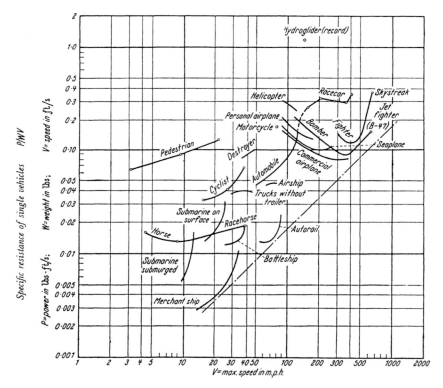

FIGURE 4.11:4 Specific resistance of single vehicles according to von Kármán and Gabrielli (1950).

vehicles and sketched a continuous curve to represent the minimum specific resistance as a function of the maximum speed for each vehicle. When these empirical curves are plotted together, they obtained the result as shown in Fig. 4.11:4. The limiting line shown in the figure is

$$\left(\frac{P}{WV}\right)_{\text{max speed}} = 1.75 \times 10^{-4} V_{\text{max}}(\text{mph}) = 3.92 \times 10^{-4} V_{\text{max}}(\text{m/sec}) \quad (7)$$

which represents a kind of limit that human technology has achieved so far. Horse and man are worse by an order of magnitude.

4.12 Cell Movement

The movements of ameba in fluid, leukocytes in blood and blood vessel walls, and neuron and smooth muscle cells in tissue have been observed. To understand these movements attention has to be given to the internal structure of the cells, and adhesion between cells. This subject is extremely important to the understanding of clinical problems of atherosclerosis, myocardial infarction, hemorrhagic death, etc., as well as basic biology. It is beyond the scope of this book to discuss this rapidly developing field. Interested readers may consult the following references. On cell *motility and Chemotaxis*: Berg and Brown (1972), Dembo et al. (1984), Dembo and Harlow (1986), Gallin and Quie (1978), Goldman et al. (1976), Huxley et al. (1982), Oster and Perelson (1987), Zhu et al. (1988), Zigmoid (1978). On *cytoskeletal networks*: Nossal (1988), Stossel (1982), Taylor and Condeelis (1979). On *actin filaments*: Pollard and Cooper (1986), Sato et al. (1985), Smith (1988), Wang (1985). On *clinical applications*: Schmid-Schönbein and Engler (1986), Wilkinson (1974).

Problems

4.1 The shark uses cartilage to form its skeleton instead of bone. Its outer shell is a cartilaginous bag. Muscles are attached to the skin. When muscles contract the cartilaginous layer is subjected to strain and stress and its elastic strain energy is varied. Imagine and describe how a shark can utilize the cartilaginous material for its locomotion; then check a book on sharks to see if your imagination is right.

The idea that animals are properly designed with respect to materials and structures is beautifully presented in the book *Mechanical Design in Organisms* by S.A. Wainwright, W.D. Biggs, J.D. Currey and J.M. Gosline, published by John Wiley and Sons, New York, 1976. It would be interesting to speculate on why the shark was successful in its adaptation through evolution.

4.2 Trees must withstand forces imposed by wind. Their swaying and bending must be related to an aerodynamic mechanism of reducing the drag force in a storm. Consider some of your favorite trees in this regard and see how well they are designed with respect to the specific environment each lives in.

4.3 At "Sea World" one can watch the whale jump out of the water to touch a ball hoisted 15 feet up in the air. You see the whale make one swoop down and jump right out. For a tank with 30 feet of water, estimate the achievable lift coefficient of the whale's tail.

4.4 Man and whale have a diving reflex. As they dive down in the ocean, their heart rates slow down. Discuss the physiological significance of this reflex.

4.5 The skin of the dolphin is compliant, and is coated by a polymer. Discuss in which way the compliance of the skin and the viscoelasticity of the polymeric fluid can affect the stability of the boundary layer on the dolphin's body, the laminar and turbulent characteristics of the boundary layer flow, and the influence of these factors on the drag force acting on a dolphin as it swims.

Immitating the dolphin, people have injected polymer into a ship's boundary layer in the hope of improving the speed of the ship. They have also used compliant paint on the hull. Fire fighters have introduced polymers into fire engines so that water can be shot out to a greater height. What success have they achieved? What possible disadvantage can there be?

4.6 When a sperm swims toward an ovum, how is its performance affected by the mechanical properties of the fluid? What characteristics of the rheological properties are important if the success of a sperm is judged by its ability to be the first to arrive at the target.

4.7 There is evidence suggesting that the swimming abilities of male and female sperms are different. Based on this observation, devise a method to separate male and female sperms in the semen for artifical insemination for sex control.

4.8 Migrating geese are always seen to line up in the form of the letter of Λ. One might suspect that there may be an energy saving in doing this as compared with a disorganized crowd. Can you offer some theoretical evidence? If an experimental approach is preferred, design an experiment. (Ref: Higdon, JJL, and Corsin, S. Induced drag of a bird flock. *Am. Naturalist* 112: 727–744, 1978.)

4.9 A chicken cannot fly very well. What is wrong with the chicken's flying machinery?

4.10 The movement of an ovum in oviduct is most likely helped by cilia lining the oviduct. Describe these cilia and the way they can help the ovum.

4.11 What are the basic factors limiting the speed of flight of a bird? For example, by diving from a very high altitude, a bird could achieve high speed. What determines the maximum speed achievable? What determines the maximum speed desirable? What ill effects will result if the speed exceeds certain critical values? What critical values are there?

4.12 Active control is undoubtedly used by birds in their flight, but not much is known about the subject. As a theoretical preliminary to an investigation, list all the important problems of flight of concern to a bird (it would be simpler if you choose a specific species) and, with a proper block diagram, describe desirable neuromuscular control systems that will enable the bird to solve each of these problems.

4.13 An airplane has to rev up the engine to produce the maximum power to take off. How does an eagle take off? One would have to consider the aerodynamic, structural, muscle dynamics, kinematics, metabolic, and energy factors.

4.14 The take-off problem for a duck on water is quite different from that of an eagle on a rock. How does the duck do it?

4.15 The locomotion methods of sea horse, clams, star fish, snails, and slugs are equally worthy of study. If you cannot find a satisfactory analysis in the literature, devise a plan of study of your own.

References

Ayman, G. (1936). *Bird Flight*, Bodley Head, London.

Bainbridge, R.A. (1958). The speed of swimming of fish as related to size and to the frequency and amplitude of the tail beat. *J. Exp. Biol.* 37: 109–133.

Bainbridge, R. (1960). Speed and stamina in three fish. *J. Exp. Biol.* 37: 129–153.

Bainbridge, R. (1963). Caudal fin and body movement in the propulsion of some fish. *J. Exp. Biol.* 40: 23–56.

Berg, H.C. and Brown, D.A. (1972). Chemotaxis in Escherichia coli analyzed by three-dimensional tracking. *Nature* London 239: 500.

Blake, J.R. and Sleigh, M.A. (1975). Hydromechanical aspects of ciliary propulsion. In *Swimming and Flying in Nature* (T.Y. Wu, C.J. Brokaw and C. Brennen, eds.), Vol. 1, Plenum Press, New York, pp. 185–209.

Bone, Q. (1975). Muscular and energetic aspects of fish swimming. In *Swimming and Flying in Nature* (T.Y. Wu, C.J. Brokow and C. Brennen, eds.), Vol. 2, Plenum Press, New York, pp. 493–528.

Breder, C.M. (1926). The locomotion of fishes. *Zoologica* 4: 159–297.

Brett, J.R. (1963). The energy required for swimming by young sockeye salmon with a comparison of drag force on a dead fish. *Trans. Roy. Soc. Can.* 1: Sec. IV, 441–457.

Brokaw, C.J. and Gibbons, I.R. (1975). Mechanisms of movement in flagella and cillia. In *Swimming and Flying in Nature* (T.Y. Wu, C.J. Brokaw and C. Brennen, eds.), Vol. 1, Plenum Press, New York, pp. 89–126.

Brown, R.H.J. (1953). The flight of birds. II. Wing function in relation to flight speed. *J. Exp. Biol.* 30: 90–103.

Childress, S. (1981). *Mechanics of Swimming and Flying.* Cambridge Univ. Press, Cambridge.

Chwang, A.T. and Wu, T.Y. (1971). A note on the helical movement of micro-organisms. *Proc. Roy. Soc. B.*, 178: 327–346.

Chwang, A.T. and Wu, T.Y. (1974, 1975, 1976). "Hydromechanics of low-Reynolds-number flow. Part I: Rotation of axisymmetric prolate bodies. *J. Fluid Mechanics* 63: 607–622, 1974. Part II: Singularity method for Stokes flows. *ibid.* 67: 787–815, 1975. Part III: (Chwang alone) Motion of spheroidal particle in quadratic flows. *ibid.* 72: 17–34, 1975. Part IV: Translation of spheroids. *ibid* 75: 677–689, 1976.

Dalton, S. (1975). *Borne on the Wind, The World of Insects in Flight.* E.P. Dutton and Co., New York.

Dembo, M. and Harlow, F. (1986). Cell motion, contractile network and the physics of interpenetrating reactive flow. *Biophys. J.* 50: 109–121.

Dembo, M., Harlow, F. and Alt, W. (1984). The biophysics of cell motility. In *Cell Surface Dynamics: Concepts and Models.* (A.S. Perelson, C. DeLisi and F.W. Wiegel, eds.), Marcel Dekker, New York, pp. 495–541.

Duncker, H.R. (1972). The structure and function of bird's lung. *Respiration Physiol.* 14: 44–63.

Ellington, C.P. (1975). Non-steady-state aerodynamics of the flight of Encarsia formosa. In *Swimming and Flying in Nature* (T.Y. Wu, C. Brokow and C. Brennen, eds.), Plenum Press, pp. 783–796.

Fung, Y.C. (1965). *Foundations of Solid Mechanics*, Prentice-Hall, Englewood Cliffs, N.J.

Fung, Y.C. (1977). *A First Course in Continuum Mechanics*, 2nd edn., Prentice-Hall, Englewood Cliffs, N.J.

Fung, Y.C. (1981). *Biomechanics: Mechanical Properties of Living Tissues*, Springer-Verlag, New York.

Gallin, J.I. and Quie, P.G. (eds.) (1978). *Leukocyte Chemotaxis: Methods, Physiology and Clinical Applications*, Raven Press, New York.

Goldman, R., Pollard, T., and Rosenbaum, J. (eds.) (1976). *Cell Motility. Cold Pring Harbor Conferences on Cell Proliferation.* Cold Spring Harbor Press, New York.

Goldspink, G. (1977). Design of muscle in relation to locomotion. In *Mechanics and Energetics of Animal Locomotion* (R.McN. Alexander and G. Goldspink, eds.), Chapman and Hall, London.

Gray, J. (1939). Aspects of animal locomotion. *Proc. Roy. Soc. London, B* 128: 28–62.

Gray, J. (1953). *How Animals Move.* Cambridge Univ. Press, Cambridge.

Gray, J. (1958). The movement of the spermatozoa of the bull. *J. Exp. Biol.* 35: 96–108.

Gray, J. (1968). *Animal Locomotion*, W.W. Norton, New York; Weidenfeld and Nicolson, London.

Gray, J. and Hancock, G.J. (1955). The propulsion of sea-urchin spermatozoa. *J. Exp. Biol.* 32: 802–814.

Gray, J. and Lissmann, H.W. (1964). The locomotion of nematodes. *J. Exp. Biol.* 41: 135–154.

Hancock, G.J. (1953). The self-propulsion of microscopic organisms through liquids. *Proc. Roy. Soc. A* 217: 96–121.

Huxley, H.E., Bray, B. and Weeds, A.G. (eds.) (1982). *Molecular Biology of Cell Locomotion.*

Phil. Trans., Roy. Soc. London, **B299**: 145–327.

Jameson, W. (1958). *The Wandering Albatross*. Hart-Davis, London.

Kuethe, A.M. (1975a). On the mechanics of flight of small insects. In *Swimming and Flying in Nature* (T.Y. Wu, C.J. Brokaw, and C. Brennen, eds.), Plenum Press, New York. pp. 803–813.

Kuethe, A.M. (1975b). Prototypes in nature. The carry-over into Technology. *Technium*, Engineering Review. 1975, Univ. of Michigan.

Kuethe, A.M. and Chow, C.-Y. (1986). *Foundations of Aerodynamics*. 4th ed. John Wiley, New York.

Lighthill, J. (1969). Hydromechanics of aquatic animal propulsion—a survey. *Ann. Rev. Fluid Mech.* 1: 413–446.

Lighthill, J. (1975). *Mathematical Biofluiddynamics*, Soc. Indus. Appl. Math. Philadelphia.

Lillienthal, O. (1889). *Der Vogelflug als Grundlage der Fliegekunst*. R. Oldenbourg, Berlin.

Maxworthy, T. (1981). The fluid dynamics of bird flight. *Ann. Rev. of Fluid Mechanics*, (M. Van Dyke and J.V. Wehausen, eds.), Annual Reviews, Palo Alto, California.

McAlister, K.W., Carr, L.W., and McCroskey, W.J. (1978). Dynamic stall experiments on the NACA 0012 airfoil. *NASA Tech. Paper* 1100.

Nachtigall, W. (1974). *Insects in Flight* (Trans. by H. Oldroyd et al). McGraw-Hill, New York.

Newman, J.N. (1973). The force on a slender fish-like body. *J. Fluid Mech.* 58: 689–702.

Newman, J.N. and Wu, T.Y. (1973). A generalized slender-body theory for fish-like forms. *J. Fluid Mech.* 57: 673–693.

Norberg, R.A. (1975). Hovering flight of the dragonfly *Aeschna Juncea L.*, kinematics and aerodynamics. In *Swimming and Flying in Nature* (T.Y. Wu, C.J. Brokaw, and C. Brennen, eds.), Plenum Press, New York, pp. 763–781.

Nossal, R. (1988). On the elasticity of cytoskeletal networks. *Biophys. J.* 53: 349–359.

Oberbeck, A. (1876). Ueber stationäre Flüssigkeitsbewegungen mit Berücksichtigung der inneren Reibung. *Crelle* 81, 62–80.

Oster, G.F. and Perelson, A.S. (1987). The physics of cell motility. *J. Cell Sci.* 8: 35–54.

Pennycuick, C.J. (1968). A wind-tunnel study of gliding flight in the pigeon *Columba livia*. *J. Exp. Biol.* 49: 509–526.

Pennycuick, C.J. (1972). *Animal Flight*. Edward Arnold, London.

Peterson, R.T. (1963). *The Birds*. Time Inc., New York.

Pollard, T.D. and Cooper, J.A. (1986). Quantitative analysis of the effect of acanthamoeba profilin on actin filament nucleation and elongation. *Biochem..* 23: 6631–6641.

Prandtl, L., and Tietjens, O.G. (1934). *Applied Hydro-and-Aeromechanics*. McGraw-Hill, New York. (Translated from the German edition, Springer, Berlin/Heidelberg, 1931).

Pringle, J.W.S. (1975). *Insect Flight*. Oxford University Press, London and New York.

Rayleigh, Lord., (J.W. Strutt). (1883). The soaring of birds. *Nature* 27: 534–535.

Sato, M., Leimbach, G., Schwartz, W.H., and Pollard, T.D. (1985). Mechanical properties of actin. *J. Biol. Chem.* 260: 8585–8592.

Schmid-Schönbein, G.W. and Engler, R.L. (1986). Granulocytes as active participants in acute myocardial ischemia and infarction. *Am. J. Cardiovasc. Patho.* 1: 15–30.

Scheid, P., Slama, H., and Piiper, J. (1972). Mechanisms of unidirectional flow in parabronchi of avian lungs. Measurements in duck lung preparations. *Respiration Physiol.* 14: 83–95.

Sleigh, M.A. (ed.) (1974). *Cilia and Flagella*. Academic Press, London and New York.

Smart, J., and Hughes, N.F. (1972). In *Insect/Plant Relationships*: Sympos. R. Entomol. Soc. London No. 6, pp. 143–155.

Smith, X. (1988). Neuronal cytomechanics: the actin-based motility of growth cones. *Science* 242: 708–715.

Stossel, T.P. (1982). The structure of the cortical cytoplasm. *Phil. Trans. Roy. Soc. London B* 299: 275–289.

Taylor, D.L. and Condeelis, J.S. (1979). Cytoplasmic structure and contractility in amoeboid cells. *Int. Rev. Cytology* 56: 57–144.

Taylor, G.I. (1951). Analysis of swimming microscopic organisms. *Proc. Roy. Soc. London Ser. A*, 209: 447–461.

Taylor, G.I. (1952). Analysis of the swimming of long and narrow animals. *Proc. Roy. Soc. London,* A, **214**: 158–183.

Tucker, V.A. (1968). Respiratory exchange and evaporative water loss in the flying budgerigar. *J. Exp. Biol.* **48**: 67–87, Company of Biologists Ltd.

Tucker, V. and Parrott, G.C. (1970). Aerodynamics of gliding flight of falcons and other birds. *J. Exp. Biol.* **52**: 345–368, Company of Biologists Ltd.

Tucker, V.A. (1975). Aerodynamics and energetics of vertebrate fliers. In *Swimming and Flying in Nature* (T. Wu, C. Brokaw and C. Brennen, eds). Plenum Press, New York, pp. 845–865.

Von Holst, E. and Küchemann, D. (1974). Motion of animals in fluids. In *J. Royal Aeronautical Soc.* **43**: 39–56.

Von Kármán, T. and Gabrielli, G. (1950). What price speed? Specific power required for propulsion of vehicles. *Mech. Eng.* **72**: 775–781.

Wang, Y.-L. (1985). Exchange of actin subunits at the leading edge of living fibroblasts: possible role of treadmilling. *J. Cell Biol.* **101**: 597–602.

Webb, P.W. (1975). Hydrodynamics and energetics of fish propulsion. *Bull. Fish Res. Bd. Can.* **190**: 1–158.

Weis-Fogh, T. and Jensen, M. (1956). Biology and physics of locust flight. I. Basic principles in insect flight. A critical review. *Phil. Trans., Roy. Soc. London,* B **239**: 415–458.

Weis-Fogh, T. (1956). Biology and physics of locust flight. II. Flight performance of desert locust (Schistocera gregaria). *Phil. Trans. Roy. Soc. London* B, **239**: 459–510.

Weis-Fogh, T. (1960). *J. Exp. Biol.* **37**: 889–907. *J. Exp. Biol.* **59**: 169–230.

Weis-Fogh, T. (1964). VIII. Lift and metabolic rate of flying locusts. *J. Exp. Biol.* **41**: 257–271.

Weis-Fogh, T. (1967). Energetics of hovering flight in hummingbirds and in *Drosophila. J. Exp. Biol.* **56**: 79–104.

Weis-Fogh, T. (1973). Quick estimates of flight fitness in hovering animals, including novel mechanisms for lift production. *J. Exp. Biol.* **59**: 169–230, Company of Biologists Ltd.

Weis-Fogh, T. (1975). Flapping flight and power in birds and insects, conventional and novel mechanisms. In *Swimming and Flying in Nature* (T.Y. Wu, C.J. Brokaw, and C. Brennen, eds.), Plenum Press, New York, pp. 729–762.

Wilkinson, P.C. (1974). *Chemotaxis and Inflammation,* Churchill Livingstone, Edinburgh and London.

Wu, T.Y. (1971). Hydromechanics of swimming fishes and cetaceans. In *Advances in Applied Mechanics* (C.S. Yih, ed.), **11**: Academic Press, New York. pp. 1–63.

Wu, T.Y. (1971). Hydromechanics of swimming propulsion. Part 3. Swimming and optimum movements of slender fish with side fins. *J. Fluid Mech.* **46**: 545–568.

Wu, T.Y. and Newman, J.N. (1972). Unsteady flow around a slender flish-like body. *Proc. International Symp. on Directional Stability and Control of Bodies Moving in Water.* Institution of Mechanical Engineers, London.

Wu, T.Y., Brokaw, C.J. and Brennen, C. (eds.) (1975). *Swimming and Flying in Nature.* Vols. 1 and 2, Plenum Press, New York.

Wu, T.Y. (1976). Introduction to the scaling of aquatic animal locomotion. In *Scale Effects of Animal Locomotion.* (M.J. Lighthill and T.J. Pedley, eds.), Academic Press, London, pp. 753–766.

Wu, T.Y. and Yates, G.T. (1978). A comparative mechanophysiological study of fish locomotion with implications for tuna-like swimming mode. In *Physiological Ecology of Tuna* (G.D. Sharp and A.E. Dizon, eds.), Academic Press, New York.

Yates, G.T. (1983). Hydromechanics of body and caudal fin propulsion. Chapter 6 in *Fish Biomechanics.* (P.W. Webb and D. Weihs, eds.), Praeger Scientific, New York.

Zhu, C., Skalak, R., and Schmid-Schönbein, G.W. (1988). One-dimensional steady continuum model of retraction of pseudopod in leukocytes. *J. Biomech. Eng.* **111**: 69–77.

Zigmoid, S.H. (1978). Chemotaxis by polymorphonuclear leukocytes. (review) *J. Cell Biol.* **77**: 269–287.

CRABS by Fung Chung-Kwang (1896–1952). Ink on rice paper.

CHAPTER 5

Blood Flow in Heart, Lung, Arteries, and Veins

5.1 Introduction

The next five chapters are concerned with flow inside the bodies of man and animals. By internal flow of blood, water, and gases, the cells of the body obtain water, oxygen, and nutrients. To understand the health and disease of these organisms, it is necessary to know the mechanics of internal flow.

A great variety of things happen in the circulatory and respiratory systems. To observe them we use a variety of tools. Most important are our eyes. To help our eyes we use x-rays, cinematography, CAT scan, NMR, ultrasound imaging, etc. For smaller things, we use optical microscopes, x-rays, electron microscopes, scanning tunneling microscopes. At different levels of scale, different objects come into view. Together they reveal the phenomenon of living. It is the function of mechanics to analyze and integrate the phenomena at different scales. If the mechanics of a phenomenon at one level of scale is called macroscopic, and that at a smaller level of scale is called microscopic, then in biomechanics one often attempts to connect the microscopic to macroscopic mechanics. For example, the dimension of an endothelial cell is much smaller than the diameter of the aorta. The mechanics of the endothelial cells is microscopic relative to the mechanics of the aorta. But they are connected when atherosclerosis is concerned. Again, the diameters of the collagen and elastin fibers are much smaller than the diameter of the arteries, but fiber mechanics and arterial mechanics are connected.

In the following, we shall first study blood flow at the scale of the heart and large arteries. Then (in Chapter 6) we focus on small blood vessels and attempt to clarify the connection between the micro- and macrohemodynamics. In Chapter 7 we do the same for the phenomena of respiration. In Chapters 8 and 9, the flow of water and other constituents from blood vessels to extravascular space, the fluid movement in the tissue space, and the fluid exchange between interstitium and cells are discussed. The presentations in Chapters 5 and 6 are rather condensed. Interested readers are referred to the author's book *Biodynamics: Circulation* (Springer-Verlag, 1984) for a more thorough treatment. The material in Chapters 7, 8 and 9 is presented at a more leisurely pace.

5.2 The Geometry of the Circulation System

Every animal's circulation system is special, but we shall consider typical features of man, dog, and cat. In these animals, blood flows from the superior and inferior vena cava into the right atrium, then through the tricuspid valve into the right ventricle, then through the semilunar valves into the pulmonary artery, the lung, the pulmonary veins, the left atrium, the mitral valve, the left ventricle, and finally through the aortic valve into the aorta. The peripheral circulation begins with the aorta, perfuses various organs, and returns to the right atrium. In each organ, flow begins in the arteries, perfuses the microcirculatory bed, then drains into veins. The vena cava collect blood from various organs, and send it to the heart.

Figure 5.2:1 shows the heart in greater detail. The two thin-walled atria are separated from each other by an interatrial septum. The two ventricles are separated by an interventricular septum. The left ventricle is thick-walled. In systolic condition, the pressure of blood in the left ventricle is higher than that in the right ventricle; hence the interventricular septum bulges out toward the right ventricle. The left ventricle can be represented as an ellipsoid; the right ventricle can be represented as a bellow. The four valves are seated in a plane and their bases are connected into an integrated structure, so that the enlargement of two opening valves is accompanied with the reduction in size of the other two closing valves. The mitral and tricuspid valves are attached to papillary muscles, which contract in systole, pulling down the valves to generate systolic pressure rapidly, and prevent the valves from inversion into the atrium.

The lung consists of three trees; see Fig. 5.2:2. The airway tree is for ventilation. The trachea is divided into bronchi which enter the lung, subdivided repeatedly into smaller and smaller branches called bronchioles, respiratory bronchioles, alveolar ducts, and alveoli. The alveoli are the smallest units of the airway. The walls of the alveoli are capillary blood vessels. Every wall of an alveolus is exposed to gas on both sides, so each wall is called an interalveolar septum. The entire lung is wrapped in a pleural membrane like a balloon. The chest wall also has a pleura. The pulmonary pleura and the visceral pleura are apposed to each other with a very small gap (a few μm thick) between them. The sealed compartment of space between the pleura is called the intrapleural space. The pressure in the intrapleural space is ordinarily lower than atmospheric. The chest wall and the transpulmonary

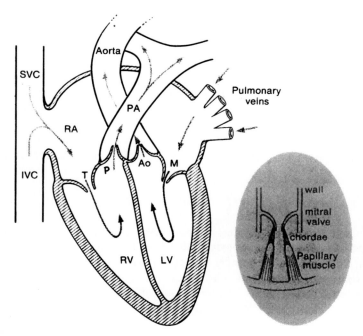

FIGURE 5.2:1 Blood flow through the heart. The arrows show the direction of blood
flow. The symbols are: SVC, superior vena cava; IVC, inferior vena cava; RA, right
atrium; RV, right ventricles; PA, pulmonary artery; LV, left ventricle. The valves are
T, tricuspid, P, pulmonary, AO, aortic, M, mitral. From Folkow and Neil (1971)
Circulation, Oxford Univ. Press, New York, p. 153, by permission.

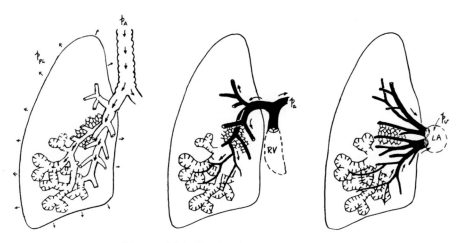

FIGURE 5.2:2 The three "trees" of the lung.

pressure (= the difference of the alveolar gas pressure and the pleural pressure) distends the lung.

The second tree is the pulmonary arterial tree. Beginning with the pulmonary artery, the tree bifurcates again and again until it form's capillary blood vessels which separate the alveoli.

The third tree is the venous tree. Beginning with the capillaries, the blood vessels converge repeatedly until they form pulmonary veins which enter the left atrium.

Such a sketch of the circulation system cannot give the needed details. Greater details can be found in Fung (1984) and other references listed therein.

5.3 The Materials of the Circulation System

The heart is a muscle. The lung is blood vessels and airways. All organs are perfused by blood via blood vessels. Blood vessels consist of smooth muscles, endothelial cells, and connective tissues. External to blood vessels are body fluids, cells, and interstitium. The circulation system also includes the lymphatic and nervous systems. The blood is a multiphase fluid composed of cells and plasma. Thus, the variety and complexity of the materials of the circulation system is truly monumental.

A detailed discussion of the chemical composition, molecular and higher structures (biochemistry, histochemistry), quantitative determination of the geometrical features of the internal structure of the tissue (morphometry, stereology, histology, anatomy), and the mechanical properties of the tissue (biomechanics) and its components (micromechanics) of any of the tissues and organs of the circulatory system would require much space, and is beyond the scope of this book. The reader is referred to the literature listed in the Bibliography at the end of this chapter. The author's book *Biomechanics: Mechanical Properties of Living Tissues* is a convenient reference. In the following sections, only the essential data required for immediate discussion are presented.

It is important to realize that the mechanical properties of many biological materials are very different from those of familiar engineering materials. For example, the incremental Young's modulus of the blood vessel wall or the relaxed muscles vary with the stress acting in the tissue; they do not remain constant as engineering materials do. For the heart, it is important to know that the maximum active tensile stress which can be generated in an isometric contraction of a cardiac muscle varies with the length of the sarcomere. See the length-tension curve in Fig. 5.3:1. If a heart normally operates at a sarcomere length marked by the point A in the figure, then when the sarcomere is lengthened, the maximum muscle tension will increase, and consequently, the systolic pressure, p_i, will increase. Since the number of sarcomeres in a heart muscle is fixed, the sarcomere length is proportional to the muscle length, and by implication, to the radius of the heart. Thus, if the radius of the

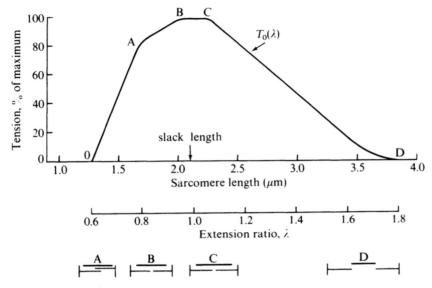

FIGURE 5.3:1 The "length-tension" curve of a skeletal muscle. The sarcomere length is plotted on the abscissa. The maximum tension achieved in isometric contraction at the length specified is plotted on the ordinate.

heart is increased, the muscle tension will increase, and so will be the systolic blood pressure. This is known as *Starling's law of the heart*. This law ceases to be valid when *A* moves off the upward-sloping leg of the curve shown in Fig. 5.3:1.

A similar length-tension curve exists for the vascular smooth muscle. Since the length of a muscle cell in a tissue depends on the strain at the place where the muscle cell is located, it becomes clear that the mechanical properties of the tissues of the circulatory system depend on the strain in the organ.

Another remarkable property of the blood vessels and the heart is the existence of large residual strains in these organs. Residual strains remain in an organ when all the external loads are removed; e.g., when the transmural pressure in a blood vessel is reduced to zero. This is discussed in Chapters 11 and 13. See Secs. 11.2 and 13.8.

5.4 Field Equations and Boundary Conditions

The basic equations of biomechanics are the equation of conservation of mass, the equation of motion, the constitutive equations specifying the mechanical properties of the materials, and, if heat and transfer and chemical reactions are involved, the energy equation and reaction rate equation. These and the equations describing the boundary conditions are all that are allowed in a theoretical analysis of circulation.

The conservation of mass is expressed by the equation of *continuity* (Sec. 1.7). For a segment of a blood vessel, it says that

$$\text{The difference of inflow and outflow} + \text{the rate of change}$$

$$\text{of the volume of the segment} = 0. \tag{1}$$

The equation of motion is a statement of Newton's law, which takes the following form when applied to a fluid or solid:

$$\text{Density} \times (\text{transient acceleration} + \text{convective acceleration})$$

$$= -\text{pressure gradient} + \text{force due to stress tensor}$$

$$+ \text{body force per unit volume}. \tag{2}$$

Here *density* refers to the density of the fluid or solid, the *transient acceleration* refers to the rate of change of the local velocity with respect to time, and the *convective acceleration* refers to the rate of change of velocity of a material particle caused by the motion of the particle from one place to another in a nonuniform velocity field. *Pressure gradient* is the rate of change of pressure versus distance. The force due to stress tensor refers to the force per unit volume due to the rate of change of stresses, taken their directions and areas into account in a proper way. See Sec. 1.7.

For the blood, the body force consists of inertial forces due to gravitational acceleration, Coriolis acceleration, and the acceleration of the body due to walking, jumping, swimming, flying, or other motion; the stress tensor consists of shear stresses caused by the viscosity of the fluid. For the blood vessel, the body forces are similar, the stress tensor is mainly caused by distension of the vessel due to blood pressure.

The constitutive equation of the blood describes the law of viscosity of the blood, which is, in fact, quite complex (see Fung, 1981). Whole blood is non-Newtonian, whose viscosity changes with the strain rate.

The constitutive equation of the blood vessel is the stress-strain relationship of the vessel wall material. It is also quite complex because it does not obey Hooke's law (see Fung, 1981, and Chapters 10 and 11 of this book).

In special situations, it is permissible to use approximate constitutive equations to simplify the analysis. For example, in large animals such as cat and man, the shear strain rate of the blood at the walls of the heart and the pulmonary and systemic arteries and veins exceeds 100 sec^{-1} so that in that region the coefficient of viscosity of blood may be regarded as a constant. The non-Newtonian feature is important only in a region near the centerline of the blood vessel. The integrated effect of the non-Newtonian feature is quite negligible so that flowing blood in these vessels may be treated as Newtonian.

For the blood vessel wall, the stress-strain relationship can be linearized (into an incremental Hooke's law) if the amplitude of deformation is very small. For pulmonary arteries and veins, the pressure-diameter relationship has been found to be linear because these vessels are embedded in an elastic medium—the lung parenchyma (see Fung, 1981, 1984, and Sec. 5.15, Figs. 5.15:4 and 5).

The boundary condition between a *viscous fluid* and a *solid* is the *no-slip condition*: there is no relative movement of material particles of the fluid at the boundary and the material particles of the solid at the same interface.

For an *ideal fluid* whose viscosity is zero, slip must be permitted, the boundary condition is then reduced to the condition that the materials on the two sides of an interface must remain contiguous: they must have the same velocity normal to the interface.

Across any surface on the boundary, the stress vectors on the two sides of the boundary surface must be equal and opposite by the condition of equilibrium.

In the analysis of blood flow in any particular blood vessel, one must not forget the two ends of the vessel. The *entry* and *exit* conditions with regard to pressure and velocity distributions at the ends must be specified.

These are the basic equations and principles. Some special problems are formulated, solved, and their physiological meaning discussed in the following sections.

5.5 Blood Flow in Heart and Through Heart Valves

The direction of blood flow in the heart is shown schematically in Fig. 5.2:1, the venous blood flows into the right atrium, through the tricuspid valve into the right ventricle, and then is pumped into the pulmonary artery and the lung, where the blood is oxygenated. The oxygenated blood then flows from the pulmonary veins into the left atrium, and through the mitral valve into the left ventricle, whose contraction pumps the blood into the aorta, and then to the arteries, aterioles, capillaries, venules, veins, and back to the right atrium.

An aortic valve with the sinus of Valsalva behind it is sketched in Fig. 5.5:1. According to model experiments by Bellhouse and Bellhouse (1969, 1972) and Lee and Talbot (1979), the flow issuing from the ventricle, immediately upon opening of the valve during systole, is split into two streams at each valve cusp, as shown in the figure. One part of the flow is directed into the sinus behind the valve cusp, where it forms a vortical flow before reemerging out of the plane of the figure, to rejoin the main stream in the ascending aorta.

When the aortic pressure rises sufficiently so that deceleration of the flow occurs, an adverse pressure gradient is produced, p_2 at the valve tip exceeds the pressure p_1 at a station upstream. The higher pressure p_2 causes a greater flow into the sinus which carries the cusp toward apposition. The peak deceleration occurs just before the valve closure. The vortical motion established earlier upon the opening of the valve has the merit of preventing the

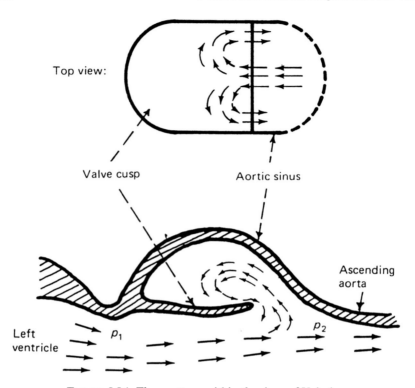

FIGURE 5.5:1 Flow pattern within the sinus of Valsalva.

valve cusp from bulging outward to contact the walls of the sinuses. The open sinus chamber thus can be supplied with fluid to fill the increasing volume behind the valve cusps as they move toward closure.

Other heart valves and the valves of the veins and lymphatics are operated by hydrodynamic forces in a similar way, although they do not have sinuses. In closing these valves, deceleration of the fluid is the essence, not backward flow.

5.6 Coupling of Left Ventricle to Aorta and Right Ventricle to Pulmonary Artery

As the heart muscle contracts periodically, blood is pumped from the left ventricle into the aorta through the aortic valve, and simultaneously from the right ventricle into the pulmonary artery through the pulmonary valve. The aorta and the pulmonary artery, being elastic, expand when they receive blood at a rate faster than the rate at which they send blood away into the peripheral organs and the lung, respectively. Expanding an elastic vessel causes an increase of the circumferential strain and stress in the vessel wall. A blood vessel with an increased circumferential stress in its wall will press harder on the blood it contains. As a result the blood pressure is increased. The increased blood pressure in the aorta acts on the aortic side of the aortic valve, tending to close it. An additional tendency to close the valves comes from the deceleration of the blood in the aorta. The deceleration occurs when the inflow exceeds the outflow. A consequence of the deceleration is the creation of a longitudinal pressure gradient through the aortic valve, again tending to close the valve. Eventually the valve is closed, blood continues to flow from the aorta into the periphery. By this mechanism the blood flow in the aorta does not have large swing of pressure as it has in the left ventricle. Similar events occur in the lung.

The process described above can be presented mathematically in various levels of generality. To be rigorous, it seems evident that the heart, aorta, arteries, and veins should be represented by three-dimensional network, and the special geometry and materials of construction of various organs must be described and incorporated in the mathematical model. In practice it is useful to consider simplified, crude models first, learn the general features, identify the important parameters, and then add details when needed. Accordingly, we shall consider the Windkessel model in this section, and the long wave, small amplitude pulse waves in the next section. Other features are added in following sections.

The *Windkessel theory* is Otto Frank's (1899) interpretation of Stephan Hale's (1733) explanation of why the pressure fluctuation in the aorta has a much smaller amplitude than that in the left ventricle. In this theory, the aorta is represented by an elastic chamber and the peripheral blood vessels are replaced by a rigid tube of constant resistance. See Fig. 5.6:1. Let $\dot{Q}$ be the inflow (cm^3/sec) into this system from the left ventricle. Part of this inflow is sent to the peripheral vessels and part of it is used to distend the elastic chamber. If p is the blood pressure in the elastic chamber (aorta), then the flow in the peripheral vessel is assumed to be equal to p/R, where R is a constant called *peripheral resistance*. For the elastic chamber, its change of volume is assumed to be proportional to the pressure. The rate of change of the volume of the elastic chamber with respect to time, t, is therefore proportional to dp/dt. Let the constant of proportionality be written as C and called *compliance*. Then, on equating the inflow to the sum of the rate of change of volume of the elastic chamber and the outflow p/R, the differential equation governing the pressure p is

$$\dot{Q} = C\frac{dp}{dt} + p/R. \tag{1}$$

FIGURE 5.6:1 The "windkessel" model of the aorta and peripheral circulation.

To solve this equation, we can use the method of integration factor. Dividing Eq. (1) by C and multiplying it by $e^{t/RC}$, we obtain

$$\frac{1}{C}\dot{Q}(t)e^{t/RC} = \frac{dp}{dt}e^{t/RC} + \frac{1}{RC}pe^{t/RC} = \frac{d}{dt}(pe^{t/RC}). \tag{2}$$

Integrating both sides from $t = 0$ to t and writing the dummy variable as τ, we have

$$p(t)e^{t/RC} = \int_0^t \frac{1}{C}\dot{Q}(\tau)e^{\tau/RC}\,d\tau + p_0, \tag{3}$$

where p_0 is the value of p at $t = 0$. If we take the instant of time when the valve opens as $t = 0$, then p_0 is the systolic pressure in the ventricle at the instant when the valve opens. Multiplying both sides with $e^{-t/RC}$, we get

$$p(t) = e^{-t/(RC)}\int_0^t \frac{1}{C}\dot{Q}(\tau)e^{\tau/(RC)}\,d\tau + p_0 e^{-t/(RC)} \tag{4}$$

which gives the pressure in the aorta as a function of the left ventricular ejection history $\dot{Q}(t)$.

An analog electric circuit can be formulated to represent the differential equation (Eq. 1). When this electric model is driven by a current $I = \dot{Q}(t)$ of the shape of an experimentally determined flow through the aortic value at the ascending aorta, the voltage V obtained is the analog of the blood pressure in the aortic arch. On comparing the analog results with an experimentally determined blood pressure curve, it is found that the actual pressure pulse deviates from the calculated results in several details: the experimental curve has a superimposed 3–6 cps oscillation apparent from midsystole throughout diastole, and a more prominent "incisura" marking aortic valve closure and a more abrupt rise, often with an "anacrotic" wave. In addition, the Windkessel model fails to explain the changes of the form of pressure wave occurring along the arterial network. These limitations of the Windkessel theory can be alleviated by an improved model such as the one presented in the next section.

The analysis also applies to the coupling of the right ventricle and pulmonary artery. The pulmonary circulation, however, is a lower pressure system. The wall of the right ventricle is thinner than that of the left ventricle; its systolic pressure is lower. The systolic and diastolic pressures in the pulmonary artery are much lower than those in the aorta. Since the flows in the aorta and pulmonary artery are about the same, the shape of $p(t)$ given by the first term on the right-hand side of Eq. (4) can be similar (except for the amplitude) only if the values of RC are approximately the same in both circuits. Hence the low pressure in pulmonary circulation must be achieved by a lower right ventricular pressure p_0, a lower resistance R, and a higher compliance C of the pulmonary circuit.

The right ventricle and the left ventricle are two pumps working in series. The flow in them must be matched perfectly, otherwise all the blood would eventually be accumulated either in the lung or in the periphery. The matching is stabilized by Starling's law of the heart (Sec. 5.3), namely, if the diastolic volume is increased, the contracting force of the muscle will increase to pump harder.

5.7 Pulsatile Flow in Arteries

The weakness of the Windkessel theory is that it allows only one degree of freedom. The pressures in the aorta and arteries are represented by a single number. It ignores the change of pressure along the vascular tree. To improve the understanding of events occurring in the arteries, we go to the next simplest model: *considering each artery as a long, isolated, circular cylindrical elastic tube, allowing an infinite number of degrees of freedom, approximating the flow to be one dimensional, and blood as a homogeneous, nonviscous, incompressible fluid.* The flow in each tube is excited at one end by the heart. The excitation is propagated in the form of elastic waves, much as an earthquake generates seismic waves. At the distal end each tube bifurcates, and the waves are partly transmitted to the daughter branches and partly reflected. This theory was originated by Euler (1775) and Young (1808, 1809), and developed by many others. It explains many things, but must be supplemented by three-dimensional theories when one wants to know the velocity profile, flow separation, stenosis, microcirculation, etc., which are important to the understanding of atherosclerosis, hypertension, etc.

To present this theory in the simplest form, *it is further assumed that the wave amplitude is small and the wave length is long* compared with the tube radius, so that the radial and circumferential velocity components are negligibly small compared with the longitudinal velocity component $u(x, t)$, which is a function of the axial coordinate x and time t only. Then the basic field equations (Sec. 3.2) are: the equation of motion,

$$\frac{\partial u}{\partial t} + u \frac{\partial u}{\partial x} + \frac{1}{\rho} \frac{\partial p_i}{\partial x} = 0 \tag{1}$$

and the equation of continuity,

$$\frac{\partial A}{\partial t} + \frac{\partial}{\partial x} (uA) = 0. \tag{2}$$

Here $A(x, t)$ is the cross-sectional area of the tube and $p_i(x, t)$ is the pressure in the tube. The relationship between p_i and A may be quite complex. For simplicity we introduce another hypothesis, that A depends on the transmural pressure, $p_i - p_e$, alone:

$$p_i - p_e = P(A), \tag{3}$$

where p_e is the pressure acting on the outside of the tube. Equation (3) is a gross simplification. In the theory of elastic shells we know that the tube deformation is related to the applied load by a set of partial differential equations and that the external load includes the inertial force of the tube wall. Hence Eq. (3) implies that the mass of the tube is ignored, and that the partial differential equations are replaced by an algebraic equation. The viscoelasticity of tube wall is ignored also.

Equation (1) is the one-dimensional case of the Eulerian equation of motion (Eq. (1.7:1)). Equation (2) can be obtained by integrating Eq. (1.7:5) over a tube. A special example of Eq. (3) is the pressure–diameter relationship of the pulmonary artery or vein (Yen et al., 1980, 1981):

$$2a_i = 2a_{i0} + \alpha p_i. \tag{4}$$

Here $2a_i$ is the vessel diameter, p_i is the blood pressure, a_{i0} and α are constants which depend on the pleural pressure p_{PL} and airway pressure p_A, but are independent of blood pressure p_i. α is the *compliance constant* of the vessel, and a_{i0} is the radius when $p_i = 0$.

Let us solve a linearized version of these equations. Consider small disturbances in an initially stationary liquid-filled circular cylindrical tube. In this case u is small and the second term in Eq. (1) can be neglected. Hence

$$\frac{\partial u}{\partial t} + \frac{1}{\rho}\frac{\partial p_i}{\partial x} = 0. \tag{5}$$

The area A is equal to πa_i^2. Substituting πa_i^2 for A in Eq. (2), remembering the hypothesis that the wave amplitude is much smaller than the wave length, so that $\partial a_i/\partial x \ll 1$, then, on neglecting small quantities of the second order, we can reduce Eq. (2) to the form

$$\frac{\partial u}{\partial x} + \frac{2}{a_i}\frac{\partial a_i}{\partial t} = 0. \tag{6}$$

Combining Eqs. (4) and (6), we obtain

$$\frac{\partial u}{\partial x} + \frac{\alpha}{a_i}\frac{\partial p_i}{\partial t} = 0. \tag{7}$$

Differentiating Eq. (5) with respect to x and Eq. (7) with respect to t, subtracting the resulting equations, and neglecting the second order term (α/a_i^2) $(\partial a_i/\partial t)\,(\partial p_i/\partial t)$, we obtain

$$\frac{\partial^2 p_i}{\partial x^2} - \frac{1}{c^2}\frac{\partial^2 p_i}{\partial t^2} = 0, \tag{8}$$

where

$$c^2 = \frac{a_i}{\rho\alpha}. \tag{9}$$

Equation (8) is the famous *wave equation*. The quantity c is the wave speed.

By direct substitution, one can verify that Eq. (8) is satisfied by the solution

$$p_i = f(x - ct) + g(x + ct), \qquad (10)$$

where f and g are arbitrary functions of the variables $x - ct$ and $x + ct$. The function $f(x - ct)$ represents a wave propagating to the right (increasing x) whereas $g(x + ct)$ represents a wave propagating to the left.

Velocity, Pressure, and Wall Displacement Waves

The velocity u is linearly related to p through Eqs. (5) and (7), and small change of the radius a is linearly related to changes in p through Eq. (4). Hence by eliminating p, it is seen that u and a are governed by the same wave equation with the same wave speed. If we write

$$p = p_0 f(x - ct) + p_0' g(x + ct),$$
$$u = u_0 f(x - ct) + u_0' g(x + ct), \qquad (11)$$

then on substituting Eqs. (11) into Eqs. (5) and (7), one sees that the amplitude p_0 and u_0 are related by the simple relationship

$$p_0 = \rho c u_0 \qquad (12)$$

for a wave that is moving in the positive x direction, and

$$p_0' = -\rho c u_0' \qquad (13)$$

for a wave which moves in the negative x direction.

Equations (12) and (13) show that *the amplitude of pressure wave is proportional to the product of wave speed and velocity disturbance and the fluid density. The pressure and velocity are in phase in an advancing progressive wave; they are out of phase in the reflected wave.*

5.8 Progressive Waves Superposed on a Steady Flow

It can be shown that the equations of Sec. 5.7 are applicable to tubes carrying a steady flow, provided that we adopt a coordinate system that moves with the undisturbed flow, and interpret u as the perturbation velocity superposed on the steady flow and c as the speed of perturbation wave relative to the undisturbed flow. The proof is as follows.

Let U be the velocity of the undisturbed flow, and u the small perturbation superposed on it. Treating u as an infinitesimal quantity of the first order, we see that the equation of motion, Eq. (5.7:1), can be linearized into

$$\frac{\partial u}{\partial t} + U \frac{\partial u}{\partial x} = -\frac{1}{\rho} \frac{\partial p_i}{\partial x}. \qquad (1)$$

This can be reduced to Eq. (5.7:6) by introducing a transformation of variables from x, t to x', t':

$$x' = x - Ut, \qquad t' = t. \tag{2}$$

From Eq. (2) we have

$$\frac{\partial}{\partial t} = \frac{\partial}{\partial t'}\frac{\partial t'}{\partial t} + \frac{\partial}{\partial x'}\frac{\partial x'}{\partial t} = \frac{\partial}{\partial t'} - U\frac{\partial}{\partial x'},$$

$$\frac{\partial}{\partial x} = \frac{\partial}{\partial t'}\frac{\partial t'}{\partial x} + \frac{\partial}{\partial x'}\frac{\partial x'}{\partial x} = \frac{\partial}{\partial x'}. \tag{3}$$

Hence, a substitution into Eq. (1) reduces it to

$$\frac{\partial u}{\partial t'} = -\frac{1}{\rho}\frac{\partial p}{\partial x'}, \tag{4}$$

which is exactly Eq. (5.7:5) in the new coordinates.

The equation of continuity, Eq. (5.7:2), now becomes

$$\frac{\partial a_i}{\partial t} + U\frac{\partial a_i}{\partial x} + \frac{a_i}{2}\frac{\partial u}{\partial x} = 0 \tag{5}$$

when πa_i^2 is substituted for A, $U + u$ is substituted for u, and the equation is linearized for small perturbations. Under the transformation Eq. (2), and using Eq. (3), Eq. (5) becomes

$$\frac{\partial a_i}{\partial t'} + \frac{a_i}{2}\frac{\partial u}{\partial x'} = 0, \tag{6}$$

which is exactly Eq. (5.7:7).

The pressure–radius relationship, Eq. (5.7:4), is independent of reference coordinates. Thus all the basic equations are unchanged. But x' and t' are the distance and time measured in the moving coordinates which translate with the undisturbed flow. Hence what we set out to prove is done.

5.9 Reflection and Transmission of Waves at Junctions

An arterial tree is composed of segments of cylindrical tubes. Consider a single junction as shown in Fig. 5.9:1 in which a tube branches into two daughters. A wave traveling down the parent artery will be partially reflected at the junction and partially transmitted down the daughters. At the junction, the conditions are: the pressure is a single-valued function and the flow must be continuous. Expressing this mathematically: p_I denote the oscillatory pressure associated with the incident wave, p_R that associated with the reflected wave, and p_{T_1} and p_{T_2} those associated with the transmitted waves in the two daughter tubes; then the single-valuedness of pressure means

$$p_I + p_R = p_{T_1} = p_{T_2}. \tag{1}$$

Similarly, let $\dot{Q}$ denote the volume–flow rate, and let the subscripts, $I, R, T_1,$

FIGURE 5.9:1 A bifurcating artery.

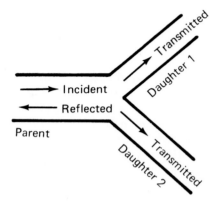

T_2 refer to the various waves as before; then the continuity condition means

$$\dot{Q}_I - \dot{Q}_R = \dot{Q}_{T_1} + \dot{Q}_{T_2}. \tag{2}$$

But $\dot{Q}$ is the product of the cross-sectional area A and the mean velocity u, which is related to p by Eqs. (12) and (13) of Sec. 5.7. Thus, the flow–pressure relationship is:

$$\dot{Q} = Au = \pm \frac{A}{\rho c} p. \tag{3}$$

Here ρ is the density of the blood and c is the wave speed. The $+$ sign applies if the wave goes in the direction of flow; the $-$ sign applies if the wave goes the other way.

The quantity $\rho c/A$ is an important characteristic of the artery, and is called the *characteristic impedance* of the tube. It is denoted by the symbol Z:

$$Z = \frac{\rho c}{A}. \tag{4}$$

Equation (3) shows that Z is the ratio of oscillatory pressure to oscillatory flow when the wave goes in the direction of flow:

$$Z = \frac{p}{\dot{Q}}, \qquad Z\dot{Q} = p, \tag{5}$$

Z has the physical dimensions $[ML^{-4}T^{-1}]$, and can be measured in units of kg m^{-4} sec^{-1}. With the Z notation, Eq. (2) can be written as

$$\frac{p_I - p_R}{Z_0} = \frac{p_{T_1}}{Z_1} + \frac{p_{T_2}}{Z_2}. \tag{6}$$

Solving Eqs. (1) and (6) for the p's, we obtain

$$\frac{p_R}{p_I} = \frac{Z_0^{-1} - (Z_1^{-1} + Z_2^{-1})}{Z_0^{-1} + (Z_1^{-1} + Z_2^{-1})} = \mathscr{R} \tag{7}$$

and

$$\frac{p_{T_1}}{p_I} = \frac{p_{T_2}}{p_I} = \frac{2Z_0^{-1}}{Z_0^{-1} + (Z_1^{-1} + Z_2^{-1})} = \mathscr{J}. \tag{8}$$

The right-hand sides of Eqs. (7) and (8) shall be denoted by $\mathscr{R}$ and $\mathscr{J}$, respectively. Hence the amplitude of the reflected pressure wave at the junction is $\mathscr{R}$ times that of the incident wave, the amplitude of the transmitted pressure waves at the junction is $\mathscr{J}$ times the incident wave. The amplitude of the reflected velocity wave is, however, equal to $-\mathscr{R}$ times that of the incident velocity wave, because the wave now moves in the negative x-axis direction, and according to Eqs. (12) and (13) of Sec. 5.7, there is a sign change in the relation between u and p depending on whether the waves move in the $+$ or $-x$-axis direction.

If the incident wave is

$$p_I = p_0 f(t - x/c_0) \tag{9}$$

and the junction is located at $x = 0$, so that x is negative in the parent tube and positive in the daughter tubes, then at the junction $x = 0$, the pressure is

$$p_I = p_0 f(t). \tag{10}$$

The reflected and transmitted waves are, therefore,

$$p_R = \mathscr{R} p_0 f(t + x/c_0),$$
$$p_{T_1} = \mathscr{J} p_0 f(t - x/c_1), \tag{11}$$
$$p_{T_2} = \mathscr{J} p_0 f(t - x/c_2).$$

Here, c_0, c_1, c_2 are the wave speeds in the respective tubes. The wave in the parent tube is

$$p = p_I + p_R = p_0 f(t - x/c_0) + \mathscr{R} p_0 f(t + x/c_0). \tag{12}$$

$$\dot{Q} = \frac{Ap_0}{\rho c_0} f(t - x/c_0) - \mathscr{R} \frac{Ap_0}{\rho c_0} f(t + x/c_0). \tag{13}$$

Equations (12) and (13) show that with a reflection, the pressure and flow wave forms are no longer equal.

5.10 Velocity Profile of a Steady Flow in a Tube

Having analyzed the aortic blood flow by lumped parameter method (Sec. 5.6), and pulse wave in arteries as one-dimensional nonstationary flow of a nonviscous fluid in an elastic tube (Secs. 5.7–5.9), we shall now consider the effect of viscosity of blood on the flow. We shall first consider blood as a Newtonian fluid.

Consider first a steady flow of an incompressible Newtonian fluid in a rigid, horizontal channel of width $2h$ between two parallel planes as shown in Fig.

FIGURE 5.10:1 Laminar flow in
a channel.

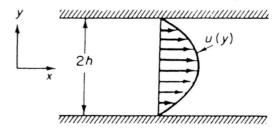

5.10:1. The channel is assumed horizontal so that the gravitational effect (a body force) may be ignored.

We search for a flow,

$$u = u(y), \qquad v = 0, \qquad w = 0, \tag{1}$$

which satisfies the no-slip conditions on the boundaries $y = \pm h$:

$$u(h) = 0, \qquad u(-h) = 0. \tag{2}$$

The function u must satisfy the Navier–Stokes equation and the equation of continuity (Sec. 1.7). It is seen that the equation of continuity is satisfied exactly. The Navier–Stokes equation is simplified to:

$$0 = -\frac{\partial p}{\partial x} + \mu \frac{d^2 u}{dy^2}, \tag{3a}$$

$$0 = \frac{\partial p}{\partial y}, \tag{3b}$$

$$0 = \frac{\partial p}{\partial z}. \tag{3c}$$

Equations (3b) and (3c) show that p is a function of x only. If we differentiate Eq. (3a) with respect to x and use Eq. (1), we obtain $\partial^2 p/\partial x^2 = 0$. Hence $\partial p/\partial x$ must be a constant. Equation (3a) then becomes

$$\frac{d^2 u}{dy^2} = \frac{1}{\mu} \frac{dp}{dx}, \tag{4}$$

which has a solution

$$u = A + By + \frac{1}{\mu} \frac{y^2}{2} \frac{dp}{dx}. \tag{5}$$

The two constants A and B can be determined by the boundary conditions, Eq. (2), to yield the final solution

$$u = -\frac{1}{2\mu}(h^2 - y^2)\frac{dp}{dx}. \tag{6}$$

Thus, the velocity profile is a parabola.

Next consider a flow through a horizontal circular cylindrical tube of radius a. Using polar coordinates, it is easy to show that the solution is [see Fung (1984), p. 83]

$$u = -\frac{1}{4\mu}(a^2 - r^2)\frac{dp}{dx}. \tag{7}$$

This is the famous parabolic velocity profile of the *Hagen–Poiseuille flow*.

From the solution (7) we obtain the *rate of flow* through the tube by integration:

$$\dot{Q} = 2\pi \int_0^a ur\,dr = -\frac{\pi a^4}{8\mu}\frac{dp}{dx}. \tag{8}$$

This classical solution has been subjected to innumerable experimental validation. It was found to be invalid near the entrance to a tube. It is satisfactory at a sufficiently large distance from the entrance but is again invalid if the tube is too large or too long if the velocity is too high. The difficulty at the entry region is due to the transitional nature of the flow in that region so that our assumption $v = 0$, $w = 0$, is not valid. The difficulty with too large a Reynolds number, however, is of a different kind: the flow becomes turbulent.

Osborne Reynolds demonstrated the transition to turbulent flow in a classical experiment in which he examined an outlet from a large water tank through a small tube. At the end of the tube there was a stopcock used to vary the speed of water through the tube. The junction of the tube with the tank was nicely rounded, and a filament of colored fluid was introduced at the mouth. When the speed of water was slow, the filament remained distinct through the entire length of the tube. When the speed was increased, the filament broke up at a given point and diffused throughout the cross-section. Reynolds identified the governing parameter $u_m d/v$—the Reynolds number— where u_m is the mean velocity, d is the diameter, and v is the kinematic viscosity. The point at which the color diffuses throughout the tube is the transition point from laminar to turbulent flow in the tube. Reynolds found that transition occurred at Reynolds numbers between 2,000 and 13,000, depending on the smoothness of the entry conditions. When extreme care is taken, the transition can be delayed to Reynolds numbers as high as 40,000. On the other hand, a value of 2,000 appears to be about the lowest value obtainable on a rough entrance. Turbulence is one of the most important and difficult problems in fluid mechanics.

5.11 Steady Laminar Flow in an Elastic Tube

If the tube is elastic (Fig. 5.11:1), then the high-pressure end would distend more than the low-pressure end. The diameter of the tube is, therefore, nonuniform (if it were uniform originally) and the degree of nonuniformity depends on the flow rate.

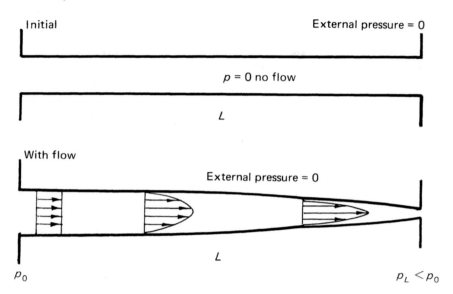

FIGURE 5.11:1 Flow in an elastic tube of length L.

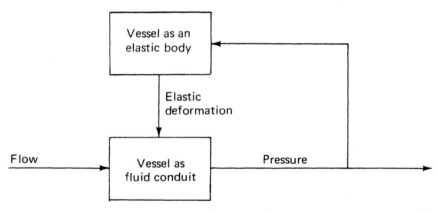

FIGURE 5.11:2 A hemoelastic system analyzed as a feedback system of two functional units: an elastic body, and a fluid mechanism.

If we wish to determine the pressure–flow relationship for such a system, we may break down the problem into two familiar components. This is illustrated in Fig. 5.11:2. In the lower block, we regard the vessel as a rigid conduit with a specified wall shape. For a given flow, we compute the pressure distribution. This pressure distribution is then applied as loading on the elastic tube, represented by the upper block. We then analyze the deformation of the elastic tube in the usual manner of the theory of elasticity. The result of the calculation is then used to determine the boundary shape of the hydrodynamic problem of the lower block. When a consistent solution is obtained, the pressure distribution corresponding to a given flow is determined.

In Sec. 5.10 we derived Poiseuille's formula under the assumption of a laminar steady flow in a circular cylindrical tube of constant radius. If the radius $a(x)$, as a function of the axial coordinate x, is not a constant, but the slope da/dx is small, then the fluid dynamic problem can be solved by perturbation method as a power series of the small parameter da/dx. Under an additional assumption that the Reynolds number is so small that the inertial force term $\rho u \, \partial u / \partial x$ is negligible in the zeroth order equations, the solution is the Poiseuille's formula

$$\frac{dp}{dx} = -\frac{8\mu}{\pi a^4} \dot{Q} \tag{1}$$

in which $\dot{Q}$ is the volume–flow rate and is a constant for the whole tube, $a(x)$ is the local radius, dp/dx is the local pressure gradient. For an elastic tube, the radius a is a function of pressure. Hence we can rewrite Eq. (1) as

$$a^4(p)\frac{dp}{dx} = -\frac{8\mu}{\pi} \dot{Q} = \text{const.} \tag{2}$$

This is very easy to integrate if the function $a(p)$ is known. For the pulmonary arteries and veins it is known that the pressure–radius relationship is given by Eq. (5.7:4)

$$a = a_0 + \frac{\alpha}{2} p, \tag{3}$$

where a_0 is the radius when p is zero, and α is the compliance constant. Substituting Eq. (3) into Eq. (2), and integrating, we obtain

$$a^4 \frac{dp}{da}\frac{da}{dx} = \frac{2}{\alpha} a^4 \frac{da}{dx} = -\frac{8\mu}{\pi} \dot{Q}. \tag{4}$$

Since the right-hand side term is a constant independent of x, we obtain the integrated result

$$[a(x)]^5 = -\frac{20\mu\alpha}{\pi} \dot{Q}x + \text{const.} \tag{5}$$

The integration constant can be determined by the boundary condition that when $x = 0$, $a(x) = a(0)$. Hence the constant $= [a(0)]^5$. Then by putting $x = L$, we obtain from Eqs. (5) and (3) the elegant result (Fung, 1984)

$$\dot{Q} = \frac{\pi}{20\mu\alpha L}\{[a(0)]^5 - [a(L)]^5\}$$

$$= \frac{\pi}{20\mu\alpha L}\left[\left(a_0 + \frac{\alpha}{2}p_0\right)^5 - \left(a_0 + \frac{\alpha}{2}p_L\right)^5\right]. \tag{6}$$

Thus the flow varies with the difference of the fifth power of the tube radius at the entry section ($x = 0$) minus that at the exit section ($x = L$). If the ratio $a(L)/a(0)$ is $\frac{1}{2}$, then $[a(L)]^5$ is only about 3% of $[a(0)]^5$ and is negligible by comparison. Hence when $a(L)$ is one-half of $a(0)$ or smaller, the flow varies directly with the fifth power of the tube radius at the entry, whereas the radius (and the pressure) at the exit section has little effect on the flow.

This analysis applies very well to the lung, in which the phenomenon just described has an important effect. See Sec. 6.8 infra.

For flow in large blood vessels with Reynolds number much greater than 1, the inertial force terms must be added. Let us consider the case of a steady flow of Newtonian fluid in an elastic tube whi which is initially a circular cylinder. Assume that the flow is predominantly one-dimensional. Let q represent the average velocity in the tube. The convective inertial force is $\rho q(\partial q/\partial x)$. The pressure drop due to blood viscosity is given by Eq. (1) even if the flow is turbulent, provided that the coefficient of viscosity μ is reinterpreted as the "apparent" coefficient of viscosity which is a function of the Reynolds number, see Sec. 5.13 infra. Then the equation of motion is

$$\rho q \frac{dq}{dx} = -\frac{dp}{dx} - \frac{8\mu}{\pi a^4} \dot{Q}. \tag{7}$$

Here ρ is the density of the blood, x is the axial coordinate, p is the pressure, $\dot{Q}$ is the volume flow rate, and μ is the apparent coefficient of viscosity of the blood corrected for turbulence, secondary flow, or entrance effect, i.e. it is a function of Reynolds number. Finally, a is the radius of the tube, which is a linear function of pressure as given by Eq. (3). When the transmural pressure is zero the tube is assumed to be cylindrical, $a_0 = $ const. The equation of continuity is

$$\pi a^2 q = \text{const} = \dot{Q}. \tag{8}$$

By differentiation, one obtains

$$a^2 dq + 2qada = 0. \tag{9}$$

On solving Eq. (8) for q, and substituting into Eq. (9) multiplied by q/a^2, we have

$$q\, dq = -\frac{2\dot{Q}^2}{\pi^2 a^5} da. \tag{10}$$

Substituting Eq. (10) into Eq. (7) and reducing, one obtains

$$\left(a^4 - \frac{\rho\alpha\dot{Q}^2}{\pi^2 a}\right)\frac{da}{dx} = -\frac{4\mu\alpha}{\pi}\dot{Q}. \tag{11}$$

Integration yields

$$a^5 - \frac{5\rho\alpha\dot{Q}^2}{\pi^2}\ln a = -\frac{20\mu\alpha}{\pi}\dot{Q}x + \text{const}. \tag{12}$$

The boundary condition $a = a(0)$ when $x = 0$ yields the integration constant. On putting this constant into Eq. (12), and then letting $x = L$ where $a = a(L)$, we obtain

$$\dot{Q} - \left[\frac{\rho}{4\mu\pi L} \ln \frac{a(L)}{a(0)} \right] \dot{Q}^2 = \frac{\pi}{20\mu\alpha L} \{ [a(0)]^5 - [a(L)]^5 \}. \tag{13}$$

This is a modification of Eq. (6). The effect of inertial force is embodied in the second term. If the elastic deformation is small, $a(L) \doteq a(0)$, then the second term tends to zero, and the solution is the same as Eq. (6) except that the apparent viscosity μ must now be considered as a function of the Reynolds number. If the elastic deformation is significant, in that $a(L)$ differs considerably from $a(0)$, then the second term must be considered. For a given set of values $a(L)$, $a(0)$, we now have two solutions of $\dot{Q}$. Conversely, for a given $\dot{Q}$ we now have multiple solutions of $a(L)$, $a(0)$. This is possible at high Reynolds number, because the inertial force and the viscous dissipation influence the pressure gradient in opposite ways.

5.12 Velocity Profile of Pulsatile Flow

To obtain the velocity profile of nonstationary flow in a blood vessel, one must solve the equations of motion and continuity of both the blood and the blood vessel wall, and boundary conditions that match the displacements, velocities, and stresses. The calculation is usually lengthy. References to the literature can be found in Fung (1984), McDonald (1974), Patel and Vaishnav (1980), Pedley (1980). In the following, a simple example is given.

Assume that the fluid is homogeneous, incompressible, and Newtonian; the vessel wall is rigid, circular, and cylindrical; the motion is laminar, axisymmetric, and parallel to the longitudinal axis of the tube. A pressure gradient drives the flow, the vessel is horizontal, and gravitation has no effect on the flow. Then the field equations are the Navier–Stokes equations, and the equation of continuity. They are simplified to the following under the conditions named above:

$$0 = -\frac{\partial p}{\partial r}, \tag{1}$$

$$0 = -\frac{\partial p}{\partial \Theta}, \tag{2}$$

$$\rho \frac{\partial u}{\partial t} = -\frac{\partial p}{\partial x} + \mu \left(\frac{\partial^2 u}{\partial r^2} + \frac{1}{r} \frac{\partial u}{\partial r} \right), \tag{3}$$

$$\frac{\partial u}{\partial x} = 0. \tag{4}$$

The boundary conditions are the axisymmetry condition at the center and no-slip on the wall, at radius a:

$$\frac{\partial u}{\partial r} = 0 \quad \text{when} \quad r = 0, \tag{5}$$

$$u = 0 \quad \text{when} \quad r = a. \tag{6}$$

Here p stands for pressure; (x, r, Θ) are cylindrical polar coordinates with x in the axial direction; u is the velocity component in the direction of x; t is time. According to Eqs. (1) and (2), p is a function of x and t only. According to (4), u is a function of r and t. On differentiating Eq. (3) with respect to x, one obtains

$$\frac{\partial}{\partial x}\left(\frac{\partial p}{\partial x}\right) = 0. \tag{7}$$

This shows that the pressure gradient must not vary with x. It can be a function of t. For a general periodic motion, one can write

$$\frac{\partial p}{\partial x} = \sum_{n=0}^{N} C_n e^{in\omega t}. \tag{8}$$

On substituting into Eq. (3), one obtains

$$\rho \frac{\partial u}{\partial t} = -\sum_{n=0}^{N} C_n e^{in\omega t} + \mu\left(\frac{\partial^2 u}{\partial r^2} + \frac{1}{r}\frac{\partial u}{\partial r}\right). \tag{9}$$

The term $n = 0$ corresponds to a steady pressure gradient investigated in Sec. 5.10; the solution is given by Eq. (5.10:7). To the other terms in (7), we can try $u(r, t)$ in the form

$$u = \sum_{n=0}^{N} v_n(r) e^{in\omega t} \tag{10}$$

which is periodic. Substituting Eq. (10) into Eq. (9) we see that the resulting equation is satisfied if we set

$$i\rho n\omega v_n = -C_n + \mu\left(\frac{d^2 v_n}{dr^2} + \frac{1}{r}\frac{dv_n}{dr}\right). \tag{11}$$

The boundary condition, Eq. (6), is satisfied if

$$v_n = 0 \quad \text{when} \quad r = a,$$

$$\frac{\partial v_n}{\partial r} = 0 \quad \text{when} \quad r = 0. \tag{12}$$

The general solution of Eq. (11) is

$$v_n(r) = A_n J_0\left(\alpha \frac{r}{a} n^{1/2} i^{3/2}\right) + B_n Y_0\left(\alpha \frac{r}{a} n^{1/2} i^{3/2}\right) + \frac{iC_n}{\rho n\omega}. \tag{13}$$

Here $J_0(kr)$ is the Bessel function of the first kind of order zero of kr, $Y_0(kr)$ is the Bessel function of the second kind of order zero of kr, k being a constant. A_n, B_n are arbitrary constants, and α is a dimensionless quantity known as the Womersley number (Sec. 5.13):

$$\alpha = a\sqrt{\frac{\omega}{v}}. \tag{14}$$

To determine A_n, B_n, the boundary conditions, Eq. (12), are used. As r approaches zero, the derivative J_0' approaches zero and Y_0' approaches infinity. Hence B_n must vanish, and the first of Eq. (12) requires

$$A_n J_0(\alpha n^{1/2} i^{3/2}) + \frac{iC_n}{\rho n \omega} = 0. \tag{15}$$

Solving this equation for A_n and substituting into Eq. (13) together with $B_n = 0$, we obtain

$$v_n(r) = \frac{iC_n}{\rho n \omega}\left[1 - \frac{J_0(\alpha\frac{r}{a} n^{1/2} i^{3/2})}{J_0(\alpha n^{1/2} i^{3/2})}\right]. \tag{16}$$

The problem is solved by substituting Eq. (16) into Eq. (10). The velocity profile depends on Womersley number α. An illustration is given in Fig. 5.12:1.

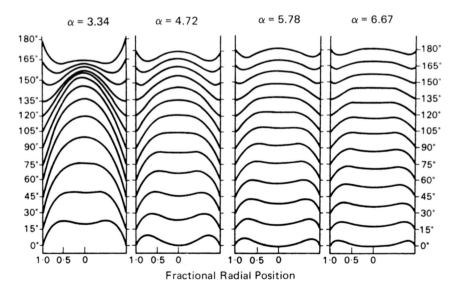

FIGURE 5.12:1 Theoretical velocity profiles of a sinusoidally oscillating flow in a pipe, with pressure gradient varying like $\cos \omega t$. α is the Womersley number. Profiles are plotted for phase angle steps of $\Delta \omega t = 15°$. For $\omega t > 180°$, the velocity profiles are of the same form but opposite in sign. Reproduced with permission from D.A. McDonald, *Blood Flow in Arteries*, copyright © 1974, the Williams & Wilkins Co., Baltimore.

5.13 The Reynolds Number, Stokes Number, and Womersley Number

The general equations of hemodynamics appear formidable. Some essential features can be identified when different terms are compared. The Navier–Stokes equation

$$\rho \frac{\partial u_i}{\partial t} + \rho \left(u_1 \frac{\partial u_i}{\partial x_1} + u_2 \frac{\partial u_i}{\partial x_2} + u_3 \frac{\partial u_i}{\partial x_3} \right)$$

$$= X_i - \frac{\partial p}{\partial x_i} + \mu \left(\frac{\partial^2}{\partial x_1^2} + \frac{\partial^2}{\partial x_2^2} + \frac{\partial^2}{\partial zx_3^2} \right) u_i, \tag{1}$$

represents the balance of four kinds of forces. Term by term, they are

$$\begin{array}{ccccc} \text{transient} & \text{convective} & \text{body} & \text{pressure} & \text{viscous} \\ \text{inertial} + & \text{inertial} & = \text{force} + & \text{force} + & \text{force} . \\ \text{force} & \text{force} & & & \end{array}$$

Not all the forces are important all the time. In a steady flow the transient inertial force vanishes. In an ideal fluid the viscous force vanishes. In hydrostatic equilibrium all but the body and pressure forces vanish. Simplifications are recognized for these cases.

Compare the transient inertial force term with the viscous force term. To make an estimate, let U be a characteristic velocity, ω a characteristic frequency, and L a characteristic length. Then the first term in Eq. (1) is of the order of magnitude $\rho \omega U$, and the last term is of the order of magnitude $\mu U L^{-2}$. The ratio is

$$\frac{\text{transient inertial force}}{\text{viscous force}} = \frac{\rho \omega U}{\mu U L^{-2}} = \frac{\rho \omega L^2}{\mu} = \frac{\omega L^2}{v}. \tag{2}$$

This is a dimensionless number. If it is large, the transient inertial force dominates. If it is small, the viscous force dominates.

The dimensionless number $\omega L^2 / v$ is a *frequency parameter*, and is called the Stokes' number because its significance was pointed out by George Stokes in 1840. It is better known by its square root,

$$N_W = L \sqrt{\left(\frac{\omega}{v} \right)}, \tag{3}$$

which is called Womersley number in honor of J.R. Womersley, who made extensive calculations on pulsatile blood flow in the 1950's. If L is taken to be the radius of the blood vessel, then Womersley's number is often written as α:

$$\alpha = N_W = \frac{D}{2} \sqrt{\left(\frac{\omega}{v} \right)}, \tag{4}$$

D being the blood vessel diameter. In large arteries of all but the smallest mammals, the value of α, calculated from the circular frequency of the heartbeat in rad/sec, is considerably larger than 1. For example, a typical value of α in the aorta of man is 20, in a dog it is 14, in a cat 8, and in a rat 3. Hence in these aortas the inertial force dominates in pulsatile flow.

If α is large, the effect of the viscosity of the fluid does not propagate very far from the wall. In the central portion of the tube the transient flow is determined by the balance of the inertial forces and pressure forces (first and fourth terms in Eq. 1), and the elastic forces in the wall (through the boundary conditions), as if the fluid were nonviscous. We, therefore, expect that when the Womersley number is large, the velocity profile in a pulsatile flow will be relatively blunt, in contrast to the parabolic profile of the Poiseuillean flow, which is determined by the balance of viscous and pressure forces. This is shown in Fig. 5.12:1.

Now compare the convective inertial force term with the viscous force term. With characteristic velocity U and characteristic length L, the order of magnitude of the inertial force is ρU^2, that of the viscous force is $\mu U/L$. The ratio is

$$\frac{\text{inertial force}}{\text{viscous force}} = \frac{\rho U^2}{\mu U/L} = \frac{\rho UL}{\mu} = \text{Reynolds number.} \tag{5}$$

A large Reynolds number signals a preponderant inertial effect. A small Reynolds number signals a predominant viscous force effect. In aorta of man the Reynolds number based on vessel diameter can be 2,000–3,000, large enough to cause possible turbulence (Sec. 5.10). In the capillary blood vessels, the Reynolds number is in the order of 0.001 to 0.01, so small that it suggests complete insignificance of the inertial force.

The occurrence of turbulence in a pulsatile flow in the aorta could be transient. Even when the condition of flow favors the transition of a laminar flow into turbulent, the actual transition into turbulence would require a certain amount of time to develop. Hence if the flow velocity fluctuates too fast, the turbulence may not develop. Similarly, if a flow is turbulent but the condition has changed to favor a transition into laminar flow, the actual transition may lag behind for a while.

Quantitative studies of the laminar–turbulent transition may seek to express the critical Reynolds number as a function of the Womersley number. Experimental results can be plotted as shown in Fig. 5.13:1. The ordinate is the peak Reynolds number. The stippled area indicates the conditions under which the flow is stable and laminar.

In the experiments whose results are shown in Fig. 5.13:1, the wide variations of velocity and heart rate were obtained with drugs and nervous stimuli in anesthetized dogs. In normal, conscious, free-ranging dogs the peak Reynolds number usually lies in an area high above the stippled area of Fig. 5.13:1. This suggests that some turbulence is generally tolerated in deceleration of systolic flow in the dog.

Turbulence in blood flow implies fluctuating pressure acting on the arterial wall, and fluctuating, increased shear stress. These stresses are implicated in murmurs, post-stenotic dilation, and atherogenesis.

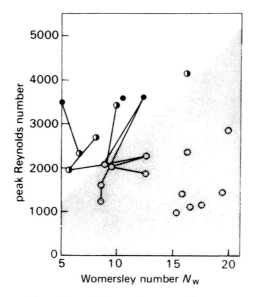

FIGURE 5.13:1 The stability of blood flow in the descending aorta of anesthetized dogs is influenced by the peak Reynolds number and the Womersley number. Points joined by the lines refer to the same animal. Open circles: laminar flow; filled circles: turbulent flow; half-filled circles: transiently turbulent flow. From Nerem, R.M., and Seed, W.A. (1972), by permission.

5.14 Equation of Balance of Energy and Work

According to the principle of conservation of energy, the rate of gain of energy of a material system (the sum of the kinetic, potential and internal energies) must be equal to the sum of the rate at which work is done on the system and heat transported in. Apply this principle to a body of blood contained in a blood vessel between two arbitrary cross sections, 1 and 2, perpendicular to the vessel axis, as illustrated in Fig. 5.14:1. Let p denote the pressure, u denote the axial component of the velocity of flow, q denote the magnitude of the velocity vector, $\dot{Q}$ denote the volume rate of flow. Let dA denote a small element of area in a cross section. At the left end, section 1, the outward normal vector of the cross section points to the left, the force acting on the area dA due to the pressure, pdA, points to the right. The positive direction of the axial velocity u point to the right. The rate of work done by the force due to pressure is $pudA$. The total work done by the force over the entire cross section is, therefore

$$\int_{A_1} pu\,dA \tag{1}$$

where the integral is taken over the area A_1 of the cross section No. 1.

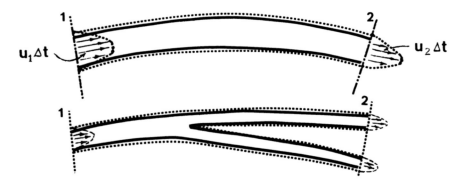

FIGURE 5.14:1 Two arbitrary cross sections, 1 and 2, of a blood vessel. At an instant of time t, a control volume of blood is bounded by the plane sections 1 and 2 and the wall of the blood vessel shown by solid lines. An infinitesimal time Δt later, the boundary of the control volume becomes that shown by the dotted line: consisting of two paraboloidal surfaces at sections 1 and 2, and a distended blood vessel wall. The equations of motion, continuity, and energy are written for the fluid in the control volume. The symbols u_1 and u_2 represent axial component of the velocity at stations 1 and 2, respectively.

The kinetic energy per unit volume of blood is $\frac{1}{2}\rho q^2$ where ρ is the density of the blood. The total kinetic energy of the blood contained in the volume between the sections 1 and 2 is, at time t,

$$\int_V \frac{1}{2}\rho q^2 \, dv. \tag{2}$$

A short time Δt later, the same body of fluid would occupy a slightly different volume which is bounded by the dotted lines shown in Fig. 5.14.1. The side wall distends a little because of vessel wall elasticity. The fluid particles composing the cross section 1 are displaced by a distance $u\Delta t$ to the right. The plane cross section No. 1 becomes curved and bulges to the right. The particles at Section 2 are also displaced to the right by the amount $u\Delta t$. During the time interval Δt, therefore, the total kinetic energy of the blood is changed by the amount

$$\int_{V'} \frac{\partial}{\partial t}\left(\frac{1}{2}\rho q^2\right)\Delta t \, dV - \int_{A_1} \frac{1}{2}\rho q^2 u\Delta t \, dA + \int_{A_2} \frac{1}{2}\rho q^2 u\Delta t \, dA. \tag{3}$$

Where V' is the volume bounded by the dotted lines. The rate of change of kinetic energy is obtained by dividing the quantity above with Δt. The $-$ and $+$ signs in the expression (3) should be carefully noted.

A similar consideration should be given to the work done by force imposed on the blood by the blood vessel wall, the potential energy change due to gravitation, the internal energy change due to temperature change, the heat transported through the boundary, and the rate of heat generation due to internal friction equal to the sum of all the products of stress components and the corresponding strain rates. The last term, the heat dissipation, is denoted by $\mathscr{D}$:

$$\mathscr{D} = \int_V \sigma_{ij} V_{ij} \, dv. \tag{4}$$

Where σ_{ij} is the stress tensor and V_{ij} is the strain rate tensor.

Now, on equating the change of energy with the work done, and dividing through by the volume flow rate $\dot{Q}$:

$$\dot{Q} = \int_A u \, dA \tag{5}$$

we obtain, if gravitational effect and heat transfer were negligible, the following equation:

$$\widehat{p_1} - \widehat{p_2} = \frac{1}{2}\rho\widehat{q_2^2} - \frac{1}{2}\rho\widehat{q_1^2} + \rho g h_2 - \rho g h_1 + \frac{\mathscr{D}}{\dot{Q}} + \frac{1}{\dot{Q}}\int_V \frac{\partial}{\partial t}\left(\frac{1}{2}\rho q^2\right) dv. \tag{6}$$

Here the velocity-weighted pressure $\widehat{p}$ and the square of velocity $\widehat{q^2}$ are defined by dividing Eqs. (1) and (2) by $\dot{Q}$:

$$\widehat{p} = \frac{1}{\dot{Q}}\int_A pu \, dA, \qquad \widehat{q^2} = \frac{1}{\dot{Q}}\int_A q^2 u \, dA. \tag{7}$$

Note that $\widehat{p}$ and $\frac{1}{2}\rho\widehat{q^2}$ have the dimensions of pressure.

The energy equation (6) was given by Pedley et al. (1977) and derived in full detail in Fung (1984), pp. 15–20. It is used frequently in this book.

5.15 Systemic Blood Pressure

If we apply the results derived in the preceding sections to a circuit of blood vessels beginning at the aortic valve and ending in the right atrium, take the average of the pressure-flow relationship of every segment over a period of time which is long compared with a single heart beat, and synthesize the segments into a whole circuit, then we obtain the result:

$$\text{Average pressure at aortic valve} - \text{average pressure}$$
$$\text{at right atrium} = \text{integrated frictional loss.} \tag{1}$$

This is often written as:

$$\text{Systemic arterial pressure} = \text{flow} \times \text{resistance.} \tag{2}$$

Here the systemic arterial pressure is the difference between the pressure at the aortic valve and that at the vena cava at the right atrium, the flow is the *cardiac output*, and the resistance is the *total peripheral vascular resistance.* Hence, writing in greater detail, we have

Pressure at aortic valve − pressure at right atrium

= (cardiac output) × (total peripheral vascular resistance),   (3)

where

$$\text{Total peripheral vascular resistance} = \frac{\text{integrated frictional loss}}{\text{cardiac output}}. \qquad (4)$$

The last term in Eq. (3) represents the sum of the pressure drops due to the friction loss along all segments of blood vessels. Since there are millions of capillary blood vessels in the body, there are millions of pathways along which one can integrate the equation of motion to obtain Eq. (3), so the final result Eq. (3) is useful only if the pressures at the aortic valve and right atrium are uniform no matter which path of integration is used. Fortunately, this is the case.

The integrated frictional loss is the sum of frictional losses in all segments of vessels of the circuit. To compute the frictional loss of a segment, let us first consider a steady laminar flow (i.e., one that is not turbulent) in a long, rigid, circular, cylindrical vessel. To such a flow, Poiseuille's formula, Eq. (5.10:8) applies. Let the vessel length be L and the vessel diameter be d, then

$$\dot{Q} = -\frac{\pi d^4}{128} \frac{\Delta p}{\mu L}. \qquad (5)$$

Here μ is the coefficient of viscosity of the fluid, and Δp is the pressure drop. Equation (5) can be written as

Δp = (laminar resistance in a tube) × (flow in the tube),   (6)

from which we obtain the resistance of a steady laminar flow in a circular cylindrical vessel:

$$\text{laminar resistance in a tube} = \frac{128 \mu L}{\pi d^4}. \qquad (7)$$

If the nth generation of a vascular tree consists of N identical vessels in parallel, then the

Pressure drop in the nth generation of vessels

= (resistance in N parallel tubes) × (total flow in N tubes)

$$= \frac{(\text{resistance in one tube})}{N} \times (\text{cardiac output}). \qquad (8)$$

Note that according to Eq. (7) the laminar flow resistance is proportional to the coefficient of viscosity μ and the length of the vessel L, and inversely proportional to the fourth power of the diameter d. Obviously the vessel diameter d is the most effective parameter to control the resistance. A reduction of diameter by a factor of 2 raises the resistance 16-fold, and hence leads to a 16-fold pressure loss. In peripheral circulation the arterioles are muscular and they control the blood flow distribution by changing the vessel diameters through contraction or relaxation of the vascular smooth muscles.

Equation (7) gives the resistance to a Poiseuillean flow in a pipe and for a given flow rate $\dot{Q}$ is the minimum of resistance of all possible flows in the pipe. If the flow becomes turbulent, the resistance increases. If the blood vessel bifurcates, the local disturbance at the bifurcation region raises resistance. In these deviations from the Poiseuillean flow the governing parameter is the Reynolds number. If a flow is turbulent, then

Resistance of a turbulent flow in a vessel

$$= \text{(laminar resistance)} \cdot (0.005\ N_R^{3/4}). \tag{9}$$

Thus, if the Reynolds number is 3,000, the resistance of a turbulent flow is over two times that of the laminar resistance. In the ascending and descending aorta of man and dog the peak Reynolds number does exceed 3,000. The energy loss that occurs at points of bifurcation, entry flow, flow separation, etc., are also functions of Reynolds number. In these cases one writes the pressure–flow relationship as

$$\dot{Q} = \frac{1}{Z(N_R)} \frac{\pi d^4}{128} \frac{\Delta p}{\mu L}, \tag{10}$$

where $Z(N_R)$ is a dimensionless function of the Reynolds number. Equation (9) shows that for a turbulent flow $Z(N_R)$ is equal to $0.005\ N_R^{3/4}$. Other examples are given in Sec. 7.2, especially Eqs. (7.2:4)–(7.2:6). All the results of fluid mechanics research on flow resistance in pipes can be packed into the function $Z(N_R)$.

Equation (3) or (8) shows the basic factors that control the systemic blood pressure. The resistance is proportional to the blood viscosity. Hence lowering the coefficient of viscosity will promote the flow. Hemodilution is thus a practical clinical tool. The resistance is sensitive to the diameter of the blood vessel. The diameter is controlled by the vascular smooth muscle. Hence the control of smooth muscle behavior is the key to the treatment of hypertension.

5.16 The Veins and Their Collapsibility

Veins are similar to arteries in size and construction, but veins have valves and smaller wall thickness to diameter ratio. In fact the wall thickness of veins is often quite nonuniform around the circumference. Because veins have thinner walls, they are more compliant than the arteries. Because the blood pressure is low in the veins, they are more sensitive to external pressure. If the external pressure exceeds the internal pressure by an amount known as the *critical buckling pressure*, then a vein will collapse. Normally, 80% of a man's blood is in the veins. For this reason veins are said to be capacitance vessels. The capacitance is sensitive to internal and external pressures. Thus raising one's leg or moving the leg muscles will reduce the blood volume in the legs, pushing blood to the heart and circulating to other parts of the body.

The collapsibility of the vein gives the venous blood flow some unique features. In dynamic condition the *transmural pressure* (Δp = internal − external pressures) acting on a vein may be a) positive throughout, b) negative and exceeding the critical buckling pressure throughout, or c) positive at the entry section, but negative and exceeding the critical buckling pressure at the exit section. Then in condition a) the vein is patent, in b) it is collapsed, whereas in c) something special will happen. If in condition c) the exit end is collapsed, then the flow would stop, the pressure drop would become zero, the whole tube would have a Δp equal to that of the entry section, the condition of a) would then prevail, and flow would start again. But if flow starts, the pressure will drop along the tube, and the exit section may be choked again. This may lead to a dynamic phenomenon of "flutter,", or to a limiting steady flow controlled by a narrowed section. In the last case, the actual value of Δp at the exit section is quite immaterial as long as the cross sectional area at the exit section is much reduced. An analogy may be drawn between this and the waterfall in our landscape, or sluicing in industry or flood control: The volume flow rate in a waterfall depends on the conditions at the top of the fall, and is independent of how high the drop is. Thus, the phenomenon of flow in case c) is described as the "waterfall" phenomenon, or as sluicing.

The waterfall phenomenon occurs in a number of important organs: the lung, the vena cava, etc. It occurs in thoracic arteries during resuscitation maneuvers, and in brachial arteries while measuring blood pressure by cuff and Korotkov sound. The same phenomenon also occurs in male and female urethra in micturition, and in manmade instruments such as the blood pump and the heart−lung machine.

Since so much depends on the collapsibility, let us consider the mechanical property of blood vessels at negative transmural pressure in greater detail.

Moreno et al. (1970) measured the change of the cross-sectional area of dog's vena cava when the transmural pressure was varied. Shapiro (1977) measured the same in latex tubing. The characteristics of the vessel and tube deformations are similar. Shapiro's results are shown by the solid curve in Fig. 5.16:1. If the tube were circular cylindrical and of Hookean elastic material when the transmural pressure is zero, then the elastic stability of the tube is amenable to mathematical analysis. The theoretical results of Flaherty et al. (1972) are shown in Fig. 5.16:1 by the curve with long dashes. Theoretically, the pressure-area curve has a sudden change of slope at each critical transmural pressure. The deformation pattern changes when the transmural pressure exceeds the critical value. If one defines the dimensionless variables

$$\tilde{p} = \frac{12(1 - v^2)R^3}{Eh^3}(p - p_e) \quad \text{and} \quad \alpha = \frac{A}{\pi R^2} \tag{1}$$

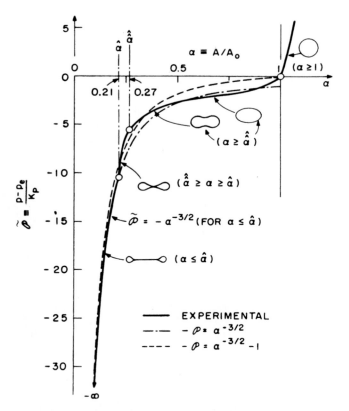

FIGURE 5.16:1 Behavior of a collapsible tube. Dimensionless transmural pressure difference, $\tilde{p}$, versus dimensionless area ratio, α. Solid curve shows a typical experimental curve for thin-walled latex tube, and adjacent to it, typical cross-sectional shapes for the different ranges of α. Dot-dash curve represents Eq. (8), coincides with solid curve for $\alpha < \hat{\alpha}$. Dashed curve represents Eq. (10). Curve with long dashes represents the theoretical result given by Flaherty et al. for cylinders whose cross sections are perfectly circular when $\tilde{p} = 0$. Point contact occurs at $\alpha = \hat{\hat{\alpha}}$, and line contact occurs at $\alpha = \hat{\alpha}$. From Shapiro (1977), by permission.

in which p represents internal pressure, p_e is the external pressure, E is the Young's modulus of the tube wall material, v is its Poisson's ratio, R is the tube radius at midwall, h is the tube wall thickness, A is the cross-sectional area of the lumen, then Flaherty et al. showed that the buckling occurs when $\tilde{p} < -3$. When $\tilde{p} = 5.247$, the opposite walls touch each other at the midpoint. Upon further increase in external pressure, the contact area increases and the open portion of the cross section is reduced in size but remains similar in shape. For this "self-similar" type of deformation Flaherty et al. obtained the relationship

$$-\tilde{p} = \alpha^{-3/2}. \tag{2}$$

Noticing the difference between the experimental curve and the theoretical curve, Shapiro (1977) proposed an empirical formula

$$-\bar{p} = \alpha^{-3/2} - 1. \tag{3}$$

Now let us see something different. Figure 6.6:5 on p. 209 shows the thickness of the pulmonary capillary sheet in the interalveolar septa as a function of the transmural pressure. These pulmonary capillary blood vessels form a dense network whose "thickness" varies with the blood pressure, whereas the dimension in the plane of the interalveolar septa is unaffected by

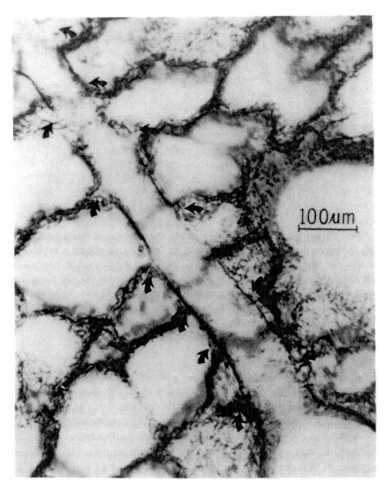

FIGURE 5.16:2 The connection between a pulmonary vein and interalveolar septa in cat's lung. Courtesy of Dr. Sidney Sobin.

the blood pressure. The difference between the blood pressure and the airway pressure is defined as the transmural pressure. Fung et al. (1972) have shown that the thickness drops very rapidly to zero when the transmural pressure changes from positive to negative values.

On the other hand, Fig. 6.6:4 on p. 208 shows the diameter versus transmural pressure relationship of pulmonary veins (Yen and Fappiano, 1981). It is seen that the relationship can be represented by straight lines. The slope of the straight line does not change when Δp changes from the positive to negative value. These veins would not collapse under negative transmural pressure in the physiological range (Fung et al., 1983).

Thus the elastic stability characteristics of the pulmonary capillaries is similar to that of the vena cava, but that of the pulmonary vein is entirely different from that of the vena cava. Not all veins are alike! The difference is actually easily explained. The vena cava was tested as an isolated tube. The pulmonary veins were, however, tested intact, embedded in the lung parenchyma which was in tension. The lung parenchyma provides an elastic support to the pulmonary veins.

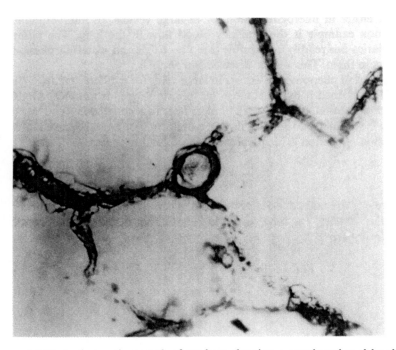

FIGURE 5.16:3 A photo micrograph of cat lung showing a venule tethered by three interalveolar septa. Vasculature perfused with a catalyzed silicone elastomer and hardened; gelatin-embedded; cresyl violet stained. From Fung et al. (1983). Reproduced by permission.

This elastic support can be seen from the photomicrographs of the lung parenchyma. In Figure 5.16:2 is shown a histological section of a cat lung parenchyma containing a large blood vessel. The lacy tissue tethering the outer wall of the blood vessel is the alveolar structure, which is in tension in an inflated lung. Thus the blood vessel is embedded in a foam-rubber-like material. In Figure 5.16:3, at a larger magnification, is shown a histological section of cat lung parenchyma containing a pulmonary arteriole. The diameter of the arteriole is about 25 μm, which is small compared with the dimension of the alveoli of the cat, about 10 μm. See that the arteriole is pulled by three interalveolar septa. These septa are in tension in an inflated lung, they tend to distend the arteriole. When the alveolar gas pressure outside the arteriole exceeds the blood pressure in the arteriole, the tendency for the vessel to collapse is resisted by the tension in the interalveolar septa attached to the outer wall of the blood vessel.

5.17 Flow in Collapsible Tubes

In circulatory physiology, flow in blood vessels in collapsible condition may occur either in microcirculation, or in large vessels. In microcirculation, a common example is the capillary blood flow in the lung. The pulmonary capillaries are readily collapsible (see Fig. 6.6:5) and waterfall phenomenon occurs in them. This will be discussed in Sec. 6.8.

Waterfall phenomenon occurs in large veins for a different reason. Shapiro (1977) explained it by an analogy (at infinite Reynolds number). Consider a one-dimensional, unsteady, frictionless flow in a collapsible tube, a gas flow in a wind tunnel, and a liquid flow in a uniform, horizontal open channel. The equation of motion is identical for each of the three cases:

$$\frac{\partial u}{\partial t} + u \frac{\partial u}{\partial x} = -\frac{1}{\rho} \frac{\partial p}{\partial x}, \tag{1}$$

where ρ is the mass density of the fluid, p is the pressure in the flowing fluid, u is the velocity, t is time, and x is longitudinal distance. The equations of continuity are

For a collapsible tube: $\quad \dfrac{\partial A}{\partial t} + \dfrac{\partial}{\partial x}(Au) = 0; \tag{2}$

For the gas flow: $\quad \dfrac{\partial \rho}{\partial t} + \dfrac{\partial}{\partial x}(\rho u) = 0; \tag{3}$

For the channel flow: $\quad \dfrac{\partial h}{\partial t} + \dfrac{\partial}{\partial x}(hu) = 0. \tag{4}$

Here A is the cross-sectional area in the first case, ρ is the mass density in the second case, and h is the height of the free surface above the bottom in the third case. The phase velocity of propagation of small perturbations, c, is, in the three cases:

$$c^2 = \frac{A}{\rho} \frac{d(p - p_e)}{dA};$$ (5)

$$c^2 = \left(\frac{dp}{d\rho}\right) \quad \text{at constant entropy;}$$ (6)

$$c^2 = \frac{h}{\rho} \frac{dp}{dh} = gh;$$ (7)

where g is the gravitational acceleration. Thus the analog is seen. Those readers who are familiar with gas dynamics may recall the shock waves, the supersonic wind tunnel, the Laval nozzle for steam turbine, the convergent section to accelerate the fluid in subsonic regime, the sonic throat, and the divergent section to accelerate the fluid in supersonic regime. Those familiar with the open channel flow may recall the flow over a dam, and the hydraulic jump. One could anticipate the existence of analogous phenomena in blood flow in collapsible vessels. One anticipates also, of course, that similar phenomena occur in air flow in the airways, Korotkov sound in arteries, urine flow in urethra, etc.

For a flow from a reservoir with a fixed total pressure head p_0 into a collapsible tube, the flow rate depends on the suction pressure p downstream and the pressure outside the tube p_e. The rate of change of flow with respect to the suction pressure p is given by the equation (to be derived below):

$$\frac{d\dot{Q}}{d(p - p_e)} = \frac{A}{\rho u}\left(\frac{u^2}{c^2} - 1\right).$$ (8)

Here $\dot{Q}$ is the flow rate, p is the internal pressure, u is the mean speed of flow, and c is the speed of the flexural wave. Note that $d\dot{Q}/dp$ depends on the ratio u/c. If the flow speed u is smaller than c, then decreasing internal pressure increases the flow. If u is larger than c, then the reverse is true. Thus the condition $u = c$ signifies the maximum flow obtainable with decreasing internal pressure. This maximum is $Q_{max} = Ac$. At this condition, the maximum flow depends neither on the upstream pressure, nor on the downstream pressure. It is an exact analog of the sonic throat of the supersonic wind tunnel. The ratio

$$S = \frac{u}{c}$$ (9)

is called the *speed index* of Shapiro (1977). It plays a central role in liquid flow through a collapsible tube as the Mach number does in gas dynamics.

Thus the condition $u = c$ signifies a flow limitation. The upstream and downstream pressures matter only to the extent of getting this condition established, just as a supersonic wind tunnel has to have suitable conditions to get it started.

The derivation of the intriguing Eq. (8) is as follows. Consider laminar flow in an elastic tube at a large Reynolds number so that Bernoulli's equation holds:

$$p + \tfrac{1}{2}\rho u^2 = p_0, \tag{10}$$

where p_0 is the stagnation pressure, p is the static pressure, ρ is fluid mass density, and u is velocity. The volume flow rate, $\dot{Q}$ is

$$\dot{Q} = Au = A\sqrt{\tfrac{2}{\rho}(p_0 - p)} = A\sqrt{\tfrac{2}{\rho}[(p_0 - p_e) - (p - p_e)]}. \tag{11}$$

A is the cross-sectional area which is a function of $p - p_e$. Although p varies with distance down the tube, $\dot{Q}$ remains constant, of course. Differentiating $\dot{Q}$ with respect to $p - p_e$, we obtain, after some reduction and using Eq. (10),

$$\frac{d\dot{Q}}{d(p - p_e)} = -A\frac{du}{dp} + \frac{dA}{d(p - p_e)}u = -\frac{A}{\rho u} + \left[\frac{\rho}{A}\frac{dA}{d(p - p_e)}\right]\frac{A}{\rho}u. \tag{12}$$

The factor in the bracket of the last term is $1/c^2$. Thus Eq. (8) is obtained.

Shapiro (1977) analyzed a number of cases in which these equations apply. Experience shows that this one-dimensional analysis is adequate to deal with the flow leading to the sonic throat. But beyond the sonic throat, the recovery of flow to the subsonic condition seems to be a three-dimensional phenomenon beyond the reach of the simplifed approach.

5.18 Pulse Wave as Message Carrier for Noninvasive Diagnosis

A subsonic flow is influenced by conditions on all of its boundary. The flow field, governed by the equations of motion and continuity and the constitutive equations, is determined by the conditions on the boundary. Anything happening anywhere on the boundary will be felt everywhere in the flow field. In fluid mechanics and the theory of partial differential equations, this is a feature of potential flow or elliptic differential equations, as distinguished from supersonic flow or hyperbolic differential equations. Now, blood flow is subsonic. Therefore, if we have the full, detailed mathematical solution of the flow field, then, in principle, by examining the flow at a given region, one should be able to tell any disturbances occurring anywhere on the boundary. Extending this concept to the diseases of the blood vessels and organs, we can anticipate that the pulse waves in a given region of an artery should carry information about stenosis, aneurysm, or atherosclerosis at distant places.

The object of studying the messages carried in arterial pulse waves is similar to the use of seismic waves to detect oil reserves underground. The mathematical problem has not been solved yet, but anecdotal, empirical information exists.

In the traditional Chinese medicine, physicians use fingers to feel the pulse waves of the radial artery on the forearm at the wrist. Through empirical information accumulated over the years, they have developed an art of diagnosis which is often marvelous but not well understood. The use of pulse waves for diagnosis was discussed extensively in one of the most ancient classics of medicine: the *Nei Jing*, i.e., the *Internal Classic of Huangti* (for Chinese references see Fung, 1984, pp. 14 and 157; and Xue and Fung, 1989) which is believed to have been written in the Warring Period (475–221 BC). In essence, the idea is that all disturbances in the function of any organ can be detected by changes in the pulse waves in the radial artery. The sensation felt by the fingers when they press on the radial artery at specified points with varying degrees of pressure is used as diagnostic criterion.

The pulse wave diagnosis method is studied intensively in China clinically, experimentally, and theoretically. In older literature the waves are treated as axisymmetric motion in circular cylindrical elastic tubes. Recent literature has included articles treating non-axisymmetric motion, including lateral oscillation of the centerline of the blood vessel. Dai et al. (1985) tested the hypothesis that a disturbance of blood flow at one place can be detected in the arterial pulse wave at a distant site. They transiently occluded blood flow in a leg and recorded the pulse waves in both radial arteries. They asked whether the right and left radial arterial waves can differentiate a disturbance in the right leg from that in the left leg. The results show that the right and left radial arterial waves do respond to the disturbances in the right and left legs differently, but the discrimination is not very strong. Xue and Fung (1989) tried to explain it on the basis of fluid mechanics. They created an unsymmetrical entry condition by blocking off one-half of the entry section of a circular cylindrical tube. As the distance from the entry section increases, the flow tends to become axisymmetric, but there is an asymmetric component which persists in propagating downstream with slowly damped amplitude. This suggests that the asymmetric flow condition from the legs may reach the arms, but whether the suggestion is quantitatively meaningful or not is entirely unknown. This remains a fascinating problem.

Problems

5.1 An energy balance equation for blood flow is desired. Consider all the arteries between two planes, for example, one plane cutting a renal artery, the other plane cut through the kidney supplied by that renal artery. Identify the rate of gain of energy of the blood in these arteries (the sum of the kinetic, potential, and internal energies) and the rate at which work is done on the blood in this system. The energy balance requires that the rate of gain of energy must be equal to the sum of the rate of work done on the system and the heat transported in. Express this energy equation in terms of pressure and velocities in the system. Cf. Fung (1984) pp. 15–20.

5.2 One of the great achievements of man in the twentieth century is the mechanical heart. What is the present status of the art in this field? What do you think needs to be done in order to make this device really available to more people at an affordable price?

5.3 What effect does a stenosis in a large artery have? To study the effect, laser-doppler velocimeter may be used. Describe the principle of this instrument. Can it be used for an unsteady flow?

5.4 Describe the theoretical criterion for the velocity distribution in the boundary layer when boundary separation from a solid wall occurs. (Cf. Yih (1977) pp. 352, 360.)

5.5 Consider the pulsatile flow in the aortic arch, part of which is highly curved as a torus. Because of the curvature of the vessel, secondary flow exists and the boundary layer thickness is a function of time and space. For the consideration of atherogenesis, we need to know the shear stress on the arterial wall. Give a qualitative discussion on the nature of variation of the flow and shear stress in the aortic arch. (Cf. Jayaraman, G., Singh, M.P., Padmanabhan, N. and Kumar, A. (1984). "Reversing flow in the aorta: a theoretical model", *J. Biomechanics*, **17**: 479–490.)

5.6 An aorta has an aneurysm, which is a sac formed by the dilatation of the wall of the artery. From the point of view of fluid mechanics, discuss the pulsatile velocity distribution and pressure in the aneurysm and its contiguous parts, and the possible sound emission (bruit, aneurysmal bruit). From the point of view of solid mechanics, discuss the stress distribution in the vessel wall. From the general biological relation between stress and growth or resorption, discuss possible reasons for the creation of the aneurysm and possible direction of its development. (Cf. Chapters 10–13 infra.)

5.7 Discuss the stress distribution in the endothelium, the intima, and the adventitia in the region of arterial bifurcation. Delve into further detail, considering the stresses acting in the endothelial cells, smooth muscle cells, collagen fibers of various kinds, elastin fibers, fibronectin, and ground substances in the vessel wall. Again, precise data are lacking. Develop a research proposal to clarify this problem. Again, cf. Chapters 10–13, and biological points of view as mentioned in Prob. 5.6.

5.8 Looking at the stenosis problem of 5.3 from the point of view of solid mechanics and biology as mentioned in Prob. 5.6, discuss the possible remodeling of the blood vessel wall when a stenosis develops.

5.9 The place where an artery branches off from the aorta is often the site of athero-sclerosis. Discuss qualitatively the velocity distribution, fluid pressure, and wall shear stress on the endothelium in this region. Develop a plan of research to gain a better understanding of these features.

References

Bellhouse, B.J. and Bellhouse, F.H. (1969). Fluid mechanics of model normal and stenosed aortic valves. *Circ. Res.* **25**: 693–704.

Bellhouse, B.J. and Bellhouse, F.H. (1972). Fluid mechanics of a model mitral valve and left ventricle. *Cardiovasc. Res.* **6**: 199–210.

Dai, K., Xue, H., Dou, R., and Fung, Y.C. (1985). On the detection of messages carried in arterial pulse waves. *ASME J. Biomed. Eng.* Vol. 107, pp. 268–273.

Euler, L. (1775). Principia pro motu sanguins per arterias determinado. *Opera posthuma mathematica et physica*. Petropoli, Vol 2, pp. 814–823.

Flaherty, J.E., Keller, J.B., and Rubinow, S.I. (1972). Post buckling behavior of elastic tubes and rings with opposite sides in contact. *SIAM J. Appl. Math.* **23**(4): 446–455.

Folkow, B. and Neil, E. (1971). *Circulation*, Oxford Univ. Press, New York.

Frank, O. (1899). Die grundform des arteriellen pulses. Erste Abhandlung, *Mathematische Analyse. Z. Biol.* **37**: 483–526.

Fung, Y.C. (1981). *Biomechanics: Mechanical Properties of Living Tissues*. Springer-Verlag, New York.

Fung, Y.C. (1984). *Biodynamics: Circulation*. Springer-Verlag. New York.

Fung, Y.C. and Sobin, S.S. (1972). Pulmonary alveolar blood flow. *Circ. Res.* **30**: 470–490.

Fung, Y.C., Sobin, S.S., Tremer, H., Yen, M.R.T., and Ho, H.H. (1983). Patency and compliance of pulmonary veins when airway pressure exceeds blood pressure. *J. Appl. Physiol.* **54**: 1538–1549.

Hales, S. (1733). *Statistical Essays. II. Haemostaticks*. Innays and Manby, London, Reprinted by Haffner, New York. p. 23.

Lee, C.S.F. and Talbot, L. (1979). A fluid mechanical study on the closure of heart valves. *J. Fluid Mech.* **91**: 41–63.

McDonald, D.A. (1974). *Blood Flow in Arteries*. Williams and Wilkins, Baltimore, MD.

Moreno, A.H., Katz, A.I., Gold, L.D., and Reddy, R.V. (1970). Mechanics of distension of dog veins and other very thin-walled tubular structures. *Circ. Res.* **27**: 1069–1079.

Nerem, R.M., Seed, W.A., and Wood, N.B. (1972). An experimental study of the velocity distribution and transition to turbulence in the aorta. *J. Fluid Mech.* **52**: 137–160.

Patel, D.J. and Vaishnav, R.N. (eds.) (1980). *Basic Hemodynamics and Its Role in Disease Process*, University Park Press, Baltimore, MD.

Pedley, T.J. (1980). *The Fluid Mechanics of Large Blood Vessels*. Cambridge University Press, London.

Shapiro, A.H. (1977). Steady flow in collapsible tubes, *J. Biomech. Eng.* **99**: 126–147. The American Society of Mechanical Engineers, New York.

Sobin, S.S., Fung, Y.C., Tremer, H., and Rosenquist, T.H. (1972). Elasticity of the pulmonary interalveolar microvascular sheet in the cat. *Circ. Res.* **30**: 440–450.

Winter, D.C., and Nerem, R.M. (1984). Turbulence in pulsatile flows. *Ann. Biomed. Eng.* **12**: 357–369.

Womersley, J.R. (1957). The mathematical analysis of the arterial circulation in a state of oscillatory motion. Wright Air Development Center, Technical Report WADC-TR-56-614. 1–123.

Xue, H. and Fung, Y.C. (1989). Persistence of asymmetry in nonaxisymmetric entry flow in a circular cylindrical tube and its relevance to arterial pulse wave diagnosis. *J. Biomech. Eng.* Vol. 111, pp. 37–41.

Yen, R.T. and Foppiano, L. (1981). Elasticity of small pulmonary veins in the cat. *J. Biomech. Eng.* **103**: 38–42.

Yen, R.T., Fung, Y.C., and Bingham, N. (1980). Elasticity of small pulmonary arteries in the cat. *J. Biomech. Eng.* **102**: 170–177.

Yih, C.S. (1977). *Fluid Mechanics*. West River Press, Ann Arbor, MI.

Young, T. (1808). Hydraulic investigations, subservient to an intended Croonian lecture on the motion of the blood. *Phil. Trans. Roy. Soc. London*, **98**: 164–186.

Young, T. (1809). On the function of the heart and arteries. *Phil. Trans. Roy. Sec. London*, **99**: 1–31.

CHAPTER 6

Micro- and Macrocirculation

6.1 Introduction

In physiology, capillary blood flow is identified with microcirculation. Flow in small blood vessels supplying and draining the capillaries, the arterioles and venules, respectively, are also included in microcirculation, but the question of how many orders are to be included in microcirculation is sometimes debated, because different organs seem to demand different answers. From fluid mechanical point of view, the distinction between micro and macro circulation can be based on the *Reynolds number, VD/v*, and *Womersley number, $(D/2)\sqrt{\omega/v}$* (Sec. 5.16), where V represents the mean velocity of flow in the vessel, D is the vessel diameter, v is the kinematic viscosity of the blood, ω is the circular frequency of oscillation of the blood velocity fluctuations. If the Reynolds number and Womersley numbers are both much smaller than 1, then the inertial force can be ignored, and the flow is said to be microcirculation. If both numbers are much greater than 1, then the fluid viscosity can be ignored, and the flow is said to be macrocirculation. In between these limits the fluid mechanical equations are harder to solve, and it is immaterial whether you classify them as micro or macro circulation.

Some of the general features of microcirculation are the following: The blood has a lower hematocrit (i.e., the volume fraction of blood cells in whole blood) than that in the heart. The apparent viscosity of blood decreases because of the decreasing concentration of red cells. The chance for activated white blood cells to stick to the blood vessel wall increases because of closer contact with the endothelium. The proportion of smooth muscle in vessel wall increases in small vessels. In systemic arterioles 70–80% of the vessel wall is smooth muscle cells. The contraction of the smooth muscle controls the vessel diameter, flow resistance, pressure gradient, and thus, eventually, the distribution of blood to various organs, as well as the systemic blood pressure.

One can get an intuitive feeling about what is going on in circulation by imagining oneself as a red blood cell. In the left ventricle, the cell is so small that the whole blood may be considered as a homogeneous fluid. A complex, unsteady flow takes place in the ventricle and aorta. The hydrodynamic force opens and closes the aortic valve periodically. The aortic diameter is about 2000 times larger than the red cell. Soon the cell gets to a bifurcation point of the channel. It is swept into one of the branches; then another bifurcation, then another. Finally, the cell gets into an arteriole, whose diameter is about 10 times larger than the red cell. After several generations of arterioles, the red cell enters a capillary blood vessel which is so narrow that its wall scrapes the red cell membrane. The cells are now flowing in single file. The plasma fluid sticks to the vessel wall and the cell membrane. There is no slip, but there is some leakback at the wall. Some fluid seeps through the endothelium. Some dissolved gas leaves the fluid and goes into the vessel wall and the tissue cells beyond. Other gases come back in reverse direction. Some bigger molecules move across the wall via tiny channels between the membranes of the endothelial cells.

After a winding trip in the capillaries, the red cell enters the venules and slows down. From venules it goes to the veins. The veins have valves which prevents backflow in unfavorable conditions. In vena cava, there may be a "sluicing gate" where the flow velocity is equal to the velocity of pulse wave. This is the choking point of the so-called "waterfall" phenomenon.

Then the red cell enters the right atrium, right ventricle, pulmonary arteries, pulmonary capillaries, pulmonary veins, left atrium, and returns to the left ventricle.

Mechanics can help clarify the understanding of these phenomena. As usual, to understand mechanics we must begin with anatomy and rheology.

6.2 Anatomy of Microvascular Beds

The microvascular beds of all organs are not the same. There are common features; there are also special features of each organ. To study an organ one must know its vasculature. Hence morphometry has been a recognized activity of bioengineers. However, it is beyond the scope of this book to describe every organ. The reader is referred to the bibliography at the end of the chapter for general references. For special organs, a search in the library is necessary; or a research project has to be undertaken.

Let us describe some common features. The smallest blood vessels are the *capillaries*. The wall of a capillary blood vessel consists of a single layer of endothelial cells lying on a basement membrane which occasionally splits to enclose the pericytes which are thought to have the potential to become smooth muscle cells. There is a large number of vesicles in the endothelial cells. These vesicles are believed to be transporters of materials.

The endothelial cells of the capillary wall are apposed to each other. In electron microscopy, there appears to be a gap of 10 to 20 nm between neighoring endothelial cell membranes. At certain points, these membranes and the adjacent cytoplasm appear darker; the intercellular clefts are sealed by tight junctions or *maculae occludens,* which are formed by close apposition or fusion of the external leaflets of plasmalemma. In certain areas (e.g., in the brain) these junctions form an uninterrupted seal, i.e., *zonulae occludens,* preventing the passage of molecules with radius of 2.5 nm or larger. These tight junctions are like spot welding (maculae) and seam welding (zonula) in industrial metal construction. They connect the endothelial cells together to form a continuous barrier and play an important role in determining the permeability of the endothelium to water and other molecules.

The appearance of the endothelial cell lining of blood vessels might be different in different organs. Generally there are three major types. In the *continuous* type, the endothelial cells are joined tightly together. In vessels of striated muscle, the cells may be quite flat and thin. In postcapillary venules, they may be cuboidal and form a thick layer. In the *fenestrated type,* the endothelial cells are so thin that at some spots the opposite surfaces of their membranes become so close together as to form small circular areas known as *diaphragms of fenestration* approximately 25 nm thick and 100 nm across. Adjacent endothelial cells are still tightly joined. This type of vessel has been described in three groups or organs: (1) endocrine gland, (2) structures engaged in the production or absorption of fluids (e.g., renal glomerulus, choroid plexus of the brain, ciliary body of the eye, intestinal villus), and (3) retia mirabilia (e.g., renal medulla, fish swim bladder).

The third type of endothelium is the *discontinuous* type in which there are distinct intercellular gaps and discontinuous basement membrane. These occur in those vessels commonly called sinusoids. They are common in organs whose primary functions are to add or extract whole cells as well as large molecules and estraneous particles, e.g., liver, spleen, and bone marrow, from the blood.

The topology of the arterioles that supply the capillaries and venules that drain them is usually special for each organ. Some are organized as trees; others are organized into arcades. Each organ is unique.

Lying next to arterioles are *lymphatic capillaries* which have a single layer of endothelium surrounded by a basement membrane, but lack smooth muscle in its walls. The walls of lymphatic capillaries are more porous than those of capillaries so that larger molecules can pass through them.

The lymphatic capillaries are blind sacs. They merge to form *collecting lymphatics* which transport fluid to the vein. Collecting lymphatics have *valves* to assure *unidirectional movement* of lymph.

The complement to the space occupied by the blood vessels, lymphatics, and cells of a tissue is called the *interstitial space* of that tissue. It is mainly connective tissue, containing collagen, elastin, hyaluronic acid, and other substances, either bathed in some kind of fluid, or embedded in a gel.

Sympathetic nerve fibers invest the aorta, large and small arteries and veins, and to a variable degree the networks of the arteriolar vessels and muscular venules in each organ. There appears to be no direct innervation of the capillary blood vessels and collecting venules, although fibers may be found in the capillary region. Sympathetic fibers are usually superimposed on the smooth muscle layer of the blood vessel wall, but do not make direct synaptic contact with the vascular smooth muscle cells.

6.3 Major Features of Microcirculation

In a typical microvascular bed of the cat mesentery, the arterial to venous distribution of intravascular pressure and velocity is shown in Fig. 6.3:1. The pressure decreases rapidly in arterioles of diameters in the range of 10–35 μm. The decline in pressure within the true capillaries and postcapillaries is much more gradual.

When the pressure profiles of normal, hypertensive, and hypotensive cats

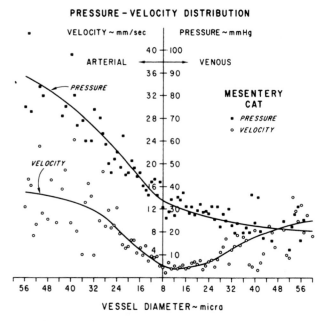

FIGURE 6.3:1 Arterial to venous distribution of intravascular pressure and velocity in the mesentery of the cat. Vessel diameter (abscissa) is taken to be representative of the functional position of each vessel in the microvascular network. Each data point represents the average value of three to five individual measurements at the abscissa (diameter) value. The solid curves are piece-wise cubic spline fits of the data and are statistically representative of the arterial to venous trends. From Zweifach and Lipowsky (1977, p. 386). Reproduced by permission of the American Heart Association, Inc.

are compared, it is found that in the hypertensive cats the pressure drop in arterioles is larger than in the normals, whereas that in the hypotensive cats is smaller. Pressures in the capillaries and postcapillary vessels are similar in the three groups, the difference between the mean pressures being only 3–5 mmHg. Thus it appears that the arterioles control the blood pressure in such a way that the capillary pressure is maintained in the normal range while the central arterial pressure can fluctuate.

In the skin, it is observed that pressures in about 10% of the venules are sometimes higher than expected (approximately 70 mmHg compared with the usual value of 30 mmHg). This is taken as evidence of the existence of *shunts* (such as *thoroughfare channels*) in the microvascular bed. These shunts are more direct low-resistance pathways connecting arterioles to venules.

Pressure and velocity pulsation are detectable in microvessels, in which the pressure oscillates with an amplitude of 1–2 mmHg normally, and 2–4 mmHg if the precapillary sphincter were dilated. There is also a random-fluctuation of a period in the order of 15–20 sec, with an amplitude of 3–5 mmHg. A third type of pressure variation is more substantial and lasts longer, in the order of 10 mmHg and 5–8 min, followed by a return to the steady-state condition in about 2–3 min. See Zweifach (1974).

It is not surprising that cardiac pulses must cause ripples in the capillaries. These waves are attenuated in the direction of propagation, reflecting the fact that in the capillaries the viscous stress dissipates the pressure fluctuations. The wave speed in the capillary is estimated to be approximately 7.2 cm/sec.

So far we have considered the pressure in the blood vessels. Of equal importance of course, is the pressure outside the blood vessels, because trans-mural pressure (i.e., internal − external pressures) is relevant in most phenomena involving blood vessels. For example, the collapse of the vein (Sec. 5.19) occurs when the external pressure exceeds internal pressure by a certain amount. The pulse wave speed depends on the Young's modulus of the blood vessel wall, the Young's modulus depends on the stress level, the stress level depends on the transmural pressure. Coming to microcirculation, the pressure outside the capillary blood vessels plays an important role in mass transport. This is because capillary walls are semipermeable: it can let water move across it, but is an impediment to ions and molecules. As a semipermeable membrane, the rate of movement of water across the capillary wall is commonly assumed to obey Starling's hypothesis (see Sec. 8.5):

$$\dot{m} = K[p_b - p_t - \sigma(\pi_b - \pi_t)] \tag{1}$$

where p stands for hydrostatic pressure, π stands for osmotic pressure, the subscript b stands for blood, t stands for tissue, $\dot{m}$ represents the volume flow rate per unit area of the membrane, K is the *permeability constant*, and σ is the *reflection coefficient*. In order to calculate the fluid transfer rate $\dot{m}$, we must know all six quantities, p_b, p_t, π_b, π_t, σ, and K; they are equally important. Since fluid movement in the tissue space is very important to health (too much or too little fluid in the tissue means edema or dehydration), the measurement of p_t and π_t has engaged attention of physiologists for a long time.

We can decrease the tissue pressure p_t in our skin by rising in a balloon. We can increase it by diving into the ocean. We can increase tissue pressure on an arm by applying a tourniquet, the pressure in abdomen by tightening abdominal muscle as in weight lifting. We can increase blood colloidal osmotic pressure by adding dextran into plasma, decrease blood colloidal osmotic pressure by adding saline. Hence p_b, p_t, π_b are variable to some extent. The regulation of p_b, p_t, π_b, π_t, and K in life is one of the central problems in physiology.

The velocity and hematocrit distributions in any capillary bed are very nonuniform and change from one instant of time to another. If you watch a microcirculatory bed in vivo, you see red cells flow and ebb.

The most unique feature of microcirculation is, of course, the prominence of red blood cells. In small blood vessels, individual red cells are so big that they may fill the vessel from wall to wall. Flowing in the capillaries, the shape of the red cells has been described as slipper, parachute, bullet, or crepe suzette. Figure 6.3:2 shows the shape of red blood cells in a capillary, computed by Zarda et al. (1977) on the basis of a constitutive equation of the red cell membrane proposed by Skalak et al. (1973) and Evans and Skalak (1979). The cell and the capillary are both assumed axisymmetric. In the unstressed state the red cell diameter is assumed to be larger than that of the tube. The values

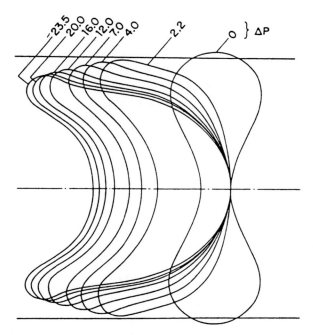

FIGURE 6.3:2 Shapes computed for red blood cells flowing axisymmetrically in capillary. The unstressed radius of the red cell is 3.91 μm, and is 5% larger than radius of the tube. From Zarda et al. (1977). Reproduced by permission.

of Δp on top of the figure is proportional to the total force applied to the red cell; it is proportional to the velocity of the cell relative to the vessel wall. The natural cross-sectional shape of a stationary red cell is shown in $\Delta p = 0$. The deformed shapes of the cell are shown for successive values of Δp. The faster the cell moves, the larger the gap between the cell and vessel wall becomes. The resistance of each cell to the flow of the blood is proportional to Δp.

6.4 The Rheological Properties of Blood

Blood is a mixture of plasma and blood cells. Plasma is a mixture of water, proteins, enzymes, and other substances. When plasma is tested in a viscometer, it is found that its coefficient of viscosity is a constant independent of the rate of strain. Hence it is a Newtonian fluid by definition, with a coefficient of viscosity about 1.2 centipoise.

Inside of the red cell is a hemoglobin solution which has a coefficient of viscosity of about 6 centipoise. Hence it is also a Newtonian fluid. Thus blood consists of one Newtonian fluid wrapped in small packages that float in another Newtonian fluid.

In large blood vessels or in viscometers whose dimensions are much larger than the diameter of the red blood cell, the mixture appears as a homogeneous fluid. The shear stress—shear strain rate relationship of the mixture can be determined by viscometry. The ratio of the shear stress to the shear strain rate is the coefficient of viscosity, see Eq. (1.7:6). Experimental results show that blood viscosity is non-Newtonian (see Fung, 1981, Ch. 3). However, as we have discussed in Sec. 5.4, in large blood vessels the non-Newtonian feature of blood is unimportant, and blood can be treated as a Newtonian fluid with a constant coefficient of viscosity without causing a significant error in the pressure-flow relationship.

In the arterioles and venules, there is a well-known *Fahreus-Lindquist effect*: the effective coefficient of viscosity decreases as the diameter of the blood vessel becomes smaller. This is explained by decreasing hematocrit in the smaller vessels. If many small blood vessels were attached to a reservoir, then the hematocrit in smaller vessels is lower. Lower the hematocrit, smaller is the viscosity.

In capillary blood vessels, the blood has to be treated as a two-phase fluid consisting of the plasma and the blood cells. The pressure-flow relationship of such a two-phase fluid in a capillary bed depends very much on the geometric characteristics of the bed. In long cylindrical capillaries such as those in muscles, retina, and mesentery, the relationship is nonlinear, especially in tightly fitting capillaries (Zarda et al., 1977). On the other hand, in the pulmonary capillaries, which are organized into sheets (see Figs. 6.6:2, 6.6:3, and 6.6:6), it has been shown that the pressure-flow relationship is linear (see Sec. 6.6). Hence an "effective" coefficient of viscosity can be determined by fitting experimental data with a theoretical formula, and this effective coefficient of viscosity is a constant (independent of velocity of flow) in the pulmonary capillaries.

6.5 General Equations Describing Flow in Microvessels

Having identified the constitutive equations of the materials, we can now write down the field equations and boundary conditions that govern a problem of interest. For example, if one is interested in the pressure–flow relationship in a capillary blood vessel, we have the Stokes equation for the plasma in the vessel:

$$0 = \frac{1}{\rho} \frac{\partial p}{\partial x_i} + v\nabla^2 u_i. \tag{1}$$

The acceleration term on the left-hand side of the equation can be set to zero because the Reynolds number and Womersley number (Sec. 5.13) are both much smaller than 1. Here u_i is the velocity vector that refers to a set of rectangular Cartesian coordinates fixed in the blood vessel. A similar equation applies to the hemoglobin solution inside the red cell. The boundary conditions for the fluids on the red cell membrane and the endothelium of the blood vessel are the no-slip condition for the velocity and continuity of normal displacements and stresses at the interfaces. The field equations for the blood vessel wall are the same as those of Sec. 1.7. The field equations for the cell membrane are the thin shell equations described in books on thin shells.

Now, if the pressure distribution and other stresses at the entry and exit sections are prescribed, then one can compute the flow in the vessel.

6.6 An Example of Analysis: Pulmonary Blood Flow

The principle of mechanics can be applied to any organ. An infinite variety of questions may be asked. To pick one for illustration, let us discuss the blood flow in the lung.

The lung is coupled to the heart. Its vascular system is illustrated in Fig. 5.2:2. Blood flows from the right ventricle to the pulmonary artery, then to pulmonary capillary blood vessels, to pulmonary veins, and finally to the left atrium of the heart. Figure 6.6:1 shows a schematic diagram of a lobule of the lung drawn by Miller (1947), p. 75. It illustrates the relation of the blood vessels to the air spaces. Note that the arteries are adjacent to bronchi, whereas the veins stand alone. Bronchial vessels and lymphatics are also shown in the figure.

The geometry of the pulmonary arterial and venous trees can be described by the *Strahler system*. In this system the capillaries are counted as vessels of order 0. The smallest arterioles are called vessels of order 1. Two order 1 vessels meet to form a larger vessel of order 2, and so on. But if an order 2 vessel meets a vessel of order 1, the order number of the combined vessel remains as 2. A similar counting is done for the venous tree. The ratio of the number

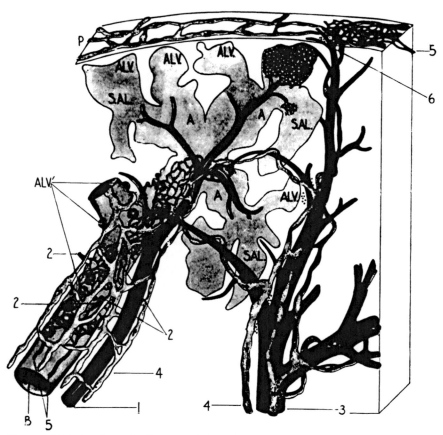

FIGURE 6.6:1 A schematic diagram of the pulmonary blood vessels, bronchial blood vessels, and lymphatic vessels in the lung and in the pleural surface. From William S. Miller, *The Lung*, 1947. Courtesy of Charles C. Thomas, Publisher, Springfield, Illinois. *B* is bronchiole, leading to two alveolar ducts one of which is shown here. *A* = atrium, ALV, ALV′ = alveoli, S.AL = alveolar sac. *P* = pulmonary pleura. 1 = pulmonary artery, dividing into capillaries. 2 = branches of pulmonary arteries distributed to bronchioles and ducts and then broken up into capillaries which unite with capillaries derived from the bronchial arteries. 3 = pulmonary vein. 4 = lymphatics, 5 = bronchial artery and capillaries. 5′ = bronchial arterial supply in the pleura (in animals with a thick pleura). 6 = pulmonary venule. 7, 8, 9, 10 are situations in which lymphoid tissue is found.

of vessels of order *n* to that of order *n* + 1 is called the *branching ratio*. The ratio of the diameter of the vessels of order *n* to that of *n* + 1 is called the *diameter ratio*. Similarly, a length ratio is defined. Yen et al. (1981, 1983) have measured the branching pattern of the pulmonary arteries and veins of the cat. The results for the arteries are presented in Table 6.6:1. Similar results for pulmonary veins are presented in Fung (1984), p. 338.

TABLE 6.6:1 Morphometric data of pulmonary arteries of the cat measured at transpulmonary pressure $p_A - p_{PL} = 10$ cm H_2O

Order	Number of branches in right lung N_n	Diameter[1] D_{on} (cm)	Length L_n (cm)	Apparent viscosity coefficient μ_n(cp)	Compliance[2] α (10^{-4} cm p_a^{-1})	Compliance[2] β ($10^{-4} p_a^{-1}$)
1	300,358	0.0024	0.0116	2.5	0.00463	1.928
2	97,519	0.0044	0.0262	3.0	0.00848	1.928
3	31,662	0.0073	0.0433	3.5	0.01407	1.928
4	9,736	0.0122	0.0810	4.0	0.02352	1.928
5	2,925	0.0192	0.151	4.0	0.02154	1.122
6	774	0.0352	0.272	4.0	0.02802	0.796
7	202	0.0533	0.460	4.0	0.03807	0.714
8	49	0.0875	0.819	4.0	0.09818	1.122
9	12	0.1519	1.426	4.0	0.4045	2.663
10	4	0.2486	1.187	4.0	0.6620	2.663
11	1	0.5080	2.500	4.0	1.353	2.663

[1] Yen et al. (1983) gives the diameter data of vessels of orders 1–4 measured at $p - p_A = -7$ cm H_2O, whereas those of orders 5–11 were measured at $p - p_{PL} = 3$ cm H_2O. In this Table, D_{on} are the diameters at zero "transmural" pressure, defined as $p - p_A = 0$ for orders 1–4, and $p - p_{PL} = 0$ for orders 5–11, and are computed from the data of Yen et al. (1980) according to the equations $D = D_o[1 + \beta(p - p_A)]$ for orders 1–4 and $D = D_o[1 + \beta(p - p_{PL})]$ for orders 5–11.

[2] Compliance was computed by fitting Eq. (1) to experimental data. Yen et al. (1980) listed data according to vessel diameters, which were interpolated to obtain data on compliance vs. vessel order. In Eq. (1), $\alpha = \beta D_{on}$. Yen et al. (1980), however, fitted data with a parabola. Their α is our β. Their β is zero in Eq. (1).

In the lung, the smallest unit of air space is the alveolus, which is bounded by networks of capillary blood vessels. The walls of each alveolus are shared by neighboring alveoli, and are called *interalveolar septa*. The overriding fact that determines the topology of the capillary blood vessels is that all pulmonary alveolar septa in adult mammalian lungs are similar. Each septum contains one single sheet of capillary blood vessels and is exposed to air on both sides.

The function of the lung is to oxygenate the blood and to remove CO_2. Nature chooses to do this by the principle of diffusion; and for this purpose, blood is spread out into very thin layers or sheets so that the blood–gas interfacial area becomes very large. In an adult human lung with a pulmonary capillary blood volume on the order of 150 ml, the pulmonary capillary blood–gas exchange area is of the order of 70 m^2, so that the average computed thickness of the sheets of blood in the pulmonary capillaries is only about 4 μm. The thin membrane that separates the blood from the air is less than 1 μm thick; it consists of a layer of endothelial cells, an interstitium, and

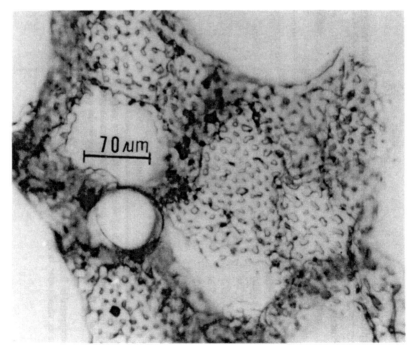

FIGURE 6.6:2 Cat lung. Flat view of interalveolar wall with the microvasculature filled with a silicone elastomer. This photomicrograph illustrates the tight mesh or network of the extensively filled capillary bed. The circular or elliptical enclosures are basement membrane stained with cresyl violet and are the nonvascular posts. Frozen section from gelatin-embedded tissue; glycerol-gelatin mount. The insert shows a detail from the region indicated by the arrow. From Fung and Sobin (1969), by permission.

a layer of epithelial cells. Each interalveolar septum is a sheet of blood. Several billion septa form a space structure which may be compared to the honeycomb (Malpighi, 1661), or a bowl of soap bubbles. Each bubble is an alveolus that is ventilated to the atmosphere through a branching airway system.

The dense network of the capillary blood vessels in an alveolar wall of the cat's lung is shown in Fig. 6.6:2. To characterize the geometry of such a network, we idealize the vascular space as a sheet of fluid flowing between two membranes held apart by a number of fairly equally spaced "posts"; see Fig. 6.6:3. In the plane view (a) this is a sheet with regularly arranged obstructions. The plane may be divided into hexagonal regions, with a circular post at the center of each hexagon. The "sheet–flow" model is therefore characterized by three parameters: L, the length of each side of the hexagon; h, the average height of thickness of the sheet; ε, the diameter of the posts.

We shall define a *"vascular-space-tissue ratio"* (VSTR) as the ratio of the vascular lumen volume to a circumscribing volume between surfaces T and B. VSTR represents the fraction of the blood volume over a sum of the volumes of the vascular space and the posts. Morphometric data of cat, dog, and human lungs show an average VSTR of 0.91 ± 0.02. Hence in the interalveolar septa, 91% of the space is occupied by blood.

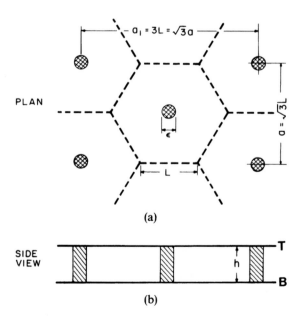

(a)

(b)

FIGURE 6.6:3 Sheet-flow model. (a) Plane view. (b) Cross section through X-X of (a). The cylindrical elements (circular in (a) and rectangular in (b)) are the "post." The space between the top and bottom walls is the flow channel. For bounding surfaces only, T, top, and B, bottom. Sheet thickness, h. C and D, contact of posts with endothelial surface at T and B. From Fung and Sobin (1969). Reprinted by permission of the American Heart Association, Inc.

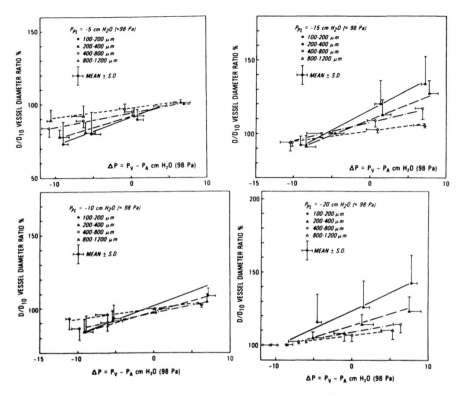

FIGURE 6.6:4 Percentage change in diameter of pulmonary veins of the cat as a function of the blood pressure, p_v. The airway pressure, p_A, is zero. The values of the pleural pressure, p_{PL}, is listed in the figure. The vessel diameter is normalized against the nominal diameter when $p_v - p_A$ is 10 cm H_2O at which the vessel cross section is circular. From Yen and Foppiano (1981). Reproduced by permission.

The distensibility of pulmonary blood vessels has been measured. A typical example is shown in Fig. 6.6:4. The empirical formula

$$D = D_0 + \alpha(p - p_A)$$
$$= D_0[1 + \beta(p - p_A)] \tag{1}$$

applies to pulmonary arteries and veins quite well. Here D is the lumen (inner) diameter of the vessel, p is the blood pressure, p_A is the pressure of alveolar gas, D_0 is the lumen diameter when $p = p_A$, and α is the *compliance constant*. In the cat's lung, this formula applies for Δp in the range of -10 to $+20$ cm H_2O ($\sim -1{,}000$ to $2{,}000$ N/m²). It is valid even for pulmonary veins subjected to negative transmural pressure. See Fig. 6.6:4. In Sec. 5.15 we have explained why the pulmonary veins will not collapse when they are subject to negative transmural pressure because of the support they receive from the tension in alveolar septa which are externally attached to them. The same tethering by interalveolar septa explains the linearity of the pressure–diameter curve: because the vessels are embedded in a linearly elastic cushioning material. Table 6.6:1 gives the values of α and β for the cat's pulmonary arteries. Similar data for the veins and the variation of the compliance constant with the transpulmonary pressure (= airway pressure minus pleural pressure) are given in Fung (1984), p. 339.

The elasticity of the pulmonary capillary sheet is exhibited in Fig. 6.6:5. The thickness h varies with the pressure difference Δp (equal to the static pressure of blood minus the pressure of the alveolar air) as follows,

$h = 0$ if Δp is negative, and smaller than $-\varepsilon$, where ε is a small number about 1 cm H_2O. $\quad(2)$

$h = h_0 + \alpha\Delta p$, if Δp is positive and smaller than a certain limiting value, p_u. $\quad(3)$

h tends to a limiting value h_∞ if $\Delta p > p_u$, and increases beyond the limiting value. $\quad(4)$

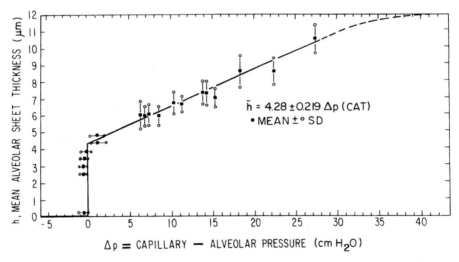

FIGURE 6.6:5 A. Sheet thickness-pressure relationship of the cat. Experimental data can be approximated by a discontinuous curve that is composed of four line segments: a horizontal line $h = 0$ for Δp negative, which jumps to $h = h_0$ at Δp slightly greater than 0, then continues as a straight line for positive Δp until some upper limit is reached, beyond which it bends down and tends to a constant thickness. B, The elastic deformation of the alveolar sheet is sketched for three conditions: $\Delta p < 0$, $\Delta p > 0$. The relaxed thickness h_0 of the sheet is equal to the relaxed length of the posts. Under a positive internal (transmural) pressure, the thickness of the sheet increases; the posts are lengthened; the membranes deflect from the planes connecting the ends of the posts; and the mean thickness becomes h. From Fung and Sobin (1972). Reproduced by permission of the American Heart Association, Inc.

TABLE 6.6:2 The compliance constant, α, and the thickness at zero transmural pressure, h_0, of the pulmonary alveolar vascular sheet at specified values of transpulmonary pressure, $p_A - p_{PL}$ (p_A = alveolar gas pressure, p_{PL} = pleural pressure)

Animal	h_0 (μm)	α ($\mu m/cm\ H_2O$)	Reference	$p_A - p_{PL}$ (cm H_2O)
Cat	4.28	0.219	Fig. 6.5:1	10
Dog	2.5	0.122	Fung, Sobin (1972)	10
	2.5	0.079	Fung, Sobin (1972)	25
Man	3.5	0.127	unpublished	10

In the small range $-\varepsilon < \Delta p < 0$, h increases from 0 to h_0. A rough approximation is $h = h_0 + (h_0/\varepsilon)\Delta p$. (5)

Here α, h_0, h_∞ and ε are constants independent of Δp. The parameter h_0 is the sheet thickness at zero pressure difference when the pressure decreases from positive values. The parameter α is called the *compliance coefficient* of the pulmonary capillary bed. The thickness h is understood to be the mean value averaged over an area that is large compared with the posts, but small relative to the alveoli. The known values of h_0 and α are given in Table 6.6:2.

Furthermore, if A_0 represents the area of a certain region of an alveolar septa when the static pressure of the blood is some physiologic value p_0, then the area of the same region A when blood pressure is changed to p is

$$A \doteq A_0. \tag{6}$$

In other words, the area is unaffected by the blood pressure.

Now, turn attention to the blood. In pulmonary capillaries, red blood cells flow in single file just as they flow in systemic capillaries. The cells deform in response to fluid pressure and shear stress as a consequence of confinement of the vessel wall and crowding of the cells. In the investigation of overall pressure-flow relationship of the lung, however, the details of cell deformation are of little interest. Only the relationship between the average pressure gradient and average velocity is of concern. These quantities can be averaged over the cross-sectional area of the blood vessels, and over a small area of the interalveolar septum. If a local frame of reference with coordinates x, y is attached to an interalveolar septum, Fig. 6.6:6, then we are interested in the gradients of the blood pressure, grad p, with components $\partial p/\partial x$, $\partial p/\partial y$; and the velocity of flow U, with components U_x, U_y. A dimensional analysis would enable us to write

$$\text{pressure gradient} = -\frac{U}{h^2}\mu kf, \tag{7}$$

where U is the mean velocity of flow, h is the thickness of blood sheet (lumen of the pulmonary alveolar septum), μ is the *apparent viscosity*, k is a dimensionless function of the sheet width to sheet thickness ratio, around 12 by value,

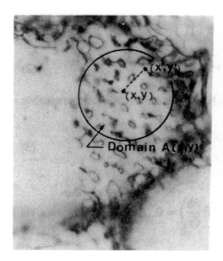 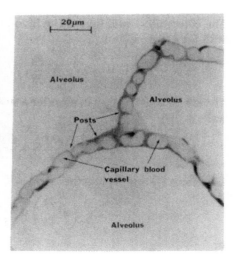

FIGURE 6.6:6 Left: A plan view of an interalveolar septum of cat lung. Right: A cross-sectional view of three interalveolar septa. In the left panel, a domain of averaging around a point (x, y) in an alveolar sheet is shown. From Fung and Sobin (1969). Reprinted by permission.

f is a dimensionless function of the sheet geometry, the ratio of the sheet thickness and the post diameter, the ratio of the interpostal distance to post diameter, and the VSTR (vascular space to tissue space ratio) defined earlier. The value of the geometric factor f lies between 1.5 to 5.0, see Lee (1969), Yen and Fung (1973). The apparent viscosity μ is a function of the coefficient of viscosity of the blood plasma μ_0, the hematocrit H (ratio of blood cell volume to whole blood volume), the cell diameter D_c, the Young's modulus of the cell membrane E_c, and the sheet thickness h. An expression for the apparent viscosity derived from a dimensional analysis is

$$\mu = \mu_0 F'\left(\frac{D_c}{h}, \frac{\mu_0 U}{E_c h}, H\right). \tag{8}$$

Details of F' are given by Yen and Fung (1973), and Fung (1981).

6.7 Pulmonary Capillary Blood Flow

Consider the local pressure gradient and local velocity defined by averaging over a small volume of an interalveolar septum, as discussed in the preceding section. The constitutive equation of the blood vessel is given by Eq. 6.6:2 in the range $0 < p - p_A <$ an upper limit:

$$h = h_0 + \alpha(p - p_A). \tag{1}$$

The constitutive equation of the blood is given by Eq. (6.6:6). Combining it with Eq. (1), we obtain

$$U = -\frac{1}{kf\mu\alpha}h^2\frac{\partial h}{\partial x}, \qquad V = -\frac{1}{kf\mu\alpha}h^2\frac{\partial h}{\partial y}. \tag{2}$$

The law of conservation of mass for a steady flow in a sheet with impermeable walls is

$$\frac{\partial(\rho h U)}{\partial x} + \frac{\partial(\rho h V)}{\partial y} = 0. \tag{3}$$

Equations (2) and (3) together yield

$$\frac{\partial}{\partial x}\left(h^3\frac{\partial h}{\partial x}\right) + \frac{\partial}{\partial y}\left(h^3\frac{\partial h}{\partial y}\right) = 0, \tag{4}$$

or

$$\left(\frac{\partial^2}{\partial x^2} + \frac{\partial^2}{\partial y^2}\right)h^4 = 0. \tag{5}$$

Thus the fourth power of h is governed by a Laplace equation. Expressed in terms of pressure, we have, on defining

$$\Phi = h^4 = [h_0 + \alpha(p - p_A)]^4, \tag{6}$$

the result

$$\left(\frac{\partial^2}{\partial x^2} + \frac{\partial^2}{\partial y^2}\right)\Phi = 0. \tag{7}$$

Solution of the Simplest Case: h Independent of y

If h is a function of x only, then Eq. (5) becomes

$$\frac{d^2 h^4}{dx^2} = 0. \tag{8}$$

The general solution is

$$h^4 = c_1 x + c_2, \tag{9}$$

where c_1, c_2 are arbitrary constants. To determine c_1, c_2, we notice the boundary conditions: (a) at the "arteriole," $x = 0$, the thickness of the "sheet" is h_a; (b) at the "venule," $x = L$, the thickness of the "sheet" is h_v. Hence,

$$h_a^4 = c_2, \qquad h_v^4 = c_1 L + h_a^4. \tag{10}$$

Solving for c_1, c_2 and substituting into Eq. (9), we obtain

$$h^4 = h_a^4 - (h_a^4 - h_v^4)x/L, \tag{11}$$

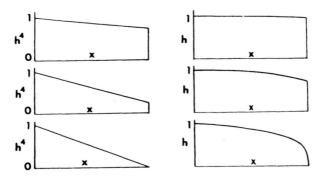

FIGURE 6.7:1 Plot of Eq. (12) showing the variation of h^4 and h with x. With an appropriate choice of units, we assume that the thickness h_a is 1. h_v^4 at $x = L$ is assumed to be 0.75, 0.25, and 0 for the three cases shown in the figure. The corresponding values of h_v are 0.931, 0.707, and 0, respectively.

or

$$h = [h_a^4 - (h_a^4 - h_v^4)x/L]^{1/4}. \tag{12}$$

For various combinations of h_a and h_v, the distribution of the thickness h is shown in Fig. 6.7:1. It is seen that the exponent $\frac{1}{4}$ makes h rather flat near the arteriole, and constricts rather rapidly toward the venule if h_v is small.

From Eqs. (1) and (2), we obtain the flow per unit width:

$$\dot{Q} = hU = -\frac{1}{4kf\mu\alpha} \frac{\partial h^4}{\partial x}. \tag{13}$$

But $\dot{Q}$ is a constant on account of conservation of mass. Hence $\partial h^4/\partial x$ is a constant, and by Eq. (11) we can write (13) as

$$\dot{Q} = \frac{1}{4kf\mu\alpha L}(h_a^4 - h_v^4). \tag{14}$$

This is the flow per unit width in a rectangular sheet. For a rectangular sheet of length L in the streamwise direction, the total width is equal to the projected area divided by L. Hence the total flow in the sheet is equal to the product of $\dot{Q}$ and sheet area divided by L.

Flow in an Interalveolar Septum

The velocity field in an interalveolar septum can be described by streamlines. The flow in a streamtube (a tube whose wall is composed of streamlines) is one-dimensional to which the analysis above applies with x identified as the coordinate along the axis of the streamtube. Thus, the flow in each streamtube is equal to the product of $\dot{Q}$ given by Eq. (14) and the ratio (projected area/length L).

The flow in a whole septum can be obtained by summing up the flow in all the streamtubes. Let A be the area of the septum and S be the vascular-space-tissue ratio; then the total area of the vascular space is SA, and the total flow in the septum can be written as

$$\text{Flow} = \frac{SA}{4\mu kf\overline{L}^2\alpha}[(h_0 + \alpha\Delta p_{\text{art}})^4 - (h_0 + \alpha\Delta p_{\text{ven}})^4], \tag{15}$$

where $\overline{L}$ is an average length of the stream tubes defined by the relation

$$\frac{SA}{\overline{L}^2} = \sum \frac{\text{area of streamtube}}{(\text{length of streamtube})^2} \tag{16}$$

in which the summation covers all individual streamtubes of the sheet whose area is SA. Expressed in terms of sheet thickness, Eq. (15) is

$$\text{Flow} = \frac{1}{C}[h_{\text{art}}^4 - h_{\text{ven}}^4], \tag{17}$$

where

$$C = \frac{4\mu kf\overline{L}^2\alpha}{SA} \tag{18}$$

$$h_{\text{art}} = h_0 + \alpha(p_{\text{art}} - p_{\text{alv}}) \tag{19}$$

$$h_{\text{ven}} = h_0 + \alpha(p_{\text{ven}} - p_{\text{alv}}). \tag{20}$$

Effect of Gravitation and Anatomy. Regional Differences in the Lung

Equation (15) shows that the total flow in a group of alveoli supplied by an arteriole and drained by one or more venule is controlled by the blood pressure at the arteriole and venule, p_{art} and p_{ven}, respectively. Now, p_{art} is equal to the blood pressure at the pulmonary valve in the pulmonary artery, minus the integrated product of resistance times flow from the pulmonary valve to the arteriole, and minus the product of the specific weight of blood times the height of that arteriole above the pulmonary valve. Similarly, p_{ven} is the difference of three terms: the left atrium pressure, minus the pressure loss due to flow, minus the product of specific weight and the height of that venule above the left atrium. Specific weight is the product of gravitational acceleration and the density of the fluid. Hence, p_{art}, p_{ven}, and the flow are influenced by the height which determines hydrostatic pressure, and the anatomy which determines the pressure loss due to resistance to flow.

Because of the height and anatomical differences, regional differences exist in the lung. For alveolar blood flow, it is convenient to define three zones as follows:

$$\text{Zone 1:} \quad p_{\text{ven}} - p_{\text{alv}} < p_{\text{art}} - p_{\text{alv}} < 0 \tag{21}$$

$$\text{Zone 2:} \quad p_{\text{ven}} - p_{\text{alv}} < 0 < p_{\text{art}} - p_{\text{alv}} \tag{22}$$

$$\text{Zone 3:} \quad 0 < p_{\text{ven}} - p_{\text{alv}} < p_{\text{art}} - p_{\text{alv}}. \tag{23}$$

In Zone 1, the capillaries are collapsed; there will be very little flow, if any.

In Zone 3, Eq. (15) applies when the blood pressure is modestly high so that the pressure-thickness relationship of the interalveolar septa is linear.

If the blood pressure in Zone 3 is so high that the thickness tends to a constant, [Eq. (6.6:3)], then the compliance α decreases as shown in Fig. 6.6:5. Then, since the compliance is very low, the thicknesses at the arteriole and venule are almost equal. Hence Eq. (17), which can be rewritten in the following form,

$$\text{Flow} = \frac{1}{C}(h_{art} - h_{ven})(h_{art}^3 + h_{art}^2 h_{ven} + h_{art} h_{ven}^2 + h_{ven}^3), \tag{24}$$

can be simplified. First, the four terms in the last parenthesis are almost equal. Then, using Eqs. (18)–(20), we can reduce Eq. (24) into

$$\text{Flow} \doteq \frac{SA}{\mu k f \overline{L}^2}(p_{art} - p_{ven})h_{art}^3. \tag{25}$$

This formula gives the flow in alveoli located low in zone 3, where the hydrostatic pressure is so high that interalveolar septa lose their compliance.

The flow in zone 2 is discussed in the following sections.

6.8 Waterfall Phenomenon in Zone 2

The inequalities

$$p_{ven} < p_{alv} < p_{art} \tag{1}$$

defines the zone 2 condition. In a standing man, a region sufficiently high above the heart is zone 2. In this zone, the capillary sheets tend to collapse at the venous end. Some sheets will be collapsed, others will remain open. In the open sheets the thickness at the venule will be small. If h_{ven} is smaller than h_{art}, then the last term in Eq. (6.7:17) is negligible and it becomes

$$\text{Flow} \doteq \frac{1}{C}h_{art}^4 \doteq \frac{1}{C}[h_0 + \alpha(p_{art} - p_{alv})]^4. \tag{2}$$

The flow is then independent of the downstream condition, in analogy with a waterfall.

The waterfall phenomenon in the lung was discovered by Banister and Torrance (1960) and Permutt et al. (1962). In earlier literature there was a debate as to whereabout on the vascular tree the "waterfalls" or "sluicing gates" are located. At that time the prevailing concepts were that capillaries were rigid, veins were collapsible, and arterioles were muscular and vasoactive. Hence the sluicing gates were searched for in veins and arterioles. Fung and Sobin (1972), however, showed that pulmonary capillaries are collapsible, whereas Fung et al. (1983) showed that pulmonary arteries and veins (and venules) will not collapse under negative transmural pressure, at least to -23 cm H_2O. Hence the sluicing gates must be located at the venular ends of the capillaries, and Eq. (2) holds.

Fung and Yen (1984) pointed out that one of the solutions of Eq. (6.7:5) is

$$h = 0 \tag{3}$$

which can occur if an area of the alveolar sheet is collapsed. Furthermore, the multifaceted alveolar structure of the lung is such that areas of collapsed alveolar sheets can be embedded in other alveolar sheets which are open and perfused. The boundary conditions for the solution (Eq. (3)) are satisfied. In the opened area there is flow, and Eq. (6.7:17) applies. The total perfused area S in the constant C of Eq. (6.7:17) must be reduced by the collapsed area. Hence to determine the flow in zone 2 condition, we must estimate the alveolar sheet area that is closed. This is done in the next section.

6.9 Open and Closed Capillary Sheets in Zone 2

The estimation of the area of the open capillary sheets is helped by several pieces of new information:

(1) The details of flow through the sluicing gate, taking into account the effect of tension in the alveolar wall and the local curvature of the membrane, as well as the Stokes flow equation have been investigated by Fung (1984) and Fung and Zhuang (1986). It is shown that as p_{ven} is decreased below p_{alv}, the flow in the gate is increased while the gate narrows further.

(2) The stability of a partially collapsed interalveolar septum illustrated in Fig. 6.9:1 (b) has been investigated by Fung and Yen (1986). The sheet shown in Fig. 6.9:1 (b) is open at the left end, closed on the right end, and partially closed in part of the sheet. Let the strain energy of the bent wall and the compressed posts be denoted by W_D, the work done by the alveolar gas and blood in creating a partially collapsed sheet be denoted by W_p, whereas the free energy of the surface be denoted by W_{CB}, (the subscript CB stands for "chemical bond"). Then if a small increment of the area of contact (adhesion), δA_c, occurs, small changes in W_{CB}, W_D, W_p would occur. If the sum $\delta W_{CB} + \delta W_D + \delta W_p$ were negative, there will be a tendency for the contact area to increase. At equilibrium, we have

$$\frac{\partial}{\partial A_c}(W_{CB} + W_D + W_p) = 0 \tag{1}$$

by which the area of contact A_c can be determined. Furthermore, the collapsed area is *stable* if

$$\frac{\partial^2}{\partial A_c^2}(W_{CB} + W_D + W_p) > 0. \tag{2}$$

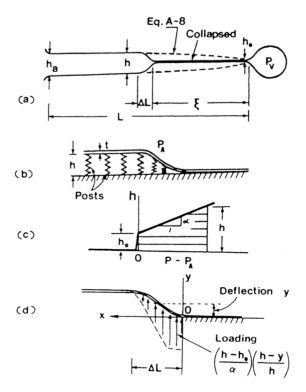

FIGURE 6.9.1 Schematic drawings for analysis of deflection pattern of collapsed inter-alveolar septa. (a) *dotted lines*, walls of sheet before collapsing, *solid lines*, walls of collapsed sheet; h is sheet thickness; E is collapsed region; L is length of sheet; ΔL is length of transition zone. Posts are drawn as springs, which balance the transmural pressure, $P - P_A$. (c) elastic characteristics of sheet described by Eq. (6.7:1). a is compliance constant. (d) out-of-balance spring forces causing deflection of wall when right-hand side of sheet is collapsed. Wall deflection measured from middle line of sheet is denoted by y. From Fung and Yen (1986), reproduced by permission.

It is *unstable* if

$$\frac{\partial^2}{\partial A_c^2}(W_{CB} + W_D + W_p) < 0. \tag{3}$$

When the details were worked out, Fung and Yen (1986) found that Eq. (3) prevails. Hence if an interalveolar septum in zone 2 starts to collapse, it will continue to collapse until the whole septum is collapsed. The result is illustrated in Fig. 6.9:2. In this figure, the left end of the collapsed septum is adjoined to two open septa. If one of the open septum started to collapse, it would continue to do so until the whole septum is collapsed, and so on.

This theoretical result is corroborated by histological evidence obtained earlier by Warrell et al. (1972), reproduced in Fig. 6.9:3. Dog's lung at zone 2

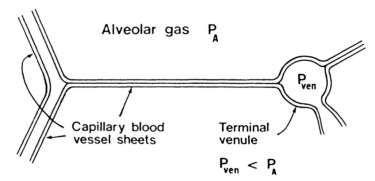

FIGURE 6.9:2 A pulmonary capillary sheet is collapsed and the collapse is arrested by two open septa at the left end. From Fung and Yen (1986), reproduced by permission.

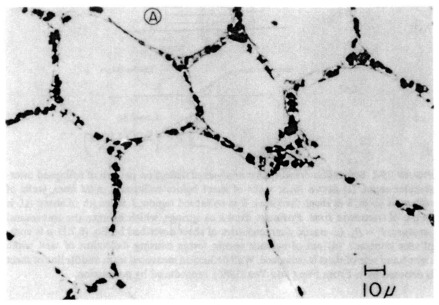

FIGURE 6.9:3 Histological micrographs of dog's lung in zone 2 condition. Specimen obtained by quick freezing, then fixed and dried at critical temperature. From Warrell et al. (1972). Reproduced by permission.

flow condition was quickly frozen by pouring liquid freon cooled to liquid nitrogen temperature. Specimens were taken within several mm below the pleura, freeze dried, and processed to preserve the geometry. In Fig. 6.9:3, it is seen that several interalveolar septa are open, several are collapsed, trapping a few red blood cells in them. Note that those septa that are collapsed did so as a whole.

Summarizing, it is found that in zone 2 condition, a number of interalveolar septa connected to the venules may be collapsed, while the remaining septa are open. Sluicing occurs in the open septa, in which the gates are located at the junctions with the venules.

A given interalveolar septum may be open or may be closed. To reopen a closed sheet requires additional energy to recreate the free surface. To make sure that every sheet is open in a lung, one should impose a large flow while the lung is in zone 3 condition. If the venous pressure is continuously reduced from zone 3 condition to zone 2 condition, more sheets will be collapsed as the venous pressure is lowered. Since the collapsing is stochastic, Fung and Yen (1986) assumed a normal probability

$$\frac{A_c}{A} = F(1 - e^{-\Delta p^2/2\sigma^2}) \tag{4}$$

to relate the collapsed area, A_c, with the pressure $\Delta p = p_{\text{ven}} - p_A$ when $\Delta p < 0$. See Figure 6.9:4. Here A is the total anatomical alveolar sheet area, F and σ are constants. This formula was validated by experimental results which will be illustrated later in Fig. 6.11:2. The physical meaning of F is the largest fraction of the alveolar sheet that will be collapsed when the pulmonary venous pressure is lowered indefinitely. This number is theoretically predictable when the relationship between the arteries, veins, alveolar ducts and alveoli is known. A theoretical model of the pulmonary alveolar duct has been proposed by Fung (1988) and validated. Fung and Yen (1986) predicted an F derivation of F based on a pentagonal dodecahedral model of the lung

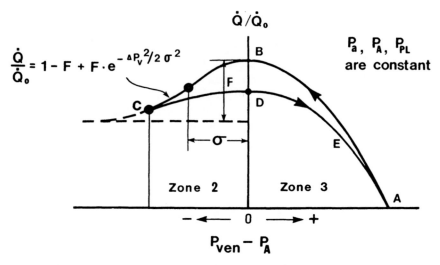

FIGURE 6.9:4 Reduction of area of perfused alveolar sheets due to collapse of capillary blood vessels and adhesion of endothelial cells in zone 2 condition is directly related to reduction of flow through lung. A mathematical expression for reduced flow and reduced alveolar area in region BC (zone 2) is given in Eq. (5). From Fung and Yen (1986), reproduced by permission.

parenchyma equal to 0.156. Experimental data yield $F = 0.104 \pm 0.016$ (SE), $\sigma = 4.45 \pm 0.45$ (SE).

Since in zone 2 blood flow exists in those capillary sheets that are *open*, we see that Eqs. (15) and (17) of Sec. 6.7 remain valid if the sheet area A in those equations is replaced by the *open* area $A - A_c$. It follows that in zone 2

$$\text{Flow} = \left(1 - \frac{A_c}{A}\right) \frac{SA}{4\mu k f L^2 \alpha} (h_{\text{art}}^4 - h_{\text{ven}}^4), \tag{5}$$

with A_c/A given in Eq. (4).

6.10 Synthesis of Micro- and Macrocirculation in the Lung

The circulation of the whole lung is the sum of flow through all of its parts. But an analysis of a circuit can be done only if one knows the circuit. Anatomic data in the form of Strahler system (Sec. 6.6) do not specify the circuit completely. It is, however, consistent with anatomical observation to assume that the vascular tree can be treated with the successive generations in series, but within each generation several possible topological arrangements can be specified. Using this simplifying assumption, Zhuang, Fung, and Yen (1983) presented a detailed analysis of flow in zone 3 condition; and Fung and Yen (1986) presented a detailed analysis of flow in zone 2 condition. Experimental validation of these results were published by Yen et al. (1984, 1986).

The theoretical procedure of synthesis begins with the capillaries, in which the pressure-flow relationship are given by Eqs. (15)–(25) of Sec. 6.7, and Eq. (5) of Sec. 6.9. Now consider the first generation of the arterioles and venules, in which the Reynolds and Womersley numbers are smaller than unity. These vessels are elastic, so the analysis presented in Sec. 5.11 applies and Eq. (5.11.6) can be used. In larger vessels with Reynolds number greater than 1, the effect of kinetic energy change on the pressure drop must be considered. The result is given in Eq. (5.11:13). In general, the needed correction for kinetic energy is quite negligible in pulmonary arteries and veins.

With these equations, we can synthesize all the segments into a circuit. At the junctions of the vessels of successive orders a finite jump of velocity occurs because of a step change in cross-sectional area. The average velocity q_n in a pulmonary artery of order n is greater than that in the next order q_{n-1} (the order number being counted from capillary upward). Hence, according to Bernoulli's equation, there is a sudden jump of pressure equal to

$$\tfrac{1}{2}\rho q_n^2 - \tfrac{1}{2}\rho q_{n-1}^2. \tag{4}$$

Similarly, at a junction of pulmonary veins of orders $n - 1$ and n, there is a corresponding sudden drop of pressure. In addition, there is the effect of entry flow and exit flow, which, however, can be taken into account by correcting the apparent coefficient of viscosity of the blood.

Calculations using detailed anatomic and rheological data following the

method just outlined were done by Zhuang et al. (1983). Some results of the
cat's lung are illustrated in the next section.

6.11 Validation of the Flow Model

Yen et al. (1984, 1986), Fung and Yen (1986) perfused cat's lung and obtained
the experimental results shown in Figs. 6.11:1 and 6.11:2. Figure 6.11:1 refers
to zone 3 condition. Figure 6.11:2 refers to zone 2 condition. In these experiments the pressures in the airway, p_A, and the largest artery, p_a, were fixed,
whereas the pressure in the left atrium, p_v, was varied, first from a higher value
downwards, reaching a minimum, then back upward. The lungs were "pre-
conditioned" by giving it a few cycles of large flow in zone 3 condition to open
up all vessels. Note that in zone 2 condition (Fig. 6.11:2), the flow reached a
peak at a certain value of p_{ven}, then decreased with further decrease of p_{ven}. A
hysteresis loop exists under cyclic change of p_{ven}.

These experimental results are compared with theoretical predictions in
Figs. 6.11:1 and 6.11:2. In the theoretical calculation, it is necessary to specify
the total area of the alveolar sheets. This area was measured from histological

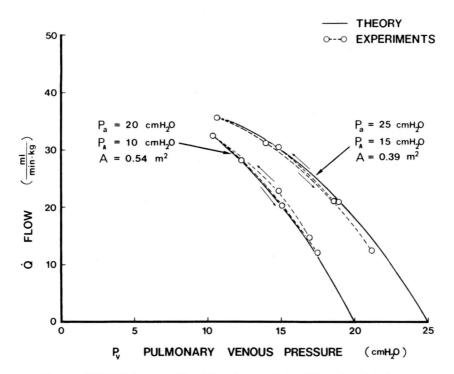

FIGURE 6.11:1 Pulmonary blood flow in zone 3 condition. Cat right lung.

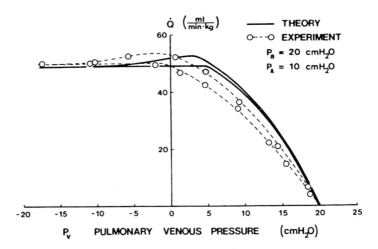

FIGURE 6.11:2 Comparison of theoretical and experimental results. Theoretical curves were based on data given in Fung (1984), and $A = 0.84$ m^2 for half lung. In zone 2 condition, Eqs. (6.9:4 and 5) were used. Curve fitting yields $\sigma = 5.3$ and $F = 0.093$. In return stroke, it is assumed that the collapsed alveolar sheets are not reopened, unless the pulmonary arterial pressure is increased under zone 3 condition. From Fung and Yen (1986), reproduced by permission.

sections by stereological methods, but the standard deviations of the measured values were large. The origin of the large standard deviation is partly due to biological variations from one animal to another and from one location in the lung to another, and partly due to finite sampling, and distorsion caused by histological preparations, especially shrinking by fixation agents. Since the available data of the total area [A in Eqs. (6.7:18) and (6.9.5)] are not very precise, we have selected a value of A that makes the theoretical prediction coincide with an experiment result at one point on the p-$\dot{Q}$ curve as noted in Figs. 6.11:1 and 6.11:2. This value of A was found to lie in the range of experimentally determined values. If A is varied, the theoretical curve would move up and down since the flow $\dot{Q}$ is directly proportional to A. Thus it is seen that the experimental trend is well predicted by the mathematical model.

In the above, the pressure-flow relationship is validated in steady flow. Of course, pulmonary blood flow is pulsatile. The features of wave propagation discussed in Ch. 5 have been shown by Bergel and Milnor (1965), Wiener et al. (1966), Milnor (1972). Waves are significant in large pulmonary arteries and veins. In pulmonary capillaries pulsatile flow is recognizable but can be treated as quasi-static. If we consider the time-averaged pressure–flow relationship, then we obtain an equation that relates the static and dynamic pressure, gravitational potential, and energy dissipation. The same equation is obtained by considering a steady flow. Hence, although steady flow does not occur in mammalian lung in life, it is still meaningful to consider it theoretically and experimentally.

In a recent book edited by Will et al. (1987), other successful approaches to the analysis of the dynamics of blood flow are reviewed. Among them are the electric circuit analog and occlusion methods by C.A. Dawson, J.H. Linehan, T.A. Bronikowski, and D.A. Rickaby, the starling resistor interpretation by W. Mitzner and I. Huang, gravity nondependent distribution by T.S. Hakim, R. Lisbona, and G.W. Dean, and models of active regulation by B. and C. Marshall. This reference serves as a convenient reference to recent literature.

Problems

6.1 *Effective diffusivity of platelets in blood with red blood cells.* The concept of molecular diffusion can be extended to fluid containing larger particles. In blood, the local fluid motion generated by individual red cell rotation will lead to greater random excursion of platelets and thus enhance its diffusivity. Consider a red cell of radius a. The angular speed ω will be proportional to the macroscopic shear rate γ. A small amount of nearby fluid will be carried along with the cell movement (due to the no-slip boundary condition at the red cell surface), leading to increased mixing, and therefore an increase of effective diffusivity. Write the effective diffusivity, D_e, as $D_e = D + D_p$, where D is the Brownian molecular diffusion coefficient and D_p is the rotation induced "diffusion" coefficient. Use dimensional analysis, show that D_p may be written as $D_p = Ca^2\dot{\gamma}$ in which C is a constant, a is the red cell radius, γ is the shear rate. (Cf. Keller, K.H. (1971), Effect of fluid shear on mass transport in flowing blood. *Fed. Proc.*, **30**, 1591.) Keller estimated that for $\dot{\gamma} = 500 \text{ sec}^{-1}$, D_p is about $10^{-5} \text{ cm}^2/\text{sec}$ in normal whole blood. This may be compared with the values of molecular diffusivity D of the following blood components:

$$O_2 \qquad D \sim 10^{-5} \text{ cm}^2/\text{sec}$$
$$\text{Protein} \qquad D \sim 10^{-7} \text{ cm}^2/\text{sec}$$
$$\text{Platelet} \qquad D \sim 10^{-9} \text{ cm}^2/\text{sec}.$$

Thus the effect of red cells is small for O_2, large for protein, very large for platelet.

6.2 Assuming a Poisseuille velocity profile and using the equation suggested in Prob. 6.1, determine the distribution of effective diffusivity of blood with platelets over the cross section of the blood vessel (i.e., as a function of the radial distance).

6.3 In a flowing bloodstream, the red blood cells and platelets are distributed differently over the cross-section of the blood vessel if the vessel diameter is much larger than the red cell diameter. Explain why? (Cf. Eckstein, E.C. (1982) Rheophoresis—broader concept of platelet dispersivity. *Biorheology* **19**: 717.)

6.4 Flow of a fluid carrying solid particles (such as coal slurry) may be an economical way of transporting the solid over a long distance. The economy depends on the rheology of the slurry. Discuss the factors that are important to the economy of transporting solids this way.

6.5 For a long time it has been controversial whether the pressure and blood flow in capillary blood vessels of most organs are significantly pulsatile or not inspite of their being driven by the heart. Modern evidence is that they are. The pulse wave velocity alters from a value in the order of 1 m/sec in large arteries to a value of the order of 1 cm/sec in microvessels. Can you ascertain this from a theoretical point of view? (Cf. Salotto, A.G., Muscarella, L.F., Melbin, J., Li, J.K.J., Noordergraaf, A.: Pressure Pulse Transmission into Vascular Beds. *Microvasc. Res.*, **32**: 152–163, 1986.)

6.6 To study coronary blood flow, a logical approach is to take steps to identify the geometry of the vascular system (branching pattern, branching number ratio, diameters, lengths, and wall thicknesses of successive generations), the rheological properties of the blood vessels and blood (constitutive equations, zero-stress state, hematocrit, blood viscosity,), the governing field equations, the boundary conditions, the mathematical solutions, and the experimental validation. Make a literature survey to find out what information is missing and what is the present status of the science of coronary blood flow. Outline a plan to improve the present status.

6.7 The study of coronary blood flow discussed in Prob. 6.6 should be extended to other vital organs such as the brain, eye, ear, kidney, liver, stomach, spleen, and bone. Make a choice and develop a plan of research.

References

Banister, J and Torrance, R.W. (1960). The effect of the tracheal pressure upon flow: Pressure relations in the vascular bed of isolated lungs. *Quart. J. Exp. Physiol.* **45**: 352–367.

Bergel, D.H. and Milnor, W.R. (1965). Pulmonary vascular impedance in the dog. *Circ. Res.* **16**: 401–415.

Branemark, P.-I. and Lindström, J. (1963). Shape of circulating blood corpuscles. *Biorheology* **1**: 139–142.

Evans, E. and Fung, Y.C. (1972). Improved measurements of the erythrocyte geometry. *Microvasc. Res.* **4**: 335–347.

Evans, E.A. and Hochmuth, R.M. (1976). Membrane viscoplastic flow. *Biophysical J.* **16**: 13–26.

Evans, E.A. and Skalak, R. (1979). Mechanics and Thermodynamics of Biomembranes. *Critical Reviews in Bioengineering*. Vol. 3, issues 3 and 4 (in 2 Vols). CRC Press, Boca Raton, Fl.

Fitzgerald, J.M. (1969). Mechanics of red-cell motion through very narrow capillaries. *Proc. Roy. Soc. London B*, **174**: 193–227.

Fung, Y.C. and Sobin S.S. (1972). Pulmonary alveolar blood flow. *Circ. Res.* **30**: 470–490.

Fung, Y.C. (1981). *Biomechanics: Mechanical Properties of Living Tissues*. Springer-Verlag, New York.

Fung, Y.C. (1984). *Biodynamics: Circulation*. Springer-Verlag, New York. In press.

Fung, Y.C. (1988). A model of the alveolar ducts of lung and its validation. *J. Applied Physiol.* **64**: 2132–2141.

Fung, Y.C. and Sobin, S.S. (1969). Theory of sheet flow in lung alveoli. *J. Applied Physiol.* **26**: 472–488.

Fung, Y.C., Sobin, S.S., Tremer, H., Yen, M.R.T., and Ho, H.H. (1983). Patency and compliance of pulmonary veins when airway pressure exceeds blood pressure. *J. Appl. Physiol. Resp. Envir. Exercise Physiol.* **54**: 1538–1549.

Fung, Y.C. and Yen, R.T. (1986). A new theory of pulmonary blood flow in zone 2 condition. *J. Appl. Physiol.* **60**: 1638–1650.

Fung, Y.C. and Zhuang, F.Y. (1986). An analysis of the sluicing gate in pulmonary blood flow. *J. Biomech. Eng.* **108**: 175–182.

Horsfield, K. and Gordon, W.I. (1981). Morphometry of pulmonary veins in main. *Lung* **159**: 211–218.

Lee, J.S. (1969). Slow viscous flow in a lung alveoli model. *J. Biomech.* **2**: 187–198.

Malpighi, M. (1661). De Pulmonibus. Letters addressed to A. Borelli. Translated by J. Young (1930), *Proc. Roy Soc. Med.* **23**: Part 1, 1–14.

Miller, W.S. (1947). *The Lung.* Thomas, Springfield, IL.

Milnor, W.R. (172). Pulmonary hemodynamics. In: *Cardiovascular Fluid Dynamics* (D.H. Bergel, ed.), Vol. 2, Academic Press, New York, Ch. 18, pp. 299–340.

Permutt, S., Bromberger-Barnea, B., and Bane, H.N. (1962). Alveolar pressure, pulmonary venous pressure, and the vascular waterfall. *Med. Thorac.* **19**: 239–260.

Schmid-Schönbein, G., Fung, Y.C., and Zweifach, B.W. (1975). Vascular endothelium-leukocyte interaction: sticking shear force in venules. *Circ. Res.* **36**: 173–184.

Skalak, R., Tozeren, A., Zarda, R.P., and Chien, S. (1973). Strain energy function of red blood cell membranes. *Biophysical J.* **13**: 245–264.

Sobin, S.S., Tremer, H.M., Fung, Y.C. (1970). Morphometric basis of the sheet-flow concept of the pulmonary alveolar microcirculation in the cat. *Circ. Res.* **26**: 397–414.

Sobin, S.S., Fung, Y.C., Tremer, H.M., and Rosenquist, T.H. (1972). Elasticity of the pulmonary alveolar microvascular sheet in the cat. *Circ. Res.* **30**: 440–450.

Warrell, D.A., Evans, J.W., Clarke, R.O., Kingaby, G.P., and West, J.B. (1972). Pattern of filling in the pulmonary capillary bed. *J. Appl. Physiol.* **32**: 346–356.

Wiener, F., Morkin, E., Skalak, R., and Fishman, A.P. (1966). Wave propagation in the pulmonary circulation. *Circ. Res.* **19**: 834–850.

Will, J.A., Dawson, C.A., Weir, E.K., and Buckner, C.K. (editors). (1987). *The Pulmonary Circulation in Health and Disease.* Academic Press, New York.

Yen, M.R.T. and Fung, Y.C. (1973). Model experiments on apparent blood viscosity and hematocrit in pulmonary alveoli. *J. Appl. Physiol.* **35**: 510–517.

Yen, M.R.T., Fung, Y.C., and Bingham, N. (1980). Elasticity of small pulmonary arteries in the cat. *J. Biomech. Eng., Trans. ASME* **102**: 170–177.

Yen, M.R.T. and Foppiano, L. (1981). Elasticity of small pulmonary veins in the cat. *J. Biomech. Eng., Trans. ASME* **103**: 38–42.

Yen, M.R.T., Zhuang, F.Y., Fung, Y.C., Ho, H.H., Tremer, H., and Sobin, S.S. (1983a). Morphometry of the cat's pulmonary venous tree. *J. Appl. Physiol. Resp. Envir. Exercise Physiol.* **55**: 236–242.

Yen, M.R.T., Zhuang, F.Y., Fung, Y.C., Ho, H.H., and Sobin, S.S. (1983b). Morphometry of the cat's pulmonary arteries. *J. Biomech. Eng.* In press.

Yen, M.R.T., Fung, Y.C., Zhuang, F.Y., and Zeng, Y.J. (1984). Comparison of theory and experiments of blood flow in cat's lung. In: *Biomechanics in China, Japan, and USA* (Y.C. Fung, E. Fukada, and J.J. Wang, eds.), Science Press, Beijing, China. pp. 240–253.

Yen, M.R.T. and Sobin, S.S. (1986). Pulmonary blood flow in the cat: correlation between theory and experiment. In: *Frontiers in Biomechanics* (G.W. Schmid-Schönbein, S.L.-Y. Woo and B.W. Zweifach, eds.), Springer-Verlag, New York. pp. 365–376.

Zarda, P.R., Chein, S., and Skalak, R. (1977) Interaction of a viscous incompressible fluid with an elastic body. In *"Computational Methods for Fluid-Structure Interaction Problems"* (Belytschko, T. and Geers, T.L., eds.) New York: American Society of Mechanical Engineers, pp. 65–82.

Zhuang, F.Y., Fung, Y.C., and Yen, M.R.T. (1983). Analysis of blood flow in cat's lung with detailed anatomical and elasticity data. *J. Appl. Physiol. Respir. Envir. Exercise Physiol.* **55**(4): 1341–1348.

Zhuang, F.Y., Yen, M.R.T., Fung, Y.C., and Sobin, S.S. (1985). How many pulmonary alveoli are supplied by a single arteriole and drained by a single venule. *Microvasc. Res.* **29**: 18–31.

Zweifach, B.W. (1974). Quantitative studies of microcirculatory structure and function. I. *Circ. Res.* **34**: 843–857, II. *Cir. Res.* **34**: 858–868.

Zweifach, B.W. and Lipowsky, H.H. (1977). Quantitative studies of microcirculatory structure and function. III. *Circ. Res.* **41**: 380–390.

CHAPTER 7

Respiratory Gas Flow

7.1 Introduction

This chapter is focused on the flow of gas into and out of the mammalian lung. We study the airway tree shown in Fig. 5.2:2. In the airway, the mixing of gases is given particular attention. In alveoli, the exchange of O_2 and CO_2 between alveolar gas and red blood cells is discussed. The effectiveness of this exchange depends on the matching of ventilation and circulation.

In Sec. 7.2 energy loss in various segments of the airway is considered. This is a major part of the cost man pays to get the gas into and out of the lung. Although a healthy person does not need to worry about this cost, a sick man may find the demand excessive.

The subject of gas exchange is treated in detail in Secs. 7.3–7.5. The effect of convection and diffusion on the gas concentration profile is discussed in Sec. 7.3. The chemical reactions taking place in the red blood cells are taken up in Sec. 7.4. The integral equation governing the distribution of ventilation/perfusion ratio in the lung is given in Sec. 7.5.

In Sec. 7.6 we discuss the pulmonary function tests that are often seen in clinics. In Sec. 7.7 we consider the analysis of the dynamics of the lung by computer simulation. Finally, in Sec. 7.8, artificial respiration, especially the high-frequency-low-tidal-volume ventilation, is discussed.

Topics such as the neural and chemical control of breathing and disease of the respiratory system, however, are omitted. The reader must consult other works for these topics, e.g. the *Handbook of Physiology* (Fenn and Rahn, eds., 1964; Fishman, Macklem, Mead, eds., 1986).

The principal variables and their notations will now be explained. A sketch of the human chest and lung is shown in Fig. 7.1:1. The lung is enveloped in a membrane—the *pulmonary* (or *visceral*) *pleura*. The internal surface of the

FIGURE 7.1:1 A sketch of human chest
and lung.

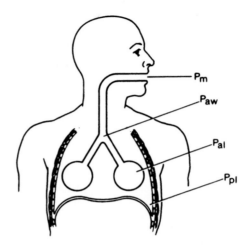

P_m

P_{aw}

P_{al}

P_{pl}

thoracic cavity is lined with another membrane—the *parietal pleura*. These
two membranes are joined together at their edges and enclose a space known
as the *pleural cavity*. Normally, the volume of the pleural cavity is small: the
spacing between the pulmonary and parietal pleura is only a few μm. This
space is filled with a fluid whose pressure is the *pleural pressure*, p_{PL}. The
pulmonary pleura is acted on one side by p_{PL}, and on the other side by the
alveolar gas pressure,* p_{ALV}, and the stress from the lung tissue σ. See the
free-body diagram in Fig. 7.1:2. The difference of the pressures, $p_{ALV} - p_{PL}$, is
responsible for inflating the pulmonary pleura as a balloon, and is called the
transpulmonary pressure. The tendency to inflate is resisted by the stresses in
pleura and *lung parenchyma*. To understand lung inflation, we must know the
stress–strain relationships of the pleura and the parenchyma (Chap. 10). On
the other hand, the movement of the gas from the mouth to the alveoli and
vice versa depends on the difference of the pressure at the mouth p_M and the
pressure in the alveoli p_{ALV}. The pressure p_M is usually atmospheric. This
leaves p_{PL} and p_{ALV} as the principal variables for ventilation. In normal
breathing, p_{PL}, p_{ALV} and the flow vary with time as illustrated in Fig. 7.1:3.

Although in Fig. 7.1:3 the pleural pressure p_{PL} is represented by a constant
at any given time, it actually varies from place to place. The spatial variation
of the intrapleural pressure can be tolerated because the intrapleural space is
very narrow, hence forming a high resistance channel for intrapleural fluid
motion. p_{PL} is generally negative (relative to the atmospheric pressure, which
is, by convention, considered to be zero) but is more so in inspiration than in
expiration. In forced expiration p_{PL} may take on large positive values.

---

* We use the standard notations p_A for airway pressure, p_a for arterial pressure, p_v for venous
pressure, p_{PL} for pleural pressure, p_{ALV} for alveolar gas pressure. Following the trend of the
literature, we use cm H_2O as a measure of pressure. 1 cm H_2O at 4°C is approximately 98 Nm^{-2}.

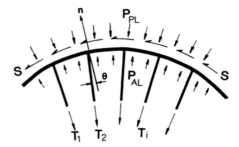

FIGURE 7.1:2 The forces acting on an element of the pleura of unit area. p_{PL} is the intrapleural pressure. p_{ALV} is the alveolar gas pressure. S is the shear stress due to relative motion of the pulmonary and parietal pleurae. T is the tensile force per unit length of the interalveolar septa. σ is the sum of the septal tension per unit area of the pleura, i.e., the normal stress in the lung tissue acting on the pleura in the direction normal to the pleura. $\sigma = (\Sigma T \cos \theta \, dL)/\text{area}$, wher dL is the length of the alveolar septa intersecting the pleura, θ is the angle between T and the normal vector of the pleura, and the product $T \cos \theta \, dL$ is summed over the intercept of the interalveolar septa on the pleura in the area named.

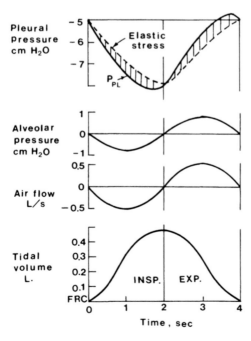

FIGURE 7.1:3 The variation of pressures in the pleura (p_{PL}) and alveoli (p_{ALV}) and flow in the airway with time in normal breathing.

p_{PL} is often measured in the esophagus which is also exposed to the pleural cavity. In making such a measurement, an elongated balloon is inserted into the esophagus. The balloon is inflated to such a degree that both the balloon and the esophagus remain buckled (collapsed) circumferentially so that the elastic resistance to inflation is negligible, and the internal and external pressures balance each other. Helium is used to inflate the balloon in order to minimize the inertial effects of the gas and improve the frequency response characteristics of the instrumentation system.

The alveolar gas pressure is also nonuniform in the lung, but normally the degree of nonuniformity is small. The alveolar gas pressure cannot be measured with transducers in life. Therefore, it is usually inferred by calculation. In normal breathing the alveolar gas pressure is practically the same as the atmospheric pressure; the difference $p_M - p_{ALV}$ does not exceed a few cm H_2O. In breath holding, or in sneezing, p_M and p_{ALV} can be very different from the atmospheric pressure.

The study of the relationship between the gas pressure p, the volume flow rate $\dot{V}$ in different parts of the airway, the distribution of p_{PL} over the pleura, the distribution of the alveolar gas pressure p_{ALV}, the partial pressures of CO_2, O_2, and N_2, and the parenchymal stress σ in various parts of the lung tissue is the main objective of respiratory mechanics.

7.2 Gas Flow in the Airway

The airway begins at the mouth and nose, extends through the larynx, trachea, bronchi, and ends at the alveoli. Detailed measurements of the geometry of the bronchial system have been made by Weibel, Cumming, and their associates. Weibel (1963) presented measurements of the length, diameter, and area of successive segments of the airway of five normal human lungs. He showed that each branch gives rise to two daughter branches, which may vary considerably in length, but less significantly in diameter. Weibel presented his data in two forms, a model A, with regularized dichotomy, and a model B, that takes into account the irregular dichotomy. In model A, Weibel took certain statistical mean values of length and diameter of each generation and presented data of the human airway which are reproduced in Table 7.2:1. The trachea is called generation 0, the main bronchi generation 1, and so on. Generations 0–16 are called *conducting airways*. Airways of generation 16 are called the *terminal bronchioles*. Those of generations 17–23 have alveoli on their walls, and constitute the so-called *respiratory zone*. Generation 23 terminates in alveoli.

The total cross-sectional area of each generation (i.e., the total area of all branches of a given generation) varies with the generation number n. When $n > 3$ the area increases with increasing n (see Table 7.2:1), and therefore increases very rapidly as the distance from the mouth increases. As a consequence, the velocity of flow decreases very rapidly toward the alveoli. The

TABLE 7.2:1 Archiecture of the human lung according to Weibel's (1963) model A, with regularized dichotomy

	Generation	Number	Diam. (mm)	Length (mm)	Cum. Length (mm)	Area (cm²)	Vol. (ml)	Cum. Vol. (ml)	At flow of 1 l/sec Speed (cm/sec)	At flow of 1 l/sec Reynolds No.
Trachea	0	1	18	120.0	120	2.6	31	31	393	4,350
Main bronchus	1	2	12.2	47.6	167	2.3	11	42	427	3,210
Lobar bronchus	2	4	8.3	19.0	186	2.2	4	46	462	2,390
	3	8	5.6	7.6	194	2.0	2	47	507	1,720
Segmental bronchus	4	16	4.5	12.7	206	2.6	3	51	392	1,110
Bronchi with cartilage in wall	5	32	3.5	10.7	217	3.1	3	54	325	690
	6	64	2.8	9.0	226	4.0	4	57	254	434
	7	128	2.3	7.6	234	5.1	4	61	188	277
	8	256	1.86	6.4	240	7.0	4	66	144	164
	9	512	1.54	5.4	246	9.6	5	71	105	99
	10	1,020	1.30	4.6	250	13	6	77	73.6	60
Terminal bronchus	11	2,050	1.09	3.9	254	19	7	85	52.3	34
Bronchioles with muscle in wall	12	4,100	0.95	3.3	257	29	10	95	34.4	20
	13	8,190	0.82	2.7	260	44	12	106	23.1	11
	14	16,400	0.74	2.3	262	70	16	123	14.1	6.5
Secondary lobule	15	32,800	0.66	2.0	264	113	22	145	8.92	3.6

Terminal bronchiole	16	65,500	0.60	1.65	266	180	30	175	5.40	2.0
resp. bronchiole	17	131×10^3	0.54	1.41	267	300	42	217	3.33	1.1
resp. bronchiole	18	262×10^3	0.50	1.17	269	534	61	278	1.94	0.57
resp. bronchiole	19	524×10^3	0.47	0.99	270	944	93	370	1.10	0.31
alveolar duct	20	1.05×10^6	0.45	0.83	271	1,600	139	510	0.60	0.17
alveolar duct	21	2.10×10^6	0.43	0.70	271	3,200	224	734	0.32	0.08
alveolar duct	22	4.19×10^6	0.41	0.59	272	5,900	350	1,085	0.18	0.04
alveolar sac	23	8.39×10^6	0.41	0.50	273	12,000	591	1,675	0.09	—
alveoli, 21 per duct		300×10^6	0.28	0.23	273		3,200	4,800		

(Rows 16–19: Terminal bronchiole / resp. bronchiole. Rows 20–23 through alveoli bracketed as **Primary lobule or acinus**.)

Notes: (1) Area = total cross sectional area, (2) cum. = cumulative, (3) Dead space, approx. 175 ml + 40 ml for mouth.

estimated order of magnitude of the velocity of flow and the Reynolds number are shown in Table 7.2:1 for a flow rate of 1,000 ml/sec.

With Strahler's method (see Sec. 6.6), Cumming and Semple (1973) obtained 17 orders of branches between the terminal bronchiole and the trachea. Taking the terminal bronchiole as order 1, they obtained a branching ratio of 2.74, and a diameter ratio of 1.37 between orders 1–17.

The pressure and velocity in the airway vary from point to point. Often the secondary flow is very strong. An approximate way to analyze the flow in such a system is to use energy equation (5.14:6). The principal variables are $\hat{p}$, the velocity-weighted average pressure, $\widehat{q^2}$, the velocity-weighted average of the square of the speed, $\dot{Q}$, the flow rate, and $\mathscr{D}$, the *dissipation function*, defined in Eqs. (5.14:7), (5.14:5), and (5.14:4) respectively. In application to the lung, $\dot{Q}$ is the gas flow rate, which will be written as $\dot{V}$ below. Of the terms in Eq. (5.14:6), the terms representing the potential energy due to the weight of the gas, $\widehat{\rho g h_2} - \widehat{\rho g h_1}$, can be neglected in the lung. The kinetic energy terms $\frac{1}{2}\rho \widehat{q_1^2} - \frac{1}{2}\rho \widehat{q_2^2}$ are important at the bifurcation points because significant velocity changes take place from one generation to another; but the last term, representing the transient change of kinetic energy within a segment, is of minor significance. Hence, in application to the airway, Eq. (5.14:6) can be simplified to

$$\widehat{p_1} - \widehat{p_2} = \frac{1}{2}\rho \widehat{q_2^2} - \frac{1}{2}\rho \widehat{q_1^2} + \frac{\mathscr{D}}{\dot{V}} \tag{1}$$

where

$$\widehat{p_1} = \frac{1}{\dot{V}} \int_A pu\,dA, \qquad \widehat{q^2} = \frac{1}{\dot{V}} \int_A q^2 u\,dA \tag{2}$$

$$\mathscr{D} = \int_V \sigma_{ij} V_{ij}\,dv, \qquad \dot{V} = \int_A u\,dA,$$

the integrals being taken over the cross sectional area A and volume of segment V. Note that $\widehat{p}$, $\widehat{q^2}$ are averages weighted with axial velocity component u. They are not equal to the average pressure and velocity $\bar{p}$ and $\bar{u}$ defined by

$$\bar{p} = \frac{1}{A} \int_A p\,dA, \qquad \bar{u} = \frac{1}{A} \int_A u\,dA. \tag{3}$$

The latter averages $\bar{p}$, $\bar{u}$ are used in Sec. 7.7 where attention is given to the airway elasticity in simulating the whole airway system. $\mathscr{D}$ is the dissipation function. In order to understand the pressure drop in the airway, major attention must be focused on the dissipation function.

The dissipation function $\mathscr{D}$ is the volume integral of the product of the coefficient of viscosity μ and the square of the shear strain rate. For a Poiseuillean flow (laminar flow) in a long circular cylindrical tube, the rate of energy dissipation $\mathscr{D}$ in a segment of length L is (see Sec. 5.10):

$$\mathscr{D}_p = 8\pi\mu\bar{u}^2 L, \tag{4}$$

where $\bar{u}$ is the mean velocity of flow, Eq. (3). (For a laminar flow, $\hat{u}^2 = 2\bar{u}^2$. See Problem 7.4 infra.) For other flows, one may represent the dissipation function as the product of $\mathscr{D}_p$ and a factor Z:

$$\mathscr{D} = Z\mathscr{D}_p \tag{5}$$

which defines Z. This Z is the same $Z(N_R)$ of Eq. (5.15:10). Examples are given below.

Entry Flow Into a Tube

Pedley et al. (1977) computed $\mathscr{D}$ for a flow entering into a circular cylindrical pipe from a reservoir on the basis of boundary layer theory, and obtained

$$Z_{\text{entry}} = \frac{\alpha_1}{16}\left(\frac{d}{L}N_R\right)^{1/2}, \qquad \text{where } N_R = \frac{\bar{u}d}{v}. \tag{6}$$

The variable d is the diameter of the pipe, v is the kinematic viscosity, $\bar{u}$ is the mean speed of flow, N_R is the Reynolds number, and α_1 is a numerical constant which is equal to 1 if the velocity profile in the boundary layer is linear and 4/3 if it is parabolic.

Turbulent Flow in a Tube

When the flow rate $\dot{V}$ exceeds a certain value, the flow becomes unstable and fluctuation develops. Further increase in flow rate leads to turbulence. In a fully developed turbulent flow, a well-known empirical formula is

$$Z_T = \frac{\mathscr{D}_{\text{turbulent}}}{\mathscr{D}_p} = 0.005N_R^{3/4}. \tag{7}$$

The critical Reynolds number for transition from laminar to turbulent flow in a pipe depends on the frequency of the pulsatile flow. At normal breathing frequency the transition Reynolds number is about 2,300. Data shown in Table 7.2:1 suggest the existence of turbulence in upper airway of man in normal breathing. In inspiration, turbulent flow is expected also in smaller airways where N_R is less than 2,300 because disturbances will be carried down with the flow, decaying rather slowly.

Branching Tubes

Experiments on branching tubes have shown the existence of secondary flow. Pedley et al. (1977) observed that for inspirational flow the new boundary layer that develops on the flow divider remains thin for some distance downstream, and extends around at least half the circumference of the daughter tube. Most of the dissipation takes place in that boundary layer. They suggest that the dissipation mechanism is similar to that of the entry flow; hence

$$Z = \gamma\left(\frac{d}{L}N_R\right)^{1/2}, \tag{8}$$

where γ is a numerical constant, L is the length of the tube, d is the diameter, and N_R is the Reynolds number. The average value of γ was found to be 0.33 for laminar flow at Reynolds numbers between 100 and 700. If the formula yields a value of Z less than 1, it should be ignored and replaced by $Z = 1$ because Poiseuillean flow yields the minimum dissipation in any tube.

Upper Airways

The flow in the nose, mouth, larynx, and trachea is very complex. The narrowest passage in the upper airway is the laryngeal constriction. About half of the viscous resistance to breathing arises in the upper airway; half of that resistance is from the laryngeal constriction. Upon inspiration the flow will separate from the larynx and form a turbulent jet in the trachea. With expiration, separation occurs again and causes a confused flow in the mouth and nasal passages. The critical Reynolds number at which the inspiratory flow becomes turbulent in the trachea is about 500, whereas for an expiratory flow it is about 1,500.

The laryngeal constriction is elastic and its aperture varies with lung volume, flow rate, and the frequency of breathing. When the lung volume is increased, the aperture increases and the resistance to flow decreases. Panting enlarges the larynx aperture and reduces the flow resistance.

Jaeger and Matthys (1968/69) expressed their experimental results on pressure drop between mouth and trachea with the formula

$$\Delta p = C\dot{V}^a, \tag{9}$$

where Δp is the pressure drop, $\dot{V}$ is the volume flow rate, and a and C are constants given in Table 7.2:2. It is interesting to note that except for the very dense gas mixture, the exponent a is close to 1.5, which is predicted for the entry flow by use of Eqs. (6), (5), and (4). Pedley et al. (1977) suggested that the higher exponent (close to 2) for the dense gas sulfur fluoride SF_6 is due to the higher Reynolds number obtained with such a gas at normal flow rate, so that most of the energy dissipation stems from the turbulent jet issuing from the larynx. Then Z_T of Eq. (7) applies, and an exponent of 2 is obtained by use of Eqs. (5), (7), and (4).

TABLE 7.2:2 Densities, viscosities, kinematic viscosities, and best fit values of a and C, in the experiments of Jaeger and Matthys (1968/69)

Gas mixtures	ρ (g cm^{-3})	μ (g cm^{-1} s^{-1})	v (cm^2 s^{-1})	a	C
He	0.45×10^{-3}	2.05×10^{-4}	0.46	1.42	0.35
O_2	1.1×10^{-3}	2.07×10^{-4}	0.18	1.55	0.74
Ne	0.88×10^{-3}	2.92×10^{-4}	0.33	1.36	0.82
SF_6	4.2×10^{-3}	1.7×10^{-4}	0.04	1.92	1.77

Reprinted from T.J. Pedley, R.C. Schroter, and M.F. Sudlow, Gas flow and mixing in the airways, in *Bioengineering Aspects of the Lung* (J.B. West, ed.), pp. 163–265, by courtesy of Marcel Dekker, Inc., 1977.

Further along the airway tree, the bronchioles divide into terminal bronchioles, then into respiratory bronchioles, alveolar ducts, and alveoli. In
bronchioles of diameters less than 0.05 cm, the Reynolds number of flow is
smaller than 1 even at high levels of ventilation. Then the inertial force of the
fluid becomes negligible and the pressure drop is proportional to the flow rate.
The contribution of this region to the overall airway resistance is negligible.

Static Pressure Variation Along the Airway Tree

Pedley et al. (1977) calculated the pressure distribution in the entire airway
during inspiration. They used Eq. (8) with an average value of $\gamma = 0.33$ for all
generations of the airway from bronchi to alveoli, and Eq. (9) for the trachea,
assuming a fully developed turbulent flow there. They calculated the term $\mathscr{D}/\dot{V}$
in Eq. (1), called it the *viscous pressure drop*, and denoted it by Δp_v. They also
calculated the kinetic energy term involving $\hat{q}^2$ in order to compute the actual
static pressure at every location on the bronchial tree. They assumed Weibel's

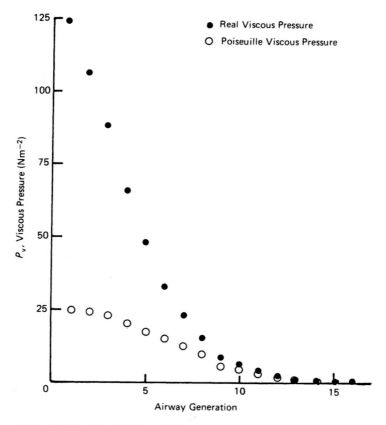

FIGURE 7.2:1 Predicted variation of viscous pressure down the bronchial tree at an
inspired flow rate of 1.67 Ls^{-1}. From Pedley, T.J. et al. (1971), by permission.

symmetric model (Table 7.2:1), and approximate equality of q and the velocity component in the axial direction u. Then, from Eq. (2),

$$\widehat{u^2} = \frac{1}{\dot{V}} \int u^3 \, dA. \tag{10}$$

By examining the empirical data, they took

$$\widehat{u^2} = 1.7\bar{u}^2. \tag{11}$$

Figure 7.2:1 shows their result on the viscous pressure drop p_v at the downstream end of each generation along the bronchial tree during inspiration, plotted against the generation number. The pressure in the alveoli is set at zero. The figure is plotted for a pulmonary flow rate of 100 liter/min. Note that the pressure drop Δp_v is equal to $\mathscr{D}/\dot{V}$, and $\mathscr{D} = Z\mathscr{D}_p$; hence when Z is given by Eq.(8), Δp_v is proportional to $\dot{V}^{3/2}$ (because $\bar{u} \sim \dot{V}$, $\mathscr{D}_p \sim \dot{V}^2$, $\mathscr{D} \sim \dot{V}^{5/2}$, and $\Delta p_v = \mathscr{D}/\dot{V}$). The hypothetical Poiseuillean flow results are also shown.

When the static pressure $\hat{p}$ is computed from Eq. (1) with the kinetic energy terms taken into account, the result is shown in Fig. 7.2:2.

A much more complex analysis is required for asymmetric branching. The flow in each branch then depends on the resistance in every other branch, both upstream and downstream.

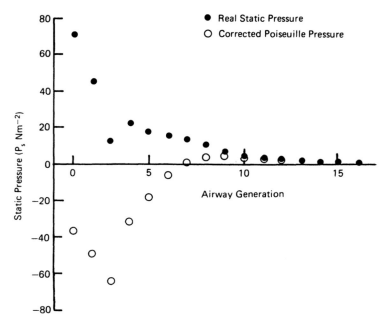

FIGURE 7.2:2 Predicted variation of static pressure down the bronchial tree at an inspired flow rate of 1.67 Ls^{-1}. From Pedley et al. (1971).

Recapitulation

Among all possible flows of a given rate in a circular cylindrical tube, the Poiseuille flow has the least energy dissipation. The energy loss due to any other flow of the same rate can be expressed as a product of a factor Z times the Poiseuillean loss. Z is a function of the pipe geometry, the entry condition, the branching pattern, the Reynolds number, the Stokes number (or its square root, the Womersley number), and the turbulence level.

The use of the Z factor outlined above is no more than a simple way of summarizing the results of some detailed fluid mechanical analysis. Obviously, more definitive detailed theoretical and experimental studies of the flow in the airway are needed. Secondary flow, separation, converging and diverging flows of expiration and inspiration in real models of the airway, the effect of flexible walls, the effect of smooth muscle contraction, mucus layer, etc., must be studied. What is known is minuscule compared to what needs to be known. This is an excellent field for computational fluid mechanics.

7.3 Interaction Between Convection and Diffusion

In each breathing cycle, we draw in fresh air and expel lung gas through the same airway. The interface of the old and new gases is somewhat blurred because of diffusion and convection. The nature of this interface is considered below.

First consider a steady laminar flow of a gas A in a long straight tube, as depicted in Fig. 7.3:1. According to Poiseuille, the velocity profile is parabolic. Imagine that at a certain time, a material B is added to A to the left of a certain section x_0 and that the interface between A and B is flat at this instant of time. If the solubility of B in A is zero, then at a later time this interface will be stretched into a paraboloid, with a vertex moving at twice the average velocity of the fluid, while the fluid in contact with the wall does not move because of the no-slip boundary condition (see Sec. 1.7). For this nondiffusible case the stretched interface at successive instants of time t_1, t_2, t_3 is shown by dashed lines in Fig. 7.3:1, and the concentration of B, averaged over a cross section, would vary with the distance x as shown by the dashed line in the graph at the bottom of Fig. 7.3:1. If, however, the material B is diffusible (soluble) in A, then molecular diffusion will take place to modify the interface. At time t_1 (see figure) the concentration gradient of B is essentially pointing in the longitudinal direction. At t_2 the concentration gradient of B in the radial direction becomes significant on the interface and diffusion in the radial direction takes place. At t_3 the average concentration of B will have a spatial distribution as shown by the solid line in the graph at the bottom of Fig. 7.3:1. The movement of the diffusible material B from the fast flowing core fluid to regions of lower velocity changes the stratification of material B, and the effective front of the material B is not located at the vertex of the paraboloid defined by convection.

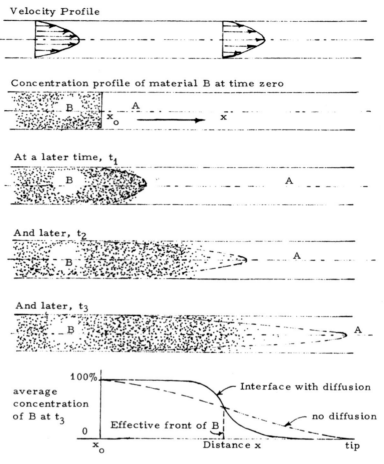

FIGURE 7.3:1 The progressive dispersion of a diffusible material B injected suddenly at time $t = 0$ into a long tube in which a fluid A flows steadily. Successive drawings show the concentration profile of material B as time increases from $t = 0$ to t_1, t_2, t_3. The last figure shows the variation of the concentration of the diffusible substance B with the distance along the tube. The "tip" indicated on the horizontal axis is the tip of the convective flow profile. The concentration shown is the average value in each cross section.

In fact, it is shown by Taylor (1953) that at large t the effective front of the material B moves at a velocity equal to the mean velocity of flow (i.e., at one-half of the velocity on the center line). Regarding the concentration distribution of the material B about this effective interface, Taylor shows that the resulting longitudinal mixing can be described as a process of diffusion, with an *apparent diffusivity* $\mathcal{D}_{app}$ (sometimes called the *dispersion coefficient*) which is very much larger than the molecular diffusivity D. He finds that $\mathcal{D}_{app}$ varies inversely with the molecular diffusivity (D), and directly with the square of the product of the average velocity (U) and the tube diameter (d); thus

$$\mathcal{D}_{app} = \frac{(Ud)^2}{192D}, \quad \text{(circular tube).} \tag{1}$$

The effective diffusivity is increased as molecular diffusivity decreases because the radial mixing is less rapid if molecular diffusivity is small. $\mathcal{D}_{app}$ also increases rapidly with an increasing flow rate and tube diameter.

This analysis also holds qualitatively for turbulent flow, because turbulent mixing is somewhat similar to molecular diffusion. Applying this concept to the gas flow in the airways, we see that the O_2 and CO_2 concentrations at the interface between the newly inhaled and alveolar gases are gently stratified in the airway during inspiration. As this stratified gas flows into smaller and smaller airways (bronchioles), U and d decrease rapidly, then the apparent diffusivity $\mathcal{D}_{app}$ becomes very small, and the stratification pattern becomes frozen. Beyond the terminal bronchiole the effect of convection is negligible, and molecular diffusion takes over as the mechanism to spread O_2 and CO_2 toward the flowing blood (Sec. 7.4).

Mathematical Analysis

The qualitative analysis above is exciting, but it may appear hard to believe at first sight. Thus, consider the component A in the mixture $A + B$ to the left of the interface. On the centerline of the tube, the mixture is moving at twice the velocity of the interface. The component A mixed with B becomes purified again on passing through the diffusion front. How is this done? To remove any feeling of mystery, it is best to reproduce Taylor's mathematical analysis here. Let us consider the case of a sudden injection at one station of a solute B into a steady flow in a channel—a situation similar to that shown in Fig. 7.3:1 with the exception that at time zero the solute B is concentrated at the section $x = 0$ (i.e., the concentration of B is a delta function of x), see the third panel of Fig. 7.3:2. When this case is solved, we can obtain the response to a step function injection as shown in Fig. 7.3:1 by a convolution integration.

To obtain the solution, we first need the basic equations governing the movement of the solute B. Let c be the concentration of the solute, D be the coefficient of molecular diffusion of the solute in the fluid A, ∇ be the gradient vector, and $\mathbf{u}$ be the velocity vector. Then the solute flux vector (mass flow rate of the solute) which is due to molecular diffusion and convection, is

$$\text{flux} = -D\nabla c + \mathbf{u}c. \tag{2}$$

The equation of balance of solute flow, describing the conservation of mass of the solute, can be derived by considering the influx and outflux through the surfaces of a control volume, and the rate of change of the mass of the solute in the volume (see Fung, 1977, Chap. 10, p. 250). The result is

$$-\frac{\partial c}{\partial t} = \nabla \cdot [-D\nabla c + \mathbf{u}c]. \tag{3}$$

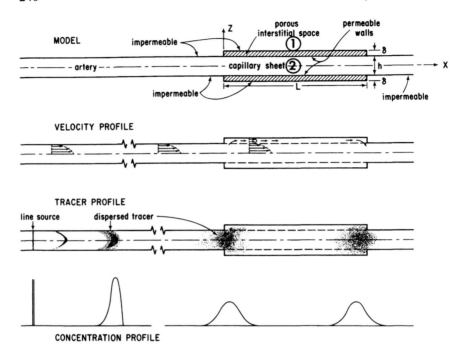

FIGURE 7.3:2 Impulsive injection of a diffusible material at station $x = 0$ at time $= 0$.

If the coefficient of diffusion D is a constant, independent of both the concentration and the spatial coordinates, and if the fluid is incompressible so that

$$\mathbf{V} \cdot \mathbf{u} = 0, \tag{4}$$

then Eq. (3) becomes

$$\frac{\partial c}{\partial t} + \mathbf{u} \cdot \mathbf{V}c = D\nabla^2 c. \tag{5}$$

Let us first solve this equation for the two-dimension channel illustrated in Fig. 7.3:1 or 7.3:2. Assume that the walls are impermeable to both the fluid A and the solute B. Let the coordinate axis x be the center line of the channel and the z axis be perpendicular to it. Let the components of velocity in the directions of x and z be u and w, respectively, whereas the third component perpendicular to the paper is zero. Then Eq. (5) can be written as

$$\frac{\partial c}{\partial t} + u \frac{\partial c}{\partial x} + w \frac{\partial c}{\partial z} = D\left(\frac{\partial^2 c}{\partial x^2} + \frac{\partial^2 c}{\partial z^2}\right). \tag{6}$$

In the case illustrated in Fig. 7.3:2, a diffusible solute B is injected at the station $x = 0$ impulsively at time $t = 0$ and uniformly over the cross section. Assume the flow to be steady except for the diffusion of the solute. We shall look for an asymptotic solution of the solute distribution after the elapse of a sufficiently long period of time following the injection. As the asymptotic condition is approached the velocity component u consists of a constant mean velocity U plus a deviation $u'(z)$ which is a function of z only; whereas the transverse velocity w vanishes. The solute B will have spread over a considerable length of the tube. Since the tube is long and narrow the rate of change of the solute concentration in the longitudinal direction will be much smaller than that in the transverse direction: hence $\partial^2 c/\partial x^2$ is negligible compared with $\partial^2 c/\partial z^2$. Hence at the asymptotic condition Eq. (6) is approximated by

$$\frac{\partial c}{\partial t} + [U + u'(z)]\frac{\partial c}{\partial x} = D\frac{\partial^2 c}{\partial z^2}. \tag{7}$$

To remove U from this equation we introduce a set of moving coordinates ξ and ζ with the origin moving at velocity U:

$$x = \xi + Ut, \qquad z = \zeta. \tag{8}$$

Then Eq. (7) becomes

$$\frac{\partial c}{\partial t} + u'(z)\frac{\partial c}{\partial \xi} = D\frac{\partial^2 c}{\partial z^2}. \tag{9}$$

Now let the average value of the concentration on the centerline $z = 0$ be denoted by $\bar{c}(\xi, t)$, which is not a function of z. Let the difference between $c(\xi, z, t)$ and $\bar{c}(\xi, t)$ be c':

$$c(\xi, z, t) = \bar{c}(\xi, t) + c'(\xi, z, t). \tag{10}$$

Substituting this into Eq. (9) yields

$$\frac{\partial}{\partial t}(\bar{c} + c') + u'(z)\frac{\partial}{\partial \xi}(\bar{c} + c') = D\frac{\partial^2 c'}{\partial z^2}. \tag{11}$$

Now Taylor argues that c' is essentially a function only of z whereas $\bar{c}$ is a function only of ξ, so that the equation above is reduced to

$$u'(z)\frac{\partial \bar{c}}{\partial \xi} = D\frac{\partial^2 c'}{\partial z^2}. \tag{12}$$

The validity of this assumption, i.e.,

$$\frac{\partial c'}{\partial \xi} \ll \frac{\partial \bar{c}}{\partial \xi}, \qquad \frac{\partial \bar{c}}{\partial t}, \frac{\partial c'}{\partial t} \ll D\frac{\partial^2 c'}{\partial z^2} \tag{13}$$

can be verified a posteriori. The boundary conditions for Eq. (12) are that $c' = 0$ when $z = 0$ (by definition of $\bar{c}$) and $\partial c'/\partial z = 0$ at an impermeable boundary ($z = \pm h/2$ in Fig. 7.3:2). Hence the integration of Eq. (12) gives

$$\frac{\partial c'}{\partial z} = \int_{-h/2}^{z} \frac{1}{D}\frac{\partial \bar{c}}{\partial \xi}u'(z')\,dz', \tag{14}$$

which satisfies the boundary condition at $z = -h/2$. A second integration gives

$$c' = f(z) \frac{\partial \bar{c}}{\partial \xi}, \tag{15}$$

where

$$f(z) = \int_0^z \left(\int_{-h/2}^{z''} \frac{1}{D} u'(z') \, dz' \right) dz''. \tag{16}$$

The rate at which mass is transported through a cross section moving at the mean velocity is

$$\dot{M} = A \overline{u'c'} \tag{17}$$

in which A is the cross-sectional area and the over bar indicates a cross sectional mean. Use of Eq. (15) in Eq. (17) yields

$$\dot{M} = A \overline{u'f} \frac{\partial \bar{c}}{\partial \xi}. \tag{18}$$

This is of the form

$$\dot{M} = -A \mathcal{D}_{app} \frac{\partial \bar{c}}{\partial \xi}, \tag{19}$$

which shows that the dispersion follows Fick's law, with a *coefficient of apparent diffusivity*:

$$\mathcal{D}_{app} = -\overline{u'f}. \tag{20}$$

Now consider a small segment of the tube of length dx. The inflow of the solute from the left is $\dot{M}$. The outflow from the right is $\dot{M} + (\partial \dot{M}/\partial \xi) \, d\xi$. The net outflow per unit time is

$$\frac{\partial \dot{M}}{\partial \xi} \, d\xi = -A \mathcal{D}_{app} \frac{\partial^2 \bar{c}}{\partial \xi^2} \, d\xi.$$

By the principle of conservation of mass this must be equal to the rate of decrease of the mass of the solute in the segment, $(-\partial \bar{c}/\partial t) A \, d\xi$. Hence the mean concentration is governed by the equation

$$\frac{\partial \bar{c}}{\partial t} = \mathcal{D}_{app} \frac{\partial^2 \bar{c}}{\partial \xi^2}, \tag{21}$$

or, by Eq. (8),

$$\frac{\partial \bar{c}}{\partial t} + U \frac{\partial \bar{c}}{\partial x} = \mathcal{D}_{app} \frac{\partial^2 \bar{c}}{\partial x^2}. \tag{22}$$

From this equation we obtain the solution which represents the concentration of the solute on the centerline following an impulsive input of unit total mass at $x = 0$ and $t = 0$:

$$\bar{c} = I(x, t) = \frac{1}{2} \frac{1}{h\sqrt{\pi \mathcal{D}_{app} t}} \exp\left[-\frac{(x - Ut)^2}{4\mathcal{D}_{app} t} \right]. \tag{23}$$

The apparent diffusivity, $\mathcal{D}_{app}$, can be computed from Eqs. (16) and (20). For a two-dimensional channel of width h with a parabolic flow profile, the velocity at a distance z from the center line is

$$u = u_0\left(1 - \frac{4z^2}{h^2}\right), \tag{24}$$

where u_0 is the maximum velocity at the axis. Hence the mean velocity U and the deviation u' are

$$U = \tfrac{2}{3}u_0, \qquad u' = \tfrac{1}{2}U\left(1 - \frac{12z^2}{h^2}\right). \tag{25}$$

On substituting into Eq. (16), integrating, and then substituting into Eq. (20), integrating again, and dividing by h, one obtains the apparent diffusivity

$$\mathcal{D}_{app} = \frac{h^2 U^2}{210D} \quad \text{(channel)}. \tag{26}$$

This result refers to a two-dimensional channel. If a corresponding problem of injecting a solute into a flow in a circular cylindrical tube of radius a is solved, the velocity u at a distance r from the central line is

$$u = u_0(1 - r^2/a^2), \qquad U = u_0/2, \tag{27}$$

and one obtains the apparent diffusivity:

$$\mathcal{D}_{app} = \frac{a^2 U^2}{48D} \quad \text{(circular tube)}. \tag{28}$$

The spatial distribution of the solute at time t after an impulsive injection at $t = 0$, $x = 0$ is given by the same $I(x, t)$ of Eq. (23).

Arbitrary Injection

Since Eq. (22) is linear, the principle of superposition holds. Hence, if one injects a solute into a tube at a rate of $f(x, t)$ per unit time per unit length, then in a small interval of time between τ and $\tau + d\tau$ and space between ζ and $\zeta + d\zeta$ the amount of solute $f(\zeta, \tau)\, d\zeta\, d\tau$ may be considered impulsively injected. The concentration distribution due to this injection is

$$f(\zeta, \tau)\, d\zeta\, d\tau\, I(x - \zeta, t - \tau).$$

By taking all such intervals into consideration, we see that the solute concentration at time t and place x is

$$c(x,t) = \int_{-\infty}^{t} d\tau \int_{-\infty}^{x} d\zeta\, f(\zeta,\tau) I(x - \zeta, t - \tau). \tag{29}$$

Further Extensions

Taylor's asymptotic solution must be supplemented by an investigation of its range of validity with regard to the conditions assumed in Eqs. (13). Obviously it is not valid at places too close to the site of injection, nor too soon after injection. Taylor (1953) has presented an extensive theoretical and experimental investigation of these questions. In Taylor (1954) the analysis is extended to turbulent flows in pipes which are smooth or rough, straight or curved. Further research has found the method valid for flow in rivers, and is used for practical determination of river flow. In physiology, use of solute injection to determine flow rate and arrival time of blood was initiated by G.N. Stewart (1893, 1900), who assumed that the tracer flows with the mean speed of flow of blood (which is only one half of the velocity on the center line of the blood vessel in a Poiseuillean flow). This is plausible, but the correct explanation was not known until Taylor presented it in 1953.

Diffusion in Alveolar Duct and Alveoli

In the respiratory bronchioles, alveolar ducts, and alveoli, the Reynolds number is less than 1, the role of convection becomes insignificant, and molecular diffusion is the mechanism that spreads O_2, CO_2, N_2 (or other gases) in the alveolar space.

Between the terminal bronchiole and the alveoli of human lung, the total distance is only about 4 mm. The volume available for the gas, however, increases very rapidly as an exponential function of the generation number, from 175 cm^3 to 4,800 cm^3 (see Table 7.2:1). In this space, the O_2 and CO_2 concentrations must change from their values in the terminal bronchiole to their values on the alveolar wall. On the alveolar wall, rapid exchange of alveolar gas with blood occurs, (Sec. 7.4). The question of stratification of O_2 and CO_2 concentration in the pulmonary alveolar duct has been investigated by Chang et al. (1973).

7.4 Exchange Between Alveolar Gas and Erythrocytes

With air brought to the alveoli and blood flowing in the capillary blood vessels, an exchange between blood and alveolar gas takes place. Now we must consider the diffusion of alveolar gas to the red blood cells, the chemical reactions in the red cell, and the spatial and temporal distribution of O_2 and CO_2.

Diffusion Across a Membrane

Analysis of diffusion across the alveolar wall can be based on the equations derived in Sec. 7.3. Let c be the concentration (mass per unit volume) of a gas in the wall, and D be the coefficient of molecular diffusion of that gas in the wall. The wall is incompressible and the convection in the wall can be ignored. For a wall (Fig. 7.4:1) of uniform thickness h exposed to a gas of concentration c_1 on one side and c_2 on the other (c_1 and c_2 being constants), Eq. (7.3:5) is reduced to the following:

$$D\frac{d^2c}{dx^2} = 0, \tag{1}$$

and the boundary conditions are

$$c = c_1, \quad \text{at } x = 0; \qquad c = c_2, \quad \text{at } x = h; \tag{2}$$

x being the coordinate perpendicular to the membrane. A solution of Eq. (1) that satisfies the boundary conditions (2) is

$$c = c_1 + \frac{c_2 - c_1}{h}x. \tag{3}$$

The flux vector, pointing in the direction of the x-axis, has a magnitude:

$$\text{flux} = -D\nabla c = -D\tfrac{1}{h}(c_2 - c_1). \tag{4}$$

The rate of mass flow across a membrane of area A is

$$\dot{m} = A \cdot \text{flux}. \tag{5}$$

In pulmonary physiology, it is customary to express the gas concentration in a tissue in terms of its *partial pressure* p and the *solubility* λ of that gas in the tissue:

$$c \equiv \lambda p. \tag{6}$$

Hence, from Eqs. (4), (5), and (6) we have the rate of diffusion across the membrane and the rate of volume flow $\dot{V}$:

$$\dot{m} = -\frac{DA\lambda}{h}(p_2 - p_1), \qquad \dot{V} = \frac{\dot{m}}{\rho}. \tag{7}$$

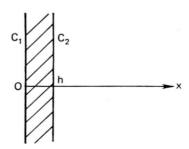

FIGURE 7.4:1 A diffusion barrier.

D is inversely proportional to the square root of molecular weight:

$$D \sim \frac{1}{\sqrt{\text{molecular weight}}}. \tag{8}$$

If we compare the diffusion of O_2 and CO_2 in a membrane, we see that CO_2 will move about 20 times faster because its solubility is much larger whereas its molecular weight is not very different from that of O_2.

Diffusion from Alveolar Gas to Red Blood Cells in a Capillary Blood Vessel (Fig. 7.4:2)

If a foreign gas such as carbon monoxide arrives at an alveolus, the gas will diffuse through the capillary blood vessel wall, the plasma, and the red cell membrane, and react with the hemoglobin. A certain amount of CO will be left in molecular form, while the rest is combined with hemoglobin into a compound. The gaseous CO will be carried by the red cell to the tissues. Hence we ask: How rapidly does the partial pressure of CO in the red blood cell rise?

The answer is illustrated in Fig. 7.4:3. The partial pressure rises slowly, because of the high chemical affinity of CO to hemoglobin.

Now let us assume that the foreign gas that arrived at the alveolus is nitrous oxide. When this gas diffuses into the red blood cell, no combination with hemoglobin takes place. As a result the partial pressure of N_2O in the red blood cell can rise rapidly. In the example shown in Fig. 7.4:3 the partial pressure of N_2O in the red cell has virtually reached the alveolar level after the cell traversed approximately one-quarter of the path along the capillary. After this point, practically no additional N_2O is transferred. Thus the amount of N_2O taken up by the blood depends entirely on the amount of blood flow.

Now, consider oxygen, which reacts with hemoglobin to form oxyhemoglobin:

$$O_2 + Hb \rightleftarrows HbO_2. \tag{9}$$

The reaction has a finite speed; in the red cell it takes about $\frac{1}{5}$ sec to complete.

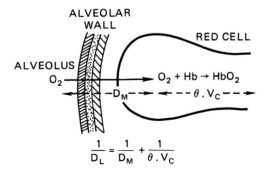

FIGURE 7.4:2 Diffusion of gas from alveolus to red blood cell in a capillary blood vessel.

FIGURE 7.4:3 CO, N_2O, and O_2 uptake by red blood cells as they travel in the pulmonary capillary. From J.B. West, *Respiratory Physiology—The Essentials*, copyright © 1974, the Williams & Wilkins Co., Baltimore.

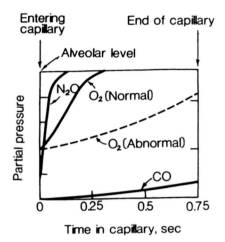

When a red blood cell enters a capillary in the alveolus it normally has an O_2 partial pressure of about 40 mmHg. The partial pressure of O_2 in the alveolus may be about 100 mmHg. The two compartments are separated by a distance of only about $\frac{1}{2}$ μm. The diffusion and chemical reaction are therefore initiated immediately in the red cell. This dynamic process continues for about $\frac{1}{4}$ sec in a normal lung until the O_2 pressure in the red cell approaches the alveolar O_2 pressure. This is depicted in Fig. 7.4:3. In abnormal circumstances, in which the diffusion properties of the lung are impaired, for example, by thickening the alveolar wall, the rate of increase of O_2 pressure in the red cell will be slowed, and it is possible that the blood p_{O_2} at the end of the capillary will not reach the alveolar value.

Another situation in which O_2 transfer may become diffusion limited occurs when the time of transit of blood through the capillary is greatly shortened (e.g., in vigorous exercise), or when the oxygen pressure in the

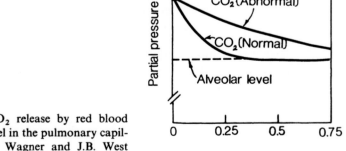

FIGURE 7.4:4 CO_2 release by red blood cells as they travel in the pulmonary capillary. After P.D. Wagner and J.B. West (1972).

atmosphere is very low (e.g., on high mountains or at high altitude while flying in an airplane with an unpressurized cabin.) Then the alveolar p_{O_2} and the blood p_{O_2} are both reduced.

Finally, consider CO_2. Its partial pressure history in a pulmonary capillary blood vessel is sketched in Fig. 7.4:4. In a normal lung there is no difficulty in remvoing blood CO_2 and reducing its partial pressure to that of the alveolar level, but in diseased states with thickening of blood–gas barrier, this elimination may have some difficulty.

Measurement of Diffusion Capacity

Equation (7) shows that to analyze diffusion across a membrane, we must know the constants D, A, λ, and h. In a real lung, there is no way to measure these constants during life. To absorb them in a simplified equation that is applicable to a whole lung, Krogh (1914) writes

$$\dot{V} = D_L(p_1 - p_2). \tag{10}$$

The constant D_L is called the *diffusion capacity of the lung*. This equation can be applied to each gas. Thus, if O_2 is considered, $\dot{V}$ is $\dot{V}_{O_2}$, the rate of transfer of oxygen from alveoli to red cells measured in terms of volume at standard temperature and pressure (0°C, 760 mm Hg, no water vapor) per unit time. p_1 is $p_{A_{O_2}}$, the partial pressure of O_2 in alveoli. p_2 is partial pressure of O_2 in the red blood cell. D_L is $D_{L_{O_2}}$, the diffusion capacity of the lung for O_2. Similarly, subscripts CO, CO_2, etc., are used when Eq. (10) is applied to these gases.

A further separation of the right-hand side of Eq. (10) into two terms would clarify the role of chemical reaction in gas transfer in the lung; see Fig. 7.4.2:

$$\dot{V}_{O_2} = D_{L_{O_2}}[(p_{O_2} \text{ in alveoli} - p_{O_2} \text{ in plasma})$$
$$+ (p_{O_2} \text{ in plasma} - p_{O_2} \text{ in red cell})]. \tag{11}$$

In analogy with Eq. (7) we write

$$\dot{V}_{O_2} = D_{M_{O_2}}(p_{O_2} \text{ in alveoli} - p_{O_2} \text{ in plasma}), \tag{12}$$

where $D_{M_{O_2}}$ is called the *membrane diffusion capacity of* O_2. On the other hand, to account for chemical reaction, Roughton and Forster (1957) write:

$$\dot{V}_{O_2} = \theta_{O_2} V_c(p_{O_2} \text{ in plasma} - p_{O_2} \text{ in red cell}). \tag{13}$$

Here V_c is the blood volume in the capillary blood vessels of the lung, and θ_{O_2} is a *reaction rate*, which is usually expressed in ml of O_2 per min per mm Hg per cm^3 of blood. Combining Eqs. (11)–(13) and omitting the subscripts O_2, we have

$$\frac{1}{D_L} = \frac{1}{D_M} + \frac{1}{\theta V_c}. \tag{14}$$

If the inverses of D_L, D_M, θV_c are interpreted as resistances to gas transport, then Eq. (14) expresses the idea that the total resistance is the sum of the "membrane" resistance and the resistance due to a finite reaction rate.

If carbon monoxide is used to measure the diffusion capacity, p_1 is the partial pressure of CO in alveoli and p_2 is that in the red blood cell. But we have seen in Fig. 7.4:3 that p_2 of CO is negligible compared with p_1. Hence the diffusion capacity of the lung for carbon monoxide is

$$D_{L_{CO}} = \frac{\dot{V}_{CO}}{p_{ACO}}. \tag{15}$$

The normal value of $D_{L_{CO}}$ of human depends on sex, age, and height; for adult man at rest it is about 25 cm^3 per min per mm Hg, and it increases to two or three times this value with exercise. θ_{CO} is very large, and D_{MCO} is essentially equal to $D_{L_{CO}}$.

For oxygen, D_{MO_2} and $\theta_{O_2} V_c$ are approximately equal. Hence a reduction of capillary blood volume V_c by disease is effective in reducing the diffusion capacity of the lung, $D_{L_{O_2}}$. For an adult man $D_{L_{O_2}}$ is about $1.23 D_{L_{CO}}$, and $D_{L_{CO_2}}$ is about $24.6 D_{L_{CO}}$, based on molecular weights and solubilities of these gases. Methods of measurements are discussed in detail in Bates et al. (1971).

7.5 Ventilation/Perfusion Ratio

An ideal condition for the lung to work is to have an exactly right amount of oxygen delivered to the alveoli in time to oxygenate all the blood that is delivered to the capillary blood vessels in the alveolar walls. This calls for a certain ideal ventilation/perfusion ratio. In a normal lung, the ideal ratio is approximately 1 to 1, i.e., the rate of fresh air delivery is roughly equal to that of blood delivery. The exact value of the ideal ventilation/perfusion ratio depends on a number of factors, such as the hematocrit, the atmospheric pressure, the altitude, the CO_2 level, and the humidity. Whatever the ideal ventilation/perfusion ratio may be, however, it cannot be achieved in the entire lung because both the ventilation and the perfusion are nonuniform in the lung. One cause of this nonuniformity is gravity. For a man in an upright position, the weight of the blood creates a hydrostatic pressure in the blood vessels, whereas the weight of the air in the airways is so small that it's negligible. The hydrostatic pressure in the blood vessels distends them. As a result those blood vessels near the base of the lung are distended more than those near the apex of the lung. The resistance to flow is thus smaller in the basal region than in the apex region. Hence the blood flow per unit volume of the lung is larger toward the base of the lung. In contrast, the alveolar size is the largest toward the apex of the lung because of the elastic deformation of the alveolar structure due to gravity. Since the lung parenchyma is very compliant, its weight pulls the alveoli in the apex region to a larger size while compressing the alveoli in the region of the base. The net result is that the ventilation/perfusion ratio is larger near the apex and smaller toward the base.

That is for the normal lung. In a diseased lung, some airways can be closed by mucus or by tumor growth. Then the alveoli downstream of these airways will not be ventilated, and the ventilation/perfusion ratio will be reduced to zero in that region. Also, a blood vessel can become obstructed by thrombi or other causes; in that case perfusion is reduced and the ventilation/perfusion ratio becomes large. This suggests that if we know how to assess the distribution of the ventilation/perfusion ratio in the lung, we may obtain an effective tool for diagnosis. The search for means to do this leads to an interesting integral equation given by Wagner and West (1972), which we shall derive and discuss.

A method to measure ventilation/perfusion distribution was proposed by West and Wagner (1977). They took a mixture of several inert gases* and equilibrated them with normal saline or isotonic dextrose in water. Then they injected the solution into a peripheral vein, and, after a steady state was established in the lungs, simultaneously collected samples of the mixed venous blood and the mixed expired gas. From the samples they measured the concentration of each gas and the solubility of the gas in blood by gas chromatography. The total pulmonary blood flow and the minute ventilation were also measured.

To analyze the situation, consider first a single alveolus and a single inert gas; see the schematic in Fig. 7.5:1. The mixed venous blood containing the inert gas flows into the capillary blood vessel, while the inspired gas does not contain that inert gas. Assuming that at a steady state the diffusion equilibrium is complete so that the partial pressure of the gas in the end-capillary blood (p_{ec}) and that in the alveolus (p_A) are equal. Let $p_{\bar{v}}$ be the partial pressure of the gas in the mixed venous blood, $d\dot{V}_A$ be the expiratory ventilation rate, and $d\dot{Q}$ be the blood flow rate in this small exchange unit. Then $p_A d\dot{V}_A$ is the rate at which that gas is expired from the alveolus to the atmosphere, whereas $d\dot{Q}(c_{\bar{v}} - c_{ec}) = d\dot{Q}\lambda(p_{\bar{v}} - p_{ec})$ is the rate at which that gas is evolved from the blood to the alveolus. Here the symbol c denotes the concentration of the inert gas, and λ *denotes the solubility of the gas in blood.* Equating these two quantities on the basis of conservation of mass, we have

$$d\dot{Q}\,\lambda(p_{\bar{v}} - p_{ec}) = p_A\,d\dot{V}_A.$$

Noting that $p_{ec} = p_A$ as assumed, and solving for p_A, we obtain

$$p_A = p_{ec} = p_{\bar{v}}\frac{\lambda}{(\lambda + d\dot{V}_A/d\dot{Q})}. \tag{1}$$

This applies to one alveolus. Now let the lung be considered as a collection of a large number of alveoli in parallel. Each alveolus has a value of $d\dot{V}_A/d\dot{Q}$. Let this value be denoted by x. In the whole lung the value of x is statistically distributed among the avleoli. On the other hand, the ventilation and perfusion of each alveolus are also statistically distributed among the alveoli.

---

* Inert in the sense that the solubility is constant and the dissociation curve is linear, not in the sense of pharmacological effects.

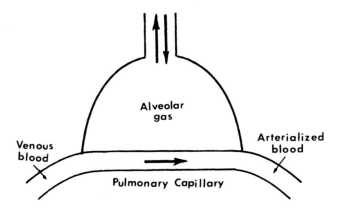

FIGURE 7.5:1 Schematic drawing of the distribution of an inert gas in a small control volume of pulmonary capillaries and alveoli. The tube below represents capillary blood vessels; the bellshaped unit above represents the alveoli. C = concentration of the inert gas (mass/vol). p = partial pressure of the inert gas. In liquid and solid the concentration is equal to the product of the solubility (λ) and the partial pressure: $c = \lambda p$. The subscripts atm, A, a, c, $\bar{v}$ denote atmospheric, alveolar, pulmonary arterial, end capillary, and pulmonary venous. $\dot{V}$ = volume rate of gas flowing out (expiration) of this small control volume. $\dot{Q}$ = volume rate of blood (perfusion) flowing into the same control volume.

Let $\Delta \dot{V}_A(x)$ be the sum of the ventilation of all alveoli whose $\dot{V}_A/\dot{Q}$ ratio lies in the range between x and $x + \Delta x$. Then the inert gas coming from these alveoli is expired at a rate $p_A(\lambda, x)\Delta \dot{V}_A(x)$. The rate of expiration of the inert gas from the entire lung is therefore the sum of $p_A(\lambda, x)\Delta \dot{V}_A(x)$ over all values of x. If the probability distribution of x is continuous we shall write $\Delta \dot{V}_A(x)$ as $\dot{V}_A f_{\dot{V}}(x) dx$, where $\dot{V}_A$ represents the total ventilation of the lung, and $f_{\dot{V}}(x)$ signifies the probability frequency function of the ventilation: the probability of finding the ventilation to come from alveolar units with $d\dot{V}_A/d\dot{Q}$ values lying between x and $x + dx$. The rate of expiration of the gas is therefore

$$\sum p_A(\lambda, x)\Delta \dot{V}_A(x) = \dot{V}_A \int_0^\infty p_A(\lambda, x)f_{\dot{V}}(x)\,dx. \tag{2}$$

Substituting from Eq. (1) and denoting the result as $\dot{G}_{\dot{V}}$, we obtain

$$\dot{G}_{\dot{V}}(\lambda) = p_{\bar{v}}\dot{V}_A \int_0^\infty \frac{\lambda}{\lambda + x} f_{\dot{V}}(x)\,dx. \tag{3}$$

By an identical reasoning, we let $f_{\dot{Q}}(x)$ denote the probability frequency of the blood flowing into those alveoli whose $d\dot{V}_A/d\dot{Q}$ ratio is x, and find the total amount of gas leaving the lung in the flowing blood, $\dot{G}_{\dot{Q}}(\lambda)$:

$$\dot{G}_{\dot{Q}}(\lambda) = p_{\bar{v}}\dot{Q} \int_0^\infty \frac{\lambda}{\lambda + x} f_{\dot{Q}}(x)\,dx. \tag{4}$$

West and Wagner (1977) call the ratio $\dot{G}_{\dot{V}}(\lambda)/p_{\bar{v}}\dot{V}_A$ the *excretion* of the inert gas and $\dot{G}_{\dot{Q}}(\lambda)/p_{\bar{v}}\dot{Q}$ the *retention* of that gas. For mathematical reasons it is more convenient to divide retention and excretion by λ, and denote the results by $r(\lambda)$ and $e(\lambda)$, respectively. They then obtain the basic integral equations:

$$r(\lambda) = \int_0^\infty \frac{1}{\lambda + x} f_{\dot{Q}}(x)\,dx, \tag{5}$$

$$e(\lambda) = \int_0^\infty \frac{1}{\lambda + x} f_{\dot{V}}(x)\,dx. \tag{6}$$

The contention of Wagner et al. is that $r(\lambda)$, $e(\lambda)$ can be measured at several values of λ by choosing suitable inert gases. The problem is to solve for the unknown functions $f_{\dot{Q}}(x)$ and $f_{\dot{V}}(x)$.

Solution of the Integral Equation

The equation

$$r(\lambda) = \int_0^\infty \frac{1}{\lambda + x} f(x)\,dx, \qquad \lambda \geqslant 0, \tag{7}$$

is a regular Fredholm integral equation of the first kind. If $f(x)$ is continuous and tends to zero as $x \to \infty$ as fast as $x^{-\mu}$, where $\mu > 0$, then the integral exists in Riemannian sense; and the function $r(\lambda)$ so defined is a continuous function of λ, except possibly at $\lambda = 0$. If $f(0)$ is finite, then $r(\lambda)$ tends to $-f(0)\log\lambda$ as $\lambda \to 0$. If $f(x)$ tends to zero as fast as x^ν as $x \to 0$, where $\nu > 0$, then $r(x)$ is finite at $\lambda = 0$ and continuous for $\lambda \geqslant 0$.

The theory of the regular Fredholm integral equation is well known. Since the homogeneous equation obtained from Eq. (7) by putting $r(\lambda)$ equal to zero has no solution other than $f(x) = 0$, the nonhomogeneous equation (7) has a unique solution.

This equation was exhaustively investigated by Stieltjes, and several solutions are given by Van der Pol and Bremmer (1950). For example, the inversion may be expressed as an integral

$$f(x) = \frac{i}{2\pi^2} \int_{c-i\infty}^{c+i\infty} dp\, x^p \sin(\pi p) \int_0^\infty du\, \frac{r(u)}{u^{p+1}}, \qquad (x > 0;\, -1 < c < 0), \tag{8}$$

or as in infinite product

$$f(x) = -x \frac{d}{dx} \prod_{n=-\infty}^{\infty}{}' \left(1 + \frac{x}{n}\frac{d}{dx}\right) r(x), \qquad (x > 0), \tag{9}$$

in which the prime at the product sign means that the term $n = 0$ should be deleted.

These exact solutions are quite useless in practice, because experimental determination of the retention $r(\lambda)$ as a function of the solubility λ is limited.

For example, in West and Wagner's extensive research on this subject, $r(\lambda)$ is determined only for six values of λ (with six specifically chosen gases). Hence the function $r(\lambda)$ is known only for six points. The question of solving Eq. (7) is transformed into the following: What can be said about the function $f(x)$ if $r(\lambda)$ is known for six values of λ? Can some major features of $f(x)$ (such as how many peaks does it have, and at what values of x are these peaks located), be determined from such scanty data? West, Wagner and their colleagues' answer is affirmative, and very elaborate theoretical investigations have been done to assess this question. These theories are not very easily understood. See West and Wagner (1977) for details.

7.6 Pulmonary Function Tests

Any measurement of the function of the respiratory system is a pulmonary function test. Pulmonary function tests help physicians to make diagnostic and therapeutic decisions. The most common pathophysiologic patterns of pulmonary diseases are obstructive (70%), restrictive (20–25%), and vascular (5–10%). Dynamic flow measurements are most useful to evaluate obstructive diseases whereas static elasticity measurements are most useful to evaluate restrictive diseases. In the following we consider the principles of some of these tests.

Volume Measurements

The volume of gas moved by normal breathing is called the *tidal volume*. When a person takes a maximal inspiration and follows this by a maximal expiration, then the exhaled volume is called the *vital capacity* (VC). Some gas remains in the lung after a maximal expiration; this is the *residual volume* (RV). The volume of gas in the lung after a normal expiration is the *functional residual capacity* (FRC). The sum of the residual volume and the vital capacity is the *total lung capacity* (TLC).

A spirometer is used to measure the tidal volume and vital capacity. It will not yield, however, the residual volume and functional residual capacity.

In order to measure absolute gas volume in the lung (TLC, FRC, and RV), three methods may be employed: inert gas dilution or wash out, radiological techniques (see Bates et al., 1971, pp. 13–15), and whole body plethysmography (Dubois, 1964, p. 454). To measure *regional lung volume* (i.e., the volume of identifiable parts of the lung), the ^{133}xenon gas-dilution method can be used. The radioactive xenon is inhaled and detected externally. The scintilation count rate of ^{133}xenon in the gas in a closed spirometer circuit is compared with the count rate after a single breath or during equilibration. Analysis is based on the principle of indicator dilution (Sec. 9.9).

Since the lung deforms under its own weight, the regional distribution of lung volume is affected by gravity (Sec. 11.6). If breathing takes place at lung volumes lower than FRC, the regional volume decreases more rapidly in the direction of the gravity. The minimal regional volume (regional residual volume) is about 20% of the regional TLC. In a young person seated upright, the regional minimal volume is reached first in the lowest zone when the total lung volume is at about 40% TLC, and in the midzone when the lung is at about 30% TLC. With loss of elastic recoil in older persons, these minimal volumes will be reached at higher fractions of the TLC. It appears liklely that the minimal volume is reached by closure of airways (Bates et al., 1973, p. 46). Airway closure beings at the lowest part of the lung which is exposed to more positive pleural pressure, and progresses upward in the lung. This phenomenon of airway closure at low volume has, of course, major physiological significance. In conditions in which the lung volume is reduced, as in gross abdominal obesity, airway closure may be responsible for the low arterial oxygen tension. In older people, ventilation is diminished immediately on the assumption of recumbency, presumably as a result of airway closure.

The airway space consists of two compartments: the *anatomical dead space* (V_D) in which there is no respiratory alveoli and hence no gas exchange with blood, and the *alveolar space* in which there is gas exchange with blood. It is of interest to measure the anatomical dead space. Such a measurement is quite easy if one makes the *assumption* that *as gas is moved into and out of the lung the velocity and concentration profiles are flat* (i.e., the velocity of flow and the concentration of any component of the gas are uniform in any cross section of the airway), but although this assumption is practically always made (tacitly) in medical literatures, it is certainly wrong (see Sections 7.2 and 7.3.). The error is generally negligible in ordinary circumstances, except in high frequency ventilation. Under this assumption, if a volume of gas V_E that is expired at the mouth contains alveolar gas, then, it must consist of a sum of the dead space and a volume V_A coming from the alveolar space:

$$V_E = V_D + V_A \tag{1}$$

because all the gas in the dead space must first come out before alveolar gas can. (If V_E is smaller than V_D then V_A must be zero.) Equation (1) is not valid if this assumption is not made. For example, if the alveolar gas fluctuates and diffuses out like a jet in the center of the bronchi and trachea, then the expired gas can indeed contain the alveolar gas while the expired volume is smaller than the dead space. Something like this actually happens in high-frequency low-tidal-volume ventilation (see Sec. 7.8).

Under the assumption named above, the anatomical dead space can be measured by simultaneous registration of expired nitrogen concentration and volume of flow, following a deep inspiration of a nitrogen-free gas. If the N_2 concentration is plotted against the volume of gas expired (Fig. 7.6:1) and the concentration profile can be approximated by a step function, then the expired gas volume before the rise of the step is the anatomical dead space.

Another method of measuring the dead space is based on CO_2 exchange. If one inhales fresh air and then collects a complete expiration in a bag, the

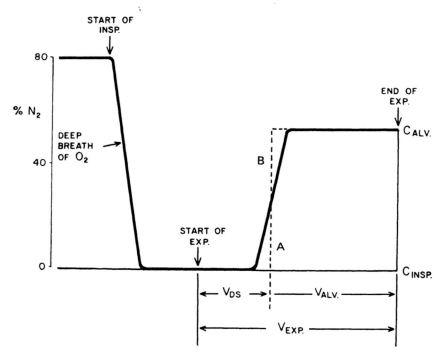

FIGURE 7.6:1 Fowler's method of measuring the anatomical dead space with a rapid N_2 analyzer. The dead space is the expired volume up to the vertical broken line which makes the areas A and B equal. From Bates et al. (1971)

amount of CO_2 in the bag is equal to the fractional concentration of CO_2 times the volume of expiration, i.e. $F_{ECO_2} \times V_E$. This amount of CO_2 is made up of two distinct portions, (1) a volume of CO_2 from the nonexchanging dead space V_D in which the concentration is the same as that in the inspired air (F_{ICO_2}), plus (2) the volume of CO_2 derived from alveolar gas, $F_{ACO_2} \times [V_E - V_D]$. Hence

$$F_{ECO_2} V_E = F_{ICO_2} V_D + F_{ACO_2}(V_E - V_D) \qquad (2a)$$

or

$$V_D = \frac{(F_{ACO_2} - F_{ECO_2})V_E}{F_{ACO_2} - F_{ICO_2}}. \qquad (2b)$$

If an assumption is made that the partial pressure of CO_2 in the arterialized blood, p_{aCO_2}, is equal to that in the alveoli, p_{ACO_2} (see Sec. 7.4), and that the inspired air contains no CO_2, then on replacing the fractional concentrations by partial pressures, Eq. (2b) becomes

$$V_{DCO_2} = \frac{V_E(p_{aCO_2} - p_{ECO_2})}{p_{aCO_2}}. \qquad (3)$$

One can thus measure V_{DCO_2} by measuring V_E, p_{aCO_2}, and p_{ECO_2}. V_{DCO_2} is called the *physiological dead space*. It is equal to the anatomical dead space if the assumption $p_{ACO_2} = p_{aCO_2}$ is valid.

Ventilation

The minute ventilation ($\dot{V}_E$) is defined as the quantity of air expired per minute. By differentiating Eq. (1) with respect to time, we obtain

$$\dot{V}_E = \dot{V}_D + \dot{V}_A, \tag{4}$$

where a dot indicates the rate of change with respect to time. The minute ventilation ($\dot{V}_E$) can be measured with a recording spirometer, a pneumotachometer, or a collecting bag. If V_D is measured then $\dot{V}_D$ is equal to the product of V_D and the frequency of breathing; and $\dot{V}_A$ can be computed according to Eq. (4).

Substituting Eq. (3) into Eq. (1), one obtains an important equation

$$V_A = \frac{V_E p_{ECO_2}}{p_{aCO_2}}, \qquad \dot{V}_A = \frac{\dot{V}_{CO_2}}{p_{aCO_2}}. \tag{5}$$

Equation (5) describes the elimination of CO_2 in health and its retention in disease. For a given metabolic need ($\dot{V}_{CO_2}$ fixed) and a decreased $\dot{V}_A$ (alveolar

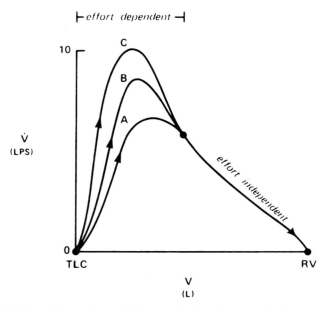

FIGURE 7.6:2 Flow ($\dot{v}$) vs. volume (V) curves in forced expiration between total lung capacity (TLC) and residual volume (RV) with graded efforts after a full inspiration. From Bates et al. (1971)

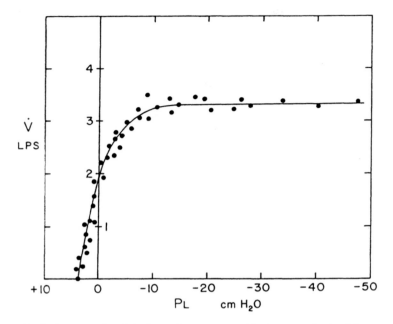

FIGURE 7.6:3 An example of the iso-volume pressure-flow curve obtained in a normal subject at 33 percent of vital capacity. From Bates, Macklem, and Christie (1971), by permission.

hypoventilation), the arterial partial pressure of CO_2 (p_{aCO_2}) will rise (hypercapnia); conversely, for an increased $\dot{V}_A$ (alveolar hyperventilation) p_{aCO_2} will decrease (hypocapnia).

A popular test is the *forced expiration* after a full inspiration. If a person is asked to perform a series of graded expiration efforts, the recorded relationship between the flow rate ($\dot{V}$) and lung volume (V) will be different for each effort. In Fig. 7.6:2, curve A refers to a small effort, curve B is for intermediate effort, curve C is for maximal effort. Note that all three curves merge at a point and thereafter follow a common pathway to the residual volume. This common portion of the curve is said to be *effort independent*. Hyatt, Schilder, and Fry (1958) plotted the forced expiration velocity at a fixed lung volume versus the transpulmonary pressure and obtained a result similar to that shown in Fig. 7.6:3. It is seen that the expiration velocity is independent of the transpulmonary pressure. Later, this was recognized by Permutt, Mead, and others as the "waterfall phenomenon." The analysis presented in Sec. 5.17 applies.

Resistance and Compliance

With the idea that the flow in the airway is driven by the difference of pressures at the alveoli and the mouth, the *airway resistance* to flow R is conventionally defined by the equation

$$R = \frac{p_{\text{mouth}} - p_{\text{alveoli}}}{\dot{V}}. \tag{6}$$

Similarly, with the thought that the lung volume is determined principally by the pressure difference across the pulmonary pleura (i.e. the difference between the alveolar pressure, p_{alv}, and the pleural pressure, p_{PL}), the *lung compliance* is defined by the equation

$$C = \frac{dV}{d(p_{\text{alv}} - p_{\text{PL}})}, \tag{7}$$

where V is the lung volume and $\dot{V}$ is dV/dt. C is approximately constant over the tidal volume range, whereas at the end of quiet expiration (FRC), $p_{\text{alv}} - p_{\text{PL}}$ is approximately 5 cm H_2O. Therefore, as an approximation

$$p_{\text{alv}} - p_{\text{PL}} = (p_{\text{alv}} - p_{\text{PL}})_{\text{FRC}} + \frac{\text{tidal volume}}{C}. \tag{8}$$

Hence, since the *transpulmonary pressure* is defined as

$$p_{\text{mouth}} - p_{\text{PL}} = (p_{\text{mouth}} - p_{\text{alv}}) + (p_{\text{alv}} - p_{\text{PL}}), \tag{9}$$

we have, from Eqs. (6) and (8),

$$p_{\text{mouth}} - p_{\text{PL}} = (p_{\text{alv}} - p_{\text{PL}})_{\text{FRC}} + \frac{\text{tidal volume}}{C} + R\dot{V}. \tag{10}$$

Quantities in these equations can be measured or calculated. p_{PL} is often approximated by measurements using an esophageal balloon. p_{mouth} and $\dot{V}$ can be measured by a pneumotachometer. Tidal volume can be measured by a spirometer. Dubois (1964, pp. 453–454) has shown how to calculate p_{alv} from Boyle's law using measurements from a body plethysmograph. From such measurements, R and C, which are functions of lung volume, can be computed.

7.7 Dynamics of the Ventilation System

Respiratory flow in the airway is a three-dimensional phenomenon of great complexity. Its analysis awaits future development in computational fluid mechanics. In Sec. 7.2 we mentioned a number of basic problems of interest. Among them is the system behavior of the branching tree. Flow in one segment affects the flow in the entire tree. (The same is true also, of course, of circulation.) Simulation of the entire tree has been done only under strong simplifying assumptions so far. A few words about this simplified approach would be worth while because it points out the advantage of introducing the adjoint system and formulating a variational principle.

A method presented by Seguchi et al. (1984, 1986) simulates the ariway as

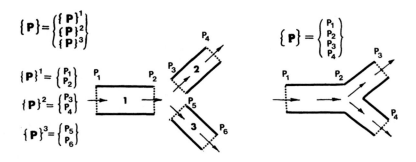

$$\{\mathbf{P}\} = \left\{ \begin{matrix} \{\mathbf{P}\}^1 \\ \{\mathbf{P}\}^2 \\ \{\mathbf{P}\}^3 \end{matrix} \right\}$$

$$\{\mathbf{P}\}^1 = \left\{ \begin{matrix} P_1 \\ P_2 \end{matrix} \right\}$$

$$\{\mathbf{P}\}^2 = \left\{ \begin{matrix} P_3 \\ P_4 \end{matrix} \right\}$$

$$\{\mathbf{P}\}^3 = \left\{ \begin{matrix} P_5 \\ P_6 \end{matrix} \right\}$$

$$\{\mathbf{P}\} = \left\{ \begin{matrix} P_1 \\ P_2 \\ P_3 \\ P_4 \end{matrix} \right\}$$

(a) **Element Vector**   (b) **System Vector**

FIGURE 7.7:1 Simulation of an airway tree by representing each segment of the airway as an element. At a branching point three elements meet. The modal vector of the elements are assembled into a system vector.

a tree (Fig. 5.2:2). Each branch of the tree is called an element, Fig. 7.7:1. Variables such as the pressure, p, are calculated at a number of nodal points and listed as nodal vectors. The value of the variables between neighboring nodal points are specified by certain interpolation formulas. The nodal vectors of the elements are assembled into a system vector, as illustrated in Fig. 7.7:1. The objective of the mathematical simulation is to calculate the system vector according to the basic equations.

The basic equations for the gas in the airways are the equations of motion, Eq. (1.7:8), continuity, Eq. (1.7:4), state (gas law), and boundary conditions. Seguchi et al. (1984) used the one-dimensional approximation of Streeter and Wylie (1967). We must first make clear what the approximation means, and generalize it as much as possible without introducing unwelcome mathematical complications.

Approximate Equations

The Navier–Stokes equation (1.7:8), reads

$$\frac{\partial u_i}{\partial t} + u_j \frac{\partial u_i}{\partial x_j} = +\frac{1}{\rho}\frac{\partial p}{\partial x_i} + \frac{\mu}{\rho}\frac{\partial^2 u_i}{\partial x_j \partial x_j} + X_i. \tag{1}$$

Here u_i is the velocity vector with components u_1, u_2, u_3 (or u, v, w) referred to a rectangular cartesian frame of reference with coordinates x_1, x_2, x_3 (or x, y, z); t is time, ρ is the gas density, p is the pressure, X_i is the gravitational force per unit volume.

Let the axial distance along an airway be represented by $x = x_1$. Consider a cross section of the airway perpendicular to the x axis. On multiplying Eq. (1) by an infinitesimal element of area $dA = dy\,dz$ and integrating the product over the cross section A, the first equation ($i = 1$) becomes

$$\int_A \frac{\partial u}{\partial t} dA + \int_A \frac{1}{2} \frac{\partial u^2}{\partial x} dA + \int_A \left(v \frac{\partial u}{\partial y} + w \frac{\partial u}{\partial z} \right) dA$$

$$= -\int_A \frac{1}{\rho} \frac{\partial p}{\partial x} dA + \frac{\mu}{\rho} \int_A \frac{\partial^2 u}{\partial x^2} dA + \frac{\mu}{\rho} \int_A \left(\frac{\partial^2 u}{\partial y^2} + \frac{\partial^2 u}{\partial z^2} \right) dA. \qquad (2)$$

To simplify this equation, we introduce the following specific approximations:

1) The pressure p is uniform over the cross section. Hence, in isothermal or adiabatic process, the density ρ is also uniform.
2) The third and the fifth integrals in Eq. (2) are zero.
3) The effect of the gravitational force is negligible.
4) The other two equations ($i = 2$ and 3) can be ignored.

With these hypotheses, we introduce the following notations for the average velocity, and transient and convective accelerations and two "conversion factors" c_1 and c_2:

$$\bar{u} \equiv \frac{1}{A} \int_A u \, dA, \qquad \overline{\frac{\partial u}{\partial t}} \equiv \frac{1}{A} \int_A \frac{\partial u}{\partial t} dA \equiv c_1 \frac{\partial \bar{u}}{\partial t}$$

$$\frac{1}{2} \overline{\frac{\partial u^2}{\partial x}} \equiv -\frac{1}{A} \int_A \frac{1}{2} \frac{\partial u^2}{\partial x} dA \equiv c_2 \frac{1}{2} \frac{\partial \bar{u}^2}{\partial x}. \qquad (3)$$

We introduce also a factor Z to replace the last term in Eq. (2) by

$$-8\pi \frac{\mu}{\rho} \bar{u} Z = \frac{\mu}{\rho} \int_A \left(\frac{\partial^2 u}{\partial y^2} + \frac{\partial^2 u}{\partial z^2} \right) dA. \qquad (4)$$

Equation (4) is motivated by a consideration of energy dissipation presented in Sec. 7.2, where the same Z is introduced to correct the dissipation function computed by the Poiseuillean formula. Poiseuille flow is presented in Sec. 5.10 under the assumption that the tube is circular, the axial velocity distribution is axisymmetric, and the flow is laminar (not turbulent). Poiseuille flow satisfies all the simplifying assumptions 1)–4) above, and Eq. (4) with $Z = 1$. This can be verified by dividing $\dot{Q}$ by πa^2, a being the radius of the airway, to obtain $\bar{u}$ using Eq. (5.10:8), substituting the result into Eq. (5.10:7), and using the transformation

$$\frac{\partial^2 u}{\partial y^2} + \frac{\partial^2 u}{\partial z^2} = \frac{1}{r} \frac{\partial}{\partial r} \left(r \frac{\partial u}{\partial r} \right) + \frac{1}{r^2} \frac{\partial^2 u}{\partial \theta^2}$$

from rectangular coordinates x, y, z to cylindrical polar coordinates x, r, θ.

With these approximations, Eq. (1) becomes

$$\rho c_1 \frac{\partial \bar{u}}{\partial t} + \frac{1}{2} \rho c_2 \frac{\partial \bar{u}^2}{\partial x} = -\frac{\partial p}{\partial x} - \frac{8\pi \mu Z}{A} \bar{u}. \qquad (5)$$

Next, apply the same procedure to the equation of continuity (1.7:4)

$$\frac{\partial \rho}{\partial t} + \frac{\partial(\rho u_j)}{\partial x_j} = 0. \tag{6}$$

Under the same approximations named above, we obtain

$$\frac{\partial(\rho A)}{\partial t} + \frac{\partial(\rho \bar{u} A)}{\partial x} = 0. \tag{7}$$

Carrying out the differentiation, regrouping terms, using the notations of "material" derivative (Sec. 1.7)

$$\dot{A} \equiv \frac{DA}{Dt} \equiv \frac{\partial A}{\partial t} + \bar{u}\frac{\partial A}{\partial x}, \qquad \dot{\rho} \equiv \frac{D\rho}{Dt} \equiv \frac{\partial \rho}{\partial t} + \bar{u}\frac{\partial \rho}{\partial x} \tag{8}$$

and dividing through by ρA, we obtain

$$\frac{\dot{A}}{A} + \frac{\dot{\rho}}{\rho} + \frac{\partial \bar{u}}{\partial x} = 0. \tag{9}$$

This equation can be simplified by taking the gas law and the area-pressure relationship of the tube into account. If the cross sectional area A is a function of the transmural pressure $p - p_e$, where p is the pressure in an airway and p_e is the pressure external to it, then the speed of pulse wave in the airway, c, is given in Sec. 5.7 (with a change of notation):

$$c = \left[\frac{A}{\rho}\frac{d(p - p_e)}{dA}\right]^{1/2}. \tag{10}$$

Hence,

$$\frac{\dot{A}}{A} = \frac{1}{A}\frac{dA}{d(p - p_e)}(\dot{p} - \dot{p}_e) = \frac{1}{\rho c^2}(\dot{p} - \dot{p}_e). \tag{11}$$

The other term $\dot{\rho}/\rho$ is related to pressure. By the gas law $p/\rho = RT$, where R is the gas constant and T is the absolute temperature, we have $p/\rho =$ constant if the process is isothermal, or $p/\rho^\gamma =$ constant if the process is adiabatic, γ being another constant. In either case, ρ is a function of p. Hence

$$\frac{\dot{\rho}}{\rho} = \frac{1}{\rho}\frac{d\rho}{dp}\dot{p} \equiv \frac{1}{K}\dot{p}. \tag{12}$$

Here $K = dp/(d\rho/\rho)$ is the bulk modulus of the gas. Note that $dp/(d\rho/\rho) = -dp/(dv/v)$, v being the specific volume of the gas, $v = 1/\rho$, and dv/v is the volumetric strain.

Combining Eqs. (9), (11), and (12), we obtain

$$\left(\frac{1}{\rho c^2} + \frac{1}{K}\right)\dot{p} - \frac{1}{\rho c^2}\dot{p}_e + \frac{\partial \bar{u}}{\partial x} = 0. \tag{13}$$

Equations (5) and (13) are the two basic equations in two variables p and $\bar{u}$ for each segment of the airway. If we assume $c_1 = c_2 = Z = 1$, then Eq. (5) is reduced to the equation used by Streeter and Wylie (1967) and Seguchi et al. (1984). The introduction of the constants c_1, c_2, Z permits us to take secondary flow and turbulence into account as partially illustrated in Sec. 7.2.

Since the number of branches in the airway system is large, a practical way to solve this large set of equations is by finite difference or finite element method. The procedure can be based on a variational principle outlined below.

Variational Principle

Seguchi et al. (1984) extended the Lagrangian multiplier method to formulate a variational principle for the airway system. For each airway branch of length L, we multiply Eq. (5) by $\lambda_1(x, t)$, Eq. (13) by $\lambda_2(x, t)$, sum, and integrate over the domain $(0 \leqslant x \leqslant L, 0 \leqslant t \leqslant T)$ to obtain a functional J:

$$J = \int_0^T \int_0^L \left\{ \lambda_1(x, t) \left[\rho c_1 \frac{\partial \bar{u}}{\partial t} + \rho \frac{c_2}{2} \frac{\partial \bar{u}^2}{\partial x} + \frac{\partial p}{\partial x} + \frac{8\pi\mu Z}{A} \bar{u} \right] \right.$$
$$+ \lambda_2(x, t) \left[\left(\frac{1}{\rho c^2} + \frac{1}{K} \right) \left(\frac{\partial p}{\partial t} + \bar{u} \frac{\partial p}{\partial x} \right) \right.$$
$$\left. \left. + \frac{\partial \bar{u}}{\partial x} - \frac{1}{\rho c^2} \left(\frac{\partial p_e}{\partial t} + \bar{u} \frac{\partial p_e}{\partial x} \right) \right] \right\} dx \, dt, \tag{14}$$

where $\lambda_1(x, t)$ and $\lambda_2(x, t)$ are Lagrangian multipliers. We seek the necessary conditions for J to assume a stationary value when the functions $p(x, t)$, $\bar{u}(x, t)$, $\lambda_1(x, t)$ and $\lambda_2(x, t)$ are allowed to vary independently and arbitrarily. p_e is considered an external forcing function and is not varied for the problem of gas flow in the airway. The first variation of J, after successive integration by parts, takes the form:

$$\delta J = \int_0^T \int_0^L \left\{ \delta\lambda_1 \left[\rho c_1 \frac{\partial \bar{u}}{\partial t} + \rho c_2 \bar{u} \frac{\partial \bar{u}}{\partial x} + \frac{\partial p}{\partial x} + \frac{8\pi\mu Z}{A} \bar{u} \right] \right.$$
$$+ \delta\lambda_2 \left[\left(\frac{1}{\rho c^2} + \frac{1}{K} \right) \dot{p} - \frac{1}{\rho c^2} \dot{p}_e + \frac{\partial \bar{u}}{\partial x} \right]$$
$$+ \delta\bar{u} \left[-\frac{\partial \lambda_1 \rho c_1}{\partial t} - \frac{\partial \lambda_1 \rho c_2 \bar{u}}{\partial x} + \frac{\lambda_1 8\pi\mu Z}{A} - \frac{\partial \lambda_2}{\partial x} + \lambda_2 \left(\frac{1}{\rho c^2} + \frac{1}{K} \right) \frac{\partial p}{\partial x} \right]$$
$$\left. + \delta p \left[-\frac{\partial \lambda_1}{\partial x} - \frac{\partial}{\partial t} \left(\frac{\lambda_2}{\rho c^2} + \frac{\lambda_2}{K} \right) - \frac{\partial}{\partial x} \left(\frac{\lambda_2 \bar{u}}{\rho c^2} + \frac{\lambda_2 \bar{u}}{K} \right) \right] \right\} dx \, dt$$
$$+ \int_0^L \left\{ \lambda_1 \rho c_1 \delta\bar{u}|_0^T + \lambda_2 \left(\frac{1}{\rho c^2} + \frac{1}{K} \right) \delta p \Big|_0^T \right\} dx$$
$$+ \int_0^L \left\{ \left[\lambda_1 + \lambda_2 \left(\frac{1}{\rho c^2} + \frac{1}{K} \right) \bar{u} \right] \delta p \Big|_0^L + [\lambda_1 \rho c_2 \bar{u} + \lambda_2] \delta\bar{u}|_0^L \right\} dt. \tag{15}$$

For the functional J to assume a stationary value, the necessary condition is that δJ vanishes for arbitrary $\delta \lambda_1$, $\delta \lambda_2$, $\delta \bar{u}$, and δp. Thus the brackets multiplying $\delta \lambda_1$, $\delta \lambda_2$, $\delta \bar{u}$, adn δp in the double integral in Eq. (15) must vanish, and we obtain the Euler equations:

$$\rho_1 c_1 \frac{\partial \bar{u}}{\partial t} + \rho_2 c_2 \bar{u} \frac{\partial \bar{u}}{\partial x} + \frac{\partial p}{\partial x} + \frac{8\pi\mu Z}{A} \bar{u} = 0, \tag{16}$$

$$\left(\frac{1}{\rho c^2} + \frac{1}{K}\right) \dot{p} - \frac{1}{\rho c^2} \dot{p}_e + \frac{\partial \bar{u}}{\partial x} = 0, \tag{17}$$

$$\frac{\partial(\lambda_1 \rho c_1)}{\partial t} + \frac{\partial(\lambda_1 \rho c_2 \bar{u})}{\partial x} - \frac{\lambda_1 8\pi\mu Z}{A} + \frac{\partial \lambda_2}{\partial x} - \lambda_2 \left(\frac{1}{\rho c^2} + \frac{1}{K}\right) \frac{\partial p}{\partial x} = 0, \tag{18}$$

$$\frac{\partial \lambda_1}{\partial x} + \frac{\partial}{\partial t}\left(\frac{\lambda_2}{\rho c^2} + \frac{\lambda_2}{K}\right) + \frac{\partial}{\partial x}\left(\frac{\lambda_2 \bar{u}}{\rho c^2} + \frac{\lambda_2 \bar{u}}{K}\right) = 0. \tag{19}$$

The initial and boundary conditions must render the line integrals in Eq. (15) as zero; i.e.,

$$\left[\lambda_1 \rho c_1 \delta \bar{u} + \lambda_2 \left(\frac{1}{\rho c^2} + \frac{1}{K}\right) \delta p\right]\Big|_0^T = 0, \tag{20}$$

$$\left\{\left[\lambda_1 + \lambda_2\left(\frac{1}{\rho c^2} + \frac{1}{K}\right)\bar{u}\right]\delta p + [\lambda_1 \rho c_2 \bar{u} + \lambda_2]\delta \bar{u}\right\}\Big|_0^L = 0. \tag{21}$$

The conditions Eqs. (20) and (21) can be satisfied by proper combinations of the following three sets of conditions:

Foced boundary conditions at $t = 0$ or T and $x = 0$ or L:

$$\begin{aligned}
\delta \bar{u}(x, 0) = 0, \qquad \delta \bar{u}(x, T) = 0, \qquad \delta p(x, 0) = 0, \qquad \delta p(x, T) = 0, \\
\delta \bar{u}(0, t) = 0, \qquad \delta \bar{u}(L, t) = 0, \qquad \delta p(0, t) = 0, \qquad \delta p(L, t) = 0,
\end{aligned} \tag{22}$$

i.e. the values of $\bar{u}$ and p specified at $x = 0$ or L for all t, or at $t = 0$ or T for all x.

Natural initial conditions at $t = 0$ or T:

$$\left[\lambda_1 \rho c_1 \bar{u} + \lambda_2 \frac{1}{\rho c^2} + \frac{1}{K}\right] p = 0. \tag{23}$$

Natural boundary conditions at $x = 0$ or L:

$$\left[\lambda_1 + \lambda_2\left(\frac{1}{\rho c^2} + \frac{1}{K}\right)\bar{u}\right] p + [\lambda_1 \rho c_2 \bar{u} + \lambda_2]\bar{u} = 0. \tag{24}$$

It is clear that Eqs. (16) and (17) are our basic equations (5) and (13), respectively. The initial and boundary conditions given by Eqs. (22)–(24) are physically realizable. Thus the variational principle $\delta J = 0$ is consistent with our system of differential equations and boundary conditions. The solution of ventilation dynamics consists of finding p, u, λ_1, and λ_2 which renders J a stationary value.

Physical Meaning of J

If we give λ_1 a physical dimension of [length^3]/[time], and λ_2 a physical dimension of [force], then J defined by Eq. (14) has the dimension of [force] · [length] or energy. We shall call J the *adjoint energy integral* and λ_1, λ_2 the *adjoint state variables*. The equations (5) and (13) are *nonself-adjoint*. The use of Lagrange multipliers or adjoint variables establishes a functional J which is stationary when the state variables obey the equations of motion and continuity. But whether J is a maximum or a minimum is uncertain. Rigorous mathematical study of this type of variational principle for nonself-adjoint systems remains to be a challenge. In a study of a non-self-adjoint system of a column loaded by a force which is always tangential to the axis of the column, Prasad and Herrmann (1969, 1972) showed that the adjoint method is more efficient for calculation than the Galerkin's method or its extension using weighted averages. Whittle (1971) has discussed variational principles of this nature in a general way.

Discretization and Computer Simulation

Computer simulation of breathing based on the variational equation $\delta J = 0$ has been presented by Seguchi et al. (1984, 1986). The airway system is subdivided into a number of subregions. See Fig. 7.7:1. In a subregion i the values of the variables p, $\bar{u}$, λ_1, and λ_2 are computed at specific points called *nodes* and listed as *nodal vectors* $\{p\}_i^e$, $\{\bar{u}\}_i^e$, $\{\lambda_1\}_i^e$, and $\{\lambda_2\}_i^e$, respectively. Polynomial interpolation matrices $[N]$ are then used to obtain p, $\bar{u}$, λ_1, λ_2 as continuous functions of x:

$$\{p\}_i = [N_p]\{p\}_i^e, \qquad \{\bar{u}\}_i = [N_u]\{\bar{u}\}_i^e,$$
$$\{\lambda_1\}_i = [N_{\lambda_1}]\{\lambda_1\}_i^e, \qquad \{\lambda_2\}_i = [N_{\lambda_2}]\{\lambda_2\}_i^e. \tag{25}$$

The overall airway system is assembled from the subregions. The global vectors can be written as

$$\{p\} = [B_p]\{p_1, p_2, \ldots, p_n\} \tag{26}$$

etc., if we denote $\{p\}_i$ by p_i. The functional J of Eq. (14) can then be evaluated as functions of $\{p\}_i^e$, etc. By setting $\delta J = 0$, a set of linear, simultaneous equations is obtained for $\{p\}_i^e$ and $\{\bar{u}\}_i^e$, which can be solved numerically. The equations for p and $\bar{u}$ are decoupled from those for the adjoints λ_1, λ_2. A key observation that will greatly simplify the analysis is that the airway of each generation has nearly constant diameter, and that the diameter, the compliance, the total cross-sectional area, and the velocity of the flow have stepwise changes at the points of bifurcation where the generation number changes. Hence according to Bernoulli's equation, the blood pressure has a stepwise change at the bifurcation point. These changes should be incorporated in the boundary conditions of the successive generations.

7.8 High-Frequency Low-Tidal-Volume Ventilation

Bohn et al. (1980) have shown that a dog can survive ventilation with a tidal volume smaller than the dead space of the airway at frequencies ranging from 5–30 Hz. This feat was anticipated by Henderson et al. (1915) who supported the idea with an admirably simple experiment. They blew tobacco smoke down a tube and found that it formed a long thin spike, concluding that the "quicker the puff, the thinner and sharper the spike." When the puff stopped "the spike breaks instantly, everywhere, and the tube is seen to be filled from side to side with a mixture of smoke and air." Later, Briscoe et al. (1954) showed that, in man, inspired volumes less than the dead space can reach the alveoli; and they speculated that this might explain why "some patients can live despite the fact that they are breathing very small tidal volumes." Lee (1984), using laser to measure the density of cigarette smoke in a tube, showed that in high frequency ventilation the *effective diffusivity* (see Sec. 7.3) of the smoke can be hundreds or thousands of times larger than the coefficient of molecular diffusion. Hence high frequency ventilation can be effective to send some aerosol drugs down to the lung.

Mechanism of Smoke Transport

The mechanism of the smoke transport in Lee's experiment is as follows: On forward stroke the smoke moves forward with the flow, with higher concentration in the core and lower concentration at the wall. Simultaneously the smoke diffuses (by molecular diffusion or by turbulent mixing) from the core into the wall region. This radial spread of smoke decreases the concentration of the smoke in the core region. In the reverse stroke the same process continues, but the core concentration is lower. Hence in a whole cycle there is a net movement of the smoke to the right. This net movement is repeated in each subsequent cycle; and by a reasoning similar to that used in Sec. 7.3, the concentration of smoke is described by the differential equation

$$\frac{\partial c}{\partial t} = D_{\text{eff}} \frac{\partial^2 c}{\partial x^2} \tag{1}$$

in which c is the concentration of the smoke, t is time, and x is the axial distance in the direction of motion. D_{eff} is a constant called the *effective diffusivity*.

Mechanism of Gas Transport in a Tube

Similar to the smoke transport in Lee's experiment, the basic reason why oxygen can get to the alveoli at a tidal volume smaller than the dead space is because the velocity and concentration profiles in the airway are not flat. The velocity is faster and the concentration is higher in the core region, so that by a mechanism similar to the Taylor diffusion (see Sec. 7.3) the effective diffusivity is augmented. Taylor's solution is, however, an asymptotic solution of a steady flow. To generalize Taylor's solution to high frequency ventilation of the lung, one must take into consideration the periodic oscillations, the branching pattern of the airway, and the fundamental difference in convergent and divergent flows of a viscous fluid in bifurcating tubes. The velocity and concentration profiles are affected by these factors.

Some features of oscillatory flows in a circular tube have been discussed in Chap. 5, Secs. 5.7, 5.12. In oscillatory flow the velocity profile is no longer parabolic, but is more blunt, and the maximum velocity occurs at a radial position which varies with time. Depending on the Womersley number, the high velocity core spreads much closer to the wall. The Taylor diffusion mechanism is enhanced by this change of velocity profile.

For oscillatory flow in a circular cylindrical tube of infinite length, Chatwin (1975) has solved the problem of dispersion. When the Womersley number is greater than 1 and the *Schmidt number* ($\eta \equiv v/D_{mol}$) is of order 1 or greater, Chatwin finds the effective diffusivity to be

$$D_{eff} = D_{mol} + \frac{v^{0.5}G^2}{4^{3.5}f^{3.5}d^2}[F_1(\eta) + F_2(\eta)(\cos 4\pi f t - \sin 4\pi f t)], \qquad (2)$$

where D_{mol} is the coefficient of molecular diffusion, v is the kinematic viscosity, f is frequency in Hz, d is diameter in cm, t is time, $F_1(\eta)$, $F_2(\eta)$ are functions of the Schmidt number η, and G is proportional to the amplitude of the pressure gradient.

$$G = -\frac{1}{\rho}\frac{dp}{dx}. \qquad (3)$$

Chang et al. (1984), using fluid transmission line theory for a sinusoidal excitation (Brown, 1962), write

$$G = 2\pi f \mathscr{Z} V_T, \qquad (4)$$

where V_T is the tidal volume in ml, and $\mathscr{Z}$ is the impedance. For air, and with $j = (-1)^{1/2}$, they found

$$\mathscr{Z}(f) = \frac{8fj}{d^2}\left\{1 + 0\left[\left(\frac{2v}{\pi f d^2}\right)^{0.5}\right]\right\} \qquad (5)$$

in which the symbol 0 means "of the order of the argument in the bracket." For high frequency ventilation of man the argument $2v/(\pi f d^2)$ is much smaller than 1, hence, on omitting the small term involving 0 in Eq. (5) and substituting Eqs. (3), (4), and (5) into Eq. (2), Chang et al. obtain

$$D_{eff} = D_{mol} + \frac{64V_T^2 f^{0.5}v^{0.5}}{\pi^{1.5}d^6}g(\eta), \qquad (6)$$

where $g(\eta)$ is the quantity in the bracket in Eq. (2). In high frequency ventilation the second term on the right-hand side of Eq. (6) is much larger than the first; and we see that according to Chatwin's theory the effective diffusivity increases with the square of the tidal volume, the square root of the frequency, and the inverse sixth power of the diameter:

$$D_{eff} \propto \frac{V_T^2 f^{0.5}}{d^6}. \tag{7}$$

Hence for the same amount of ventilation $(V_T f)$, it is more effective to increase the tidal volume V_T than to increase the frequency; and the effective diffusivity is much larger in the small airways than in large ones. Experimental validation of this formula has been done by Chang et al. (1984). The maximum values of D_{eff} reached $180 \, cm^2/sec$ in Chang's experiments as compared to the molecular diffusivity of CO_2 in air, $0.16 \, cm^2/sec$ at $20°C$.

Gas Transport in Airway Tree

The airway is an assemblage of tubes which end on a common alveolar space. Since the total cross-sectional area of the airway increases with the generation number beyond the third bronchial generation, the inspiratory flow is divergent at the points of bifurcation whereas the expiratory flow is convergent at these points. The secondary flow of a divergent flow is very different from that of a convergent flow, and this will influence the mass transport characteristics. The boundary layer thickness grows faster in a divergent flow and the velocity profile peaks in the central region. The boundary layer is thinner in a convergent flow and the velocity profile is blunter in the middle. Scherer et al. (1982) have paid attention to this feature. Using a flow visualization method they have measured the transport characteristics of small opaque beads (diameter $\sim 40\mu$) in a glycerine-water solution flowing in a physical model of bronchial tree. With the empirical results they constructed a theory of gas transport in high frequency ventilation.

While the Taylor mechanism dominates the scene from mouth to bronchioles, beyond the respiratory bronchioles the molecular diffusion mechanism takes over. Some authors (see Chang, 1984), however, emphasize the convective effect of the direction of flow relative to the angles of branches at the points of bifurcation. This last mechanism may persist at the level of alveolar ducts.

Many authors performed animal experiments (e.g. Bohn et al. (1980), Slutsky et al. (1981)), whereas others investigated the effects of turbulence on effective diffusivity (Lee, 1984), and the transition between laminar and turbulent flows in oscillatory condition (Winter and Nerem, 1984). Chang (1984) explored high frequency ventilation by vibrating the chest and abdominal wall. Clinical applications have been vigorously pursued.

Problems

7.1 Draw a free-body diagram of the lung. Analyze the condition of equilibrium of the lung in the thoracic cavity without neglecting the weight of the lung, and describe a method to determine how the pleural pressure varies with the height.

7.2 It is common practice to assume the mean stress in the lung parenchyma to be roughly equal to the transpulmonary pressure $p_{ALV} - p_{PL}$. Assess the accuracy of this hypothesis analytically on the basis of biomechanical principles.

7.3 Analyze the pressure in the esophagus as measured by a catheter tip balloon inserted into the esophagus and discuss whether the result is approximately equal to the pleural pressure or not.

7.4 *The Dissipation Function $\mathscr{D}$ of Eq. (2) of Sec. 7.2.* In an incompressible Newtonian viscous fluid, consider a control volume of unit volume consisting of a parallelopiped with edges parallel to the coordinates axes. The stresses acting on the surfaces of this element are σ_{ij}. The rate of change of strain is

$$\dot{e}_{ij} = \frac{1}{2}\left(\frac{\partial u_i}{\partial x_j} + \frac{\partial u_j}{\partial x_i}\right),$$

where u_1, u_2, u_3 are the velocity components. The rate at which the work is done on this element by the stresses is $dW = \sigma_{ij}d\dot{e}_{ij}$. The stress–strain relationship is given by Eq. (7) of Sec. 1.7. Hence show that $W = -p\dot{e}_{kk} + \mu\dot{e}_{ij}\dot{e}_{ij}$. This work is dissipated as heat. The dissipation $\mathscr{D}$ is obtained by integrating W throughout the volume occupied by the fluid. For a Poiseuillean flow, only one derivative of the velocity field does not vanish; namely, $\partial u/\partial r$. Deduce $\mathscr{D}_P$ for Poiseuille flow and verify Eq. (4) of Sec. 7.2.

7.5 Describe qualitatively the mechanics of gas flow in the airway when one sneezes or coughs. Write down the governing differential equations and boundary conditions.

7.6 From the point of view of boundary layer development, what is the difference between inspiring flow and the expiratory flow in the airway? What effect does this have on the resistance to flow in inspiration and in expiration? Write down the boundary layer equations and the boundary conditions that apply.

7.7 Discuss the difference between the tendency for flow separation (the failure of boundary streamlines to follow the solid wall) to occur in the airway in inspiration and that in expiration. What effect does flow separation have on the resistance to flow? Formulate this discussion mathematically on the basis of boundary layer equations.

7.8 Discuss the difference in the Z factor for inspiration from that for expiration. Why does such a difference exist? Is Z larger in inspiration than in expiration? (Cf. Pedley et al. (1977).)

7.9 To analyze the dynamics of breathing, the flow in the airway may be idealized as a one-dimensional flow. The flow is assumed to depend only on the distance x along the airways. Consider a control volume of length Δx, confined between sections x and $x + \Delta x$. Use the method of continuum mechanics, express the balance of forces acting on the control volume. Make suitable approximations to derive a differential equation governing the flow in inspiration and in expiration. (Cf. Schmid-Schönbein and Fung (1978).)

7.10 A small perturbation in pressure is introduced to the air flow at the mouth by means of an "interrupter." Derive the linearized differential equation and boundary conditions for the perturbation of flow in the entire airway system. Take notice of the elasticity of the airways. In order to completely determine the boundary conditions, it is necessary to consider the lung parenchyma, pleura, chest wall, diaphragm, abdomen, and the heart. Describe a mathematical procedure to complete the formulation of the problem. (Cf. loc. cit. Prob. 7.9)

7.11 Estimate the Mach number of the flow of air in human lung breathing at a rate of 1 liter/sec. Can the air be treated as an incompressible fluid in Problems 7.9–10? Show that the compressibility of air can be ignored.

7.12 Consider a numerical method of solving the equations obtained in Prob. 7.9. Outline the necessary steps. (Cf. Sec. 7.8)

7.13 From the solutions $\bar{c}(\xi, t) \equiv \bar{c}(x - Ut, t)$ and $c'(\xi, z, t)$ given in Eqs. (10), (15), and (23) of Sec. 7.3, reexamine the hypotheses expressed in Eqs. (13) of Sec. 7.3. Derive the conditions under which Eqs. (13) are valid.

7.14 Derive Eq. (7.3:21) from Eq. (7.3:19) by considering the balance of mass inflow and outflow of the solute in a control volume with transient change of the total mass in the volume. Derive Eq. (7.3:23) from (7.3:22).

7.15 Consider a circular cylindrical tube (Fig. 7.3:2). The right-hand side of Eq. (6) has to be replaced by (see Fung, 1977, p. 277)

$$\nabla^2 c = \frac{\partial^2 c}{\partial x^2} + \frac{1}{r}\frac{\partial}{\partial r}\left(r\frac{\partial c}{\partial r}\right),$$

where r is the radial distance. Derive the formula for $f(r)$ corresponding to $f(z)$. Show that $\mathscr{D}_{app}$ is given by Eq. (1) in this case.

7.16 Consider flow in a semicircular cylindrical open channel. The boundary condition on the free surface is that the shear stress $= 0$ and pressure $= 0$. The channel is tilted to such a degree that the fluid fills the semicircular cross section all the time. The gravitational force and the slope of the channel drives the flow. Show that the Poiseuille solution for the velocity field in an axisymmetric flow in a full cylinder satisfies all the boundary conditions here. Consider diffusion of an impulsively injected solute at a section in analogy to the cases considered above. Deduce the apparent diffusivity $\mathscr{D}_{app}$ in this case.

The solution considered in Section 7.3 is the foundation of the indicator dilution method of measuring flow rate and organ volume. The example of the open channel suggests the applicability of the analysis to measure flow rate in sewers and rivers.

7.17 With the solution $I(x, t)$ given in Eq. (7.3:23) for the response to an impulsive input, derive the concentration distribution following a step input at time $t = 0$ of the solute B of unit mass per unit axial length distributed uniformly from $x = 0$ to $-\infty$, i.e., with the input $f(x, t) = \mathbf{1}(t) \cdot \mathbf{1}(-x)$. $\mathbf{1}(t)$ is the unit-step function, which is zero when the argument is negative, and is $\mathbf{1}$ when the argument is positive.

7.18 Consider an airway tree. Hypothetically, regarding the first generation as an infinitely long tube extending from $x = 0$ to $-\infty$. When the solute B is injected as a unit step function $\mathbf{1}(t) \cdot \mathbf{1}(-x)$, derive the concentration distribution of the solute in the tree following the injection.

7.19 Extend the solution of Prob. 7.18 to the alveolar wall. Then formulate the mathematical problem of oxygen transport across the alveolar wall. Do the same for CO_2.

7.20 Using qualitative physical reasoning, explain why the ternary diffusion effect on O_2 and CO_2 transport in the alveoli will be significant if the third gas is helium.

7.21 Consider a liquid layer between two gases (Fig. 7.4:1). Show that in general the partial pressure of a diffusible gas is discontinuous at the liquid-gas interface.

7.22 The solubility of oxygen in blood is nonlinear. A curve showing the oxygen concentration in blood versus the partial pressure of O_2 is called a dissociation curve. Describe the dissociation curve of O_2 as affected by the CO_2 concentration. Similarly, describe the dissociation curve of CO_2 in blood as affected by the oxygen concentration. (Cf. West and Wagner (1977), West (1974).)

7.23 Take a breath of N_2O. Calculate the time course of uptake of N_2O by the blood. *Answer.* Let V_0 be the intake volume of N_2O, corresponding to a partial pressure p_0. Let V be the volume of N_2O transfered to blood. Then the partial pressures of N_2O in the alveoli and blood, p_1 and p_2 respectively, are

$$p_1 = p_0\left(1 - \frac{V}{V_0}\right), \qquad p_2 = \frac{V}{\lambda V_c},$$

where V_c is the capillary blood volume and λ is the solubility of N_2O in blood. Hence, the equation (7.4:10),

$$\dot{V} = D_L(p_1 - p_2)$$

is reduced to

$$\dot{V} + D_L\left(\frac{p_0}{V_0} + \frac{1}{\lambda V_c}\right)V = D_L p_0.$$

The solution of this differential equation is

$$V = Ce^{-\alpha t} + \frac{D_L p_0}{\alpha},$$

where

$$\alpha = \frac{p_0}{V_0} + \frac{1}{\lambda V_c}$$

and C is an integration constant. Since $V = 0$ when $t = 0$, we have

$$C = -\frac{D_c p_0}{\alpha}.$$

Hence the solution is

$$V = \frac{D_L p_0}{\alpha}(1 - e^{-\alpha t}).$$

7.24 Discuss the changes in the $\dot{V}/\dot{Q}$ ratio in the following circumstances: 1) obstruction in blood vessels, 2) local obstruction in airway, 3) restricted ventilation in some airways, 4) asthma, 5) edema in the lung, 6) black lung disease or pneumoconiosis of coal miners, 7) silicosis.

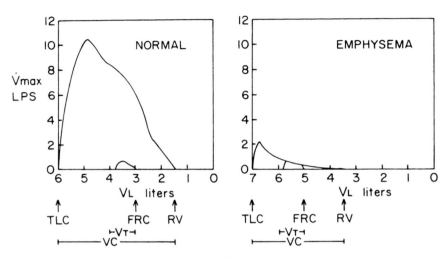

FIGURE P.7.27 Expiratory flow vs lung volume curves for normal and patients with emphsema. From Bates et al. (1971)

7.25 Explain why the airway resistance to flow varies inversely with lung volume.

7.26 In forced expiration the pleural pressure can become greater than atmospheric. Based on the pressure-diameter relationship, Eq. (1) of Sec. 5.7, and the dependence of σ_{alv} on p_{PL}, discuss the consequent change in the airway diameter.

7.27 The left panel of Fig. P7.27 shows a normal maximal expiratory flow vs volume curve together with a flow rate curve during quite breathing over the tidal volume range. The right panel shows the maximal expiratory flow vs volume curve obtained from a patient with emphysema. For the emphysema patient the flow rates during quite breathing are the maximal ones over the tidal volume range. Explain the cause of difference of the flow-volume curves of the emphysema patient in terms of the geometric and structural changes of the lung. (Cf. Bates et al. (1971), Tisi (1980).)

7.28 Name and describe briefly a few principles by which the velocity of flow of a gas can be measured.

7.29 Name a few principles by which the pressure in a gas can be measured.

7.30 Consider an airway (A) branching into two daughters, B_1 and B_2, one of which (B_1) is severely constricted. At a given frequency of ventilation, the velocity of flow into branch B_1 will be slower than that in B_2 in inspiration and the volume of the lung supplied by B_1 will increase at a rate slower than that supplied by B_2. In other words, there will be a phase lag. If the compliance C is defined by Eq. (7.6:7). Show that the compliance of the lung supplied by the airway A will be frequency dependent. (Cf. Otis et al. (1956).)

7.31 What kind of patient needs oxygen therapy? The physiologist Haldane used to say that intermittent oxygen therapy is like bringing a drowning man to the surface—occasionally! Why is it?

7.32 Design an instrument to measure the volume flow rate of breathing through the mouth.

7.33 Design an instrument which can measure velocity at a point in the trachea.

7.34 Design an instrument to measure the gas volume of inspiration and expiration.

7.35 Design an instrument by which the volume of inspired or expired gas can be recorded as a function of time during breathing.

7.36 Errors in the use of respirators is a major cause of accidents in hospitals. If you were a hospital engineer, what would you do to minimize this problem?

7.37 An effective way to measure the resistance to gas flow in the airway, and the elastic compliance of the lung is to use the small perturbation method (see Secs. 7.6, 7.7). Define "small" for the small perturbations in pressure and flow. Set a criterion to test the smallness. Design an instrumentation system which is practical for patient use.

7.38 Small perturbations of pressure that act on the chest wall but not at the mouth can be introduced by using a loud speaker in a body plethysmograph box. Small perturbations of pressure and flow that act on the flow-through the mouth, but not on the chest can be introduced by using a loud speaker in the respirator circuit. Through a speaker system one can impose perturbations of sinusoidal or other wave forms. But if a delta function perturbation is desired (so that the perturbation is very brief in time), an interrupter may be introduced in the respirator circuit or a small impact may be used on the chest. Discuss the pros and cons of these methods from the point of view of clinical use.

7.39 Reliable measurements of the partial pressure of oxygen in resting mammalian skeletal muscle show that the pO_2 is about 100 mm Hg in large arteries, 40 mm Hg in large veins, 15–19 mm Hg in the muscle tissue. Explain why the tissue pO_2 is so much *lower* than the venous pO_2? (Cf. Harris, P.D., (1986). Movement of oxygen in skeletal muscle. *News in Physiol. Sci.* 1: 147–149.)

7.40 Pulmonary edema is caused by movement of fluid from blood into the interstitial and alveolar space. One major problem when dealing with critically ill patients is the accumulation of edema fluid within the interstitial spaces of the lung. The excess interstitial fluid reduces the lung's ability to provide oxygen to the body, thus compromising the recovery of patients already at risk. List the factors that can influence edema. Discuss the clinical measures that will control edema. Consider especially the role of lymph flow and its relation to the systemic venous pressure; and of the effects of infection. (Cf. Laine, G.A., Allen, S.T., Williams, J.P., Katz, J., Gabel, J.C. and Drake, R.E. (1986). A new look at pulmonary edema. *News in Physiol. Sci.* 1: 150–153.)

References

Bates, C.V., Macklem, P.T., and Christie, R.V. (1971). *Respiratory Function in Disease.* Saunders, Philadelphia.

Bohn, D.J., Miyasaka, K., Marchak, B.E., Thompson, W.K., Froese, A.B., and Bryan, A.C. (1980).

Ventilation by high frequency oscillation. *J. Appl. Physiol.: Respir. Environ. Exer. Physiol.* **48**: 710–716.

Briscoe, W.A., Forster, R.E., and Comroe, J.H. (1954). Alveolar ventilation at very low tidal volumes. *J. Applied Physiol.* **7**: 27–30.

Brown, F.T. (1962). The transient response of fluid lines. *J. Basic Engineering* **84**: 547–553.

Chang, H.K. (1984). Mechanics of gas transport during ventilation by high frequency oscillation. *J. Appl. Physiol.* **56**: 553–563.

Chang, H.K., Cheng, R.T., and Farhi, L.E. (1973). A model study of gas diffusion in alveolar sacs. *Respiration Phys.* **18**: 386–397.

Chang, H.K., Isabey, D., Shykoff, B.E., and Harf, A. (1984). Gas mixing during high frequency oscillation. In *Biomechanics in China, Japan and USA*. (Y.C. Fung, J.J. Wang, and E. Fukada, eds.), Science Press, Peaking, pp. 264–280.

Chang, H.K., Tai, R.C., and Farhi, L.E. (1975). Some simplifications of ternary diffusion in the lung. *Respiration Phys.* **23**: 109–120.

Chatwin, P.C. (1975). On the longitudinal dispersion of passive contaminant in oscillatory flows in tubes. *J. Fluid Mech.* **71**: 513–527.

Cumming, G., Henderson, R., Horsfield, K., and Singhal, S.S. (1968). The functional morphology of the pulmonary circulation. In *The Pulmonary Circulation and Interstitial Space* (A. Fishman and H. Hecht, eds.), University Chicago Press, Chicago, pp. 327–338.

Cumming, G. and Semple, S.J. (1973). *Disorders of the Respiratory System*. Blackwell Sci. Pub., London.

Dubois, A.B. (1964). Resistance to breathing. In *Handbook of Physiology*, Sec. 3 *Respiration*, Vol. 1 (W.O. Fenn and H. Rahn, eds.). Amer. Physiol. Soc. Washington, D.C. 1964, pp. 451–462.

Fenn, W.O. and Rahn, H. (eds.) (1964). *Handbook of Physiology*: Sec. 3 *Respiration*, Vols. 1 & 2., 1696 pp. American Physiological Society, Washington, D.C.

Fishman, A., Macklem, P., and Mead, J. (eds.) (1986). *Handbook of Physiology*, Sec. 3. *Respiration*. Amer. Physiol. Soc. Washington, D.C.

Fry, D.L. and Hyatt, R.E. (1960). Pulmonary mechanics. A unified analysis of the relationship between pressure, volume and gasflow in the lungs of normal and diseased human subjects. *Amer. J. Med.* **29**: 672–689.

Fung, Y.C. (1977). *A First Course in Continuum Mechanics*, 2nd edn. Prentice-Hall, Englewood Cliffs, N.J.

Fung, Y.C. (1983). *Biodynamics: Circulation*. Springer-Verlag, New York.

Henderson, Y., Chillingworth, F.P., and Whitney, J.L. (1915). The respiratory dead space. *Amer. J. Physiol.* **38**: 1–19.

Hirschfelder, J.O., Curtis, C.F., and Bird, R.B. (1954). *Molecular Theory of Gases and Liquids*. Wiley, New York.

Hyatt, R.E., Schilder, D.P., and Fry, D.L. (1958). Relationship between maximum expiratory flow and degree of lung inflation. *J. Appl. Physiol.* **13**: 331–336.

Jaeger, M.J. and Matthys, H. (1968/69). The patten of flow in the upper human airways. *Respiration Phys.* **6**: 113–127.

Krogh, M. (1914/15). The diffusion of gases through the lungs of man. *J. Physiol (London)* **49**: 271–300.

Lee, J.S. (1984). The mixing and axial transport of smoke in oscillatory tube flows. *Annals of Biomed. Eng.* **12**: 371–383.

Lee, J.S. (1984). A transient analysis of gas transport in oscillatory tube flows. In *Biomechanics in China, Japan, and USA* (Y.C. Fung, J.J. Wang, and E. Fukada, eds.). Science Press, Peking, pp. 254–263.

Otis, A.B., McKerrow, C.B., Bartlett, R.A., Mead, J., McIlroy, M.B., Selverstone, N.J., and Radford, E.P., Jr. (1956). Mechanical factors in distribution of pulmonary ventilation. *J. Appl. Physiol.* **8**: 427–443.

Pedley, T.J., Schroter, R.C., and Sudlow, M.F. (1971). Flow and pressure drop in systems of repeatedly branching tubes. *J. Fluid Mech.* **46**: 365–383.

Pedley, T.J., Schroter, R.C., and Sudlow, M.F. (1977). Gas flow and mixing in the airways. In *Bioengineering Aspects of the Lung* (J.B. West, ed.), Marcel Dekker, New York, pp. 163–265.

Prasad, S.N. and Herrmann, G. (1969). The usefulness of adjoint systems in solving nonconservative stability problems. *Int. J. Solids and Struct.* 5: 727–735.

Prasad, S.N. and Herrmann, G. (1972). Adjoint variational methods in nonconservative stability problems. *Int. J. Solids and Struct.* 8: 29–40.

Pride, N.B., Permutt, S., Riley, R.L., and Bromberger-Barnea, B. (1967). Determinants of maximal expiratory flow from the lungs. *J. Appl. Physiol.* 23: 646–662.

Roughton, F.J.W. and Forster, R.E. (1957). Relative importance of diffusion and chemical reaction rates in determining rate of exchange of gases in the human lung. *J. Appl. Physiol.* 11: 291–302.

Scherer, P.W. and Haselton, F.R. (1982). Convective exchange in oscillatory flow through bronchial-tree models. *J. Appl. Physiol.: Respir. Environ. Exer. Physiol.* 53(4): 1023–1033.

Schmid-Schoenbein, G. and Fung, Y.C. (1978). Forced perturbation of respiratory system. (A) The traditional model. *Annals of Biomed. Eng.* 6: 194–211. (B) A continuum mechanics analysis. *ibid*, 6: 367–398.

Seguchi, Y., Fung, Y.C., and Maki, H. (1984). Computer simulation of dynamics of fluid-gas-tissue systems with a discretization procedure and its application to respiration dynamics. In *Biomechanics in China, Japan, and USA* (Y.C. Fung, E. Fukada, J.J. Wang, eds.). Chinese Science Press, Beijing, pp. 224–239.

Seguchi, Y., Fung, Y.C. and Ishida, T. (1986). Respiratory Dynamics-Compter simulation. In *Frontiers in Biomechanics* (G.W. Schmid-Schönbein, S.L.Y. Woo, B.M. Zweifach, eds). Springer Verlag, New York, pp. 377–391.

Slutsky, A.A., Drazen, J.M., Ingram, R.H., Jr., Kamm, R.D., Shapiro, A.H., Fredberg, J.J., Loring, S.H., and Lehr, J. (1980). Effective pulmonary ventilation with small-volume oscillations at high frequency. *Science* 209: 609–611.

Slutsky, A.S., Kamm, R.D., Rossing, T.H., Loring, S.H., Lehr, J., Shapiro, A.H., and Ingram, R.H., Jr. (1981). Effects of frequency, tidal volume, and lung volume on CO_2 elimination in dogs by high frequency (2–30 Hz) low tidal volume ventilation. *Clin. Invest.* 68: 1475–1484.

Stewart, G.N. (1893). Researches on the circulation time in organs and on the influences which affect it. *J. Physiol. (London)* 15: 1–30. II. Time of the lesser circulation, 15: 31–72. III. Thyroid gland, 15: 73–89. The output of the heart, 22: (1900) 159–173.

Streeter, V.L. and Wylie, E.B. (1967). *Hydraulic Transients*. McGraw-Hill, New York.

Taylor, G.I. (1953). Dispersion of soluble matter in solvent flowing slowly through a tube. *Proc. Roy. Soc. (London)*, Series A, 219: 186–203.

Taylor, G.I. (1954). The dispersion of matter in turbulent flow through a pipe. *Proc. Roy. Soc. (London)*, Series A., 223: 446–468.

Tisi, G.M. (1980). *Pulmonary Physiology in Clinical Medicine*. Williams and Wilkins, Baltimore, Maryland.

Van der Pol, B. and Bremmer, H. (1950). *Operational Calculus*, Cambridge University Press, London/New York, pp. 305–307.

Wagner, P.P. and West, J.B. (1972). Effects of diffusion impairment on O_2 and CO_2 time courses in pulmonary capillaries. *J. Applied Physiol.* 33: 62–71.

Weibel, E.R. (1963). *Morphometry of the Human Lung*. Academic Press, New York.

West, J.B. (1974). *Respiratory Physiology—The Essentials*. Williams and Wilkins, Baltimore.

West, J.B. (1982). *Pulmonary Pathophysiology—the essentials*. 2nd ed. Williams and Wilkins, Baltimore.

West, J.B. and Wagner, P.D. (1977). Pulmonary gas exchange. In *Bioengineering Aspects of the Lung* (J.B. West, ed.), Marcel Dekker, New York, pp. 361–457.

Whittle, P. (1971). *Otpimization Under Constraints: Theory and Applications of Nonlinear Programming*. Wiley-Interscience, London, New York.

Winter, D.C. and Nerem, R.M. (1984). Turbulence in pulsatile flows. *Annals of Biomed. Eng.* 12: 357–369.

CHAPTER 8

Basic Transport Equations According to Thermodynamics, Molecular Diffusion, Mechanisms in Membranes, and Multiphasic Structure

8.1 Introduction

Now we shall consider the movement of water and other fluids in our bodies, especially the exchange of fluid between blood and the extravascular tissues. Red blood cells cannot leave the blood vessel; but water, ions, and some white blood cells can. The fluid in the extravascular space moves and exchanges matter with the cells in the body. The ionic composition of the fluid in the cells is quite different from that in the extracellular space. Extracellular fluid is rich in Na^+, Cl^-, HCO_3^-, whereas the intracellular fluid is rich in K^+, Mg^{++}, phosphates, proteins, and organic phosphates. The composition of blood plasma is fairly similar to that of the extravascular fluid, except that the plasma has some 14 mEq/L of proteins while extracellular fluid has essentially none. See Table 8.1:1. To talk about mass transport in the body we must explain how this difference in composition comes about.

In man, the total body water accounts for 45–50 percent of the body weight in adult females and 55–60 percent of the body weight in adult males. Approximately 50 percent of this water is in muscle, 20 percent in the skin, 10 percent in the blood, and the remainder in the other organs. The total body water is contained in two major compartments which are divided by the cell membrane: the *cell water* and the *extracellular water* (Table 8.1:2). The extracellular compartment is subdivided into several subcompartments: the *interstitial-lymphatic* compartment, the *vascular* compartment, the *bone* and *dense connective tissue* compartment (cartilage and tendons), and the *transcellular water* compartment (epithelial secretions). In this chapter we shall mostly be concerned with the interstitial-lymphatic compartment which communicates directly with the vascular compartment.

Water and solutes are continually exchanged between these fluid compart-

TABLE 8.1:1 Ionic composition of body water compartments*

	Extravascular Extracellular fluid (mEq/L)	Intracellular fluid of skeletal muscle cell (mEq/L)	Normal blood plasma (mEq/L)
Na^+	145.1	12.0	142.0
K^+	4.4	150.0	4.3
Ca^{++}	2.4	4.0	2.5
Mg^{++}	1.1	34.0	1.1
Cations Total	153.0	200.0	149.9
Cl^-	117.4	4.0	104.0
HCO_3^-	27.1	12.0	24.0
$HPO_4^{2-}, H_2PO_4^-$	2.3	40.0	2.0
Proteins	0.0	54.0	14.0
Other	6.2	90.0††	5.9
Anions Total	153.0	200.0	149.9

* Data from D.M. Woodbury, in Howell and Fulton's *Physiology and Biophysics*, 20th Ed., T.C. Ruch and H.D. Patton (eds.). Saunders, Philadelphia, 1974.
†† This largely represents organic phosphates such as ATP.

TABLE 8.1:2 Distribution of total body water in an average 70-kg male man, according to Woodbury, loc. cit., supra

Expressed in percent body weight and approximate volume		
Cell water	36.0%	25 liter
Extracellular water		
Interstitial fluid & lymph	11.5%	8 liter
Blood plasma	4.5%	3 liter
Bone	3.0%	2 liter
Dense connective tissue	4.5%	3 liter
Transcellular water*	1.5%	1 liter
Total body water $\cong$ 60 percent body weight		

* The transcellular water consists of epithelial secretions such as the digestive secretions, sweat, and cerebrospinal, pleural, synovial, and intraocular fluids.

ments. This exchange occurs by passive and active mechanisms. The movement of particles is *passive* if it develops spontaneously and does not require a supply of metabolic energy, or *active* if it depends upon energy derived from metabolic processes. In humans, solute movement occurs by both passive and active mechanisms whereas all water movement is passive. Active transport requires a certain machinery, certain biochemical processes occurring in some specific physical space. The space is usually the cell membrane and membranes within cells. Hence mass transport in membranes has many unique features.

This extensive subject will be treated in two chapters. In this chapter we shall consider the basic equations that describe the transport phenomena: the constitutive equations and the phenomenological laws. In the next chapter, applications of these basic equations will be considered.

The phenomenological laws are empirical equations formulated with the guidance of theoretical considerations. The most general theoretical foundation is thermodynamics, with the concepts of entropy production, Onsager's principle, and molecular motion. Greater details are obtained, however, by consideration of specific molecular and structural models. Hence we begin this chapter with thermodynamics (Sec. 8.2), emphasizing chemical potential because of its central importance on mass transport (Secs. 8.2–8.4), then discuss the internal entropy production (Sec. 8.5) and phenomenological laws of diffusion, osmotic pressure, Darcy, Fick, and Starling (Sec. 8.6). We then proceed to molecular theory (Sec. 8.7), add detailed mechanisms in biological membranes (Sec. 8.8) and solid immobile matrix (Sec. 8.9).

8.2 The Laws of Thermodynamics

One of the ways to arrive at a suitable formulation of the phenomenological laws of fluid motion in interstitial space is by thermodynamics. In this section, a brief review of thermodynamics is given with a view to clarify the concept of chemical potential.

Classical thermodynamics is based on the following axioms:

1) conservation of matter,
2) an isolated system tends toward equilibrium,
3) the first law: conservation of energy,
4) the second law,
5) the third law.

By 2), thermometers are used to characterize empirical temperature, and calorimeters are used to quantify heat. For a given system, a small increment of heat (dQ) is proportional to a small change of temperature (dt), so that

$$dQ = C\,dt, \tag{1}$$

where C is the *heat capacity*. The accepted unit of heat is *calorie*, the heat which must be imparted to 1 gm of water to raise its temperature by $1°C$ at $15°C$ and atmospheric pressure. Davy, Joule, and others established the mechanical equivalent of heat: one calorie equals 4.1868 Joule. They showed that if a small amount of heat dQ is imparted on a body, the body would change its state and do work:

$$dQ = dU - dW. \tag{2}$$

The variable U is a function of the state of the body; W is the work done by the system on the surrounding. Equation (2) is a statement of the *first law of thermodynamics.*

Joule and others, in establishing the first law, experimented on many forms of work and energy: mechanical, electrical, chemical, etc. If the body is a solid and the stress in the body is σ_{jk} (free Latin indexes range over 1, 2, 3), and the body deforms so that its strain is changed by de_{jk}, then the work done by the stress is $\sigma_{jk} de_{jk}$ per unit volume, or $\sigma_{jk} de_{jk} V$ in a body of volume V.* If a force $\mathbf{F}$ acts on a material particle which is moved by a displacement $\mathbf{dx}$, then the work done is $\mathbf{F} \cdot \mathbf{dx}$.** If a muscle shortens by an amount $-\mathbf{dL}$ under a force $\mathscr{F}$, it performs work equal to $-\mathscr{F} \cdot \mathbf{dL}$. If a quantity of electricity $-de$ is given off by a system at an electric potential ψ, an electric work $\psi \, de$ is performed. If dn_i mole of the ith species of chemicals is transported into the system from the surrounding, a chemical energy $\mu_i \, dn_i$ is added to the system (the constant of proportionality μ_i is called the *chemical potential*). Thus, on denoting the change of energy by dU for a solid body containing muscle fibers and subject to a body force $\mathbf{F}$, an electric potential ψ, and a transfer of N species of chemicals No. 1, 2, ..., N into the body by the amounts $dn_1, dn_2, \ldots, dn_N$, we have, by the first law:

$$dQ = dU + \mu_\alpha \, dn_\alpha - \mathscr{F} \cdot \mathbf{dL} + \mathbf{F} \cdot \mathbf{dx} - \psi \, de + \sigma_{jk} \, de_{jk} V. \tag{3}$$

The Greek index α ranges over 1, 2, ... N. The summation over α includes all chemical species of the system. The energy dU includes kinetic, potential, and internal energies.

The *second law of thermodynamics* may be stated as follows:

There exists two single-valued functions of state T, called the absolute temperature, and S, called the entropy, with the following properties:

  I. *T is a positive number which is a function of empirical temperature only.*
 II. *The entropy of the system is equal to the sum of the entropies of its parts.*
III. *The entropy of a system can change in two distinct ways: by transfer from the surrounding and by internal changes.* Thus

$$\frac{dS}{dt} = \frac{d_e S}{dt} + \frac{d_i S}{dt}, \tag{4}$$

   where dS/dt denotes the rate of increase of entropy of the system, $d_e S/dt$ denotes the rate of transfer of entropy from the surrounding, $d_i S/dt$ denotes the rate of internal entropy production.
 IV. *The internal entropy production is never negative. If it is zero, the process is said to be reversible. If it is positive, the process is said to be irreversible.*

---

* The summation convention is used; see Sec. 1.7. The equations derived below can be used for a gas if the term $\sigma_{jk} \, de_{jk}$ is replaced by $-p \, dV$.
** Boldface letters denote vectors.

V. *If dQ is the heat absorbed by a system in a reversible process, then the entropy of the system is changed by the amount:*

$$dS = \frac{dQ}{T}. \tag{5}$$

The absolute temperature T and the entropy S are defined completely by their properties as expressed in the second law. For T, William Thomson (Lord Kelvin) has shown how to calibrate any thermometer into absolute temperature scale on the basis of the second law. For S, the law says that it is an attribute of a material body. Its units are Joules per degree Kelvin. For example, a kilogram of saturated steam at a pressure of 1 atm (101.3 kN/m^2) and a temperature of 373.16 K has an entropy of 7.358 kJ/K/kg. Water at the same pressure and temperature has an entropy of 1.307 kJ/K/kg.

The physical meaning of entropy has intrigued everybody since its introduction by Clausius in 1865. A thought that is conducive to an intuitive idea about entropy comes from statistical mechanics. Boltzmann has shown that entropy is proportional to the number of configurations Ω in which a system can be realized, according to the relation

$$S = k \ln \Omega. \tag{6}$$

Here k is the Boltzmann constant. An increase in the number of configurations that a system can assume increases its entropy, while any restriction in the modes of realization decreases entropy. Thus, mixing, disorganization, and randomization increase entropy, while organization and ordering decrease it.

The second law states how entropy of a system is changed, but it does not say how an absolute value can be assigned to entropy. The assignment of an absolute value of entropy is done by the *third law of thermodynamics*: *The entropy in the state $T = 0$ is equal to zero in every system occurring in nature.* This statement is Planck's formulation of Nernst's postulate.

Example. Entropy of a perfect gas. A perfect gas obeys the equation of state $pV = RT$, with R a constant, and has a constant specific heat C_v. Hence for one mole of gas, we have, on defining U, V, C_v, R for one mole,

$$dS = \frac{dQ}{T} = \frac{dU + p\,dV}{T} = \frac{C_v\,dT + (RT/V)\,dV}{T}.$$

An integration and use of the equation of state gives

$$\begin{aligned}
S &= C_v \log T + R \log V + S_0 \\
&= C_v \log p + C_p \log V + S_0' \\
&= C_p \log T - R \log p + S_0'',
\end{aligned} \tag{7}$$

where S_0, S_0' and S_0'' are constants.

8.3 The Gibbs and Gibbs–Duhem Equations

Combining Eqs. (3) and (5) of Sec. 8.2, we obtain the *Gibbs equation* for a tissue containing muscle fibers with tension $\mathscr{F}$, subject to external force $\mathbf{F}$, electric potential ψ, and transfer of N species of chemicals of mass $dn_1, dn_2, \ldots, dn_N$ into the tissue,

$$dU = T\,dS + \mu_\alpha\,dn_\alpha - \mathscr{F}\cdot d\mathbf{L} + \mathbf{F}\cdot d\mathbf{x} - \psi\,de + \sigma_{jk}\,de_{jk}V. \qquad (1)$$

This Gibbs equation can be integrated. A method for doing so is based on Euler's theorem on homogeneous function. A function $\Phi(x, y, z)$ is said to be homogeneous of the mth order if

$$\Phi(\lambda x, \lambda y, \lambda z) = \lambda^m \Phi(x, y, z). \qquad (2)$$

For example, $ax + by$ is homogeneous of the first order, $ax^3 + bx^2 y$ is homogeneous of the third order. For such a function, Euler showed that

$$m\Phi = x\left(\frac{\partial\Phi}{\partial x}\right)_{y,z} + y\left(\frac{\partial\Phi}{\partial y}\right)_{z,x} + z\left(\frac{\partial\Phi}{\partial z}\right)_{x,y} \qquad (3)$$

The proof is very simple. Differentiation of the left-hand side of Eq. (2) yields

$$\frac{\partial\Phi(\lambda x, \lambda y, \lambda z)}{\partial\lambda} = \left(\frac{\partial\Phi}{\partial\lambda x}\right)x + \left(\frac{\partial\Phi}{\partial\lambda y}\right)y + \left(\frac{\partial\Phi}{\partial\lambda z}\right)z. \qquad (4)$$

Differentiation of the right-hand side of Eq. (2) yields

$$\frac{\partial\lambda^m\Phi(x, y, z)}{\partial\lambda} = m\lambda^{m-1}\Phi(x, y, z). \qquad (5)$$

Equating Eqs. (4) and (5), and setting $\lambda = 1$, we obtain Eq. (3).

Now apply Euler's theorem to Eq. (1), the Gibbs equation, noting that energy, volume, entropy, and chemical quantity are all proportional to the mass of the system. Hence, if the internal energy is expressed as a function $U(S, e_{jk}V, n_1, n_2, \ldots)$, then U is homogeneous of order 1:

$$U(\lambda S, \lambda e_{jk}V, \lambda n_1, \lambda n_2, \ldots) = \lambda U(S, e_{jk}V, n_1, n_2, \ldots), \qquad (6)$$

therefore according to Euler's theorem, $U(S, e_{jk}V, n_1, n_2, \ldots)$ can be written as follows:

$$U = S\left(\frac{\partial U}{\partial S}\right) + e_{jk}V\left(\frac{\partial U}{\partial e_{jk}V}\right) + \sum_\alpha n_\alpha\left(\frac{\partial U}{\partial n_\alpha}\right) + \cdots. \qquad (7)$$

Hence, for small changes of the variables we have

$$dU = dS\left(\frac{\partial U}{\partial S}\right) + de_{jk}V\left(\frac{\partial U}{\partial e_{jk}V}\right) + \sum_\alpha dn_\alpha\left(\frac{\partial U}{\partial n_\alpha}\right) + \cdots. \qquad (8)$$

Comparing Eqs. (8) and (1), we see that

$$\left(\frac{\partial U}{\partial S}\right)_{V, n_\alpha} = T, \qquad \left(\frac{\partial U}{\partial e_{jk} V}\right)_{S, n_\alpha} = \sigma_{jk},$$

$$\left(\frac{\partial U}{\partial n_\alpha}\right)_{S, V, n_1, n_2, \ldots} = \mu_\alpha, \tag{9}$$

etc. Introducing these into Eq. (7), we obtain the *integrated form of the Gibbs equation*:

$$U = TS + \sum_\alpha \mu_\alpha n_\alpha - \mathscr{F} \cdot \mathbf{L} + \mathbf{F} \cdot \mathbf{x} - \psi e + \sigma_{jk} e_{jk} V. \tag{10}$$

Now if we differentiate Eq. (10), we obtain

$$dU = T \, dS + S \, dT + \sum_\alpha (\mu_\alpha \, dn_\alpha + n_\alpha \, d\mu_\alpha) + \cdots. \tag{11}$$

On subtracting Eq. (1) from Eq. (11), we obtain the *Gibbs–Duhem equation*:

$$\sum_\alpha n_\alpha \, d\mu_\alpha + S \, dT - \mathbf{L} \cdot d\mathscr{F} + \mathbf{x} \cdot d\mathbf{F} - e \, d\psi + e_{jk} \, d\sigma_{jk} V = 0. \tag{12}$$

If experiments were done in such a manner that T, $\mathscr{F}$, $\mathbf{F}$, ψ, and σ_{jk} remain constant, then

$$\sum_\alpha n_\alpha \, d\mu_\alpha = 0. \tag{13}$$

These equations are used extensively in mass-transport analysis (Sec. 8.5).

8.4 Chemical Potential

Let us define

$$F = U - TS \tag{1}$$

as the *Helmholtz free energy*, and

$$G = U - TS - \sigma_{jk} e_{jk} V \tag{2}$$

as the *Gibbs free energy* (or *Gibbs thermodynamic potential*). Then, on substituting Eq. (10) of Sec. 8.3 into Eqs. (1) and (2) above, taking differentials of the resulting expressions and combining them with the Gibbs equation, Eq. (8.3:1), we obtain

$$dU = T \, dS + \sum_\alpha \mu_\alpha \, dn_\alpha - \mathscr{F} \cdot d\mathbf{L} + \mathbf{F} \cdot d\mathbf{x} - \psi \, de + \sigma_{jk} \, de_{jk} V, \tag{3}$$

$$dF = dU - T \, dS - S \, dT = -S \, dT + \sum_\alpha \mu_\alpha \, dn_\alpha + \sigma_{jk} \, de_{jk} V + \cdots, \tag{4}$$

$$dG = -V e_{jk} \, d\sigma_{jk} - S \, dT + \sum_\alpha \mu_\alpha \, dn_\alpha + \cdots. \tag{5}$$

Thus we see that

$$\mu_\alpha = \left(\frac{\partial U}{\partial n_\alpha}\right)_{S, e_{jk}, L, n_1, \ldots} = \left(\frac{\partial F}{\partial n_\alpha}\right)_{T, e_{jk}, L, n_1, \ldots} = \left(\frac{\partial G}{\partial n_\alpha}\right)_{T, \sigma_{jk}, L, n_1, \ldots} \qquad (6)$$

In other words, the *chemical potential* μ_α *of the αth species of chemicals is the rate of change of internal energy U with respect to the mass of the αth species of the chemical when entropy, strain, muscle length, electric charge, and concentrations of all other chemicals,* n_β, $(\beta \neq \alpha)$ *are held constant. Similarly, it is the partial derivative of the Gibbs thermodynamic potential,* $\partial G/\partial n_\alpha$, *when the temperature, stress, etc. are held constant; or the partial derivative of the Helmholtz free energy* $\partial F/\partial n_\alpha$, *when the temperature, strain, etc. are held constant.*

Now, in chemical and biological laboratory experiments, it is relatively easy to maintain a constant temperature by a heat bath, and a constant pressure by exposure to the atmospheric pressure. If the material is a solid or liquid and is left stressfree in atmosphere at constant temperature, then the measurement of the chemical potential is best done by measuring the Gibbs thermodynamic potential under isothermal and isobaric conditions. If the material is a gas and the container volume is fixed, then it is best done by measuring the Helmholtz free energy under isothermal condition.

In the chemical literature, following G.N. Lewis, the chemical potential of a substance is often expressed in terms of *activity* defined by the following equation:

$$\mu = \mu_s + RT \log a \qquad (7)$$

in which a is called *activity*, T is the absolute temperature, R is the gas constant, and μ_s is the chemical potential of the substance at a *standard state at the same temperature* T. The standard state can be chosen for each chemical component independently and arbitrarily with respect to other components in the system. For example, the activity of liquid water is 1 at 1 atm pressure, but at 100 atm pressure it is 1.0757, 1.0728, and 1.0703 at 25°C, 37°C, 50°C, respectively.

Example 1. Perfect gas. By using Eq. (7) of Sec. 8.2 for S, and $dQ = C_p dT$ for isobaric process, show that the Gibbs thermodynamic potential G per unit mass, which, by Eq. (6), is the chemical potential of the perfect gas, is

$$\mu = \mu_s(T) + RT \log p, \qquad (8)$$

where

$$\mu_s(T) = pV + \int_0^T C_p dT - T \int_0^T \frac{C_p dT}{T} + u_0 - TS_0'', \qquad (9)$$

u_0 being a constant. Comparing Eqs. (7) and (8), we see that *the activity of a perfect gas is equal to the pressure.* In a *real* gas, experimentally determined activity replaces the role of pressure in the chemical potential.

Example 2. A mixture of perfect gases. It follows from the gas law that the pressure of a mixture of perfect gases is equal to the sum of the partial pressures of the individual gases if there is no chemical reaction between the gases. Hence the internal energy, entropy, and Gibbs free energy of each gas are the same as if that gas were alone in a vessel, and *the chemical potential of a gas species α in the mixture is given by Eq. (8)*, with the addition of a subscript α to μ, μ_s, p, C_p, u_0 and S_0'' to indicate their belonging to the species α. Let the mole number of the gas species α be n_α, and denote by

$$c_\alpha = \frac{n_\alpha}{n_1 + n_2 + \cdots + n_k}, \qquad \alpha = 1, \ldots, k \tag{10}$$

the *concentration* of the component α in the mixture. Then

$$p_\alpha = n_\alpha RT/V = c_\alpha p, \tag{11}$$

and we can write the chemical potential of the gas species α as

$$\mu_\alpha = \mu_{s\alpha}(T) + RT \log c_\alpha + RT \log p. \tag{12}$$

It is seen that the product of the concentration and the pressure is the activity. In a *real gas mixture*, experimentally determined activities replace the $c_\alpha p$ terms in the chemical potential.

Example 3. Dilute solution. For a dilute solution consisting of chemically pure substances $0, 1, 2, \ldots k$, of which the component 0 is the most plentiful, denote the mole numbers by $n_0, n_1, \ldots, n_k$, respectively. A solution is dilute if the internal energy and volume of the solution are linear functions of n_α/n_0, ($\alpha = 1, 2, \ldots, k$). It can be shown (Epstein, 1937, Chap. 9) that the Gibbs thermodynamic potential of a dilute solution is

$$G = \sum_{\alpha=0}^{k} n_\alpha(g_\alpha + RT \log c_\alpha) \tag{13}$$

in which c_α is the concentration, $g_\alpha = u_\alpha - Ts_\alpha + pv_\alpha$ is a function of p and T, independent of n_α. The role of concentration in the chemical potential is seen again. In *real* solution, activity replaces the concentration.

8.5 Entropy in a System with Heat and Mass Transfer

We now consider a system in which the thermomechanical variables are functions of time and space. The system is not necessarily in equilibrium, and heat and mass transfer take place within. To analyze such a system, *one makes the basic assumption that the Gibbs equation for internal energy remains valid locally at any time and place*, that is, the internal energy as a function of the state variables is the same whether the state variables are arrived at by reversible process or not. Thus, focusing our attention on heat and mass transfer, and ignoring for the time being th muscle contraction, body force, electric discharge, and strain energy, we have, for a small volume ΔV of the system,

$$\Delta Q = \Delta U - \Delta W = T\Delta S + \sum_\alpha \mu_\alpha \Delta n_\alpha \tag{1}$$

according to Eqs. (8.2:2) and (8.3:1). Further analysis is facilitated by defining the entropy per unit volume s_v, concentration of a chemical species c_α, and the heat per unit volume q_v, by the expressions

$$s_v = \frac{\Delta S}{\Delta V}, \qquad c_\alpha = \frac{\Delta n_\alpha}{\Delta V}, \qquad q_v = \frac{\Delta Q}{\Delta V}. \tag{2}$$

Thus Eq. (1) may be written as

$$T\,ds_v = dq_v - \sum_\alpha \mu_\alpha\,dc_\alpha. \tag{3}$$

If the system moves and has a velocity field v' which varies from place to place, then Eq. (3) can be interpreted, in the notation of continuum mechanics, Sec. 1.7, as

$$T\frac{Ds_v}{Dt} = \frac{Dq_v}{Dt} - \sum_\alpha \mu_\alpha \frac{Dc_\alpha}{Dt}, \tag{4}$$

where D/Dt denotes a derivative with respect to time of a function (such as s_v) belonging to a fixed set of material particles. D/Dt is called the *material derivative*. It can be shown (see any book on continuum mechanics, e.g. Fung, 1977) that the material derivative of a function F is

$$\frac{DF}{Dt} = \frac{\partial F}{\partial t} + \mathbf{v}' \cdot \operatorname{grad} F. \tag{5}$$

If the system has no bulk flow so that $\mathbf{v}' = 0$, then Eq. (4) is reduced to

$$T\frac{\partial s_v}{\partial t} = \frac{\partial q_v}{\partial t} - \sum_\alpha \mu_\alpha \frac{\partial c_\alpha}{\partial t}. \tag{6}$$

Analysis of Entropy Production

When heat and mass transfers occur we define heat or mass *flux* as the amount of heat or mass passing through a surface of unit area. Consider a volume V enclosed in a surface A fixed in space. The rate of increase of the mass of a chemical species α in V is due to the flow of matter through the boundary A:

$$\frac{\partial n_\alpha}{\partial t} = -\int_A \mathbf{J}_\alpha \cdot \mathbf{v}\,dA, \tag{7}$$

where $\mathbf{J}_\alpha$ is the mass flux vector of species α through a surface of area dA out of the volume V; $\mathbf{v}$ is a unit vector normal to the surface pointing outward from V. But if c_α is the concentration of species α we have, by definition,

$$n_\alpha = \int_V c_\alpha\,dV. \tag{8}$$

Furthermore, by Gauss's theorem, we have

$$\int_A \mathbf{J}_\alpha \cdot \mathbf{v}\, dA = \int_V \operatorname{div} \mathbf{J}_\alpha\, dV. \tag{9}$$

Hence, by Eqs. (8) and (9), Eq. (7) is reduced to

$$\int_V \frac{\partial c_\alpha}{\partial t}\, dV = -\int_V \operatorname{div} \mathbf{J}_\alpha\, dV. \tag{10}$$

Since the control volume V is arbitrary, we must have

$$\frac{\partial c_\alpha}{\partial t} = -\operatorname{div} \mathbf{J}_\alpha. \tag{11}$$

This equation *expresses the conservation of the mass of the chemical species* α. Similarly,

$$\frac{\partial q_v}{\partial t} = -\operatorname{div} \mathbf{J}_q \tag{12}$$

expresses the conservation of heat, $\mathbf{J}_q$ being the heat flux vector. On the other hand, according to the second law of thermodynamics, *the change of entropy in a system consists of two parts. One part is transported through the boundary; the other part is produced locally and internally.* The first part is transferred by the *entropy flux vector,* $\mathbf{J}_s$. The second part is the $d_i S/dt$ term in the second law, Eq. (8.2:4). Thus

$$\frac{\partial s_v}{\partial t} = -\operatorname{div} \mathbf{J}_s + \frac{d_i s_v}{dt}. \tag{13}$$

Substituting these expressions into Eq. (6) and dividing through by T, we obtain

$$\frac{d_i s_v}{dt} - \operatorname{div} \mathbf{J}_s = -\frac{1}{T} \operatorname{div} \mathbf{J}_q + \sum_\alpha \frac{\mu_\alpha}{T} \operatorname{div} \mathbf{J}_\alpha. \tag{14}$$

This equation may be modified by making use of the following relation which holds for a scalar a and a vector $\mathbf{b}$:

$$\operatorname{div} a\mathbf{b} = a \operatorname{div} \mathbf{b} + \mathbf{b} \cdot \operatorname{grad} a.$$

Thus Eq. (14) can be written as

$$\frac{d_i s_v}{dt} - \operatorname{div} \mathbf{J}_s = -\operatorname{div} \frac{\mathbf{J}_q}{T} + \mathbf{J}_q \cdot \operatorname{grad} \frac{1}{T} + \sum_\alpha \left(\operatorname{div} \frac{\mu_\alpha \mathbf{J}_\alpha}{T} - \mathbf{J}_\alpha \cdot \operatorname{grad} \frac{\mu_\alpha}{T} \right). \tag{14a}$$

The right-hand side can be divided into two parts: a divergence of a vector and a scalar product of two vectors. Now we shall interpret Eq. (14a) by splitting it into two parts: We equate the divergence on the two sides of the equation to obtain

$$\operatorname{div} \mathbf{J}_s = \operatorname{div} \frac{\mathbf{J}_q}{T} - \operatorname{div} \sum_\alpha \frac{\mu_\alpha \mathbf{J}_\alpha}{T}, \tag{15}$$

and we equate the entropy production with the scalar products so that

$$\frac{d_i s_v}{dt} = \mathbf{J}_q \cdot \operatorname{grad} \frac{1}{T} + \sum_\alpha \mathbf{J}_\alpha \cdot \operatorname{grad}\left(-\frac{\mu_\alpha}{T}\right). \tag{16}$$

Equation (15) shows that the flow of entropy across a surface is

$$\mathbf{J}_s = \frac{\mathbf{J}_q}{T} - \sum_\alpha \frac{\mu_\alpha \mathbf{J}_\alpha}{T} \tag{17}$$

as suggested directly by Eq. (3). Equation (16) is of the form

$$\frac{d_i s_v}{dt} = \sum_k J_k X_k, \tag{18}$$

where X_k can be called the *generalized force* "conjugate" to the *flux* J_k. Thus, by comparing Eq. (18) with Eq. (16), we obtain the important result that *the generalized force which is conjugate to the flux of a chemical species is the negative gradient of the chemical potential of that species divided by T*. This is why the chemical potential is so important in mass transfer phenomena.

Phenomenological Laws and Onsager Principle

The relationships between the fluxes $J_1, J_2, \ldots, J_k$ and the conjugate generalized forces $X_1, X_2, \ldots, X_K$, (K being the number of irreversible processes involved), are called *phenomenological laws*. If the phenomenological laws are linear we can write

$$J_k = \sum_{m=1}^{K} L_{km} X_m, \qquad k, m = 1, 2, \ldots, K, \tag{19}$$

where L_{km} are constants. Experience with known phenomenological laws in thermoelasticity, thermodiffusion, piezoelectricity, etc. has shown that the matrix of the coefficients L_{km} is symmetric, i.e.

$$L_{km} = L_{mk}, \qquad k, m = 1, 2, \ldots, K. \tag{20}$$

Onsager derived Eq. (20) from statistical mechanics, and Eq. (20) is known as *Onsager's principle* or the Onsager reciprocal relations.

If we substitute J_k from Eq. (19) into Eq. (18), we obtain the rate of entropy production:

$$\frac{d_i s_v}{dt} = \sum_{k=1}^{K} \sum_{m=1}^{K} L_{km} X_m X_k. \tag{21}$$

According to the second law of thermodynamics, *the entropy production is nonnegative*. Hence the second degree polynomial on the right-hand side of Eq. (21) is positive definite. This, together with Eq. (20), imposes restrictions on the coefficients L_{km}. For example, if L_{km} is collectively considered to be a square matrix, then the diagonal terms $L_{11}, L_{22}, ..., L_{KK}$ must be positive, the determinant of the matrix must be positive, and the sum of the principal minors of any given order must be positive (see Fung, 1965, pp. 8, 29, 30).

8.6 Diffusion, Filtration, and Fluid Movement in Interstitial Space from the Point of View of Thermodynamics

Let us now consider the movement of water and macromolecules in the interstitial space at constant temperature. We shall follow Zweifach and Silberberg (1979) to consider a simplified model of the interstitium consisting of three groups of components: a "solvent" component, a diffusible "solutes" component, and a "matrix" component. The solvent is essentially water, the solutes are principally macromolecules. (Small molecules that can move freely through the tissue space do not create sufficient driving force to move water and macromolecules and can be lumped with the solvent.) The matrix consists of collagen, hyaluronate, etc., see Fig. 8.6:1. Our formulation follows Zweifach and Silberberg (1979), but greatly simplifies it.

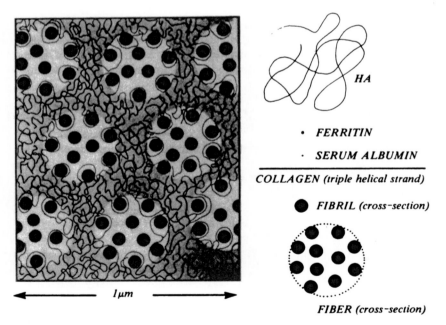

FIGURE 8.6:1 An idealized interstitium consisting of collagen fibers, hyaluronic acid molecules, globular proteins, ferritin, and serum albumin. From Zweifach and Silberberg (1979), by permission.

Let us use the subscripts 0, 1, and 2 for the three components, water, solutes, and matrix, respectively. Then according to Eqs. (8.5:16–20) and under the assumptions of constant temperature and linear phenomenological laws, we can write*

$$J_0 = -L_{00}\,\text{grad}\,\mu_0 - L_{01}\,\text{grad}\,\mu_1 - L_{02}\,\text{grad}\,\mu_2$$
$$J_1 = -L_{10}\,\text{grad}\,\mu_0 - L_{11}\,\text{grad}\,\mu_1 - L_{12}\,\text{grad}\,\mu_2 \qquad (1)$$
$$J_2 = -L_{20}\,\text{grad}\,\mu_0 - L_{21}\,\text{grad}\,\mu_1 - L_{22}\,\text{grad}\,\mu_2$$

in which L_{00}, L_{01}, etc. are constants obeying Onsager's reciprocal relations and the positive-definiteness of entropy production, Eqs. (8.5:20) and (8.5:21).

Chemical potentials of water, solute, and matrix are functions of temperature, pressure, and composition. According to Eq. (8.4:12), at constant temperature, we have

$$\mu_\alpha = \mu_\alpha^{(0)} + v_\alpha p + RT\log a_\alpha, \qquad \alpha = 1, 2, 3, \qquad (2)$$

where R is the gas constant, v_α is the molecular volume (see Eq. (13) of Sec. 8.4) of the species α, and a_α is its activity.

The chemical potentials of all the components in a system cannot vary independently. They must obey the Gibbs–Duhem equation, Eq. (8.3:12). Thus, if the temperature, pressure, and stress deviations are kept constant, then

$$\sum_{\alpha=1}^{3} c_\alpha\, d(\log a_\alpha) = 0, \qquad (3)$$

c_α being the concentration of the species α. Using Eq. (3), one of the $d(\log a_\alpha)$ terms, say, that of water, can be isolated and expressed as a linear function of all the others. Thus

$$d(\log a_0) = -\frac{c_1}{c_0} d(\log a_1) - \frac{c_2}{c_0} d(\log a_2). \qquad (4)$$

Zweifach and Silberberg (1979) define Π_1 and Π_2 by the equations

$$\Pi_1 \equiv kT \int_0^{c_1} \frac{c_1}{c_0 v_0} d(\log a_1), \qquad c_2, T = \text{const.}$$
$$\Pi_2 \equiv kT \int_0^{c_2} \frac{c_2}{c_0 v_0} d(\log a_2), \qquad c_1, T = \text{const.} \qquad (5)$$

It will be shown later that these Π's can be revealed as osmotic pressure with

---

* Here we treat the "immobile" matrix as a fluid also. This leads to greater mathematical simplicity. If the immobile matrix must be treated as a solid, be it elastic, viscoelastic or otherwise, then one must consider stress tensor in place of pressure, and a much more detailed treatment has to be given to the matrix component. This is done in Sec. 8.9, which has applications to dense connective tissues such as cartilage, spinal discs, etc.

suitable semipermeable membrane. From Eqs. (2), (4) and (5) we can write

$$d\mu_0 = v_0(dp - d\Pi_1 - d\Pi_2), \tag{6a}$$

$$d\mu_1 = v_1\left(dp + \frac{c_0 v_0}{c_1 v_1} d\Pi_1\right). \tag{6b}$$

A similar equation for $d\mu_2$ is obtained from (6b) by replacing the subscript 1 by 2. An integration of these equations gives

$$\mu_0 = v_0(p - \Pi_1 - \Pi_2) + \mu_0^{(0)},$$
$$\mu_1 = v_1\left(p + \frac{c_0 v_0}{c_1 v_1}\Pi_1\right) + \mu_1^{(0)}. \tag{7}$$

With these equations, we shall consider some simplified cases below.

Dilute Solutions

For a dilute solution, the activity a_α is equal to the concentration of the species α (Sec. 8.4). Then, with c_1 and c_2 small compared with c_0, Eqs. (2) and (4) yield

$$\mathrm{grad}\,\mu_0 = v_0\,\mathrm{grad}\,p - \frac{RT}{c_0}\,\mathrm{grad}\,c_1 - \frac{RT}{c_0}\,\mathrm{grad}\,c_2, \tag{8a}$$

$$\mathrm{grad}\,\mu_1 = v_1\,\mathrm{grad}\,p + \frac{RT}{c_1}\,\mathrm{grad}\,c_1, \tag{8b}$$

and a similar equation for μ_m by replacing the subscript 1 by 2 in Eq. (8b). In a nonuniform solution, the fluxes are given by Eq. (1).

Fick's Law of Diffusion

If there is no matrix, no pressure gradient, and no bulk flow, then the second equation of Eq. (1) is reduced to

$$J_1 = -\mathrm{const}\cdot\mathrm{grad}\,\mu_1. \tag{9}$$

For a *dilute solution*, Eq. (9) then yields

$$J_1 = -D\,\mathrm{grad}\,c_1 \tag{10}$$

which is *Fick's law of diffusion*. D is a constant called the *coefficient of diffusion*.

Osmotic Pressure

If a solution of chemical species 1 and 2 is put in equilibrium with pure water across a membrane which is not permeable to 1 and 2, but is permeable to a water, then when there is no flow of water across the membrane, the chemical potential of water on the two sides of the membrane must be balanced. On equating μ_0 on the "outside" and "inside" of the membrane, via Eq. (7) we have

$$v_0 p_{out} = v_0(p_{in} - \Pi_{1\ in} - \Pi_{2\ in})$$

so that

$$p_{in} - p_{out} = \Pi_{1\ in} + \Pi_{2\ in}. \tag{11}$$

This represents the *osmotic pressure* of the solution. The pressure difference has to be mechanically compensated, which is done in this case by the tension and curvature of the membrane. It is interesting to note that although Π_1 and Π_2 are measures of the concentrations c_1 and c_2 of species 1 and 2, respectively, they do have the units of pressure.

To reveal chemical potential as osmotic pressure a semipermeable membrane is required. One should remember this in order to avoid confusion. For example, solute concentration can be measured by freezing point depression. Solute-free water freezes at 0°C. If 1 osmol of any solute is added to 1 kg of water, the freezing point of this water will be depressed by 1.86°C. (1 osmol is defined as one gram molecular weight of any nondissociable substance and contains 6.02×10^{23} molecules.) But one must realize an important difference between the osmotic pressure of a solution and its osmolality as measured by the freezing-point depression. When the freezing-point depression is measured, all solutes contribute in relation to their concentration, including those that are ineffective osmoles. In contrast, ineffective osmoles do not contribute to the osmotic pressure of solution. For example, plasma osmolality of 280 mosmol/kg represents a potential osmotic pressure of 5404 mm Hg. However, almost all the plasma solutes (primarily Na^+ salts) are ineffective since Na^+ salts are able to cross the capillary wall separating the plasma from the interstitial fluid. The net effective plasma osmolality is only about 1.3 mosmol/kg (generating an osmotic pressure of 25 mm Hg) and is due to the plasma proteins which are restricted to the vascular space.

As another example, consider the cells. The cell membrane is permeable to Na^+ but the cell Na^+ concentration is maintained at 12 mEq/L by active transport of Na^+ out of the cell. Thus the cell membrane is effectively, though not actually, impermeable to Na^+. In this setting, an osmotic pressure is generated in the extracellular fluid, and these solutes are called *effective osmoles*. If the solute is able to cross the membrane and reach concentration in both compartments, e.g., urea, then no osmotic pressure is generated across the membrane; hence urea is an *ineffective osmole* for cells.

Starling's Law of Membrane Filtration

Consider two solutions separated by a semipermeable membrane. Each solution consists of a solute in water. Assume first that *the membrane is impermeable to the solute but is permeable to water*. The chemical potentials of the water on the two sides of the membrane (indicated as "inside" and "outside") are, according to Eq. (27),

$$\mu^{(0)}_{0\,in} + v_{0\,in}(p_{in} - \Pi_{in}), \qquad \mu^{(0)}_{0\,out} + v_{0\,out}(p_{out} - \Pi_{out}). \tag{12}$$

Here, Π stands for the sum $\Pi_1 + \Pi_2$ if an immobilized matrix exists, or Π_1 alone if such a matrix does not exist. At equal temperature $\mu^{(0)}_{0\,in}$ and $\mu^{(0)}_{0\,out}$ are equal. For dilute solutions, $v_{0\,in}$ and $v_{0\,out}$ are approximately equal. Hence, the difference of the chemical potentials of water on the two sides of the membrane is

$$v_0(p_{in} - p_{out} - \Pi_{in} + \Pi_{out}). \tag{13}$$

This drives the water across the membrane. Let the flux of water across the membrane be J_0 (volume rate of flow per unit area of the membrane). If we assume that flux is linearly proportional to the driving force, then we obtain

$$J_0 = K(p_{in} - p_{out} - \Pi_{in} + \Pi_{out}), \tag{14}$$

where K is the *permeability constant of the membrane*. This is *Starling's law of membrane filtration*.

If the membrane is also permeable to the solute, then there will be a flux of the solute across the membrane. Let us use the symbol Δ for the difference of a quantity on the two sides of the membrane. We may write, in analogy with Eqs. (1), and using Eq. (7),

$$\begin{aligned}
J_0 &= K_{11}(p_{in} - p_{out} - \Pi_{in} + \Pi_{out}) \\
&\quad - K_{22}\left[p_{in} - p_{out} - \frac{c_0 v_0}{c_1 v_1}(\Pi_{in} - \Pi_{out}) \right], \\
&= K(\Delta p - \sigma \Delta \Pi), \tag{15}
\end{aligned}$$

where K and σ are constants, $\Delta p = p_{in} - p_{out}$, and $\Delta \Pi = \Pi_{in} - \Pi_{out}$. σ is called the *reflection coefficient*: $\sigma = 1$ when the membrane is impermeable to the solute, $\sigma < 1$ when it is permeable to the solute.

Darcy's Law of Flow through Porous Media without Chemical Gradient

If the osmotic pressures can be ignored and the solutes are freely movable Eq. (1) can be reduced to the form

$$J_0 = -\frac{\rho k}{\mu} \operatorname{grad} p, \tag{16}$$

where ρ is the density of the fluid, μ is its coefficient of viscosity, and k is also called a *permeability constant*. This is *Darcy's law* of fluid flow through porous media.

One may wish to emphasize the *porosity* of the porous medium in which the fluid moves. Let us define the porosity (or the *effective pore space per unit volume of the medium*) as the volume fraction Φ_0:

$$\Phi_0 = c_0 v_0(c_0 v_0 + c_1 v_1 + c_2 v_2). \tag{17}$$

Then, we have

$$J_0 = -\kappa \Phi_0 \, grad \, p, \tag{18}$$

where κ is a constant.

Recapitulation

According to thermodynamics of irreversible prcesses, phenomenological laws should be based on a consideration of the local entropy production. From such a consideration it is shown that the gradient of the chemical potential of each chemical species in the tissue space is the driving force for the movement of that species. Phenomenological laws relate the fluxes of the chemical species to the gradients of the chemical potentials. The tissue space may be divided into several compartments. Each compartment may be constituted of a solution (a liquid) or a mixture (of gases, liquids, and solids). The chemical potential of all the components in a compartment cannot vary independently; they must obey the Gibbs–Duhem equation. Using the Gibbs–Duhem equation, we show that Fick's law of diffusion, Starling's law for membrane, and Darcy's law of flow through porous media are special cases of a general relationship based on the consideration of local entropy production and the second law of thermodynamics.

So far the matrix is considered mobile and treated as fluid. Immobile matrix is treated in Sec. 8.9.

8.7 Diffusion from the Molecular Point of View

In the preceding section we saw diffusion driven by chemical potential. Now, following Einstein (1908), let us look at it from the point of view of Brownian motion.

Consider a molecule undergoing random collisions in the x-direction. Each collision results in a step displacement d_i. After N collisions, the displacement Δx and its square are, respectively:

$$\Delta x = d_1 + d_2 + \cdots + d_N, \tag{1}$$

$$(\Delta x)^2 = d_1^2 + d_2^2 + \cdots + d_N^2 + 2d_1 d_2 + 2d_1 d_3 + \cdots + 2d_{N-1} d_N. \tag{2}$$

If the probability for the positive d's is equal to that for the negative, then when N is large Δx will tend to zero and $(\Delta x)^2$ would approach the sum of the squares of the d's. Thus, if $\overline{d^2}$ is the mean value of the square of the displacement step, then

$$\overline{(\Delta x)^2} = N\overline{d^2}. \tag{3}$$

If τ is the average time between steps, then in a time interval t, the number of steps N is equal to t/τ, and Eq. (3) becomes

$$\overline{(\Delta x)^2} = N\overline{d^2} = \left(\frac{\overline{d^2}}{\tau}\right)t. \tag{4}$$

When applied to a whole population of molecules at the origin initially, Eq. (4) describes a fundamental characteristic of diffusion.

Einstein (1908) identified the coefficient $\overline{d^2}/\tau$ with twice the coefficient of diffusion by considering the movement of molecules across a plane. Consider a plane A with a box attached to each of its sides. Each box has a length d and a cross-sectional area of 1 cm^2, d being the mean step length between collisions, i.e., the square root of $\overline{d^2}$. There are $C_1 d$ molecules in the box to the left and $C_2 d$ molecules in the box to the right, (C being the concentration of the solute, and equal to the product of the Avogadro's number and the moles of the solute per unit volume). In a time interval τ, all solute molecules will leave these boxes, but on average half of the molecules will diffuse to the left and half to the right. The number of molecules crossing the plane A from left to right is $\frac{1}{2}dC_1$, that from right to left is $\frac{1}{2}dC_2$. The solute flux J_1 (moles/sec/cm^2) through the plane A is therefore,

$$J_1 = \frac{d}{2\tau}(C_1 - C_2). \tag{5}$$

For a continuous distribution of concentration C as a function of x,

$$C_1 - C_2 = -\frac{dC}{dx}d. \tag{6}$$

Hence

$$J_1 = -\left(\frac{d^2}{2\tau}\right)\frac{dC}{dx}. \tag{7}$$

This is identical with Fick's law given in Eq. (8.6:10) if we identify $d^2/2\tau$ with the coefficient of diffusion D:

$$D = \frac{\overline{d^2}}{2\tau}. \tag{8}$$

Einstein (1908) went on to derive an expression relating the diffusion coefficient to the resistance that a solute molecule encounters when it moves in a solvent. He obtained the expression

$$D = \frac{RT}{N_A 6\pi\eta a}, \tag{9}$$

where D is the diffusion coefficient (cm^2/s), R is the gas constant (1.99 cal.$^\circ$K^{-1} mol^{-1}), T is temperature ($^\circ$K), N_A is Avogadro's number (6.023 × 10^{23} molecules/mol), η is the coefficient of viscosity (dyn. s. cm^{-2}), a is the solute radius (cm). Other authors extended Einstein's method to derive expressions for diffusion through membranes with pores (Pappenheimer et al., 1951) or in fiber-matrix, Curry and Michel, 1980), see Curry (1984) for a comprehensive review.

8.8 Transport Across Cell Membranes

Cells can retain, gain, or expel ions against gradients of chemical potential, or transport matter through their membranes at an augmented rate. The kidney can concentrate urine. The nerve cells can transmit signals. The muscle cells can rest or contract. All these are initiated through actions of their membranes.

Obviously, in these membranes there must be some specific machineries, specific structures, specific materials, each obeying physical and chemical laws, and together leading to specific functions of each organ. It is equally obvious that this is a long story, not possible to be compressed into a single section, a single chapter, a single book. We shall consider the subject only so far as to define the main topic of this chapter: the movement of fluid in the interstitial space. This space is bounded by blood vessels on one side, and cells on the other. Blood vessels and cell membranes are both semipermeable. The interstitial space is bounded by semipermeable membranes.

In the preceding sections we have discussed the transport of fluid across capillary blood vessels. For a substance (such as water) which moves according to the gradient of chemical potential, cell membranes obey the same laws. The chemical potential in the cell, however, depends on the activity of all materials in the cell. Hence we must know the transport laws of all substances in the cell. Certain substances can enter or leave cells by way of specific protein carriers of the plasma membrane. If the process consists of "pumping" a substance against a gradient of chemical potential, then it is called an *active* transport. If the process tends to equilibrate the substances across the membrane spontaneously but faster, then it is said to be a *facilitated transport.*

How to Discover the Transport Mechanisms in the Membrane

Since the cell membranes are so thin and the transport mechanisms cannot be observed directly, one can approach the problem of determining the mechanisms only by speculation and experimentation. Generally, one measures what can be measured and relies on physics and chemistry, especially mechanics, thermodynamics, and statistical mechanics, to define all the restrictions quantitatively as far as possible, and to formulate hypotheses that are not in conflict with these restrictions. If the predictions deduced from the hypotheses agree with experimental results, then one gains confidence in the hypotheses. If new experimental results contradict theoretical predictions, then some hypotheses must be abandoned. There is no unique procedure to the formulation of hypotheses. If two sets of hypotheses predict the same results, scientists usually favor the simpler, weaker, more general, less restrictive, and the least ad hoc.

Proposed Mechanisms of Facilitated Diffusion

Figure 8.8:1 shows several proposed models of *channels* in which ions can pass through a membrane. In (a) a single membrane protein, an *ionophore* is shown which has a hole down its center and is oriented in the direction of membrane thickness. The channel is lined with charged or polar amino acid residues, and filled with water. In (b) a set of rod-like proteins arranged in parallel with space between them lined with polar group is assumed. In (c) and (d) *gates* are

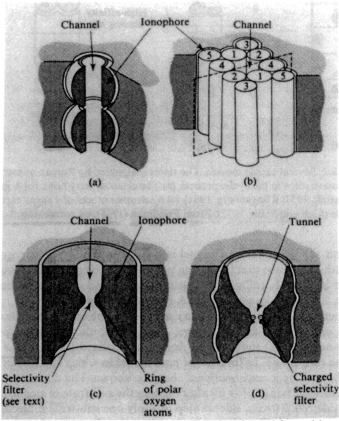

FIGURE 8.8:1 Several ion channel models. (a) Tubular channel formed by two molecules of the ionophore, the antibiotic gramicidin. (b) Rodlike ionophores proposed for the anion channel in the red blood cell (Solomon et al., 1983). The ionophore is a dimer. Each monomer consists of five rodlike helices. Three helices (*1, 2,* and *4*) from each monomer surround the pore. (c) Potassium-selective channel in nerve (Latorre and Miller, 1983). The "selectivity filter" does not pass Li or Na readily in their hydrated forms, but does if Li and Na are unhydrated. (d) Sarcoplasmic reticulum maxipotassium channel (Miller, 1982). The tunnel is occupied by no more than one ion at a time. The diffusion is relatively unrestricted at the mouth of the pore. Figure reproduced by permission from Morton H. Friedman (1986).

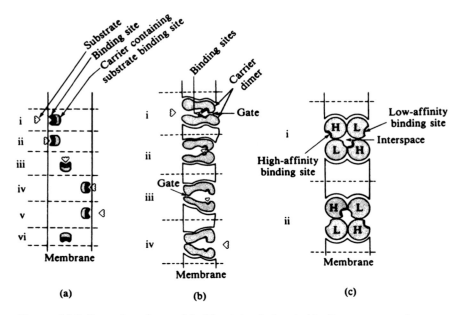

FIGURE 8.8:2 Several carrier models. The states designated by Roman numerals represent successive steps in the cyclic process. (a) The classical ferry boat. (b) A gated pore model (Patlak, 1957; Klingenberg, 1981). (c) A tetramer model of a sugar carrier in red blood cells (Lieb and Stein, 1972). From Friedman (1986), by permission.

added and the shape of the channel is defined to match the energy profile. Many authors assume that the gating process is statistical, and can be opened or closed by ions, hormones, neurotransmitters, and changes in membrane potential. Specific rules apply to each of these control agents.

Another class of assumed protein mediators in the cell membrane that can facilitate diffusions is called *carriers*. Figure 8.8:2 shows several proposed carrier models. In (a) the classical *ferry boat* is shown. The carrier diffuses across the membrane. In (b) a *gated pore* model in which the carrier accomplishes the task by a *conformational change* is shown. The dimer undergoes a conformational change (ii → iii) after binding the substrate in step ii, exposing it to the side No. 2 (*trans.* side) in step iii. Only one substrate molecule can fit between the two binding sites, so only one molecule is transferred per cycle. In (c) a *tetrameric model*, consisting of two monomers (*H*) whose binding sites have high affinity for the substrate, and two monomers (*L*) of lower affinity is shown. The monomers undergo concurrent conformational changes, such that the carrier alternates between states i and ii, exchanging solute between the ambient phases and the interspace. Solute in the interspace equilibrates alternatively between the high and low affinity inward-facing sites.

Mathematical models of these channels and carriers are fairly similar; they differ mainly on the physical meaning of the variables that enter the kinetic

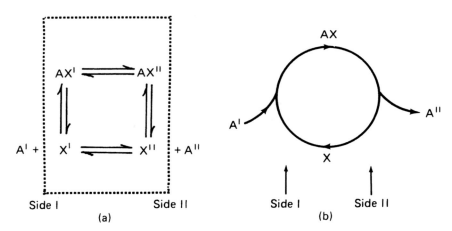

FIGURE 8.8:3 Schematic representations of the simplest carrier system. From Friedman (1986), by permission.

equations. The kinetic analysis of the simplest carrier system can be presented in a page or two. An examination of such an analysis will increase our understanding of these phenomena. The starting point is a list of hypotheses:

H1) The carrier X, transports a single solute A. Fig. 8.8:3.

H2) Each molecule of X reversibly binds a single molecule of A. A can cross the membrane only as AX.

H3) The binding/unbinding reaction $A + X \rightleftharpoons AX$ has an equilibrium constant K:

$$\frac{c_A^i c_X^i}{c_{AX}^i} = K, \qquad i = \text{I, II}, \tag{1}$$

where c is concentration (to be replaced by activity if necessary), i is the side number (I or II). The reaction is fast, and is in equilibrium on both sides of membrane. K is the same at both interfaces.

H4) The phases on both sides of the membrane are at steady state or quasi-steady state.

H5) The rate constant D', for the movement of carriers from one side of the membrane to the other is the same for X and AX in either direction. Thus the net flux of j ($j = X$ or AX) from side I to side II is

$$J_j = D'(c_j^{\text{I}} - c_j^{\text{II}}). \tag{2}$$

Under these hypotheses, we see that the *conservation of mass of the carrier X* requires that the sum of the parts be equal to the total X_T,

$$c_{AX}^{\text{I}} + c_{AX}^{\text{II}} + c_X^{\text{I}} + c_X^{\text{II}} = X_T. \tag{3}$$

At steady state, the combined transmembrane flux of X and AX must be zero because the carrier X is confined to the membrane. Hence $J_X + J_{AX} = 0$, i.e.

$$D'(c_X^I - c_X^{II} + c_{AX}^I - c_{AX}^{II}) = 0. \tag{4}$$

On the other hand, the flux of the substrate A is equl to the flux of AX because A cannot cross the membrane in any other form:

$$J_A = D'(c_{AX}^I - c_{AX}^{II}). \tag{5}$$

Solving Eqs. (2)–(5) for $c_X^I, c_X^{II}, c_{AX}^I, c_{AX}^{II}$, and J_A, using Eq. (1) as the boundary conditions, we obtain the important result

$$J_A = \frac{D'X_T}{2}\left[\frac{c_A^I}{K + c_A^I} - \frac{c_A^{II}}{K + c_A^{II}}\right]. \tag{6}$$

This equation shows that the flux of the solute A is linearly proportional to $c_A^I - c_A^{II}$ when the concentration c_A is very small compared with K; and that as c_A increases relative to K, the flux tends asymptotically to a maximum and the system is said to be *saturated*. If c_A^{II} were zero, the maximum flux attainable by the system is $V_m = D'X_T/2$.

It is easy to imagine more complex models. See Friedman (1986) for examples and for channels also.

Proposed Mechanisms of Active Transport

The difference between active transport and facilitated diffusion may be illustrated by the flux versus chemical potential difference relationship depicted in Fig. 8.8:4. In a passive transport mechanism, the flow J_i, is a function of its *conjugate*, the chemical potential difference $\Delta\mu_i$ (see Sec. 8.6), but J_i always vanishes when the conjugate chemical potential difference vanishes, i.e., $J_i = 0$ when $\Delta\mu_i = 0$. The hallmark of active transport is that there exists a flow even when the difference of chemical potentials $\Delta\mu_i$ is zero. The total flux may be written as

$$J_i = J_i^p + J_i^a \tag{7}$$

in which $J_i^p = 0$ when $\Delta\mu_i = 0$; whereas J_i^a is a result of the coupling of the ith flux to a driving force that is not its conjugate driving force. The force driving J_i^a is derived from a metabolic reaction. J_i^a is called the *metabolic contribution* to J_i, or the *rate of active transport* of the ith species.

In concept, a metabolic reaction causes a conformational change in a carrier protein, which then converts chemical energy into transport work. The driving force for active transport is the affinity A, of the metabolic reaction:

$$A = \Delta\mu_{\text{reaction}} = \sum_{\text{reactants}} v_i\mu_i - \sum_{\text{products}} v_i\mu_i. \tag{8}$$

Here v_i are the stoichiometric coefficients of the reaction, and μ_i are the chemical potentials. Many active transport systems also convert chemical energy into electrostatic potential energy by contributing to the electric poten-

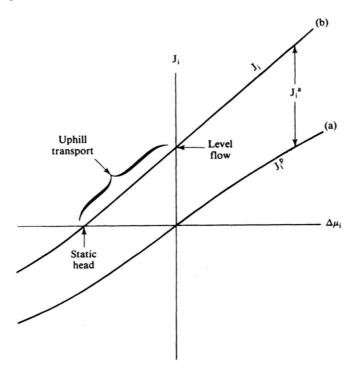

FIGURE 8.8:4 Flow vs. its conjugate driving force in (a) a passive transport, and in (b), an active transport. The terminology of an industrial pump is used to describe the properties of an active transport mechanism by analogy. From Friedman (1986), by permission.

tial difference across the membrane. The membrane potential can influence cellular and transport events.

The major metabolic step relevant to active transport conceived so far is the *hydrolysis* of *adenosine triphosphate* (ATP) to *adenosine diphosphate* (ADP) and inorganic phosphate (PO_4):

$$ATP + H_2O \rightarrow ADP + 3H^+ + PO_4. \qquad (9)$$

The reverse step, the *phosphorylation* of ADP into ATP, is endothermic and uses the energy of metabolism derived from food. Because the substrates are oxidized during the energy extraction process, it is referred to as *oxidative phosphorylation*.

ATP is the principal carrier of metabolic energy in the cell. When acted upon by an adenosine triphosphatase (*ATPase*), it undergoes hydrolysis into ADP, Eq. (9). The dephosphorylation reaction is accompained by the release of the energy stored in the terminal phosphate bond, about 7.8 ± 0.5 kcal/mol.

The specific effect of the hydrolysis of ATP is determined by the specific enzyme (ATPase) that causes the hydrolysis reaction. A specific enzyme coupled with the flux of a specific transported solute results in a *primary active transport* process. In animal cells, only three primary transport systems are known:

1. The *sodium-potassium exchange pump*, which exchanges two K^+ for three Na^+ per ATP hydrolyzed.
2. The *calcium pump* for which the stoichiometry is uncertain; either one or two Ca^{++} are transported across the cell membrane per ATP molecule hydrolyzed.
3. The *hydrogen-potassium exchange pump* in the stomach, which pumps one proton into the stomach per potassium removed.

The movement of Na^+ and K^+ ions in the Na–K exchange pump may cause other ions to move across the membrane, e.g. the use of sodium influx to accumulate a sugar in the cell. This is called a *secondary* active transport. Further, proteins may be transported across the cell membrane, the mechanism of which is still unknown although vigorously studied.

From this brief outline one can see that active transport is initiated by specific ATPase on the cell membrane. It depends on the availability of ATP. The chemical reaction also depends on the membrane potential, and the presence or absence of inhibitors and stimulators. The dynamic interaction of these carriers on the cell membrane and the chemical and electric environment of the cell creates the phenomena of nerve function, muscle contraction, urine formation, and food absorption, which however, are, outside the scope of this volume. Friedman (1986) is recommended for further reading.

8.9 Solid Immobile Matrix

So far we have treated the matrix component of the interstitial space as a fluid (Sec. 8.5). In reality, especially in dense connective tissues, its quality as a solid cannot be ignored. In a solid the state of stress must be described by a tensor. The analog of pressure gradient in a fluid is the divergence of stress tensor in a solid. In fluid we focus on flow. In solid we focus on strain rate. In fluid one usually pays little attention to the displacement of particles, whereas in solid one must. These are the points to be made in the mathematical analysis presented below. These considerations are essential to answering questions such as the bearing capacity, rigidity, and lubrication of the articular cartilage or the intervertebral disc.

Biological tissues are multi-phasic. To present a simple illustration, however, let us lump solute and solvent together and call it the "fluid" phase, and lump immobile matrix as the "solid" phase. The fluid phase has already been discussed in Secs. 8.6–8.8. We view the system as a mixture; and each of the two phases is present anywhere in the whole space (sometimes called the "equipresence" hypothesis). Hence each point in the tissue is occupied simultaneously by both phases. This concept can be expressed in terms of the density and volume fractions defined as follows (with the two phases identified by superscripts s and f):

$d^{(s)}$, $d^{(f)}$–the true density, i.e. mass per unit volume, of the solid and fluid.

$\rho^{(s)}$, $\rho^{(f)}$–the phase density (mass/tissue vol) in the mixture.

$\phi^{(s)}$, $\phi^{(f)}$–phase volume fraction, defined by

$$\phi^{(s)} = \rho^{(s)}/d^{(s)}, \qquad \phi^{(f)} = \rho^{(f)}/d^{(f)}. \tag{1}$$

According to the hypothesis of equipresence, ρ_s and ρ_f are well defined, and, since there is no void,

$$\phi^{(s)} + \phi^{(f)} = 1. \tag{2}$$

With respect to a set of rectangular cartesian coordinates, the displacement u_i, velocity v_i, acceleration a_i, strain e_{ij}, and strain rate $\dot{e}_{ij}$, $(i, j = 1, 2, 3)$ of the solid and fluid can be defined in the usual way in view of the equipresence hypothesis. The law of conservation of mass is expressed, in the absence of chemical reaction, by the usual equation of continuity

$$\frac{\partial \rho^{(s)}}{\partial t} + \frac{\partial (\rho^{(s)} v_j^{(s)})}{\partial x_j} = 0, \qquad \frac{\partial \rho^{(f)}}{\partial t} + \frac{\partial (\rho^{(f)} v_j^{(f)})}{\partial x_j} = 0. \tag{3}$$

These equations can be expressed differently if we can assume that both phases are incompressible by themselves and the densities d_s and d_f are constant. Then, by Eqs. (1), Eqs. (3) become

$$\frac{\partial \phi^{(s)}}{\partial t} + \frac{\partial (\phi^{(s)} v_j^{(s)})}{\partial x_j} = 0, \qquad \frac{\partial \phi^{(f)}}{\partial t} + \frac{\partial (\phi^{(f)} v_j^{(f)})}{\partial x_j} = 0. \tag{4}$$

Adding, and using Eq. (2), we obtain

$$\frac{\partial (\phi^{(s)} v_j^{(s)} + \phi^{(f)} v_j^{(f)})}{\partial x_j} = 0 \tag{5}$$

which can be written, by Eq. (2), as

$$\frac{\partial}{\partial x_j}(v_j^{(s)} + q_j) = 0, \qquad q_j = \phi^{(f)}(v_j^{(f)} - v_j^{(s)}). \tag{6}$$

Equation (6) says that the divergence of the velocity of the solid phase plus the filtration flux q (the volume fraction of the velocity of the fluid phase relative to the solid) vanishes.

Refinement of the Phenomenological Laws

In Sec. 8.6 we used the entropy production concept in thermodynamics to identify the driving forces conjugate to the fluxes of the materials in various phases of a mixture. The driving forces turn out to be the negative gradient of the chemical potentials divided by the absolute temperature T. A chemical species moves when it is subjected to the driving forces, and in motion it encounters resistance due to friction and momentum transfer from particle to particle. The actual mechanism of resistance is very complex. Instead of delving into this complexity, we just assumed in Sec. 8.6 that the mechanism is such that the fluxes are linearly related to the driving forces, leaving the coefficients in this relation to be determined by experiments "phenomenologically." The same idea will be used in the present section to deal with the solid matrix, except that the forces of interaction will be expressed in terms of stress tensors and the phenomenological laws will be replaced by the equations of motion.

Simplifying Assumptions

In applying the theory to living tissues, the following assumptions are sometimes acceptable.

(1) All phases are incompressible, but the volume of the phases may vary so that the tissue as a whole may be compressible (e.g. by squeezing water out or by swelling).

(2) The velocities are so small that the convective acceleration is negligible. (Rigid body motion is of no interest to the topic discussed here.)

(3) For the solid phase, a strain energy function W exists, which is a function of the strain, and symmetric with respect to the strain components $e_{ij}^{(s)}$ and $e_{ji}^{(s)}$, so that the stress is given by the equation (see Fung, 1981, p. 247):

$$\sigma_{ij}^{(s)} = \frac{\partial W}{\partial e_{ij}}. \tag{7}$$

This stress is defined with respect to the tissue at a whole. W is part of the chemical potential of the solid matrix (see Sec. 8.4).

(4) At every point in the fluid phase there is a hydrostatic pressure p, and a stress deviation tensor which represents the shear stress in the fluid caused by its moving through the solid matrix, a product of the viscosity of the fluid and the strain rate due to flow in every "pore." The *resultant of the viscous shear forces in the pores, per unit volume of tissue, can be expressed as a body force per unit volume*, ψ, *called the filtration resistance*. We shall assume that the filtration resistance is proportional to the filtration velocity, so that the ith component of ψ is

$$\psi_i = -R_{ij}q_j, \qquad i = 1, 2, 3, \tag{8}$$

where q_j is the filtration flux defined in Eq. (6), and R_{ij} is a matrix of constant coefficients, called the *resistivity tensor*. That R_{ij} are constants is suggested by the smallness of the filtration velocity so that the flow is in the Stokes regime, in which the resistance is linearly proportional to the velocity of flow (see Chap. 5 in Fung, 1984). R_{ij} depend on the size of the pores, hence are in general nonlinear functions of the deformation.

Thus, we can write the stress in the fluid as

$$\sigma_{ij}^{(f)} = -p\phi^{(f)}\delta_{ij} + \hat{\sigma}_{ij}^{(f)}, \tag{9}$$

where $\phi^{(f)}$ is the volume fraction of the fluid phase, and $\hat{\sigma}_{ij}^{(f)}$ is the stress deviation named above, which has the property that

$$\frac{\partial}{\partial x_j}\hat{\sigma}_{ij}^{(f)} = \psi_i. \tag{10}$$

(5) The structural fibers in the solid phase are stressed by strain according to Eq. (7), subjected to the same pressure p of the fluid in contact, and the shear stress in the fluid, $\hat{\sigma}_{ij}^{(f)}$, in the opposite direction. Hence the stress in the solid phase may be written as

$$\sigma_{ij}^{(s)} = \frac{\partial W}{\partial e_{ij}} - p\phi^{(s)}\delta_{ij} - \hat{\sigma}_{ij}^{(f)}. \tag{11}$$

(6) The chemical potential of the fluid consists of a part due to the pressure, $RT\ln p$, a part due to the concentration, $RT\ln c_\alpha$, of the chemical species numbered α (the concentration c_α to be replaced by "activity" in real solution), and a part due to electric charges of the molecules. See Sec. 8.4. The chemical potential of the solid has, in addition to those listed above, the strain energy function $W(e_{11}, e_{12}, \ldots)$ named in Eq. (7). However, since p and W have been named explicitly in Eqs. (9) and (11), we shall use the notations $\mu^{(f)}$, $\mu^{(s)}$ for chemical potentials other than $RT\ln p$ and W.

Equations of Motion

The familiar Eulerian equation of motion (Sec. 1.7)

$$(\rho \cdot \text{acceleration})_i = \frac{\partial \sigma_{ij}}{\partial x_j} \tag{12}$$

is derived for a homogeneous continuum. If the chemical composition is nonhomogeneous, we must add the driving forces due to chemical non-uniformity, namely, the gradients of the chemical potentials divided by the absolute temperatures T, to the right-hand side (see Sec. 8.6). Thus, in view of Eqs. (9), (11), and (12), the equations of motion of the fluid and solid phases are, respectively,

$$\rho^{(f)}a_i^{(f)} = -\phi^{(f)}\frac{\partial p}{\partial x_i} + \psi_i + \frac{1}{T}\frac{\partial \mu^{(f)}}{\partial x_i}, \tag{13}$$

$$\rho^{(s)}a_i^{(s)} = -\phi^{(s)}\frac{\partial p}{\partial x_i} + \frac{\partial}{\partial x_j}\left(\frac{\partial W}{\partial e_{ij}}\right) - \psi_i + \frac{1}{T}\frac{\partial \mu^{(s)}}{\partial x_i}. \tag{14}$$

Here a_i is the ith component of the accleration vector. $\mu^{(f)}$ and $\mu^{(s)}$ are the chemical potentials of the fluid and solid which depend on the concentrations and electric charges of the chemical species.

Variations of the chemical potentials of a system must obey the Gibbs–Duhem equation, Eq. (12) of Sec. 8.3. Under isothermal and constant stress conditions, Gibbs–Duhem equation reads, for the biphasic system, i.e.

$$d(\mu^{(f)} + \mu^{(s)}) = 0, \tag{15}$$

i.e.

$$\frac{\partial \mu^{(f)}}{\partial x_i} = -\frac{\partial \mu^{(s)}}{\partial x_i}. \tag{16}$$

This means that the effect of chemical potential on the motion of the liquid phase is equal and opposite to that on the solid phase, i.e., it is an internal interaction.

Some authors add a buoyance force

$$b_i = p \frac{\partial \phi^{(f)}}{\partial x_i} \tag{17}$$

to the right-hand side of Eq. (13) and $-b_i$ to the right-hand side of Eq. (14). This is based on molecular arguments by Muller (1968) for an intrinsic tendency of material to become homogeneous in density.

Adding Eqs. (13) and (14), and using Eq. (16), we obtain

$$\rho a_i = -\frac{\partial p}{\partial x_i} + \frac{\partial}{\partial x_j}\left(\frac{\partial W}{\partial e_{ij}}\right) \tag{18}$$

which is the normal equation of motion of the whole tissue. Of the three equations (13), (14), and (18), two are independent. Hence the solution of any problem of biphasic flow must involve chemical potential and filtration resistance.

Boundary Conditions

If a boundary of the mixture is in contact with an impermeable solid, then the normal outflow will be zero, the normal traction will be specified, and the normal velocity of the mixture will be equal to that of the solid. The tangential component of the velocity and displacement are uncertain and need further specification.

If the boundary of the mixture is in contact with another biphasic tissue, then the normal stress, flow, velocity, and displacement of the two media must be continuous at the boundary. If the boundary is a semipermeable membrane, then Π can be revealed as an osmotic pressure. The chemical potential will be continuous across the boundary, i.e., $p - \Pi$ is continuous; hence p will have a jump if Π does.

The multiphasic theory has been developed and applied to cartilage, skin, and other connective tissues by Armstrong et al. (1984), Eisenberg and Grodzinsky (1985), Lai et al. (1981), Lanir (1987), Mow et al. (1980, 1986), Myers et al. (1984), Omens et al. (1984), Urban et al. (1979). For fundamental mathematical development, one should consult Bowen (1976), Kenyon (1976), and Truesdell (1962). There are controversies about acceptable approximate simplifying assumptions for various tissues and organs: This has to be sorted out carefully. Some authors focus attention on frictional reistance to flow. Other authors speak of concentration effect or osmotic pressure effect. I believe, however, that it is best to cast the theory in terms of chemical potential. Then a unification with the classical theories of diffusion, filtration, osmosis, and permeation in porous media can be achieved.

Problems

8.1 Name two organs whose functions significantly rely on active transport of matter across membranes. Describe their anatomy, and how they work.

8.2 Generalize Fick's law, Eqs. (37), (39) of Sec. 8.5 to a solution that is not necessarily dilute.

8.3 Generalize Fick's law to a solution with n species of solutes.

8.4 Under the assumption leading to Starling's law, Eqs. (14), (15) of Sec. 8.6, write down an expression for the flux of a solute across the membrane.

8.5 Generalize Darcy's law, Eq. (16) of Sec. 8.6 to the movement of a nonhomogeneous solution in a porous medium, including the effect of the chemical potentials of various chemicals in the fluid or matrix.

    Note: Further generalization to consider the effects of electric charges on the solutes and matrix can be done. But consideration of the effect of elastic deformation of the "pores" in the media, leading to a nonlinear constitutive relation, requires an analysis of a different nature which is presented by Lew and Fung (1970).

8.6 G.S. Beavers and D.D. Joseph, in a paper entitled "Boundary Conditons at a Naturally Permeable Wall" (*J. Fluid Mech.* **30**, p. 197, 1967), have proposed that at a permeable wall the no-slip condition should be replaced by

$$\sqrt{k}\frac{\partial u}{\partial r} = -\alpha u,$$

where α is the "slip parameter", and k is the specific permeability of the porous medium. Thus they postulated the existence of slip at the boundary. They confirmed it experimentally.

    A qualitative estimation of the effect of slip is quite easy. Consider Poiseuille flow in a tube. What is the effect of slip?

    In blood circulation, could this effect be important?

8.7 Since the xylem fluid in trees is almost pure water, it will freeze in subzero temperature. On thawing again, some trees can recover, other trees will drop their frozen branches because cavitation forms in the xylem and water column cannot be formed on thawing. Explain the difference. (Ref.: Hammel, H.T. (1967) Freezing of xylem sap without cavitation. *Plant Physiol.* **42**: 55–66.)

8.8 Describe the active and passive mass transport processes taking place in the different segments of the loop of Henle in the kidney. (Cf. books such as P.C. Johnson (1978), *Peripheral Circulation*, Wiley, New York.)

8.9 The leaves of a tree in the desert look wilted in the afternoon, but become turgid by the next morning. From what you know about the negative pressure in the xylem water, how can this phenomenon be explained? Can this explain how desert plants get water from the atmosphere? What critical experiments should be done to verify your hypothesis? (Refs.: Dixon, H.H. (1914) *Transpiration and the Ascent of Sap in Plants.* 216 pp., London, MacMillan and Co. and Dainty, J. (1963) Water relations of plant cells. *Adv. Bot. Res.* **1**: 279–326.)

8.10 The beautiful flower of water lilies opens up in the sun and closes up when the sun sets. This is a mechanical event of which I know no literature. So let's speculate. Propose a possible explanation, and possible experiments, and quantitative analysis to verify your hypothesis.

8.11 Plants are well adapted to the environmental mechanics. Consider pine, bamboo, and orchid in wind. How well streamlined they are! To be convinced of the optimal design of the foliage, assume that the petiole (leaf stalk) and the leaf blade were rigid, and compute the shear force and bending moment that must act in the petiole at the stem in a wind of 30 km/hr. At what gale force would the leaf-stalk break?

References

Andreoli, T.E., Hoffman, J.F., and Fanestil, D.D. (eds.) (1980). *Membrane Physiology*. Plenum Medical, New York, London (being Parts 1, 2 and 3 of *Physiology of Membrane Disorders* by the same editors, which contains Part 4, *Specialized Cells, Tissues and Organs*, and Part 5, *Clinical Disorders of Membrane Disorders*).

Armstrong, C.G., Lai, W.M., and Mow, V.C. (1984). An analysis of the unconfined compression of articular cartilage. *J. Biomech. Eng.* **106**: 165–173.

Bird, R.B., Stewart, W.E. and Lightfoot, E.N. (1960). *Transport Phenomena*. Wiley, New York.

Bowen, R.M. (1976) Theory of mixtures. In: *Continuum Physics*, (A.C. Eringen, ed.), Vol. *III*, Academic Press, New York, pp. 1–127.

Curry, F.E. and C.C. Michel (1980). A fiber matrix model of capillary permeability. *Microvasc. Res.* **20**: 96–99.

Curry, F.E. (1984). Mechanics and thermodynamics of transcapillary exchange. *In Handbook of Physiology*, Sec. 2, *Cardiovascular System*, Vol. IV, Part 1, (E.M. Renkin and C.C. Michel, eds.). Amer. Physiol. Soc. Bethesda, MD.

Darcy, H. (1956). *Les Fontaines Publiques de la Ville de Dijon*. Dalmont.

Eisenberg, S.R. and Grodzinsky, A.J. (1985). Swelling of articular cartilage and other connective tissues: Electromechanochemical forces. *J. Orthop. Res.* **3**: 148–159.

Einstein, A. (1908). The elementary theory of the Brownian motion. *Z. Elektrochem.* **14**: 235–239. Reprint, 1956, *Investigations on the Theory of the Brownian Movement*. Dover, New York.

Epstein, P. (1937). *Textbook of Thermodynamics*. Wiley, New York.

Friedman, M.H. (1986). *Principles and Models of Biological Transport*. Springer-Verlag, New York.

Fung, Y.C. (1965). *Foundations of Solid Mechanics*. Prentice-Hall, Englewood Cliffs, N.J.

Fung, Y.C. (1977). *A First Course in Continuum Mechanics*. Prentice-Hall, Englewood Cliffs, N.J.

Fung, Y.C. (1981). *Biomechanics: Mechanical Properties of Living Tissues*. Springer-Verlag, New York.

Hargens, A.R. (ed.) (1981). *Tissue Fluid Pressure and Composition.* Williams and Wilkins, Baltimore, p. 3.

Intaglietta, M. and Johnson, P.C. (1978). Principles of capillary exchange. In: *Peripheral Circulation* (P.C. Johnson, ed.). Wiley, New York.

Katchalsky, A. and Curran, P.F. (1965). *Nonequilibrium Thermodynamics in Biophysics.* Harvard University Press, Cambridge, MA.

Kenyon, D.E. (1976). The theory of an incompressible solid fluid mixture. *Archs. Ration. Mech. Anal.* **62**: 131–147.

Klingenberg, M. (1981). Membrane protein oligomeric structure and transport function. *Nature* **290**: 449–454.

Krupka, R.M. and Deves, R. (1983). Kinetics of inhibition of transport systems. *Int. Rev. Cytol.* **84**: 303–352.

Lai, W.M., Mow, V.C., and Roth, V. (1981). Effects of nonlinear strain-dependent permeability and rat of compression on the stress behavior of articular cartilage. *J. Biomech. Eng.* **103**: 61–66.

Lanir, Y. (1987). Biorheology and flux in swelling tissue. I. Bicomponent theory for small deformation including concentration effect. *Biorheology* **24**: 173–187. II. Analysis of unconfined compressive response of transversely isotropic cartilage disc. *Biorheology* **24**: 189–205.

Lew, H.S. and Fung, Y.C. (1970). Formulation of a statistical equation of motion of a viscous fluid in an anisotropic non-rigid porous solid. *Int. J. Solids Struct.* **6**: 1323–1340.

Lieb, W.R. and Stein, W.D. (1969). Biological membranes behave as non-poroud polymeric sheets with respect to the diffusion of non-electrolytes. *Nature* **224**: 240–243.

Lieb, W.R. and Stein, W.D. (1972). Carrier and non-carrier models for sugar transport in the human red blood cell. *Biochim. Biophy. Acta* **265**: 187–207.

Lightfoot, E.N. (1974). *Transport Phenomena and Living Systems.* Wiley, New York.

Lotorre, R. and Miller, C. (1983). Condition and selectivity in potassium channels. *J. Membrane Biol.* **71**: 11–30.

Middleman, S. (1972). *Transport Phenomena in the Cardiovascular Systems.* Wiley, New York.

Miller, C, (1982). Feeling around inside a channel in the dark. In: *Transport in Biomembranes: Model Systems and Reconstitution.* (R. Antolini, G. Allessandra, and A. Gorio, eds.). Raven Press, New York, p. 99.

Mow, V.C., Kuei, S.C., Lai, W.M., and Armstrong, C.G. (1980). Biphasic creep and stress relaxation of articular cartilage: Theory and experiments. *J. Biomech. Eng.* **102**: 73–84.

Mow, V.C., Kwan, M.K., Lai, W.M., and Holmes, M.H. (1986). A finite deformation theory for nonlinearly permeable soft hydrated biological tissues. In *Frontiers in Biomechanics,* (Schmid-Schönbein, G., Woo, S.L.Y., Zweifach, B.W., eds). Springer-Verlag, New York. 153–179.

Muller, I. (1968). A thermodynamic theory of mixtures of fluids. *Arch. Rat. Mech. Anal.* **28**: 1–38.

Myers, E.R., Lai, W.M., and Mow, V.C. (1984). A continuum theory and an experiment for the ion-induced swelling behavior of articular cartilage. *J. Biomech. Eng.* **106**: 151–158.

Omens, C.W.J., van Campen, D.H., Grootenboer, H.J., and De Boer, L.J. (1984). Experimental and theoretical compression studies on porcine skin. *European Soc. of Biomechanics (ESB) Conference,* Davos, Switzerland.

Onsager, L. (1931). Reciprocal relations in irreversible processes. *Phys. Rev.* **37**: 405–426; **38**: 2265–2279.

Pappenheimer, J.R., Renkin, E.M., and Borrero, L.M. (1951). Filtration, diffusion and molecular sieving through peripheral capillary membranes. A contribution to the pore theory of capillary permeability. *Amer. J. Physiol.* **167**: 13–46.

Patlak, C.S. (1957). Contributions to the theory of active transport. II. The gate type non-carrier mechanism and generalizations concerning tracer flow, efficiency, and measurement of energy expenditure. *Bull. Math. Biophys.* **19**: 209–235.

Prigogine, I. (1955). *Introduction to the Thermodynamics of Irreversible Processes.* Thomas, Springfield, IL.

Renkin, E.M. (1977). Multiple pathways of capillary permeability. *Circ. Res.* **41**: 735–743.

Solomon, A.K., Chasan, B., Dix, J.A., Lukocovic, M.F., Toon, M.R., and Verkman, A.S. (1983).

The aqueous pore in the red cell membrane: B and 3 as a channel for anions, cations, nonelectrolytes, and water. *Ann. N.Y. Acad. Sci.* **414**: 97–124.

Starling, E.H. (1896). On the absorption of fluid from the connective tissue spaces. *J. Physiol. (London)* **19**: 312–326.

Truesdell, C. (1962). Mechanical basis of diffusion. *J. Chem. Phys.* **37**: 2336–2344.

Urban, J.P.G., Maroudas, A., Bayliss, M.T., and Fillon, J. (1979). Swelling pressures of proteoglycans at the concentration found in cartilageneous tissues. *Biorheology* **16**: 447–464.

Zweifach, B.W. and Silberg, A. (1979). The interstitial-lymphatic flow system. In: *International Review of Physiology, Cardiovascular Physiology III*, Vol. 18, (A.C. Guyton and D.B. Young, eds.). University Park Press, Baltimore.

The strain energy function $\rho_0 W$ plays a central role in biomechanics. Caligraphy and lithography by Aphrodite Sobin for the Symposium on Frontiers of Biomechanics, 1984.

CHAPTER 9

Mass Transport in Capillaries, Tissues, Interstitial Space, Lymphatics, Indicator Dilution Method, and Peristalsis

9.1 Introduction

In this chapter some applications of the basic equations derived in Chap. 8 are demonstrated. Flow across the walls of the capillary and lymph vessels is discussed in Sec. 9.2. Methods for measuring the permeability of vessel walls are presented in Sec. 9.3. A model of oxygen delivery and consumption in tissues is given in Sec. 9.4. Fluid movement in interstitial space is discussed in Sec. 9.5.

The "tissue pressure" in the very small interstitial space is very important but very difficult to measure, and very controversial. This subject is discussed in Secs. 9.6 and 9.7. Then in Sec. 9.8 the lymph flow is discussed.

In Secs. 9.9 and 9.10, the indicator dilution method for the measurement of the extravascular space is discussed. In the last section, peristalsis as a mode of mass transport is analyzed.

9.2 Fluid Movement Across Capillary Blood Vessel Wall

A simplified description of the flow through capillary walls is *Starling's law* (Sec. 8.6):

$$J_0 = K[p_c - p_t - \sigma(\pi_c - \pi_t)],$$
$$J_1 = \alpha_1(p_c - p_t) + \alpha_2(\pi_c - \pi_t), \tag{1}$$

in which J_0, J_1 represent the fluxes of water and solutes (rates of volume flow per unit area from the capillary side to the tissue side), respectively, p is static pressure, π is osmotic pressure, K, σ, α_1, α_2 are material constants, K is called *hydraulic conductivity*, σ is the *reflection coefficient*, the subscript c refers to the capillary, whereas t refers to the tissue. σ is related to the degree of permeability of the capillary wall to the solutes: $\sigma = 1$ if the wall is impermeable to the solutes; $\sigma < 1$ if it is permeable. The smaller the σ, the leakier the wall is to the solutes.

The coefficients K, σ, α_1, α_2 vary from organ to organ, and are affected by stress and strain, and the molecular sizes of the solutes.

To understand the nature of the coefficients K, σ, α_1, and α_2, a number of mechanical models of the capillary endothelium have been proposed. Renkin and Curry (1982) have examined the multiple pathways for transcapillary exchange. The pathways that have been documented in the literature are:

(1) Transport through the endothelial cells by crossing two plasma membranes and cell cytoplasm.
(2) Transport within the cell membranes by lateral diffusion in the lipid phase through intercellular junctions or vesicular channels.
(3) Transport through junctions between endothelial cells in the aqueous extracellular phase either a) via small pores or a fibrous meshwork impermeable to plasma proteins, or b) via large pores or a less-dense fibrous meshwork permeable to plasma proteins.
(4) Transport through endothelial cell fenestrae which are either specialized openings across the cells, or are areas at which two cell membranes are fused and modified, bypassing the aqueous cytoplasmic phase.
(5) Transport by vesicles in the endothelial cells that either a) move back and forth between cell surfaces, or b) communicate with one another transiently to exchange their contents or to establish temporary open channels.
(6) Transport through the basal laminar which consists of a layer of fine collagenous fibers.
(7) Transport through the cell surface coat which consists of a layer of glycoprotein strands that extends into the intercellular junctions and lines the vesicles.

Water, lipophilic solutes, hydrophilic solutes, and macromolecules may use different pathways. See Curry (1984) and Table 9.2:1.

In the following, we shall consider a few highly idealized examples to

TABLE 9.2:1 Partition of transcapillary fluxes between different pathways*

Species	Pathways
Water	1, 3, 4
Lipophillic solutes	1, 2, 3, 4
Hydrophilic solutes	3, 4
Macromolecules	3, 4, 5

*According to Curry (1984).

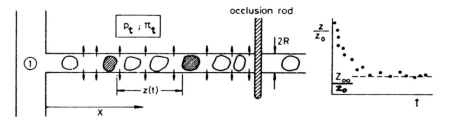

FIGURE 9.2:1 A model of a capillary blood vessel.

illustrate the nature of the mathematical problem. The capillary geometry is shown in Fig. 9.2:1. It has a length L and a constant radius R. The blood flow rate in the capillary is $Q(x)$; the filtration rate across the wall is $J(x)$, x being the coordinate along the length of the capillary, $x = 0$ at the arteriolar end, $x = L$ at the venular end. The subscripts c and t stand for capillary and tissue, respectively, whereas a and v stand for the arteriolar and venular ends of the capillary, respectively.

The amount of fluid that is filtered by the capillary per unit time is the *filtration rate*,

$$\text{filtration rate} = 2\pi \int_0^L R(x)J(x)\,dx. \qquad (2)$$

If a dimensionless variable $\xi = x/L$ is introduced, the radius is constant, and $\sigma = 1$, then

$$\text{filtration rate} = A \int_0^1 K(\xi)\{[p_c(\xi) - p_t(\xi)] - [\pi_c(\xi) - \pi_t(\xi)]\}\,d\xi, \qquad (3)$$

where $A = 2\pi RL$ is the capillary surface area.

Example 1. Assume that K, π_c, p_t, and π_t are constant, whereas

$$p_c(\xi) = p_a(1 - b\xi), \qquad b = \text{const.} \qquad (4)$$

Then

$$\text{filtration rate} = KA[p_a(1 - \tfrac{b}{2}) - p_t - \pi_c + \pi_t]. \qquad (5)$$

Example 2. (Apelblat et al., 1974; An and Salathé, 1976). Assume that K, π_c, π_t, and p_t are constant, whereas p_c obeys the momentum equation

$$\frac{dp_c}{d\xi} = -8\frac{\mu}{\pi R^4}Q(\xi), \qquad (6)$$

$$\frac{dQ}{d\xi} = -2\pi RK[p_c(\xi) - p_t - \pi_c + \pi_t], \qquad (7)$$

$$p_c(0) = p_a, \qquad p_c(1) = p_v. \qquad (8)$$

Then the filtration rate is

$$KA\left[\frac{\cosh 4\sqrt{\varepsilon} - 1}{2\sqrt{\varepsilon}\sinh 4\sqrt{\varepsilon}}\left(\frac{p_a + p_v}{2} - p_t + \pi_t\right) - \pi_c\right], \tag{9}$$

where ε is a dimensionless parameter:

$$\varepsilon = \frac{K\mu}{R}\left(\frac{L}{R}\right)^2, \tag{10}$$

where μ is blood viscosity. ε is usually <0.01. The blood viscosity μ is a function of capillary radius, tubular hematocrit, volume rate of flow, elastic properties of the red cells, and plasma protein concentration. (See Fung, 1981, Chap. 5.)

   Example 3. (Deen et al., 1972). Assume that K, p_t, and π_t are constant, and $p_c(\xi)$ is given by Eq. (4). The plasma oncotic pressure π_c is required to satisfy the equation of conservation of mass of protein in the capillary. Then the filtration rate is

$$KA\left[p_a\left(1 - \frac{b}{2}\right) - p_t + \pi_t - \int_0^1 \pi_c(\xi)\,d\xi\right] = (1 - H_F)Q_a\left(1 - \frac{c_a}{c_v}\right). \tag{11}$$

c_a, c_v are the plasma concentrations at the arterial and venous ends of the capillary, respectively. Q_a is the volume rate of flow into the capillary, H_F is the (feed) hematocrit of blood entering the capillary.

   Example 4. (Papenfuss and Gross, 1978). Assume that K, p_t, π_t are constant, while p_c, π_c satisfy the equations of motion and continuity. Consideration of mass balance leads to

$$\frac{dQ}{d\xi} = -2\pi RK[p_c(\xi) - p_t - \pi_c(\xi) + \pi_t] = -2\pi RK[p_{\text{eff}}(\xi) - \pi_c(\xi)], \tag{12}$$

where p_{eff} is defined as the *effective pressure*:

$$p_{\text{eff}}(\xi) = p_c(\xi) - p_t + \pi_t. \tag{13}$$

The gradient of p_{eff} is given by the Poiseuille equation

$$\frac{dp_{\text{eff}}}{d\xi} = -8\frac{\mu}{\pi R^4}Q(\xi). \tag{14}$$

   The "oncotic" pressure π_c of the blood plasma in the capillary is a function of the plasma protein concentration c, and may be written as

$$\pi_c = \alpha c + \beta c^2 + \delta c^3, \tag{15}$$

in which the coefficients α, β, δ have been reported by Landis and Pappenheimer (1963). Conservation of plasma protein in the capillary blood vessel leads to the equation

$$c(\xi) = c_a \frac{(1 - H_F)Q_a}{Q(\xi) - H_F Q_a}. \tag{16}$$

Here $c(\xi)$ is the plasma protein concentration at ξ, H_F is the hematocrit as in Eq. (11), Q is the flow, and the subscript a denotes the arteriolar end. The boundary conditions for Eqs. (12), (13), and (14) are $\xi = 0$, $p_c = p_a$, $Q = Q_a$, $c = c_a$. These equations can be solved numerically, or by asymptotic expansion in terms of the small parameter ε defined by Eq. (10), after the equations are properly nondimensionalized. Papenfuss and Gross (1978) have applied this theory to the rat glomerulus and intestinal muscle, and showed that it is important to consider the nonlinear terms in Eq. (15).

Discussion

None of the above examples deal with tissue pressure p_t and tissue osmotic pressure π_t; yet these two variables are very important for the health of tissue. For example, changes in p_t are associated with edema, swelling, pain, and injury. We shall address this topic in Secs. 9.5–9.7.

9.3 Experimental Determination of the Permeability Characteristics of Capillary Blood Vessel Wall

Landis (1927) first showed how to get hold of a single capillary blood vessel and measure the permeability of the vessel wall. In the 1960's and 70's, Zweifach, Michel, and others made extensive investigations with Landis's method. Other investigators, however, experimented with whole organs and then used simplified models of the organs to interpret the results, e.g., the *osmotic transient* method of Pappenheimer et al. (1948, 1951), and the *indicator dilution* method. These methods are outlined below.

Landis Micro-occlusion Technique

If the flow in a capillary is suddenly stopped with a fine needle (Fig. 9.3:1), then the pressure on the left-hand side suddenly becomes that of the arteriolar pressure, while the pressure on the right-hand side becomes venular. This upsets the equilibrium condition across the vessel wall, and fluid will move out of or into the capillary. To measure the filtration rate, Landis used the red cells in the blood vessel as markers. Assuming that the vessel radius R does not change, and that the cells sufficiently plug up the vessel so that there is no leakage of plasma between the red cells and the capillary wall, then any loss of fluid in the vessel will be seen by a shortening of the distance between neighboring red cells. The distance L between a red cell and the occluding needle is measured as a function of time. In a time interval Δt, the distance

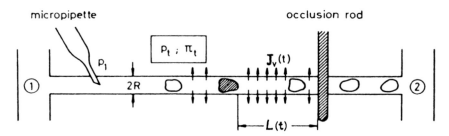

FIGURE 9.3:1 Landis's (1927) microocclusion experiment.

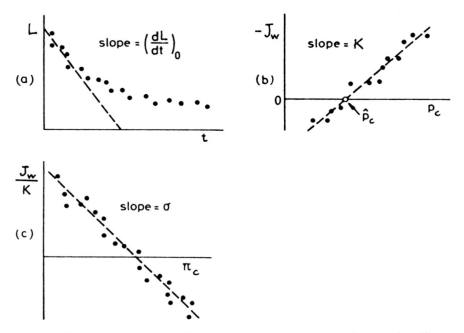

FIGURE 9.3:2 Data reduction method to determine transport coefficients of capillary wall by Landis's method. From Papenfuss and Gross (1979), by permission.

changes by ΔL, the volume of the capillary segment is changed by $\pi R^2 \Delta L$, while the amount of fluid that flows out of the vessel wall is $J_w L\, 2\pi R\, \Delta t$, J_w being the flux across the wall. Equating the outflow with the change of volume and solving for J_w, one obtains

$$J_w = -\frac{R}{2L}\left(\frac{\Delta L}{\Delta t}\right) = -K[(p_c - p_t) + \sigma(\pi_c - \pi_t)]. \tag{1}$$

When L is plotted against t as in Fig. 9.3:2(a), it is seen that the slope dL/dt varies with time. Landis extrapolated the slope and L to $t = 0$ and computed

J_w at time 0. He measured the capillary pressure p_c in the neighboring arteriole (Fig. 9.3:1). When J_w is plotted against p_c at $t = 0$, the slope of the resulting regression line yields the hydraulic conductivity K (Fig. 9.3:2(b)).

This calculation is valid only if J_w, p_t, π_c, π_t, R, σ, and K are constant along the capillary. Zweifach and Intaglietta (1968), however, have shown that this last assumption is quite wrong: in fact, K varies along a capillary by a factor of almost 3. The occluding needle certainly disturbs the radius and the tissue. The hypothesis that the red cells can serve as dividers of plasma segments without leaking is also uncertain.

Lee–Smaje–Zweifach Method

Zweifach and Intaglietta (1968) modified Landis's method by observing two red cells and measuring the distance between them $z(t)$, as a function of time following occlusion (Fig. 9.2:1). This avoids the assumption that the occluding needle does not deform the blood vessel and surrounding tissue. The theory is given by Lee et al. (1971). Assuming $\sigma = 1$, they write:

$$J_w(t) = -\frac{R}{2z(t)}\frac{dz}{dt} = -K[p_{\text{eff}} - \pi_c(t)], \tag{2}$$

where $p_{\text{eff}} = p_c - p_t + \pi_t$, and π_c is the colloidal osmotic pressure. The variation of π_c with the loss of water as z changes is analyzed and incorporated into the calculation (cf. Eq. (16) of Sec. 9.2). Equation (2) is solved as a differential equation for $z(t)$ to obtain a solution in the form of $t = f(z)$. Then a least-squares method is used to fit the data and obtain K.

Pappenheimer and Soto-Rivera Isogravimetric Method

Pappenheimer and Soto-Rivera (1948) isolated an organ and weighed it continuously while it was well supported to preserve its natural shape, environment, and state of stress. They controlled the conditions so that the weight did not change with time (isogravimetric), at which the averge osmotic pressure and static pressure across the capillary blood vessels balanced each other according to Starling's law. Now, if either the static pressure or the osmotic pressure of blood in the capillaries were changed, filtration or absorption of fluid into or from the tissue would occur, and the weight of the organ would change. By plotting the rate of increase of weight against the changes in static or osmotic pressure, they identified the slope of the curve as the *filtration coefficient* from which the permeability constants can be calculated. Arterial pressure, venous pressure, osmotic pressure of the perfusing fluid, blood flow, and temperature can be varied at will, and their influence onthe filtration-absorption equilibrium can be measured separately. But interpretation of whole organ experiments is always somewhat difficult because it is necessary to use some kind of model.

The Indicator Dilution Method

The indicator dilution method (Sec. 9.9) was first used by G. N. Stewart to measure the circulation time in organs in 1893. The theory was clarified by G. I. Taylor (see Sec. 7.3). It is now widely used as a clinical tool. In applying it to measure permeability, however, considerable difficulty exists. Issues and controversies are well documented in Crone and Lassen (1970), Aschheim (1977), Pappenheimer (1948, 1951), Renkin (1959), and Johnson and Wilson (1966).

9.4 The Krogh Cylinder as a Model of Oxygen Diffusion from Capillary Blood Vessel to Tissue

A mathematical idealization that has served as a model of capillary-tissue gas (or solutes) exchange was proposed by Krogh (1919). In this model, the capillary blood vessel is represented by a straight circular cylindrical tube of radius r_c; the tissue is represented by a stationary concentric cylindrical tube of inner radius r_c, outer radius r_t. A large number of identical tubes are closely packed to represent the tissue of an organ; therefore, the boundary condition at the outer radius r_t is no-flux.

Since diffusion is the main mechanism, the field equations are the same as those of Sec. 7.4. Let c be the concentration of the chemical species of interest (moles/volume). The flux vector (the rate of movement of the species across a cross section of unit area per unit time) consists of two parts: diffusive and convective. The former obeys Fick's law (Sec. 8.5). The latter is proportional to the product of velocity and concentration. Hence the flux

$$J_x = -D_x \frac{\partial c}{\partial x} + v_x c, \qquad J_y = -D_y \frac{\partial c}{\partial y} + v_y c, \qquad J_z = -D_z \frac{\partial c}{\partial z} + v_z c, \quad (1)$$

where D_x, D_y, D_z are the *coefficients* of *diffusion* in the x, y, z directions, respectively, and v (with components v_x, v_y, v_z) is the velocity vector of the medium. (x, y, z) is a fixed frame of reference.

The law of conservation of mass can be expressed by considering the mass transfer into a small control volume $dx\,dx\,dz$, as shown in Fig. 9.4:1. The inflow into the surface $dy\,dz$ at the left is $J_x\,dy\,dz$ per unit time. The outflow through a parallel surface at a distance dx to the right is $[J_x + (\partial J_x/\partial x)]\,dy\,dz$. Therefore, the net inflow rate through these two surfaces is $-(\partial J_x/\partial x)\,dx\,dy\,dz$. Considering all six surfaces, we obtain the net inflow rate:

$$-\left(\frac{\partial J_x}{\partial x} + \frac{\partial J_y}{\partial y} + \frac{\partial J_z}{\partial z} \right) dx\,dy\,dz. \quad (2)$$

In the meantime, the mass of the species in the volume is changing at a rate $\frac{\partial c}{\partial t}\,dx\,dy\,dz$. Further, there may be chemical reaction in the volume $dx\,dy\,dz$

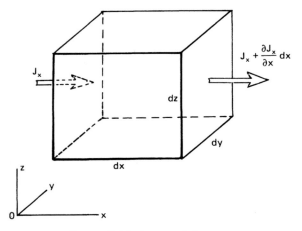

FIGURE 9.4:1 A control element.

which consumes the species at a rate of g moles per unit volume per unit time. Adding all three rates together, we obtain

$$\frac{\partial c}{\partial t} = -\left(\frac{\partial J_x}{\partial x} + \frac{\partial J_y}{\partial y} + \frac{\partial J_z}{\partial z}\right) - g. \tag{3}$$

On substituting Eqs. (1) into Eq. (3), we obtain

$$\frac{\partial c}{\partial t} = \frac{\partial}{\partial x}\left(D_x \frac{\partial c}{\partial x}\right) + \frac{\partial}{\partial y}\left(D_y \frac{\partial c}{\partial y}\right) + \frac{\partial}{\partial z}\left(D_z \frac{\partial c}{\partial z}\right) - g$$

$$- c\left(\frac{\partial v_x}{\partial x} + \frac{\partial v_y}{\partial y} + \frac{\partial v_z}{\partial z}\right) - v_x \frac{\partial c}{\partial x} - v_y \frac{\partial c}{\partial y} - v_z \frac{\partial c}{\partial z} \tag{4}$$

which is the basic equation for diffusion. If the fluid is incompressible, then

$$\frac{\partial v_x}{\partial x} + \frac{\partial v_y}{\partial y} + \frac{\partial v_z}{\partial z} = 0. \tag{5}$$

If, in addition, the coefficient of diffusion D is isotropic and is a constant, then Eq. (5) is reduced to

$$\frac{Dc}{Dt} \equiv \frac{\partial c}{\partial t} + v_x \frac{\partial c}{\partial x} + v_y \frac{\partial c}{\partial y} + v_z \frac{\partial c}{\partial z}$$

$$= D\left(\frac{\partial^2 c}{\partial x^2} + \frac{\partial^2 c}{\partial y^2} + \frac{\partial^2 c}{\partial z^2}\right) - g. \tag{6}$$

In the tissue space, let the coefficient of diffusion be D_t, the velocity of motion of the tissue be zero, and the distribution of oxygen be axisymmetric. Then, on using cylindrical-polar coordinates, Eq. (6) becomes, in the tissue region,

$$\frac{\partial c}{\partial t} = D_t \left(\frac{\partial^2 c}{\partial r^2} + \frac{1}{r} \frac{\partial c}{\partial r} + \frac{\partial^2 c}{\partial z^2} \right) - g(c), \qquad r_c \leqslant r \leqslant r_c. \tag{7}$$

It is often observed that oxygen consumption in tissue follows Michaelis–Menten kinetics (White et al., 1959), which may be expressed as

$$g(c) = \frac{Ac}{(B + c)} \tag{8}$$

with two constants A, B.

In the capillary region, let D_b be the overall coefficient of diffusion of oxygen in the blood, which varies with the hematocrit and capillary size. Let $d(c)$ represent the rate of generation of oxygen within the blood due to the dissociation of oxyhemoglobin contained in the red cells. Assume incompressibility of blood and axisymmetry in oxygen distribution. Then Eq. (6) becomes

$$\frac{\partial c}{\partial t} + \mathbf{v} \cdot \nabla c = D_b \left(\frac{\partial^2 c}{\partial r^2} + \frac{1}{r} \frac{\partial c}{\partial r} + \frac{\partial^2 c}{\partial z^2} \right) + d(c), \qquad 0 \leqslant r \leqslant r_c. \tag{9}$$

The boundary conditions are (a) symmetry about the axis of the cylinder, (b) no flux at $r = r_t$, and (c) continuity of flux at the interface $r = r_c$. Thus,

$$r = 0: \quad \partial c / \partial r = 0, \tag{9a}$$

$$r = r_c: \quad \left(D_b \frac{\partial c}{\partial r} \right)_{\text{blood}} = \left(D_t \frac{\partial c}{\partial r} \right)_{\text{tissue}}, \tag{9b}$$

$$r = r_t: \quad \partial c / \partial r = 0. \tag{9c}$$

These equations have been solved for a variety of conditions. Some comments follow:

(a) *Krogh's Steady-State Model* (1919). Assuming that 1) $g(c)$ in Eq. (6) is a constant g_0; 2) c is equal to a constant c_0 in the blood vessel; 3) the condition is steady-state; and 4) the axial diffusion of oxygen is negligible, then Eq. (6) is reduced to

$$0 = D_t \left(\frac{d^2 c}{dr^2} + \frac{1}{r} \frac{dc}{dr} \right) - g_0 \tag{10}$$

whereas Eq. (8) is reduced to $c = c_0$.

The solution, subjected to the boundary conditions (9c) and $c = c_0$ at $r = r_c$, is

$$c = c_0 = \frac{g_0}{D_t} \left(\frac{r^2 - r_c^2}{4} - \frac{r_t^2}{2} \log \frac{r}{r_c} \right). \tag{11}$$

(b) Bloch (1943) and Blum (1960) relaxed the boundary condition $c = c_0$ at $r = r_c$ in Krogh's approximation, and considered the axial variation of oxygen concentration in the capillary blood vessel. Several models of the oxygen generation function and oxygen consumption are considered. Reneau et al. (1967, 1969) presented numerical solutions of these equations in both steady-state and nonstationary cases, with particular reference to the brain.

They concluded that axial diffusion makes a significant contribution only near the entrance to the capillary.

(c) Bassingthwaighte et al. (1970) made an analysis of a Krogh tube which has a circular capillary blood vessel in the center and a tissue cylinder with hexagonal cross section on the outside. Hexagonal shape is selected because it is space-filling. The dimensions and parameters are selected for the model to represent a myocardium, and the boundary conditions used are pertinent to the problem of fractional extraction of a diffusible indicator during trans-capillary passage. It is concluded that the intratissue radial concentration gradients are probably negligible for small solutes, and that the longitudinal concentration gradients in tissue and capillary are important even for non-metabolized indicators.

(d) Niimi and Sugihara (1984), using a finite difference method, have analyzed a Krogh tube with a variable radius, $r_t(z)$. They considered counter current flow in neighboring tubes. They treated blood as a two-phase fluid, with red blood cell and plasma as two separate phases.

(e) *Other Diffusing Substance.* Oxygen is lipid-soluble. It can freely penetrate the endothelial cells and other elements of the vascular wall. Hence the boundary condition specified in Eq. (9) is appropriate. For lipid-insoluble substances such as water or albumin, the boundary condition (Eq. (9)) is inappropriate. Starling's law is considered a better approximation for trans-capillary movement of lipid-insoluble substances.

(f) *Larger Blood Vessels.* Oxygen diffusion takes place not only in the capillaries, but also in arteries and veins. Figure 9.4:2 shows a profile of

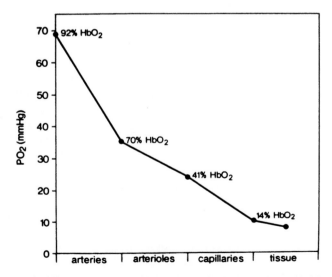

FIGURE 9.4:2 Perivascular oxygen tension and estimated oxyhemoglobin saturation in the arterial network of the hamster cheek pouch. Data from Duling and Berne (1970). Figure by Intaglietta and Johnson (1978). Reproduced by permission.

the perivascular partial pressure of oxygen (which is very nearly equal to intravascular pO_2) plotted against vessel type. The figure refers to a hamster cheek pouch. The arterialized blood from the lung contains approximately 20 ml O_2 per 100 ml blood at a partial pressure of 100 mm Hg. By the time the blood reaches the beginning of capillaries, the pO_2 is only 25 mm Hg.

9.5 Fluid Movement in the Interstitial Space

A schematic drawing of the interstitial space is shown in Fig. 9.5:1. Blood enters capillary blood vessels from the left and exits to the right. Blood and lymph vessels are embedded in a tissue, which consists of cells and an interstitial matrix. The walls of the capillaries and lymphatics are permeable to water, solutes, and some proteins. The blood inflow is controlled by arterioles and precapillary sphinctors (P.S.). The exchange of water and solutes between the capillaries, lymphatics, and the tissue is determined by the static pressures in the capillaries (p_c), lymphatics (p_L), and the tissue (p_t), and the osmotic pressures in the capillaries (π_c), lymphatics (π_L), and the tissue (π_t). The flux of the fluid from capillaries into the tissue is denoted by J_a at the arteriolar end , J_v at the venular end. J_L is the flux across the lymph vessel. In the interstitial space, fluid moves slowly as in a porous medium. The flux is measured by the rate of movement of the interstitial fluid across a cross section of the tissue per unit area of the tissue.

Equations governing the fluid movement in interstitial space have been derived in Sec. 8.5. The simplest version is Darcy's law, which ignores the effect of osmotic pressure on water movement.

Darcy's law (Eq. (18) of Sec. 8.5) states

$$\mathbf{J}_0 = -\kappa \Phi \nabla p, \tag{1}$$

where $\mathbf{J}_0$ represents the volume flow rate of the interstitial fluid, κ is a constant, Φ is the "porosity" of the tissue, i.e., the fraction of a given portion of tissue occupied by mobile interstitial fluid, ∇ is the gradient symbol, p is the pressure, and Φ and κ are functions of p. By the method detailed in Sec. 9.4, the rate of change of fluid volume in a unit volume of tissue can be written as $-\nabla \cdot \mathbf{J}_0$, which must be equal to the rate of change of the porosity, $\partial \Phi / \partial t$. Hence, the principle of conservation of mass is expressed by the equation:

$$\frac{\partial}{\partial t} \Phi(p) - \nabla \cdot \Phi(p)\kappa(p)\nabla p = 0. \tag{2}$$

For steady flow this is reduced to $\nabla \cdot \Phi \kappa \nabla p = 0$; and for constant Φ and κ, to the Laplace equation, $\nabla^2 p = 0$.

The flux of solutes and proteins is given by J_1 of Eq. (22) of Sec. 8.5. To obtain a simplified version which is consistent with Darcy's law for water flux, we assume that the solution is dilute, and neglect the term L_{21} grad μ_0 in the second equation of (8.5:22), because the conjugate term L_{12} grad μ_1 is

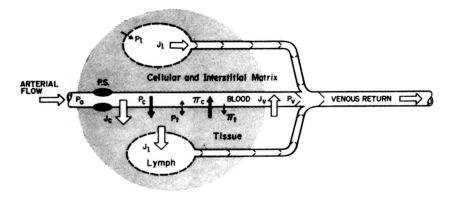

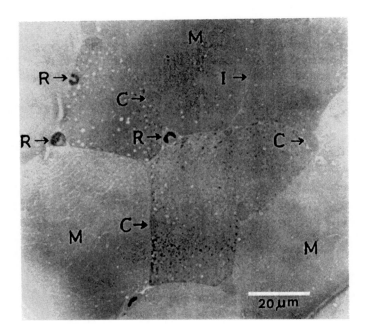

FIGURE 9.5:1 Top: Concept of balance of tissue fluid. Transcapillary pressures (closed arrows) determine the transcapillary flows (open arrows). Reproduced with permission from A.R. Hargans (Ed.), *Tissue Fluid Pressure and Composition*, p. 3, copyright © 1981, the Williams & Wilkins Co., Baltimore. Bottom: A histological photograph showing an example that in the rat gracilius muscle the interstitial space is very small. Fascia intact. Capillary blood pressure was 25 cm H_2O. M = muscle cell, C = capillary blood vessel, R = red blood cell in capillary, I = interstitium. Courtesy of Jye Lee, Ph.D., from his dissertation: "Morphometry and Mechanical Properties of Skeletal Muscle Capillaries." University of California, San Diego, 1990.

neglected in Darcy's law. Neglect also the term L_{23} grad μ_2 because the matrix is immobilized. Further, using Eq. (8.5:29), we obtain

$$\mathbf{J}_1 = -L_{22}(\Phi_1 \nabla p + kT\nabla c_1), \tag{3}$$

where Φ_1 is the volume fraction of the solute, k is the Boltzmann constant. T is the absolute temperature, and L_{22} is a constant. The first term on the right-hand side of Eq. (3) represents transport by convection, the second term represents diffusion.

The conservation of mass of the solute requires that

$$\frac{\partial c_1}{\partial t} + \nabla \cdot \mathbf{J}_1 = 0, \tag{4}$$

because $\nabla \cdot \mathbf{J}_1$ represents the rate of loss of solute in a unit volume, and $\partial c_1/\partial t$ the rate of increase in the same volume.

If the effect of osmotic pressure of the solutes or water movement is not neglected, then Darcy's law is not valid and we must use the full Eq. (22) of Sec. 8.5.

These equations are nonlinear and difficult to solve. It is clear that the interstitial pressure is a dynamic quantity, it changes both in space and time. For example, the vasomotion of the arterioles and venules will affect p significantly.

At a steady-state, Eq. (2) is reduced to

$$\nabla \cdot \Phi(p)\kappa(p)\nabla p = 0. \tag{5}$$

To solve this equation, Salathé and An (1976) introduced a new variable, ψ, which is a function of p defined by

$$\frac{d\psi}{dp} = \Phi(p)\kappa(p), \qquad \psi(p_\infty) = 0. \tag{6}$$

Then

$$\nabla\psi = \frac{d\psi}{dp}\nabla p,$$

and Eq. (5) is reduced to the Laplace equation

$$\nabla^2\psi = 0. \tag{7}$$

A solution that satisfies the condition $\psi \to 0$ as $r^2 + z^2 \to \infty$ is

$$\psi(r, z) = \int_0^L \frac{f(\zeta)\, d\zeta}{\sqrt{(\zeta - z)^2 + r^2}}, \tag{8}$$

where $f(\zeta)$ is an arbitrary function defined over the interval $0 \leqslant \zeta \leqslant L$. This solution is useful if the interstitial space is infinitely large compared with the capillaries. It states that the interstitial pressure can be represented as a distribution of sources and sinks of the function ψ distributed along the capillary axis, and reduced the determination of $p(r, z)$ to the determination of a function of only one variable $f(z)$. Salathé and An (1976) show further that Eq. (4) admits a solution of the form $c_1 = F(\psi)$, which states that the interstitial protein concentration is a function only of the interstitial fluid pressure, see Salathé (1980).

Nonuniformity and Other Idiosyncrasies of Interstitial Space

There is much evidence for nonuniformity of porosity in tissue space. Diffusible dyes (e.g., fluorescent dextran) injected into the blood vessel diffuses out into the tissue, and great nonuniformity of the concentration of the dye is seen in the tissue. Often there seems to be invisible channels carrying the dyes off. This means that either κ or Φ or both are functions of location in the interstitial space. In practical application of the mathematical analysis, the nonuniformity must be recognized, but not much is known at this time.

In skeletal muscle, the extravascular space is filled mostly by muscle cells. The cell volume of the muscle cells are closely regulated by the active pump of Na^+, K^+ ions by the cell membrane. As a result, although the cell membrane is permeable to water, the cell volume changes little, and the movement of water in the extravascular space takes place only in the extracellular space. The extracellular–extravascular space in the skeletal muscle is very small in vivo; perhaps no more than 2 or 3 percent of the muscle volume. Thus the domain in which the field equations (2), (3), (4) or (5) apply is long and narrow. Again, anatomical and rheological data are missing, and both theory and experiment remain to be developed.

In the lung, the interstitial space in the alveolar wall is normally much smaller than the vascular space. In normal conditions without interstitial edema, the thickness of the interstitium in the alveolar wall of man, dog, and cat is no more than 1 μm, whereas the thickness of the capillary blood vessels is 5–10 μm in zone 3 condition (see Chap. 6). The tissue pressure in this space depends on the structure of the interstitium, especially on the volume of the macromolecules such as the hyaluronic acid. In this case the geometrical configuration is sufficiently simple that a detailed analysis can be done. A theoretical analysis is presented in Sec. 9.10.

The interstitial spaces in tissues of other organs such as liver, kidney, arterial wall, etc., are also complex in their own way. Each organ needs a special treatment. This rich field is waiting to be cultivated.

9.6 Measurement of Interstitial Pressure

The time-averaged interstitial pressure p_t has been measured by several authors. Guyton and his associates (1963, 1981) imbedded perforated, hollow, rigid capsules (made of plastic or metal, with diameters ranging from 3–15 mm) in tissues. They let the interstitial fluid collect itself in the capsules in the course of days, then inserted needles into the capsules to measure the fluid pressure. If the pressure of the capsule fluid can be assumed to have exactly the same pressure as thatof the interstitial fluid in the tissue, then the capsule pressure yields the latter.

Wiederhielm (1969, 1972, 1979, 1981) employed micropipettes to measure the tissue pressure. These glass micropipettes with dip diameters less than 1 μm are produced by locally heating glass pipettes until they elongate and break under tension. By adjusting the tension and heating rate, the tip diameter can be controlled. Each pipette is then filled with a NaCl solution whose conductivity is different from that of the tissue fluid. The impedance of the pipette depends on the location of the interface between the saline and the tissue fluid in the tip section. If the saline pressure in lower than that of the tissue fluid, the interface will be pushed into the tube and the impedance will change. By a servo control with a pump, the back pressure in the pipette can be increased until the impedance reaches an extremum. At this condition the interface is located at the tip of the pipette. The back pressure in the pipette is then taken as the pressure of the tissue fluid.

Scholander et al. (1968) inserted cotton wicks with connective tissues of animals. The wicks are enclosed in polyethylene tubes which are perfused with saline. The wick and tube are introduced into the tissue with a trocar. When the trocar is withdrawn, $\frac{1}{2}$–1 cm of the wick is freely embedded in the tissue, the rest is enclosed in the polyethylene tube, which is connected to a horizontal pipette, then to a saline reservoir. A small air bubble is left in the horizontal pipette. The fluid in the wick is then allowed to come into equilibrium with the tissue fluid by adjusting the height of the reservoir in such a way that the bubble in the horizontal pipette remains stationary. In a few minutes equilibrium is established and the pressure in the pipette is measured and taken to be the tissue pressure.

Scholander's wick is considered to be equivalent to a large number of micropipettes by the capillary action. The principle is similar to Wiederhielm's except that the wick has soft walls and has many, many points of contact with the interstitial space. Wiederhielm's micropipette, though small, is still like "telephone poles" to the interstitial space.

The results from the three approaches do not always agree. For animals in the atmosphere Guyton and his associates have found tissue pressure in various organs to be negative (zero being atmospheric), in the order of -6 mm Hg, except in the kidney and the brain. In the kidney the tissue pressure is lower than the pressure under the tight renal capsule. In the brain, it is lower than the pressure in the cerebrospinal fluid that surrounds the brain. Scholander et al. (1968) found negative pressure in subcutaneous tissues. But most investigators using the Wiederhielm probe have found positive mean pressure, in the order of 1–3 mm Hg in subcutaneous tissue. Everybody agrees, however, that edema develops when the tissue pressure exceeds a certain limit.

These differences have not been completely resolved. Injury, healing, large disturbance of the normal state by the measuring instrument, and the changed boundary conditions are all parts of the story. The disturbed boundary conditions are especially important, as will be seen in the next section.

9.7 Pressure in an Incompressible Material

The tissue pressure measurements (Sec. 9.6) led to results that are perplexing. To gain a deeper understanding we shall consider the pressure in an incompressible fluid in general; namely, the biological tissue and body fluids.

Every student of mechanics knows that pressure in an incompressible material is not defined by strain and strain rate. As far as the constitutive equation is concerned, the pressure in an incompressible material is an arbitrary constant. It can assume any value. The pressure of a gas (considered compressible) is determined by its volume or density. The pressure of water (considered incompressible) is determined not by the water volume or density, but by its motion and boundary conditions. The pressure in a piece of steel (which is compressible) is determined by the strains in the steel (being proportional to the mean principal strains). The pressure in a piece of rubber (considered incompressible) is not determined by the deformation of the rubber, but by the equations of equilibrium (or motion) and the boundary conditions. Note that we usually consider steel as compressible but rubber as incompressible, not because rubber is harder to deform, but because its elastic modulus for volume change (the bulk modulus) is 10,000–100,000 times larger than its Young's modulus and shear modulus. For steel, the bulk modulus, Young's modulus, and shear modulus are of the same order of magnitude. Most biological materials are considered incompressible for the same reason as rubber is: they can be deformed easily, but their volumes are hard to change.

An interstitium without gas bubbles is incompressible. Therefore, the pressure in it has to be determined by the conditions of equilibrium (or motion) and the boundary conditions. This most elementary fact must be remembered, or one becomes confused. To assess the interstitial pressure at a certain point, we must begin with the external environment whose pressure we do know, and trace it step by step through each membrane and each compartment, until the point is reached.

In the following, let us make a digression from the interstitium and follow Per (Pete) Scholander (see Hammel and Scholander, 1976, p. 41 et seq.) to consider some instructive examples and provocative thoughts.

In Fig. 9.7:1 several cases considered by famous physicists are shown. To the left is Lord Kelvin (W. Thomson)'s (1871) capillary with a drop of water in its closed end. The capillary is placed in an isothermal evacuated jar with a layer of water. The vapor from the water will distill down into the capillary until the height of the meniscus reaches the same level as if the capillary were open below. Lord Kelvin derived a formula relating the curvature of the

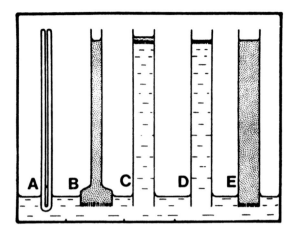

FIGURE 9.7:1 A series of experiments considered by famous physicists. (A) Kelvin's capillary. (B) Arrhenius's osmometer. (C) Noyes's osmometer. (D) and (E) Hulett's water and osmotic columns. In all these cases the chamber is isothermal and evacuated. From Hammel and Scholander (1976), reproduced by permission.

meniscus and the lowering of the water pressure immediately below the meniscus, and explained the movement of the water vapor into the capillary tube by the lowering of water pressure in the tube.

Ten years later, J. H. Poynting (1881) suggested that the vapor pressure change is in fact *caused* by the difference in water pressure. Based on the theorem of equipartition of energy, he wrote a simple equation:

$$\bar{v}_L \cdot dp_L = \bar{v}_g \cdot dp_g, \tag{1}$$

i.e., the molar volume of water times a change of its pressure is equal to the molar volume of the gas phase times its change in pressure.

In 1889, S. Arrhenius saw an important relation to osmosis in Poynting's equation. He considered an osmotic column (Fig. 9.7:1B) with its semipermeable membrane dipping in the water, and concluded that the solution would rise to a height at which the vapor pressure would match the rise from the water surface below; otherwise distillation would take place.

In 1900, A. Noyes moved the membrane to the top of Arrhenius's column, covering it with an infinitely thin layer of solution (Fig. 9.7:1C). He could then claim that the osmotic pressure times cross-sectional area is equal to the weight of the *water* beneath the free surface.

In 1902, G. Hulett gave a dynamic interpretation of the negative pressure of water below the membrane of Noyes's osmometer. He saw that the negative pressure is derived from the reflection of the solute molecules from the free surface and is analogous to a capillary lift from a membrane (Fig. 9.7:1D and E). The free surface of the solution is exposed to the vapor. The solute

FIGURE 9.7:2 Scholander's experiment on sedimented starch columnat equilibrium. Pure water gradients are measured with wick catheters. From Hammel and Scholander (1976), by permission.

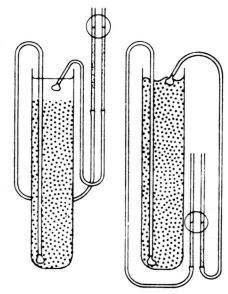

molecules tend to leave the surface, and are pulled back by the solvent, thus imparting tension to the solvent.

The last point was made clear by Scholander in his experiment on sedimented starch matrix (Fig. 9.7:2). A test tube is filled with water and some starch (left figure). The starch is thixotropic. A cotton wick in a plastic tubing is put at the bottom of the starch column, while another is put in the clear water at the top. Each one is connected with an identical glass capillary, scrupulously cleanned. The menisci are seen to stand at the same level. Now, let more starch be added until it hits the water surface, making it pasty by capillary lift. At this time, both capillaries will show negative pressure (Fig. 9.7:2, right). This shows that the interaction between the starch and water occurs at the free surface, and at that surface only. The relevance of this example to the matrix force in the interstitium shown in Fig. 9.7:1 is obvious.

Other examples are given in Hammel and Scholander (1976). A particular application of the principle to botany is very interesting. This is concerned with the question: What drives the sap of a tree from the root to the top in such tall trees as the redwood which may stand as tall as 100 m? First, it was found that the xylem water is nearly pure water, with a freezing point of 0.1°C or less. For a tree like the mangrove which can grow in sea water, the root cells that act as a semipermeable membrane must separate the xylem water from the sea water at an osmotic pressure difference of some 24 stm. To draw pure water into the xylem from the sea is a process of reverse osmosis that needs a driving pressure greater than 24 atm. The source of the driving pressure was found to be located in the parenchyma cells of the leaves on which the xylem terminates. The parenchyma cells contain a high concentration of solutes and have supporting matrix. They are exposed to atmosphere. When the leaves lose their turgor, the cells are flat and soft, so the internal pressure in the cells must be nearly atmospheric. Now, across that part of the cell membranes which are in contact with the xylem water, an osmotic pressure difference exists. If we apply Starling's law (Eq. (35) or (36) in Sec. 8.5) to the cell membrane, we find that the difference in static pressure must balance the osmotic pressure on the two sides of the membrane at equilibrium (noflow). Since the sum of osmotic and static pressure on the cell side is atmospheric, the static pressure in the xylem water must be negative because the osmotic pressure is essentially zero there. Hence it explains the negative pressure in the xylem.

Using a pressure chamber as shown in Fig. 9.7:3, Scholander et al. (1965) compressed the leaves until the cells lose their turgor and the xylem water is expressed from the cut stem of the twig. This pressure must be equal to the negative pressure in the xylem before the twig was cut. In this way, they measured the sap pressure in a variety of vascular plants, and demonstrated that all plants tested had negative sap pressure, most remarkably so in the mangroves (Rhizophora and Lagunchularia) and desert plans. See Fig. 9.7:4.

Hammel and Scholander (1976) prefer to think of water on the two sides of a semipermeable membrane as a single continuum with identical properties;

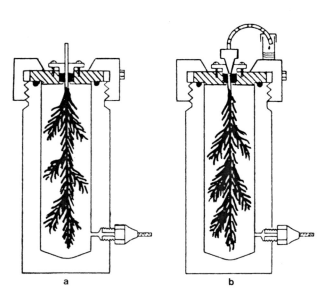

a   b

FIGURE 9.7:3 Pressure chambers used by Scholander et al. for measurement of sap tension in a twig. (a) Direct observation. (b) Step-by-step sap extrusion and pressure measurement to obtain a pressure-volume curve. From Scholander et al. (1965), by permission.

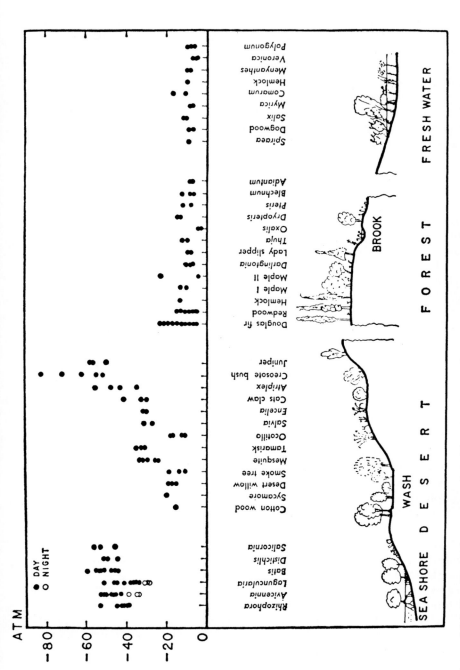

FIGURE 9.7:4 Sap pressure in a variety of vascular plants. Most measurements were taken during daytime in sunlight. Night values are likely to be less negative by several atmosphere. From Scholander et al. (1965), by permission.

in particular, at equilibrium they have the same pressure. If the pressure of water on the side of pure water is negative, then that on the solution side must be negative also and at the same value. This is Scholander's tensile water theory. He generalized it to all solutions in living organisms. To him, the explanation of osmotic pressure and surface tension has to be sought at the interface: the very boundary of the fluid, with a dynamic view of the solutes as in the molecular theory of gases and liquids. Some people challenge his view on the molecular theory of solutes. I have no doubt that the basic premies that in an incompressible fluid one can determine pressure only by tracing the balance of forces to the boundary where stress is known is correct.

9.8 Lymph Flow

There was a time when people believed that fluid enters the tissue space from the arterial ends of the capillaries and all of it is returned to the capillaries at their venular ends. The fluid balance in the tissue is thus maintained by the capillaries. This idea was known as Starling's hypothesis (Starling, 1896). More recent measurements, however, have shown that the return of fluid to the blood vessels and the prevention of edema relies on the lymph vessels (Zweifach and Prather, 1975). This is true not only for water, but also for various solutes, especially macromolecules.

The lymphatic vessel system is similar to the venous system except that the terminals (capillaries) are blind-ended. Like veins, the lymph vessels have one-way valves, and the lymph flow can be propelled by contraction of the vessels. Such contractions can be produced either by the smooth muscles in the lymphatic vessel wall, or by the motion of nearby arterioles, or by the contraction of the skeletal muscles, or movement of the organs (e.g., motion of the arms or legs, peristalsis of the intestine, or breathing of the lung). Figure 9.8:1 shows several records of pressure waves in the lymphatic vessels of the mesentery of the cat. The pulsatile nature is clearly seen. The period is long. Between a pair of valves (not necessarily consecutive), the propulsion of the lymph fluid is similar to that of the ventricles of the heart, consisting of a diastolic filling phase, and a systolic expulsion phase.

Thus, beyond the first valve, there is no difficulty in understanding the nature of lymph flow. Conceptual difficulty lies with the terminal lymphatics. How is it filled? What force drives the fluid from the tissue space into the terminal lymphatics?

To answer this question, let us first consider the structure of the terminal lymphatics. These vessels are extremely thin-walled and are much wider than the blood capillaries. They possess irregular contours, vary in diameter from 15–20 μm, and have flattened segments that are as wide as 300 μm (Cliff and Nicholl, 1970). They have different geometric patterns in different tissues. In the mesentery they conform to the modular configuration of the blood capillary network. In the skeletal muscle, lymphatics are found in the immedi-

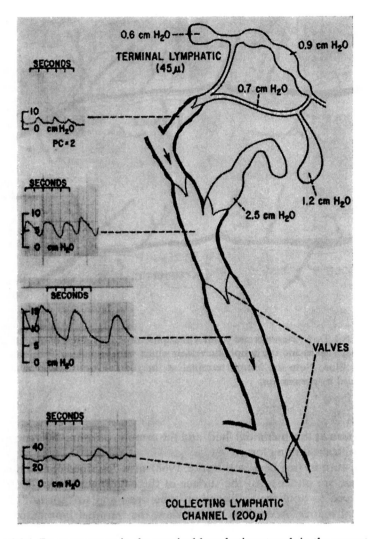

FIGURE 9.8:1 Pressure waves in the terminal lymphatic network in the mesentery of the cat. From Zweifach and Prather (1975). Reprinted by permission.

ate neighborhood of the arterioles (see Fig. 9.8:2). In the lung, the terminal lymphatics lie at the junctions of interalveolar septa. The walls of the terminal lymphatics consist of an attenuated endothelium which is similar to, but usually more attenuated and looser than the endothelium of blood capillaries, and is frequently without a basement membrane. With such a loose structure, the fluid in the terminal lymphatics seems to be in free communication with the rest of the interstitium. Thus, the lymph fluid has much the same

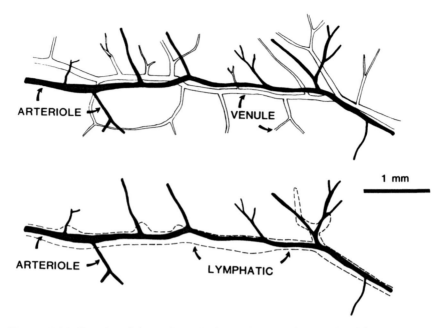

FIGURE 9.8:2 Top: Arterioles and venules in a spinotrapezius muscle of the rat. Bottom: the same arterioles and the lumphatic vessels which were visualized by microinjection of Evans Blue. Note the lumph terminal at the right. From Skalak et al. (1984). Reproduced by permission.

composition as the interstitial fluid, and the osmotic pressure difference cannot be a significant driving force.

Some authors (see Leak and Burke, 1968) have described connective tissue fibers that are attached to the surface of the endothelial cells. These fibers are believed to hold the endothelium tube open and to "anchor" it to the interstitial matrix. According to this view, the terminal lymphatic is filled because of the natural tendency of the vessel to be open.

Skalak, Schmid-Schönbein, and Zweifach (1984) have shown that, in the spinotrapezius muscle of the rat, all lymphatic vessels in the 20–200 μm diameter range lie in immediate neighborhoods of arterioles or venules. See Fig. 9.8:2. The majority of the lymphatics are in direct contact with the arterioles. When the arterioles were dilated, the contiguous lymphatics were seen to be partially or completely collapsed, whereas lymphatics around contracted arterioles were seen to be wide open. No significant deformation of the adjacent skeletal muscle cells was observed. These lymphatics have no smooth muscle; they have only a thin lining of endothelium. These results suggest that the contraction and relaxation of the vascular smooth muscle in the arterioles serves to open and close the lymphatics; thus vasomotion of the arterioles and venules appears to be the motive force for lymph transport in terminal lymphatics.

Lymph Flow During Hemorrhage

If enough blood is lost in hemorrhage so that the mean arterial blood pressure is reduced to 50–60 mm Hg, what happens to lymph flow? In such a person, extensive vasconstriction would raise peripheral resistance and the hydrostatic pressure at the capillary level may fall to 8–10 mm Hg. Under this condition, reabsorption of fluid is favored, and the uptake of fluid will reduce circulating plasma protein concentration by up to 30–50%. The hemodilution develops rapidly and approaches asymptotically to a maximum level in approximately 60 mins. During the initial response to hemorrhagic hypotension, lymph flow is drastically reduced. With the completion of hemodilution, lymph flow reappears at a reduced rate.

Baez (1960) observed that, in animals suffering severe blood loss, the frequency of spontaneous contractions of the collecting lymphatics in the intestinal mesentery increases some two- to threefold. Such heightened activity leads to a visible collapse of the terminal lymphatics, and after 20–30 mins, the spontaneous vasomotion slows and finally stops.

At a later stage in hemorrhage reaction, as tissue hypoxia develops, especially in tissue such as the small intestine, there is a marked increase in peristalsis. At the same time, an increased vascular permeability develops in these splanchnic viscera, as shown by loss of plasma protein and numerous petechial hemorrhages into the tissue. Simultaneously, spontaneous contraction of lymphatics increases and lymph flow picks up. On the other hand, lymph flow from the skeletal muscle of the extremities remains low (Arturson, 1971). These findings are suggestive of some feedback mechanism in the intestinal wall which responds to the changes in intestinal plasma protein level. A host of possible sensors and feedback loops may be suggested, but identification is a job for the future.

9.9 Measurement of Extravascular Space by Indicator Dilution Method

An indicator (or tracer) is a substance injected into a system in order to measure the distribution of some endogenous substance in the system. It must have the following properties:

1) It can be introduced into the living body.

2) Its presence does not alter any properties of the system.

3) It is neither metabolized nor sequestered in the system; all of it can be recovered.

4) It can be distinguished from other substances in the system and can be measured.

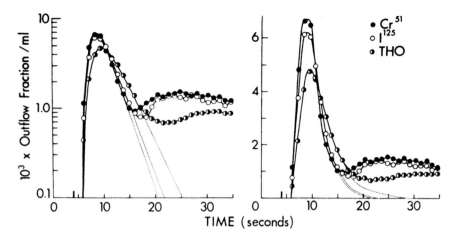

FIGURE 9.9:1 A typical set of indicator dilution curves. The ordinate shows the outflow fraction per ml, in logarithmic scale on the left, and in linear scale on the right. The abscissa shows time. The injection time and duration is shown by the black block on the abscissa. The sampling time interval is 1.09 sec. in this example. From Goresky et al. (1969). Reproduced by permission.

5) Its distribution in the system is a measure of the way the endogenous substance of interest is distributed.

6) The input-output relationship is linear; i.e., the distribution of the indicator in the system is a linear function of the input.

7) The basic system is in a steady-state.

Consider an organ fed by one artery and drained by one vein. Assume that the organ has a constant volume and that the inflow is steady. Let a tracer of amount q_0 be injected into the artery *as an impulse* at time zero. Let the concentration of the tracer measured at an outflow section in the vein be denoted by $\bar{c}(t)$. (See Fig. 9.9:1 as an example). We normalize $\bar{c}(t)$ and define an *impulse-response function* as follows:

$$h(t) \equiv \bar{c}(t) \Big/ \int_0^\infty \bar{c}(t)\,dt. \qquad (1)$$

The function $h(t)$ is nonnegative, and

$$\int_0^\infty h(t)\,dt = 1, \qquad (2)$$

therefore, it has the character of a *probability density function*. Since t is the time after injection, and $h(t)$ is the portion of the tracer that arrived at the exit section in a time interval dt after spending a "transit time" t in the system, $h(t)$ is spoken of as the *probability density function of the transit time t.*

The *precise* definition of $\bar{c}(t)$ is as follows: Assume that the vein is straight and let x, y, z be the coordinates in the vein, with the x-axis coinciding with the axis of the vein. Let the tracer be measured at an "exit" cross section located at $x = L$, where the velocity normal to the section is $u(L, y, z)$. Then the mean concentration of the tracer is

$$\bar{c}(t) = \frac{1}{F} \int \int c(L, y, z, t) u(L, y, z) \, dy \, dz, \tag{3}$$

where F is the "flow":

$$F = \int \int u(L, y, z) \, dy \, dz. \tag{4}$$

The integrals are taken over the entire cross section.

From Eq. (3) we can show that

$$\int_0^\infty \bar{c}(t) \, dt = q_0/F, \tag{5}$$

where q_0 is the total amount of the tracer injected. This is because $cu \, dt \, dy \, dz$ is the quantity of tracer that flows across $dy \, dz$ in time dt, so that the integral

$$\int \int \int cu \, dt \, dy \, dz = q_0. \tag{6}$$

Measurement of Flow

Equation (5) can be used as a basis to measure flow. For example, to measure cardiac output, one may inject, impulsively, an amount of dye q_0, into the right heart, and record the concentration of the dye as it passes through the aorta. The area under the concentration curve is the integral on the left-hand side of Eq. (5). Then Eq. (5) can be used to compute the cardiac output F.

Response to Arbitrary Injection History

Equations (1) and (5) give the impulse response function

$$\bar{c}(t) = \frac{q_0}{F} h(t). \tag{7}$$

This formula can be generalized to give the response to an arbitrary injection which has a history $q(t)$ per unit time. Regarding the input as a superposition of successive impulses of magnitude $q(t) \, dt$, we have, under the assumption of linearity between input and output, the result.

$$\bar{c}(t) = \frac{1}{F} \int_0^t q(\tau) h(t - \tau) \, d\tau. \tag{8}$$

Measurement of the Volume of the Traced System

If $q(t)$ is a *step function* with amplitude A, then the concentration at the exit section is the indicial-response function

$$\bar{c}(t) = \frac{A}{F} \int_0^t h(t - \tau)\,d\tau = \frac{A}{F} \int_0^t h(\tau')\,d\tau'. \tag{9}$$

Now, $F\bar{c}(t)$ is the rate at which the tracer leaves the system. Therefore, the difference $AH(t) - F\bar{c}(t)$ is the rate at which the tracer is accumulated in the system, $H(t)$ being the *unit-step function*. An integration gives the total amount of tracer that is accumlated in the system.

$$Q(t) = \int_0^t [AH(t) - F\bar{c}(t)]\,dt. \tag{10}$$

As $t \to \infty$, $Q(t)$ tends to a constant $Q(\infty)$. From Eq. (9),

$$Q(\infty) = A \int_0^\infty \left[1 - \int_0^t h(t)\,dt \right] dt. \tag{11}$$

Integrating by parts and remembering Eq. (2), we obtain

$$Q(\infty) = A \int_0^\infty th(t)\,dt = A\bar{t}, \tag{12}$$

where $\bar{t}$ is the mean transit time defined by the integral in Eq. (12). Now, under constant infusion, Eqs. (9) and (12) show that as $t \to \infty$, the concentration at the exit tends to a constant $c(\infty) = A/F$. *If we assume that the concentration tends to $c(\infty)$ throughout the entire system,* then $c(\infty)$ multiplied by the volume V of the system is the total tracer left in the system, $Q(\infty)$. Thus, from Eq. (12), we obtain the principal result

$$V = F\bar{t}. \tag{13}$$

Thus the flow multiplied by the mean transit time gives the volume of the system. This derivation was given by Meier and Zierler (1954) and Zierler (1963) and can be applied to measure the size of the left ventricle, the total blood volume in the lung, etc.

Zierler's hypothesis that the concentration be uniform throughout the entire system, unfortunately, is violated in many practical cases. Hence we add the remark that Eq. (13) *hold under the weaker condition that the* **averge** *concentration of the tracer over the entire system be equal to $c(\infty)$*. The proof follows Eq. (12) immediately.

Volume of Extravascular Water in an Organ

Consider a specific system, e.g., the lung. Let us simultaneously inject two tracers and continuously measure their concentrations at an exit section. Let the probability density functions of the transit time be $h(t)$ and $y(t)$ for these two tracers. The mean transit times $\bar{t}(h)$ and $\bar{t}(y)$ may be different. The difference

$$\Delta V = V(y) - V(h) = F\bar{t}(y) - F\bar{t}(h) \tag{14}$$

represents the difference of volumes traced by these two tracers.

If one of the tracers is labeled water which permeates through both vascular and extravascular spaces, whereas the other tracer is confined to the blood stream, then ΔV gives the extravascular water volume. This is a practical way to determine extravascular water in the lung and edema.

These formulas can be generalized in a simple way to cover the situation in which a tracer occupies only part of the fluid. For example, for a tracer which tags only red blood cells, or another which tags only the plasma, the corresponding flow rate will be only a fraction of the flow rate of the whole blood.

9.10 Tracer Motion in a Model of Pulmonary Microcirculation*

As an example of analysis of the motion of fluid in very narrow tissue space, the tracer motion (Sec. 9.9) in the extravascular and extracellular interstitial space of the interalveolar septa of the lung is presented. This space lies between the basement membrane of the pulmonary capillary sheet and the layer of epithelial cells. In the human lung, it is a multiply-connected sheet of the order of 100 m² in total area and 1 μm in thickness, with the total volume in the order of 100 ml. But the thickness and volume are highly variable. The total area changes with the transpulmonary pressure (airway minus pleural pressure). The thickness varies with the blood pressure, tissue pressure, osmotic pressures of the blood and the tissue, the tightness of the epithelial cell junctions, and the lymph drainage.

Figure 9.10:1 shows a sketch of an idealized capillary blood vessel sheet between an arteriole and a venule: a channel bounded by two thin layers of porous material. The channel (compartment No. 2) represents the capillary blood vessel. The porous layers (compartment No. 1) represent the tissue space. The membrances that divide these two compartments are permeable to water, but are semipermeable selectively with respect to the solutes. The external boundaries (the epithelia at $z = \pm(\frac{h}{2} + \delta)$), are impermeable to both water and solutes. Beyond the epithelia is the alveolar air space. The channel is supplied by an arteriole and drained by a venule, which are represented here by uniform channels with impermeable walls. Variables in the two compartments are distinguished by the subscripts 1 and 2.

The justification of this idealized model is discussed in Fung (1983), Chap. 6. The fluid coming from the left in the channel will permeate through the membrane into the porous layer under the influence of hydrostatic and

* This analysis was given by Tang and Fung (1975) and Fung and Tang (1975).

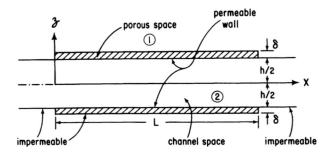

FIGURE 9.10:1 Flow of blood, interstitial fluid, and tracer in a pulmonary interalveolar septum. The upper figure shows the compartments of the blood (No. 2) and the porous interstitial space (No. 1), the boundary conditions, the coordinates, and the dimensions. The lower figure shows a sketch of the velocity distribution. From Fung and Tang (1975), by permission.

osmotic pressures. Because of fluid viscosity, the horizontal velocity component must vanish on the wall. Therefore, at the interface, the velocity must be perpendicular to the wall. In the porous layer, the fluid flows to the right and back into the channel.

The blood is assumed to be Newtonian. The interstitial fluid is assumed to obey Darcy's law (Eq. 9.5:1). The fluids and the tissue are incompressible. The Reynolds number of flow is very small ($\ll 1$). Hence with the coordinates shown in Fig. 9.10:1 and restricting our consideration to the two-dimensional case, we have the following governing equations:

In the porous layer, $\frac{h}{2} \leqslant z \leqslant \frac{h}{2} + \delta$, Darcy's law applies:

$$u_1 = -\kappa \frac{\partial p_1}{\partial x}, \qquad w_1 = -\kappa \frac{\partial p_1}{\partial z}. \tag{1}$$

On the boundaries $x = 0$, $x = L$, and $z = \pm(\frac{h}{2} + \delta)$, there is no slip:

$$u_1 = w_1 = 0. \tag{2a}$$

On the interface at $z = \pm\frac{h}{2}$, Starling's law applies:

$$w_1 = K[(p_2 - p_1) - \sigma(\pi_2 - \pi_1)]. \tag{2b}$$

Here u_1, w_1 are velocity components in the x and z directions, respectively; κ is Darcy's constant, K is the membrane permeability constant, p is pressure,

π is osmotic pressure, σ is reflection coefficient, δ is the thickness of the interstitium. The subscripts 1, 2 refer to the compartments 1, 2 respectively.

In the channel, $0 \leqslant z \leqslant \frac{h}{2}$, Stokes equation applies:

$$\frac{\partial p_2}{\partial x} = \mu\left(\frac{\partial^2}{\partial x^2} + \frac{\partial^2}{\partial z^2}\right)u_2, \qquad \frac{\partial p_2}{\partial z} = \mu\left(\frac{\partial^2}{\partial x^2} + \frac{\partial^2}{\partial z^2}\right)w_2, \tag{3}$$

where μ is the coefficient of viscosity of the fluid. The boundary conditions at the ends $x = 0$ and $x = L$ are assumed to conform to Poiseuillean flow:

$$u_2 = \frac{3}{2}U\left(1 - 4\frac{z^2}{h^2}\right), \qquad w_2 = 0. \tag{4a}$$

The boundary condition on the walls $z = \pm\frac{h}{2}$ is no-slip:

$$u_2 = 0, \qquad w_2 = w_1. \tag{4b}$$

U is the mean velocity of flow in the channel ahead of the porous walls.

The incompressibility condition is expressed by

$$\frac{\partial u}{\partial x} + \frac{\partial w}{\partial y} = 0. \tag{5}$$

Further, the distribution of each tracer (a solute or a component of blood) obeys Eq. (9.4:6). Let c be the concentration of the solute, D be its coefficient of diffusion, (u, w) be the velocity vector; then,

$$\frac{\partial c}{\partial t} + u\frac{\partial c}{\partial x} + w\frac{\partial c}{\partial z} = D\left(\frac{\partial^2 c}{\partial x^2} + \frac{\partial^2 c}{\partial z^2}\right), \tag{6}$$

and the boundary conditions are

$$D_1\frac{\partial c_1}{\partial z} - D_2\frac{\partial c_2}{\partial z} = w(c_1 - c_2), \quad \text{for a permeable wall,} \tag{7a}$$

$$D_1\frac{\partial c_1}{\partial z} = wc_1, \qquad \text{for an impermeable wall,} \tag{7b}$$

$$\frac{\partial c_1}{\partial z} = 0, \qquad \text{for a wall impermeable to water.} \tag{7c}$$

The solutions are as follows:

Flow in the Interstitium. Equation (5) is satisfied by an arbitrary stream function ψ if u, w are defined by

$$u = -\frac{\partial \psi}{\partial z}, \qquad w = \frac{\partial \psi}{\partial z}. \tag{8}$$

Using (8), Eq. (1) is reduced to

$$\left(\frac{\partial^2}{\partial x^2} + \frac{\partial^2}{\partial z^2}\right)p_1 = 0, \qquad \left(\frac{\partial^2}{\partial x^2} + \frac{\partial^2}{\partial z^2}\right)\psi_1 = 0. \tag{9}$$

It can be proven by direct substitution that the following solutions satisfy Eqs. (2a), (2b), and (9) if $\lambda_n = n\pi/L$, $n = 1, 2, 3 \ldots$, and $z \geqslant 0$:

$$\psi_1(x, z) = \sum_{n=1}^{\infty} a_n \sin \lambda_n x \sinh \lambda_n \left[\left(\frac{h}{2} + \delta \right) - z \right] \bigg/ \sinh \lambda_n \delta, \tag{10}$$

$$p_1(x, z) = \sum_{n=1}^{\infty} \frac{a_n}{K} \cos \lambda_n x \cosh \lambda_n \left[\left(\frac{h}{2} + \delta \right) - z \right] \bigg/ \sinh \lambda_n \delta + b_1, \tag{11}$$

where b_1 is an integration constant. This leaves the boundary condition Eq. (2c) at $z = \pm \frac{h}{2}$ to be matched later with the solution in the channel. For $z \leqslant 0$, ψ_1 and p_1 are mirror images of Eqs. (10) and (11).

Flow in the Channel. With Eq. (8), Eq. (3) can be reduced to the form

$$\left(\frac{\partial^2}{\partial x^2} + \frac{\partial^2}{\partial z^2} \right) p_2 = 0, \qquad \left(\frac{\partial^4}{\partial x^4} + 2 \frac{\partial^4}{\partial x^2 \partial x^2} + \frac{\partial^4}{\partial x^4} \right) \psi_2 = 0. \tag{12}$$

Let

$$u_2 = \tfrac{3}{2} U \left(1 - \frac{4z^2}{h^2} \right) + u_2', \qquad w_2 = w_2'. \tag{13}$$

Then the perturbations u_2', w_2' due to permeability of the wall must satisfy the following equations:

$$u_2' = 0, \qquad w_2' = 0, \quad \text{at} \quad x = 0, L; \tag{14}$$

$$u_2' = 0, \qquad w_2' = w_1, \quad \text{at} \quad z = \pm h/2; \tag{15}$$

$$u_2' = -\partial \psi_2'/\partial z, \qquad w_2' = \partial \psi_2'/\partial x; \tag{16}$$

$$\nabla^4 \psi_2' = 0. \tag{17}$$

According to the methods discussed in Fung (1965, p. 206 et seq.), we find the following solution which satisfies the condition of symmetry with respect to z and the boundary condition $u_2' = 0$ at $x = 0, L$:

$$\psi_2' = \sum_{n=1}^{\infty} \left(A_n \sin \lambda_n x \sinh \lambda_n z + C_n \frac{2z}{h} \sin \lambda_n x \cosh \lambda_n z \right)$$

$$+ \sum_{n=1}^{\infty} \left[G_n \left(1 - \frac{x}{L} \right) \sinh \gamma_n x \sinh \gamma_n z \right.$$

$$\left. + H_n \frac{x}{L} \sinh \gamma_n (L - x) \sin \gamma_n z \right], \tag{18}$$

where $\lambda_n = n\pi/L$, $\gamma_n = 2n\pi/h$, and $n = 1, 2, 3, \ldots$. This can be proven by substitution. The boundary condition $w_2' = w_1$ at $z = \pm \frac{h}{2}$ is satisfied by choosing

$$C_n = (a_n/\cosh \lambda_n \tfrac{h}{2}) - A_n \tanh \lambda_n \tfrac{h}{2}. \tag{19}$$

The remaining unknown constants A_n, G_n, and H_n can be determined in terms of a_n by using the yet unsatisfied conditions $u'_2 = 0$ at $z = \pm\frac{h}{2}$ and $w'_2 = 0$ at $x = 0, L$. The complete solution is obtained by matching the solutions, Eqs. (10) and (18) at the boundary, Eqs. (2b) and (15). The equations are complex, but the numerical calculations are straightforward. See Tang and Fung (1975) for details.

Distribution of the Tracer. If a tracer (such as THO) is permeable to the membrane at $z = \pm\frac{h}{2}$, then at a steady-state the concentration of that substance is uniform in both compartments. This is because a solution of Eqs. (6) and (7a) is

$$c_1 = c_2 = \text{const.} \tag{20}$$

If a substance (such a hyaluronate) in compartment 1 (interstitium) cannot penetrate into compartment 2 (blood vessel), then in a steady-state the solution must satisfy Eq. (6) with u_1, w_1 given by Eq. (10), and the boundary conditions (7b) at $z = \frac{h}{2}$, and (7c) at $x = 0$ and L. It can be verified by substitution that the solution is

$$c_1(x, z) = c_0 \exp[-\varphi(x, z)], \tag{21}$$

where c_0 is a constant and

$$\varphi(x, z) = \frac{1}{D_1} \sum_{n=1}^{\infty} a_n \cos \lambda_n x \cosh \lambda_n [(h/2 + \delta) - z]/\sinh \lambda_n \delta. \tag{22}$$

The verification is easy if one notices that $\nabla^2 \varphi = 0$ and that the functions φ and ψ_1 satisfy the Cauchy–Riemann differential equations for an analytical function $\varphi + i\psi_1$ of a complex variable $z = x + iy$.

Numerical evaluation of $\varphi(x, z)$ is given in Fung and Tang (1975). It was found that for values of L, h, δ pertaining to mammalian lungs, the values of φ are small; hence the concentration $c_1(x, z)$ is nearly equal to a constant through the interstitium.

Finally, consider a substance (such as albumin) which exists in the vascular space (compartment 2) but cannot penetrate into the interstitium. Then its concentration c_2 must satisfy Eq. (6) with u_2, w_2 given by Eqs. (13), (16), (18), and the boundary conditions (4a), (4b), and (7b). In this case the exact solution is unknown. An approximate solution can be obtained by relaxing the boundary conditions $w = 0$ at $x = 0, L$, yielding a simplified solution ψ'_2 given by Eq. (17) with $G_n = H_n = 0$. Then the coefficients A_n, C_n are proportional to a_n of Eq. (10). Alternatively, Ritz's method can be applied to obtain an approximate solution. In this method we write down a solution that satisfies all the boundary conditions and contains a set of undetermined coefficients. These coefficients are then determined by minimizing an error integral I, representing the square of the deviation from the differential equation when the approximate solution is substituted into it.

$$I = \int_0^L \int_0^{h/2} \left[u_2 \frac{\partial c_2}{\partial x} + w_2 \frac{\partial c_2}{\partial z} - D_2 \nabla^2 c_2 \right]^2 dx \, dz. \tag{23}$$

If $I = 0$, then Eq. (6) is satisfied exactly. If an exact solution is unknown, one determines the undetermined coefficients so that

$$I = \text{minimum.} \tag{24}$$

Numerical results are given in Tang and Fung (1975), and the following conclusions are reached:

1) If a tracer (such as THO) is permeable through the membrane that separates the blood from the tissue space, which in turn is limited by an impermeable wall, then at a steady-state the concentration of that tracer is uniform in both compartments.

2) If a tracer is confined to the vascular space, then, because of water movement through the porous layers, the concentration of the tracer in the channel is nonuniform, but the average concentration (averaged over the entire channel) is the same as that at the entry.

3) The hypothesis for the validity of the tracer-volume theorem, Eq. (9.9:13), or $V = F\bar{t}$, is therefore fulfilled. We conclude that the tracer-volume formula can be used for the pulmonary alveoli.

4) If a solute is confined to the porous layers, then its concentration distribution is nonuniform because of the nonuniform convection of water through the boundary. The nonuniformity, however, is negligible for practical ranges of physiological parameters of the lung. Thus solutes in the interstitial space are nearly uniformly distributed.

9.11 Peristalsis

Besides circulation, respiration, and interstitial fluid flow, there are many other physiological flows, e.g., micturition, biphasic flow in arthroidal cartilage, urine extraction in kidney, peristalsis in ureter, intestine, stomach, arterioles, venules, and lymphatics. The mathematical problem of micturition is similar to that of the collapsible tube discussed in Sec. 5.16. Effort independent limitation of flow in urination occurs because of muscle contraction at the beginning of urethra. Flow in cartilage is an example of biphasic flow considered in Sec. 8.9. Urine transport in kidney is an example of active transport (Sec. 8.8).

Peristalsis is a muscle-controlled flow akin to the flow in the heart. Since peristalsis occurs in many organs, one example, that of the ureter, is discussed in greater detail below.

Ureteral peristalsis was described by Aristotle (384–322 B.C.) in one of his books on animals (*Historia animalium*) several hundred years B.C. The ureters collect urine from the kidneys and send it to the bladder. In X-ray the waves of urine appear as shown in Fig. 9.11:1: discontinuous boluses passing down slowly one at a time. At the bladder each ureter passes through a valveless one-way valve, called the *ureterovesicular junction*, which is a Z- or U-shaped fold of the ureter in the wall of the bladder. It works by the pressure in the

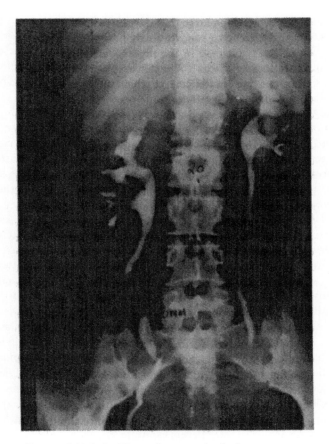

FIGURE 9.11:1 Au X-ray photograph of a human ureter.

bladder. When the bladder is full, its pressure is high, its wall is in tension in the circumferential direction. The bladder pressure presses on the Z or U-shaped fold in the bladder wall and collapses it, forming a check valve, and stopping back flow into the ureter. This valve can be opened by each bolus of urine in the ureter if the pressure in the bolus of urine in the ureter exceeds the lateral pressure imposed by the urine in the bladder and the muscle in the bladder wall. If the smooth muscle of the ureter is unable to generate such a higher pressure in the bolus of urine, or if the legs of the Z or U have inadequate length, then the ureterovesicular junction would not work properly and a disease called *hydro-ureter* results. A hydro-ureter is a swollen ureter with much increased lumen filled with urine. The reason why hydro-ureter is a disease becomes clear when one considers the hoop stress in a pressurized tube. In a tube of radius a, a tention T generated by the ureteral smooth muscle will create a pressure $p = T/a$. For a given T, p can be large if a is small. But in a hydro-ureter a becomes so large that the pressure which can be generated by the ureter is insufficient to send the urine through the ureterovesicular junction. Then urine is collected in the ureter, backs up to the kidney, and before long causes kidney disease. A weak ureterovesicular junction would also permit reflux, causing some bacteria which might exist in the bladder to reach back to the kidney.

The surgical treatment of hydro-ureter is to cut the ureter open longitudinally, remove the excess material, and suture the vessel back into a tube of smaller radius. Repair the ureterovesicular junction if it were the cause.

Thus, to understand the physiological process of ureteral peristalsis, one must study the ureteral smooth muscle, find out how it generates tension, what maximum tension can be generated, and how the tension moves the urine.

Mathematical studies of peristalsis were initiated by Shapiro et al. (1969), Fung and Yih (1968), and others. Most of the analyses are based on the Navier–Stokes equation, considering flow in a circular cylindrical tube or two-dimensional channel with a sinusoidal displacement wave traveling in its wall at a constant velocity. The objects of the studies are: (1) to determine the longitudinal pressure gradient that can be generated by the traveling wave; (2) the flow resulting from peristalsis superposed on pressure differences at the ends; and (3) conditions of reflux. To simplify the analysis, various approximations are introduced such as: (1) small amplitude of the wall displacement compared with the undeformed radius of the tube; or (2) long wave length compared with the tube radius; or (3) very small Reynolds number so that the nonlinear convective acceleration term in the Navier–Stokes equation can be neglected, or a combination of these. Mathematical techniques include the use of stream functions, resulting in Bousinesq equation or Oberbeck–Bousinesq equation, series expansion in small parameters, and numerical methods of finite differences, finite elements, and boundary integrals.

Experimental fluid mechanical studies were made by Yin and Fung (1971a) who compared theoretical predictions with experimental results. Formulation of the ureteral peristalsis problem as a muscle controlled flow was first presented by Fung (1971). Ureteral muscle mechanics was then studied by Yin and Fung (1971b), and Zupkas and Fung (1985).

One conclusion reached by these studies is that peristalsis is an effective method to move fluid only if the fluid is transported in the form of a series of isolated boluses, as seen in Fig. 9.11:1. If the amplitude of the displacement of the wall is small compared with the tube radius, very little pressure gradient can be generated by the traveling wave. Pressure gradient increases significantly when the radius of the minium section of the wave approaches zero. No wonder then, that peristaltic waves of the ureter, intestine, and lymphatics in normal conditions are in this mode. Artificial peristaltic pumps usually operate in this mode also, fully occluding the tube between boluses.

Figure 9.11:2 shows an example of the theoretical shape of the ureteral bolus and pressure distribution when the muscle contraction follows the

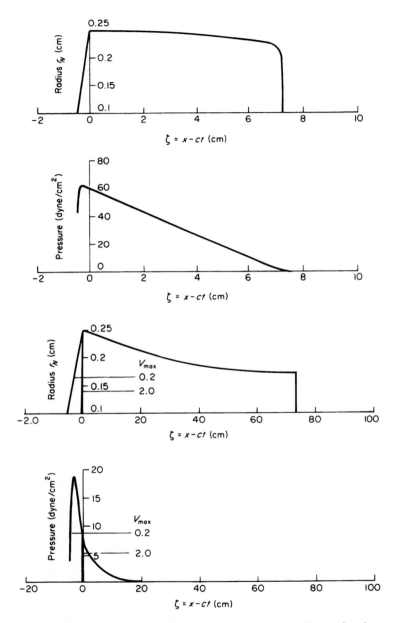

FIGURE 9.11:2 Theoretical shapes of the ureteral boluses and associated pressure distribution. From Fung (1971), by permission.

constitutive equation given by Fung (1970). The bolus moves from left to right, r_N is the radius of the bolus, x is the distance from the kidney, c is the wave speed, t is the time. It is seen that the ureter is closed in front of the bolus. The parameters of the ureter assumed in this example are given in the original paper, Fung (1971). This example shows that the pressure in the front part of the bolus is small, and that significant pressure is generated toward the end of the bolus. This is because the muscle was relaxed before the wave reached it; it generates tension gradually after stimulation. When the maximum tension is reached, the lumen of the ureter is forced closed. The closure at the rear end generates a pressure that may be sufficiently high to send the urine bolus through the ureterovesicle junction even when the bladder is full.

For a full understanding of physiological peristalsis, future research should focus on the smooth muscles, identify their constitutive equations for active contraction, and their response to drugs and other physical, chemical, and biological stimuli, as well as the muscle-controlled flow.

Problems

9.1 Design a method to measure the permeability of the capillary blood vessels of an organ. The permeability constant in Starling's law is defined for fluid transfer per unit area. Since the total area of the capillary blood vessel wall is unknown for any organ, what can you do? Is the product of permeability and the capillary wall area meaningful?

9.2 Using the indicator dilution method (Sec. 9.7), design an experiment to measure the permeability of capillary blood vessel wall to water.

9.3 Formulate a mathematical theory to study whether red blood cells can serve as markers for transcapillary fluid transfer in the method of Landis, Zweifach, and Michel. Estimate the order of magnitude of possible errors due to this source.

9.4 Many authors believe that all capillaries in an organ are not open to flow; many are closed at any given time. This leads to the concept of recruitment of capillaries. It is suggested that with the development of essential hypertension more capillaries are closed. Design an experiment to study this problem. What animal and what organ will you choose? Why? What instruments are needed? What man power? What financial resources are needed?

9.5 Formulate an experiment to evaluate the distensibility of the capillary blood vessels with respect to changes in hydrostatic pressure. How significant is the result with respect to Landis's method for measuring capillary permeability? Discuss this with reference to a specific organ of your choice.

9.6 Write a computer program using the finite element method to analyze the Krogh cylinder (Sec. 9.4). Generalize the cylinder to hexagonal cross section so that a space-filling system can be obtained. What other cross-sectional shapes should be (or can be) studied?

9.7 Solve the Krogh cylinder problem of Prob. 9.6 with the finite differences method. Discuss the pros and cons of the two methods.

9.8 What effect does vasomotion (continuous periodic contraction and relaxation) of arterioles and venules have on blood flow and on fluid and solutes transport in the skeletal muscle? Formulate a mathematical theory to investigate this problem. What are the missing pieces of information that need experimental determination? How can one get this information? Is the fact that the interstitial space in the muscle is very small (< 3% muscle volume) significant?

9.9 If the preceding question is asked with regard to the mesentery, what difference does it make?

9.10 Where are the lymph terminals located in the pulmonary alveoli? Formulate a theory for water movement in the interstitial space of the alveolar wall and its relationship to the lymphatics, breathing, and possible damage to epithelium due to drug abuse or disease.

9.11 Discuss the conditions that may cause edema in the lung.

9.12 What mechanism keeps the volume of living cells constant? Where is the source of energy? Swelling usually occurs whan a cell dies. Why?

9.13 What controls water balance in an organ? Is there a feedback control mechanism working for water balance?

9.14 What happens to a person if the oxygen tension in the blood is higher than normal? What if the partial pressure of CO_2 is higher than normal?

9.15 Select an organ. Design an experiment to determine the permeability of capillary blood vessels by the isogravimetric method.

9.16 Name some tracer materials that can be used to determine the volume of the vascular space of an organ, and the extravascular water volume in that organ.

9.17 Consider all the factors that can influence pulmonary edema. Discuss what you can do to control the respiratory distress syndrome of a young student involved in a car accident.

9.18 Fluids in the human stomach and small intestine can be characterized as a non-Newtonian viscous fluid. What kinds of fluid motion are physiologically significant in these organs?

9.19 In skeletal muscle the level of oxygen in outflowing venous blood is much higher than the level of oxygen in tissue. This implies that the blood oxygen does not have enough time to come into equilibrium with the oxygen in tissue. How can this be explained? Harris named counter current flow in arteries and veins as a reason: the diffusion between parallel and adjacent arterioles and venules as a kind of shunt from the tissue space. Discuss this idea more thoroughly. (Cf.: Harris, P.D. (1986). Movement of oxygen in skeletal muscle *News in Physiol. Sci.* 1:147–149. Whalen, W.J., Nair, P. and Ganfield, R.A. (1973). Measurements of oxygen tension in tissues with a micro-oxygen electrode. *Microvasc. Res.* 5:254–262.)

9.20 In the lung, lymph flow drains into the vein. If the systemic venous pressure is temporarily increased, how would the pulmonary lymph flow change? How would this affect pulmonary edema formation? (Cf.: G.A. Laine et al. (1986): A New Look at Pulmonary Edema, *News in Physiol. Soc.* 1:150–153.)

9.21 Peripheral nerves need nutrition and hence blood supply. Groups of nerve fibers are enclosed in a specialized, multilayered membrane called perineurium into fascicles. Small blood vessels enter each fascicle through the perineurium. If the nerve is injured one of the findings is that water accumulates in the fascicle (i.e. edema occurs), resulting in an increase of pressure and distension of the perineurium. This distension takes place mainly in the circumferential direction—the length of the nerve fascicle changes little. The penetrating blood vessel is distorted by the deformation of the perineurium: its cross section becomes more or less elliptical or kidney shaped, or biconcave; its cross-sectional area is reduced; the resistance to blood flow from outside of the fascicle to the inside through the perineurium, or vice versa, is increased; and the blood flow is reduced. This is an important problem that is relevant to nerve trauma. Formulate a mathematical model to analyze this problem. The following tentative data may be used for such an analysis (see Myers, R.R., Murakami, H., Powell, H.C.: Reduced Nerve Blood Flow in Edematous Neuropathies: A Biomechanical Mechanism. *Microvasc. Res.* **32**:145–151, 1986):

	Perineurium	*Blood Vessel*
Diameter	2,000 µm	50 µm
Wall thickness	15 µm	5 µm
Internal pressure	0.8 kPa	33 kPa
External pressure	0.0 kPa	0.0 (outside perin.)
		0.8 (inside perin.)
Elasticity (Young's mod.)	3.3 kPa	50 kPa
Poisson's ratio	0.1	0.3

References

An, K.N. and Salathé, E.P. (1976). A theory of interstitial fluid motion and its implications for capillary exchange. *Microvasc. Res.* **12**: 103–119.

Apelblat, A., Katzir-Katchalsky, A., and Silberberg, A. (1974). A mathematical analysis of capillary-tissue fluid exchange. *Biorheology*, **11**: 1–49.

Arrhenius, S. (1889). Einfache Ableitung der Beziehung zwischen osmotischem Druck und Erniedrigung der Dampfspannung. *Z. Phys. Chem.* **1**: 115–119.

Arturson, G. (1971). Effect of colloids on transcapillary exchange. In *Hemodilution: Theoretical Basis and Clinical Application*, (K. Messmer and H. Schmid-Schoenbein, eds.), pp. 84–104. S. Karger, Basel.

Aschheim, E. (1977). Passage of substances across the walls of blood vessels. In *Microcirculation*, (G. Kaley and B.M. Altura, eds.), Vol. 1, Ch. 10, University Park Press, Baltimore, pp. 213–249.

Baez, S. (1960). Flow properties of lymph: A microcirculatory study. In *Flow Properties of Blood and Other Biological Systems*, (A.L. Copley and G. Stainsby, eds), pp. 398–411, Pergamon Press, New York.

Bassingthwaighte, J.B., Knopp, T.J., and Hazelrig, J.B. (1970). A concurrent flow model for capillary-tissue exchange. In *Capillary Permeability*, (C. Crone and N.A. Lassen, eds.), Academic Press, New York.

Bloch, I. (1943). Some theoretical considerations concerning the interchange of metabolites between capillaries and tissue. *Bull. Math. Biophysics,* **5**: 1–14.

Blum. J.J. (1960). Concentration profiles in and around capillaries. *Am. J. Physiol.* **198**: 991–998.

Cliff, W.J. and Nicoll, P.A. (1970). Structure and function of lymphatic vessels of the bat's wing. *Quart J. Exp. Physiol.* **55**: 112–121.

Crone, C. and Lassen, N.A. (1970). *Capillary Permeability,* Academic Press, New York.

Curry, F.E. (1984). Mechanics and thermodynamics of transcapillary exchange. In *Handbook of Physiology,* Sec. 2., *Cardiovascular System,* Vol. IV, Part I, (E.M. Renkin and C.C. Michel, eds.) Amer. Physiol. Soc., Bethesda, MD.

Deen, W.M., Robertson, C.R., and Brenner, B.M. (1972). A model of glomerular ultrafiltration in the rat. *Amer. J. Physiol.* **223**: 1178–1183.

Duling, B.R. and Berne, R.M. (1970). Longitudinal gradients in periarteriolar oxygen tension. *Circ. Res.* **27**: 669–678.

Fung, Y.C. (1965). *Foundations of Solid Mechanics.* Prentice-Hall, Englewood Cliffs, N.J.

Fung, Y.C. (1970). Mathematical representation of the mechanical properties of the heart muscle. *J. Biomech.* **3**: 381–404.

Fung, Y.C. (1971). Peristaltic pumping: a bioengineering model. In *Urodynamics: Hydrodynamics of the Ureter and Renal Pelvis.* (S. Boyarsky, C.W. Gottschalk, E.A. Tanagho, and P.D. Zimskind, eds). Academic Press, New York, pp. 177–198.

Fung, Y.C. (1972). Theoretical pulmonary microvascular impedance. *Annals of Biomed. Eng.* **1**: 221–245.

Fung, Y.C. (1974). Fluid in the interstitial space of the pulmonary alveolar sheet. *Microvasc. Res.* **7**: 89–113.

Fung, Y.C. (1981). *Biomechanics: Mechanical Properties of Living Tissues.* Springer-Verlag, New York.

Fung, Y.C. and Yih, C.S. (1968). Peristaltic transport. *J. Appl. Mech.* **35**, Ser. E. 669–675.

Fung, Y.C. and Sobin, S.S. (1972). Elasticity of the pulmonary alveolar sheet. *Circ. Res.* **30**: 451–469.

Fung, Y.C. and Tang, H.T. (1975). Solute distribution in the flow in a channel bounded by porous layers. *J. Appl. Mech.* **42**: 531–535.

Fung, Y.C., Zweifach, B.W., and Intaglietta, M. (1966). Elastic environment of the capillary bed. *Circ. Res.* **14**: 441–461.

Goresky, C.A., Cronin, R.F.P., and Wangel, B.E. (1969). Indicator dilution measurements of extravascular water in the lungs. *J. Clin. Invest.* **48**: 487–501.

Granger, H.J. and Shepherd, A.P. (1979). Dynamics and control of the microcirculation. In *Advances in Biomedical Engineering,* Vol. 7, pp. 1–63, (J.H.U. Brown, Ed.), Academic Press, New York.

Gross, J.F. and Popel, A. (Eds.) (1980). *Mathematics of Microcirculation Phenomena,* Raven Press, New York.

Guyton, A.C. (1963). A concept of negative interstitial pressure based on pressures in implanted perforated capsules. *Circ. Res.* **12**: 399–414.

Guyton, A.C., Barber, B.J., and Moffatt, D.S. (1981). Theory of interstitial pressures. In *Tissue Fluid Pressure and Composition* (A. Hargens, ed.), pp. 11–19, Williams & Wilkins, Baltimore.

Hammel, H.T. and Scholander, P.F. (1976). *Osmosis and Tensile Solvent.* Springer-Verlag, Berlin.

Hargens, A.R. (Ed.) (1981). *Tissue Fluid Pressure and Composition,* p. 3, Williams and Wilkins, Baltimore.

Hulett, G.A. (1902). Beziehung zwischen negativm Druck und osmotischen Druck. *Z. Phys. Chem.* **42**: 353–368.

Intaglietta, M. and Johnson, P.C. (1978). Principles of capillary exchange. In *Peripheral Circulation* (P.C. Johnson, ed.), Wiley, New York.

Johnson, J.A. and Wilson, T.A. (1966). A model for capillary exchange. *Amer. J. Physiol.* **210**: 1299–1303.

Kedem, O. and Katchalsky, A. (1958). Thermodynamic analysis of the permeability of biological membranes to non-electrolytes. *Biochim. et Biophys. Acta.* **27**: 229–246.

Krogh, A. (1919). The number and distribution of capillaries in muscle with calculations of the oxygen pressure head necessary for supplying the tissue. *J. Physiol.* **52**: 409–415.

Landis, E.M. (1927). Micro-injection studies of capillary permeability. II. The relation between capillary pressure and the rate at which fluid passes through the walls of single capillaries. *Amer. J. Physiol.* **82**: 217–238.

Landis, E.M. and Pappenheimer, J.R. (1963). Exchange of substances through the capillary walls. In *Handbook of Physiology*, Sec. 2: *Circulation*, Vol. 2, Ch. 29, American Physiological Society, Washington, D.C.

Leak, L.V. and Burke, J.F. (1968). Ultrastructural studies on the lymphatic anchoring filaments. *J. Cell Biol.* **36**: 129–149.

Lee, J.S., Smaje, L.H., and Zweifach, B.W. (1971). Fluid movement in occluded single capillaries of rabbit omentum. *Circ. Res.* **28**: 353–370.

Meier, P. and Zieler, K.L. (1954). On the theory of the indicator-dilution method for measurement of blood flow and volume. *J. Appl. Physiol.* **6**: 731–744.

Michel, C.C. (1978). The measurement of permeability in single capillaries. *Arch. Int. Physiol. Biochim.* **86**: 657–667.

Michel, C.C. (1980). Filtration coefficients and osmotic reflexion coefficients of the walls of single frog mesenteric capillaries. *J. Physiol.* (London) **309**: 341–355.

Middleman, S. (1972). *Transport Phenomena in the Cardiovascular System*. Wiley-Interscience, New York.

Niimi, H. and Sugihara, M. (1984). Hemorrheological approach to oxygen transport between blood and tissue. *Biorheology* **21**: 1–17.

Noyes, A. (1900). Die genaue Beziehung zwischen osmotischem Druck und Dampfdruck. *Z. Phys. Chem.* **35**: 707–721.

Ogston, A.G., Preston, B.N., and Wells, J.D. (1973). On the transport of compact particles through solutions of chain-polymers. *Proc. Roy. Soc.* (London) **333**: 297–316.

Papenfuss, H.D. and Gross, J.F. (1978). Analytical study of the influence of capillary pressure drop and permeability on glomerular ultrafiltration. *Microvasc. Res.* **16**: 59–72.

Pappenheimer, J.R. (1953). Passage of molecules through capillary walls. *Physiol. Rev.* **33**: 337–423.

Pappenheimer, J.R. (1970). Osmotic reflection coefficients in capillary membranes. In: *Capillary Permeability*, (C. Crone and N-A. Lassen, eds.), Munksgaard, Copenhagen, pp. 278–286.

Pappenheimer, J.R. and Soto-Rivera, A. (1948). Effective osmotic pressure of the plasma proteins and other quantities associated with the capillary circulation in the hind-limbs of cats and dogs. *Amer. J. Physiol.* **152**: 471–491.

Pappenheimer, J.R., Renkin, E.M., and Borrero, L.M. (1951). Filtration, diffusion and molecular sieving through peripheral capillary membranes. A contribution to the pore theory of capillary permeability. *Amer. J. Physiol.* **167**: 13–46.

Poynting, J.H. (1981). Change of state: solid-liquid. *Phil. Mag.* **5**: 32–48.

Reneau, D.D., Jr., Bruley, D.F., and Knisely, M.H. (1967). A mathematical simulation of oxygen release, diffusion and consumption in the capillaries and tissue of the human brain. In *Chemical Engineering in Medicine and Biology*, (D. Hershey, ed.), Plenum Press, New York, pp. 135–241.

Reneau, D.D., Jr., Bruley, D.F., and Knisely, M.H. (1969). A digital simulation of transient oxygen transport in capillary-tissue systems (cerebral gray matter) *Amer. Inst. Chem. Eng. J.* **15**: 916–925.

Renkin, E.M. (1959). Transport of potassium-42 from blood to tissue in isolated mammalian skeletal muscles. *Amer. J. Physiol.* **197**: 1205–1210.

Renkin, E.M. (1977). Multiple pathways of capillary permeability. *Circ. Res.* **41**: 735–743.

Renkin, E.M. and Zaun, B.D. (1955). Effects of adrenal hormones on capillary permeability in perfused rat tissues. *Amer. J. Physiol.* **180**: 498–502.

Renkin, E.M. and Curry, F.E. (1982). Endothelial permeability: pathways and modulations. *Proc. N.Y. Acad. Sci.* **401**: 248–259.

Salathé, E.P. (1977). An analysis of interstitial fluid pressure in the web of the bat wing. *Amer. J. Physiol.* **232**: H297–H304.

Salathé, E.P. (1980). Convection and diffusion in the extravascular space. In *Mathematics of Microcirculation Phenomena* (J.F. Gross, and A. Popel, eds.), Raven Press, New York.

Salathé, E.P. and Venkataraman, R. (1978). Role of extravascular protein in capillary-tissue fluid exchange. *Amer. J. Physiol.* **234**: H52–H58.

Salathé, E.P. and An, K.N. (1976). A mathematical analysis of fluid movement across capillary walls. *Microvasc. Res.* **11**: 1–23.

Scholander, P.F., Hammel, H.T., Bradstreet, E.D., and Hemmingsen, A.E. (1965). Sap pressure in vascular plants. *Science* **148**: 339–346.

Scholander, P.F., Hargens, A.R., and Miller, S.L. (1968). Negative pressure in the interstitial fluid of animals. *Science* **161**: 321–328.

Scholander, P.F., Hammel, H.T., Hemmingsen, E.A., and Bradstreet, E.D. (1964). Hydrostatic pressure and osmotic potential in leaves of mangroves and some other plants. *Proc. Natl. Acad. Sci.* **52**: 119–125.

Shapiro, A.H., Jaffrin, M.Y. and Weinberg, S.L. (1969) Peristaltic pumping with long wave lengths at low Reynolds number. *J. Fluid Mech.* **37**, 799.

Skalak, T.C., Schmid-Schoenbein, G.W., and Zweifach, G.W. (1984). New morphological evidence for a mechanism of lymph formation in skeletal muscle. *Microvasc. Res.* **28**: 95–112.

Starling, E.H. (1896). On the absorption of fluids from the connective tissue spaces. *J. Physiol. (London)* **19**: 312–326.

Staub, N.C. (ed.) (1978). *Lung Water and Solute Exchange*, Marcel Dekker, New York.

Staverman, A.J. (1951). The theory of measurement of osmotic pressure. *Recl. Trav. Chim. Pays-Bas. Belg.* **70**: 344–352.

Stewart, G.N. (1893–1897). Researches on the circulation time in organs and on the influences which affect it. *J. Physiol.* (London) **15**: 1–30, **15**: 31–72, **15**: 78–89 (1893); **22**: 159–173 (1897).

Tang, H.T. and Fung, Y.C. (1975). Fluid movement in a channel with permeable walls covered by porous media. (A moel of lung alveolar sheet). *J. Appl. Mech.* Trans. ASME Vol. 97, Ser. E **42**(1): 45–50.

Taylor, G.I. (1953). Dispersion of soluble matter in solvent flowing slowly through a tube. *Proc. Roy. Soc.* Ser. A, **219**: 186–203.

Taylor, G.I. (1954). The dispersion of matter in turbulent flow through a pipe. *Proc. Roy. Soc.* Ser. A, **223**: 446–468.

Thomson, W. (1871). On the equilibrium of vapour at a curved surface of liquid. *Phil. Mag.* (A) **42**: 448–452.

Van't Hoff, J.H. (1886a). Une propriete general de la matiere diluee. *Svenska Vet. Akad. Handl.* **21**: 17–43.

White, A., Handler, P., Smith, E., and Stetten, D., Jr. (1959). *Principles of Biochemistry*, Chapt. 12, 2nd ed., McGraw-Hill, New York.

Wiederhielm, C.A. (1968). Dynamics of transcapillary fluid exchange. *J. Gen. Physiol.* **52**: 295–615.

Wiederhielm, C.A. (1969). The interstitial space and lymphatic pressures in the bat wing. In *The Pulmonary Circulation and the Interstitial Space*. (A.P. Fishman and H.H. Hecht, eds.), pp. 29–41. University of Chicago Press, Chicago.

Wiederhielm, C.A. (1972). The interstitial space. In *Biomechanics: Its Foundations and Objectives*. (Y.C. Fung, N. Perrone and M. Anliker, eds.) Prentice Hall, Englewood Cliffs, N.J.

Wiederhielm, C.A. (1979). Dynamics of capillary fluid exchange: a nonlinear computer simulation. *Microvas. Res.* **18**: 48–82.

Wiederhielm, C.A. (1981). *The tissue pressure controversy, a semantic dilemma*. In *Tissue Pressure and Composition* (A. Hargens, ed.), Williams & Wilkins, Baltimore, pp. 21–33.

Yin, F.C.P. and Fung, Y.C. (1971a). Comparison of theory and experiment in peristaltic transport. *J. Fluid Mech.* **47**: 93–112.

Yin, F.C.P. and Fung, Y.C. (1971b). Mechanical properties of isolated mammalian ureteral segments. *Am. J. Physiol.* **221**: 1484–1493.

Zierler, K.L. (1963). Theory of use of indicators to measure blood flow and extracellular volume and calculations of transcapillary movement of tracers. *Circ. Res.* **12**: 464–471.

Zupkas, P.F. and Fung, C.Y. (1985). Active contractions of ureteral segments. *J. Biomech. Eng.* **107**: 62–67.

Zweifach, B.W. and Intaglietta, M. (1968). Mechanics of fluid movement across single capillaries in the rabbit. *Microvasc. Res.* **1**: 83–101.

Zweifach, B.W. and Prather, J.W. (1975). Micromanipulation of pressure in terminal lymphatics in the mesentery. *Amer. J. Physiol.* **228**: 1326–1335.

Zweifach, B.W. and Silberberg, A. (1979). The interstitial-lymphatic flow system. In *International Review of Physiology, Cardiovascular Physiology* III, Vol. 18. (A.C. Guyton and D.B. Young, eds.) University Park Press, Baltimore.

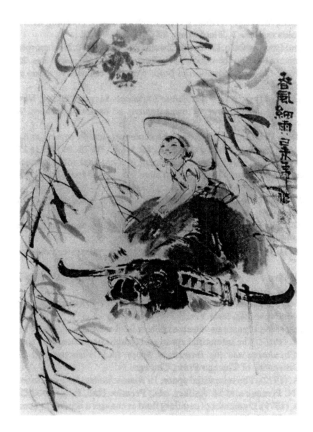

An expression of a philosophy of education: LIKE A MISTY BREEZE OF SPRING. Painted by Chen Chia Yu (陳加遊) in Wuchan, during the author's two-month course on biomechanics in 1979.

CHAPTER 10

Description of Internal Deformation and Forces

10.1 Introduction

Since living organs normally go through finite deformation, a bioengineer should know the subject of finite deformation analysis. This subject is not difficult, but it usually lies outside the common engineering curriculum. It is not simple, and considerable patience is needed to master it. In the following, a presentation of its most important aspects is given in easy to understand physical terms. There are many books and papers on this subject (see References). Fung (1965) is believed to be one of the easiest to read.

The concept of strain is reviewed in Secs. 10.2 and 10.3. The concept of stress is discussed in Sec. 10.4. The equation of motion is presented in Sec. 10.5, work and energy in Sec. 10.6, the use of the strain energy function in Sec. 10.7, and complementary energy function in Sec. 10.8, all without linearization. The separation of local rotational motion and strain in a general deformation of a continuum is discussed in Sec. 10.9.

Of these topics, the most important are the definitions of Cauchy's, Lagrange's, and Kirchhoff's stress tensors discussed in Sec. 10.4. They are denoted by σ_{ij}, T_{ij} and S_{ij}, respectively. The stresses σ_{ij}, S_{ij} are symmetric tensors, but T_{ij} is not. Then it is shown in Sec. 10.7 that if the material is elastic and the strain energy function per unit initial volume, $\rho_0 W$, is expressed in terms of the Green's strain tensor, E_{ij}, we have

$$S_{ij} = \frac{\partial(\rho_0 W)}{\partial E_{ij}}. \tag{1}$$

On the other hand, if $\rho_0 W$ is expressed in terms of the deformation gradient tensor $\partial x_i / \partial a_j$, then

353

$$T_{ji} = \frac{\partial(\rho_0 W)}{\partial(\partial x_i/\partial a_j)}. \tag{2}$$

If the strain energy per unit initial volume is expressed as a function of the Kirchhoff stress tensor S_{ij}, then it is called the complementary strain energy denoted by $\rho_0 W_c$. We have

$$E_{ij} = \frac{\partial \rho_0 W_c}{\partial S_{ij}}. \tag{3}$$

These important formulas are used frequently in biomechanics.

After these mathematical concepts are clarified, we return to biology. The first question we ask is whether living tissues are elastic. The answer, unfortunately, if no. But the concepts of pseudo-elasticity and the pseudo-strain energy function provides us with a useful approximation. Applications of these concepts are discussed in Sec. 10.8 and continued in Chap. 11.

10.2 Description of Internal Deformation

To describe the deformation of a body we need to know the position of any point in the body with respect to an initial configuration which shall be called the *reference state*. Moreover, to describe position we need a *frame of reference*. Let us choose a rectangular cartesian frame of reference, Fig. 10.2:1. Every material particle in the body at the reference state S_0 has three coordinates a_1, a_2, a_3 which can be written as a column matrix denoted by any one of the following forms:

$$\begin{pmatrix} a_1 \\ a_2 \\ a_3 \end{pmatrix} \quad \text{or} \quad \{a_1, a_2, a_3\} \quad \text{or} \quad \{a_i\} \quad \text{or} \quad a_i, \quad i = 1, 2, 3.$$

A particle P has the coordinates a_i, and a neighboring particle P' has the coordinates $a_i + da_i'$. When the body is deformed the particles P, P' are moved to Q, Q' whose coordinates are x_i and $x_i + dx_i'$, respectively. The deformation of the body is known completely if we know the relationship

$$x_i = x_i(a_1, a_2, a_3) \qquad i = 1, 2, 3 \tag{1a}$$

or its inverse

$$a_i = a_i(x_1, x_2, x_3) \qquad i = 1, 2, 3 \tag{1b}$$

for every point in the body. If we write

$$x_i = a_i + u_i \qquad i = 1, 2, 3 \tag{1c}$$

then u_i is called the *displacement* of the particle P.

In general, we do not have the luxury of knowing the transformation law (or mapping function), as expressed in Eq. (1a) or (1b), at the beginning of

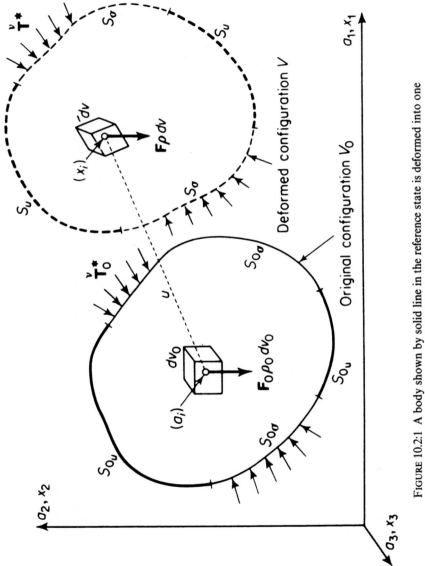

FIGURE 10.2:1 A body shown by solid line in the reference state is deformed into one bounded by the dotted line. The frame of reference for the reference state is a_1, a_2, a_3. That for the deformed state is x_1, x_2, x_3.

solving any biological or engineering problems. More often the seeking of such a transformation law is the final objective of research. In the process, however, there is a general belief (axiom) that the stress at a point in a continuum depends only on the deformation in the neighborhood of that point. Therefore, we ask what features of the transformation describe the deformation in the immediate neighborhood of a particle? The answer is that the principal features are given by the partial derivatives of Eqs. (1a–c). Usually the first derivatives provide all the information we need. These derivatives can be arranged in the form of matrices

$$\begin{bmatrix} \dfrac{\partial x_1}{\partial a_1} & \dfrac{\partial x_1}{\partial a_2} & \dfrac{\partial x_1}{\partial a_3} \\[2mm] \dfrac{\partial x_2}{\partial a_1} & \dfrac{\partial x_2}{\partial a_2} & \dfrac{\partial x_2}{\partial a_3} \\[2mm] \dfrac{\partial x_3}{\partial a_1} & \dfrac{\partial x_3}{\partial a_2} & \dfrac{\partial x_3}{\partial a_3} \end{bmatrix} \quad \text{or} \quad \begin{bmatrix} \dfrac{\partial a_1}{\partial x_1} & \dfrac{\partial a_1}{\partial x_2} & \dfrac{\partial a_1}{\partial x_3} \\[2mm] \dfrac{\partial a_2}{\partial x_1} & \dfrac{\partial a_2}{\partial x_2} & \dfrac{\partial a_2}{\partial x_3} \\[2mm] \dfrac{\partial a_3}{\partial x_1} & \dfrac{\partial a_3}{\partial x_2} & \dfrac{\partial a_3}{\partial x_3} \end{bmatrix} \tag{2a}$$

which can be written simply as

$$\partial x_i/\partial a_j, \qquad \partial a_i/\partial x_j, \; (i,j = 1, 2, 3). \tag{2b}$$

They are called the *deformation gradient* tensors.

Physically, what the deformation gradients enable us to do is to describe how the distance between two neighboring points such as $P(a_i)$, $P'(a_i + da_i)$ is changed after they are deformed to $Q(x_i)$, $Q'(x_i + dx_i)$. To measure distance between two points Q, Q' we assume that the world we live in is Euclidean and Pythagoras rule applies:

$$ds^2 = dx_1^2 + dx_2^2 + dx_3^2, \tag{3}$$

where dx_1, dx_2, dx_3 are the projections of QQ' on the three coordinate axes and ds is the length QQ'.

We may write Eq. (3) in any one of the following forms:

$$ds^2 = \sum_{i=1}^{3} dx_i\, dx_i = dx_i\, dx_i = \delta_{ij}\, dx_i\, dx_j. \tag{3a}$$

Here δ_{ij} is the *Kronecker delta*, defined as $\delta_{ij} = 1$ when $i = j$, $\delta_{ij} = 0$ when $i \neq j$. The *summation convention* is used (Sec. 1.3). In the reference state, the distance between the same two points at P, P' is ds_0: Applying Eq. (3) to the line elements $PP'(da_i)$ and $QQ'(dx_i)$, we have

$$ds_0^2 = da_i\, da_i, \qquad ds^2 = dx_i\, dx_i. \tag{4}$$

Hence we obtain the difference:

$$ds^2 - ds_0^2 = dx_i\, dx_i - da_i\, da_i = \delta_{ij}(dx_i\, dx_j - da_i\, da_j). \tag{5}$$

Now, from Eqs. (1) we have

$$dx_i = \frac{\partial x_i}{\partial a_k} da_k, \qquad da_i = \frac{\partial a_i}{\partial x_k} dx_k. \tag{6}$$

A substitution of Eq. (6) into Eq. (5) and several changes of the dummy indices yield the result:

$$ds^2 - ds_0^2 = \left(\frac{\partial x_i}{\partial a_l} \frac{\partial x_i}{\partial a_k} - \delta_{lk} \right) da_l \, da_k \tag{7a}$$

$$= \left(\delta_{lk} - \frac{\partial a_i}{\partial x_l} \frac{\partial a_i}{\partial x_k} \right) dx_l \, dx_k. \tag{7b}$$

We rewrite Eq. (7) in the form

$$ds^2 - ds_0^2 \equiv 2E_{ij} \, da_i \, da_j = 2e_{ij} \, dx_i \, dx_j \tag{8}$$

and call the coefficients E_{ij} the *Green's strain tensor* and e_{ij} the *Almansi's strain tensor*. Thus

$$E_{ij} = \frac{1}{2} \left(\frac{\partial x_k}{\partial a_i} \frac{\partial x_k}{\partial a_j} - \delta_{ij} \right) = \frac{1}{2} \left(\frac{\partial u_i}{\partial a_j} + \frac{\partial u_j}{\partial a_i} + \frac{\partial u_k}{\partial a_i} \frac{\partial u_k}{\partial a_j} \right) \quad \text{(Green)} \tag{9}$$

$$e_{ij} = \frac{1}{2} \left(\delta_{ij} - \frac{\partial a_k}{\partial x_i} \frac{\partial a_k}{\partial x_j} \right) = \frac{1}{2} \left(\frac{\partial u_i}{\partial x_j} + \frac{\partial u_j}{\partial x_i} - \frac{\partial u_k}{\partial x_i} \frac{\partial u_k}{\partial x_j} \right) \quad \text{(Almansi)}. \tag{10}$$

If the strains E_{ij} or e_{ij} are known, then according to Eq. (8), we can calculate the change of the *square* of the distance between any two neighboring points in a deformed body. (Only the square of the distance can be computed so easily.) To obtain the change in distance itself we would have to take the square roots of the righthand sides of Eqs. (4) and form the difference. The resulting formula for $ds - ds_0$ is inconvenient to use. The vanishing of E_{ij} or e_{ij} implies (according to Eq. (8)) no change of distance between any two points. Hence a rigid body motion has no strain. If the deformation is small in the sense that the values of the derivatives $\partial u_i / \partial a_j$, $\partial u_i / \partial x_j$ are infinitesimal so that their squares can be neglected in Eqs. (9) and (10), then both E_{ij} and e_{ij} are reduced to

$$\varepsilon_{ij} = \frac{1}{2} \left(\frac{\partial u_i}{\partial x_j} + \frac{\partial u_j}{\partial x_i} \right) = \frac{1}{2} \left(\frac{\partial u_i}{\partial a_j} + \frac{\partial u_j}{\partial a_i} \right), \tag{11}$$

which are called *Cauchy's infinitesimal strain tensor*.

Returning to what we said at the beginning of this section, we may assert that biomechanics problems are usually structured so as to find the stresses and strains first, then the deformation gradient, and finally, the mapping functions in Eq. (1a–c).

The following examples illustrate the meaning of the finite strain components:

Example 1. A block is deformed in such a way that

$$x_1 = \lambda_1 a_1, \qquad x_2 = \lambda_2 a_2, \qquad x_3 = \lambda_3 a_3. \tag{12}$$

Then

$$E_{11} = \tfrac{1}{2}(\lambda_1^2 - 1), \qquad E_{22} = \tfrac{1}{2}(\lambda_2^2 - 1), \qquad E_{33} = \tfrac{1}{2}(\lambda_3^2 - 1). \qquad (13)$$

If the material is *incompressible* so that the volume of the block does not change, then

$$\lambda_1 \lambda_2 \lambda_3 = 1, \qquad (14a)$$

or

$$(1 + 2E_{11})(1 + 2E_{22})(1 + 2E_{33}) = 1. \qquad (14b)$$

Example 2. Consider a rubber band. Initially its length is L_0. By pulling on it, its length is increased to L. We can define its "strain" in several ways:

$$\varepsilon = \frac{(L - L_0)}{L_0}, \qquad \varepsilon' = \frac{(L - L_0)}{L},$$

$$E_{11} = \frac{(L^2 - L_0^2)}{2L_0^2}, \qquad e_{11} = \frac{(L^2 - L_0^2)}{2L^2},$$

$$\gamma = \int_{L_0}^{L} \frac{dL}{L} = \ln \frac{L}{L_0}.$$

If $L = 2$, $L_0 = 1$, we have $\varepsilon = 1$, $\varepsilon' = \tfrac{1}{2}$, $E_{11} = \tfrac{3}{2}$, $e_{11} = \tfrac{3}{8}$, $\gamma = 0.693$. Thus for a finite deformation each definition leads to a different number. If $L = 1.01$, $L_0 = 1.00$, then $\varepsilon = 0.01$, $\varepsilon' \doteq 0.01$, $E_{11} \doteq 0.01$, $e_{11} \doteq 0.01$, $\gamma \doteq 0.01$; and all definitions lead approximately to the same number when the deformation is small. γ is called the "true" strain by its enthusiastic supporters and is often used in metallurgy and thermal stress in discussing creep and plasticity. Depending on personal preference, one or the other is called "engineering" strain. Our reasoning presented above suggests that E_{11} or e_{11} is preferred in biomechanics when dealing with finite deformation.

Example 3. The following deformation is called a *pure shear*

$$x_1 = \lambda_1 a_1, \qquad x_2 = \frac{1}{\lambda_1} a_2, \qquad x_3 = a_3. \qquad (15)$$

It preserves the volume.

Example 4 (Fig. 10.2:2). A deformation

$$x_1 = \lambda_1 a_1 + k a_2, \qquad x_2 = m a_1 + \lambda_2 a_2, \qquad x_3 = a_3 \qquad (16)$$

corresponds to the following strains according to the definition, Eq. (9).

$$E_{11} = \tfrac{1}{2}(\lambda_1^2 - 1), \qquad E_{22} = \tfrac{1}{2}(\lambda_2^2 - 1), \qquad E_3 = 0,$$

$$E_{12} = \tfrac{1}{2}(\lambda_1 k + m\lambda_2) = E_{21}, \qquad E_{23} = E_{31} = 0. \qquad (17)$$

The *maximum shear strain* is equal to the radius of the Mohr's circle (see Fung, 1977, p. 95)

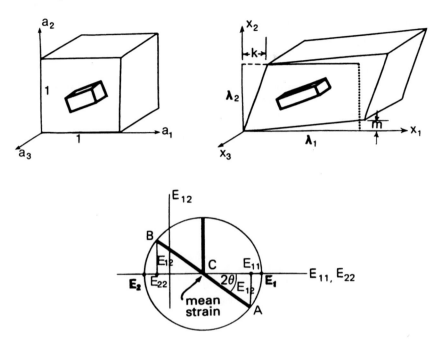

FIGURE 10.2:2 Deformation described by Eqs. (16), (17) when $\lambda_1 = 1.5$, $\lambda_2 = 0.9$, $k = 0.3$, $m = 0.1$ is shown by the figures at the top. The corresponding Mohr's circle is shown below. For the construction of the Mohr's circle, the normal strains are plotted as abscissa, the shear stress is plotted as ordinate. We plot $E_{11}, -E_{12}$ to obtain a point A, plot E_{22}, E_{12} to obtain point B. Join A, B by a straight line. AB intersects the horizontal axis (normal strain) at C. With C as center, draw a circle passing through A, B. This is the Mohr's circle. The center, C, represents the mean strain in x_1, x_2 plane. The intercepts E_1, E_2 yield the values of the principal strains. The radius is the maximum shear strain. The angle θ is the angle between the x_1 axis and the principal axis associated with E_1. The angle between CA and CE_1 is equal to 2θ. Analytically, $\tan 2\theta = E_{12}/[E_{11} - (E_{11} + E_{22})/2]$.

$$\text{max shear strain} = \left[\left(E_{11} - \frac{E_{11} + E_{22}}{2} \right)^2 + E_{12}^2 \right]^{1/2}.$$

Example 5 (Strain in Myocardium). To measure the strain in the beating heart Fenton et al. (1978), Waldman et al. (1985) inserted small lead beads into the myocardium and took x-ray cine-photographs from two different directions simultaneously to obtain the coordinates of the beads as a function of time. Any four nonplanar beads form the vertices of a tetrahedron which has 6 edges. From the measured coordinates, they compute ds^2, ds_0^2, dx_i, da_i of each edge, substitute into Eq. (8), and obtain 6 linear equations for the 6 unknowns $E_{11}, E_{22}, E_{33}, E_{12} = E_{21}, E_{23} = E_{32}, E_{31} = E_{13}$. These are the average strains in the tetrahedron. Using a standard computing program one obtains further the principal strains and principal directions. If N beads are inserted, one can thus obtain the strains in $N - 3$ nonoverlapping regions in the myocardium, because, starting with a basic tetrahedron, every addition of a new bead defines a new tetrahedron. Waldman et al. (1985) used this method to measure the strains in the heart. By creating an axisymmetric infarct in the left ventricle, the stress distribution can be computed by finite element method on the basis of an assumed form of constitutive equation. The identification of the computed and measured strains can then yield the material constants of the infarct myocardium.

10.3 Use of Curvilinear Coordinates

In the preceding section we used a fixed rectangular cartesian frame of reference to describe the location of a material particle in both the reference and deformed states. An alternative is to use curvilinear coordinates, such as cylindrical polar coordinates for a cylinder, spherical coordinates for a sphere. When the body is deformed, we can use a different set of curvilinear coordinate to describe the deformed body.

When curvilinear coordinates are used, the square of the length of a differential element must be written as

$$ds_o^2 = g_{ij}^{(0)} \, da_i \, da_j, \qquad ds^2 = g_{ij} \, dx_i \, dx_j. \tag{1}$$

The quantities $g_{ij}^{(0)}$ and g_{ij} are called *metric tensors*. When $g_{ij} \neq \delta_{ij}$, the geometry is non-Euclidean.

A particularly advantageous choice of coordinates is to consider the frame of reference as attached to the material particles. Then as the body deforms, x_i remains the same as a_i, but the square of the length between any two neighboring particles become:

$$ds_o^2 = g_{ij}^{(0)} \, da_i \, da_j, \qquad ds^2 = g_{ij} \, da_i \, da_j. \tag{2}$$

It follows that

$$ds^2 - ds_o^2 = (g_{ij} - g_{ij}^{(0)}) \, da_i \, da_j. \tag{3}$$

If we write

$$ds^2 = ds_0^2 = 2E_{ij} \, da_i \, da_j \tag{4}$$

then

$$E_{ij} = \tfrac{1}{2}(g_{ij} - g_{ij}^{(0)}). \tag{5}$$

Thus the strain tensor is just one-half of the change of the metric tensor.

The elegance of this approach is incomparable. The coordinates so chosen have been called *convective, co-moving, dragging along, embedded, or body coordinate system*. Its use requires, however, a knowledge about the general tensor analysis. In the rest of this book we limit ourselves to cartesian tensors.

10.4 Description of Internal Forces

Internal forces exist in a deformed body. To describe internal forces we imagine an arbitrary cut in the body and examine the force acting on the cut surface (Fig. 10.4:1). The magnitude of the force must vary with the area of the cut. We assume that a limiting value of the force per unit area of the cut exists, and call it a *stress vector* or *surface traction*. The surface traction varies with the orientation of the surface. To obtain a full description of the state of stress at any point in a body we imagine an infinitesimal cube enclosing that point (Fig. 10.4:2). On each surface of the cube there acts a stress vector. On the surface whose normal vector coincides with the x_1 axis the stress vector has three components σ_{11}, σ_{12}, σ_{13} in the directions of the x_1, x_2, x_3 axes respectively, (Fig. 10.4:3). We call σ_{11} a *normal stress* and define it to be *positive (tensile)* if it points in the direction of the x_1-axis. We call σ_{12} a shear stress and define it to be *positive* if it points in the direction of the x_2-axis. Similarly, σ_{13} (shear) is positive if it points in the direction of the x_3-axis. The other

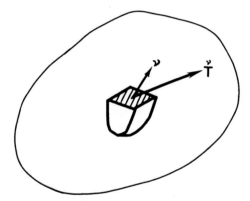

FIGURE 10.4:1 Stress vector $\overset{v}{\mathbf{T}}$ acting on a surface whose normal vector is **v**.

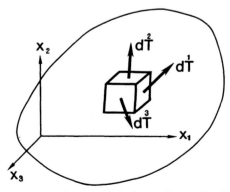

FIGURE 10.4:2 Stress vectors acting on the surfaces of a small cube with edges parallel to coordinate axes x_1, x_2, x_3.

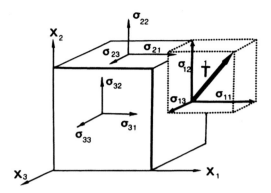

FIGURE 10.4:3 The three components of the stress vector $\overset{1}{T}$ acting on a surface whose normal vector points in the direction of the positive x_1 axis area $\sigma_{11}, \sigma_{12}, \sigma_{13}$. Similarly, $\overset{2}{T}$ has components $\sigma_{22}, \sigma_{23}, \sigma_{21}$, $\overset{3}{T}$ has components $\sigma_{33}, \sigma_{31}, \sigma_{32}$.

surface that is perpendicular to the x_1-axis has a normal pointing to the negative x_1-axis. On that surface the stresses are $\sigma_{11}, \sigma_{12}, \sigma_{13}$ also, with all the directions reversed. Similarly, on the surfaces perpendicular to the x_2 and x_3 axes we have normal stresses σ_{22}, σ_{33} and shear stresses $\sigma_{21}, \sigma_{23}, \sigma_{31}, \sigma_{32}$. By conditions of equilibrium we can show that $\sigma_{12} = \sigma_{21}, \sigma_{23} = \sigma_{32}, \sigma_{31} = \sigma_{13}$. Hence 6 stress components define the stress state at a point in the body. This was first clarified by Cauchy, hence σ_{ij} is known as *Cauchy stress*. If the body is subjected to a *body force X* (with components X_1, X_2, X_3) per unit volume of the material in the deformed state, then it can be shown that the *equation of equilibrium* of the internal forces is given by

$$\frac{\partial \sigma_{ij}}{\partial x_j} + X_i = 0. \tag{1}$$

If the internal forces are not in equilibrium, then motion ensues, and according to Newton's law, the *equation of motion* is

$$\rho \alpha_i = \frac{\partial \sigma_{ij}}{\partial x_j} + X_i. \tag{2}$$

Here ρ is the density of the material in the deformed state, α_i is the *acceleration* of the material particle.

All this is discussed in much greater detail in mechanics books, e.g., Fung (1977). In the course of analysis, however, we must relate stresses to strains. Hence, if strains were referred to the original position of particles in a continuum, it would be convenient to define stresses similarly with respect to the original configuration.

Consider an element of a strained solid as shown on the right-hand side of Fig. 10.4:4. Assume that in the original state this element has the configuration as shown on left side of Fig. 10.4.:4. A force vector dT acts on the surface $PQRS$. A corresponding force vector dT_0 acts on the surface $P_0Q_0R_0S_0$. If we

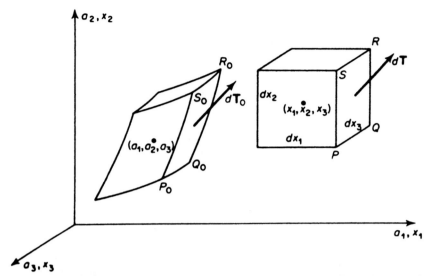

FIGURE 10.4:4 The corresponding tractions in the original and deformed states of a body.

assign a rule of correspondence between dT and dT_0 for every corresponding pair of surfaces, and define stress vectors in each case as the limiting ratios dT/dS, dT_0/dS_0, where dS and dS_0 are the areas of $PQRS$, $P_0Q_0R_0S_0$, respectively, then by the usual method (see Equations (15), (16) infra) we can define stress tensors in both configurations. The assignment of a correspondence rule between dT and dT_0 is arbitrary, but must be mathematically consistent.

The following is known as the *Lagrangian rule*:

$$dT_{0_i}^{(L)} = dT_i, \qquad (d\mathbf{T}_0^{(L)} = d\mathbf{T}). \tag{3}$$

According to this rule, the force vector is transported without change of direction and magnitude from the surface element in the deformed configuration to the same surface element in the undeformed configuration. An alternative, known as the *Kirchhoff's rule*, however, requires a change of magnitude and direction of the force vector. The Kirchhoff's rule states that

$$dT_{0_i}^{(K)} = \frac{\partial a_i}{\partial x_j} dT_j, \tag{4}$$

so that the force vectors $d\mathbf{T}_0^{(K)}$ and $d\mathbf{T}$ are related by the same rule as the transformation of line elements

$$da_i = \frac{\partial a_i}{\partial x_j} dx_j \tag{5}$$

when the body is subjected to deformation. These correspondence rules are illustrated in Fig. 10.4:5 for a two-dimensional case, in which the rectangle represents an element in the deformed body, and the parallelogram represents

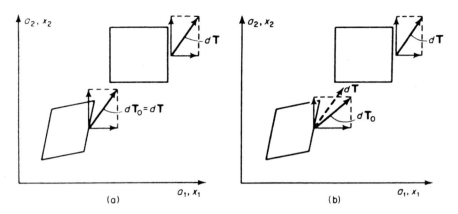

FIGURE 10.4:5 The correspondence of force vectors in defining (a) Lagrange's and (b) Kirchhoff's stresses, illustrated in a two-dimensional case.

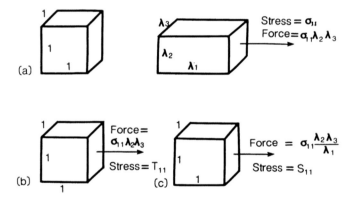

FIGURE 10.4:6 The relationship of stresses defined according to (a), Cauchy, (b), Lagrange, and (c) Kirchhoff.

the same element in the undeformed configuration. Let us illustrate these relations by a few examples.

Example 1. Let us consider the deformation described by Eq. (10.2:12). A unit cube becomes a parallelopiped with edge lengths $\lambda_1, \lambda_2, \lambda_3$. A Cauchy stress σ_{11} acts on the surface perpendicular to the x_1-axis in the deformed state. Since the area of that surface is $\lambda_2 \lambda_3$, the force is $\sigma_{11} \lambda_2 \lambda_3$ (Fig. 10.4:6a). To derive a Lagrangian stress associated with the original cube we transfer this force to the corresponding surface of the original cube (Fig. 10.4:6b) without change. The area of that surface in the cube is 1, and the Lagrangian stress, written as T_{11}, is

$$T_{11} = \sigma_{11} \lambda_2 \lambda_3. \tag{6}$$

To derive a Kirchhoff stress associated with the original cube (Fig. 10.3:6c)

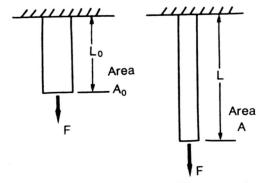

FIGURE 10.4:7 A bar in the original and deformed configurations.

we transfer the force $\sigma_{11}\lambda_2\lambda_3$ to the corresponding face on the cube according to the rule (4). The inverse deformation gradient $\partial a_1/\partial x_1$ is $1/\lambda_1$, thus the force $S_{11} = \sigma_{11}\lambda_2\lambda_3/\lambda_1$. Since the area is 1, the Kirchhoff stress is

$$S_{11} = \sigma_{11}\frac{\lambda_2\lambda_3}{\lambda_1} = \frac{T_{11}}{\lambda_1}. \tag{7}$$

If the material is incompressible, so that $\lambda_1\lambda_2\lambda_3 = 1$, then

$$S_{11} = \sigma_{11}\frac{1}{\lambda_1^2}, \quad \text{if } \lambda_1\lambda_2\lambda_3 = 1. \tag{8}$$

Figure 10.4:7 shows a usual experimental set up in which a cylindrical test specimen is clamped at the top and loaded by a force F at the bottom. If the cross-sectional area is A, the Cauchy stress is F/A. If the unloaded specimen has a length L_0 and a cross-sectional area A_0, we have $\lambda_1 = L/L_0$ and $\lambda_2\lambda_3 = A/A_0$. Hence

$$\sigma_{11} = \frac{F}{A}, \quad T_{11} = \frac{F}{A_0}, \quad S_{11} = \frac{F}{A_0}\frac{L_0}{L}. \tag{9}$$

Example 2. Consider a deformation described by the mapping function relating the coordinates of a point in the original position (a_1, a_2, a_3) to those in the deformed position (x_1, x_2, x_3):

$$x_1 = a_1 + ka_2, \quad x_2 = a_2, \quad x_3 = a_3. \tag{10a}$$

$$a_1 = x_1 - kx_2, \quad a_2 = x_2, \quad a_3 = x_3. \tag{10b}$$

A unit cube in the deformed state corresponds to a parallelopiped in the initial state, as it is shown in Fig. 10.4:8. Let the unit cube be subjected to shear stresses $\sigma_{12} = \sigma_{21}$ while all other stress components vanish. Compute the Lagrangian and Kirchhoff stresses.

Solution. Consider first the top surfaces $ACDA$ and $A'C'D'A'$. They have the same orientation and same area, hence under either Lagrangian rule or

Kirchhoff rule, the force acting on the surface $A'C'D'A'$ consists of a horizontal force σ_{21}. Therefore, we have

$$T_{22} = 0, \qquad T_{21} = \sigma_{21}, \qquad S_{22} = 0, \qquad S_{21} = \sigma_{21}. \tag{11}$$

Next consider the wedge formed by the inclined surface $A'B'$ and the surfaces OA', OB' normal to the axes a_1, a_2 respectively. Figure 10.4:8(b) shows the Lagrangian case, with the stress vectors and the areas indicated. In view of $T_{21} = \sigma_{21}$ (Eq. (11)), the conditions of equilibrium in the vertical and horizontal directions respectively yield

$$T_{12} = \sigma_{12}, \qquad T_{11} + kT_{21} = 0 = T_{11} + k\sigma_{21}. \tag{12}$$

Hence in this case $T_{12} = T_{21}$ and $T_{11} = -k\sigma_{21}$.

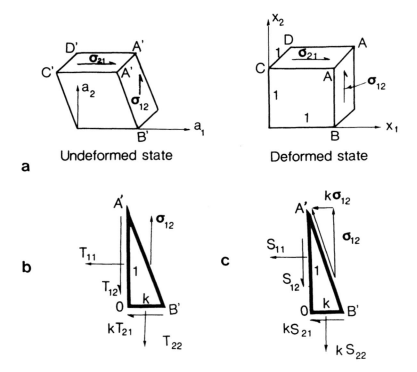

FIGURE 10.4:8 (a) The corresponding shapes of a body in the undeformed and deformed states described by Eq. (10). The Cauchy shear stress $\sigma_{21} = \sigma_{12}$ acts on the body in the deformed state which is a unit cube. The forces on the surfaces AB and AC are σ_{12} and σ_{21}, resp. since the area $= 1$. These forces are transported to the surfaces $A'B', A'C'$ in the undeformed state according to Lagrange's rule. (b) Forces acting on a wedge $A'B'O$ in the undeformed state. (c) The force that acts on the surface $A'B'$ in the undeformed state obtained from the Cauchy stress σ_{12} according to Kirchhoff's rule.

Similarly, for the Kirchhoff's case, consider the wedge shown in Fig. 10.4:8(c). According to Kirchhoff's rule Eq. (4), the force acting on the inclined surface $A'B'$ has the following components:

$$\text{vertical: } \frac{\partial a_2}{\partial x_2}\sigma_{12} = \sigma_{12}; \quad \text{horizontal: } \frac{\partial a_1}{\partial x_2}\sigma_{12} = -k\sigma_{12}.$$

Hence the conditions of equilibrium yield

$$S_{12} = \sigma_{12}, \qquad S_{11} + kS_{21} + k\sigma_{12} = 0. \tag{13}$$

But, according to Eq. (11) $S_{21} = \sigma_{21}$, and $\sigma_{21} = \sigma_{12}$. Hence

$$S_{11} = -2k\sigma_{12}, \qquad S_{12} = S_{21}.$$

Example 3. Consider the same deformation as in Example 2, but now let there be, in the deformed state, a normal stress $\sigma_{22} \neq 0$ in addition to $\sigma_{12} = \sigma_{21}$. Then according to the Lagrangian rule, there acts a force with a vertical component σ_{22} and a horizontal component σ_{21} on the top surface $A'C'D'A'$. Thus, in this case, Eqs. (11) are replaced by

$$T_{22} = \sigma_{22}, \qquad T_{21} = \sigma_{21},$$

and Eqs. (12) are replaced by

$$T_{12} + kT_{22} - \sigma_{12} = 0, \qquad T_{11} + kT_{21} = 0.$$

Hence

$$T_{12} = \sigma_{12} - k\sigma_{22}, \qquad T_{11} = -k\sigma_{21}.$$

Thus, in this case $T_{12} \neq T_{21}$, i.e. the Lagrangian stress tensor is asymmetric.

For the Kirchhoff stresses observe that according to the transformation rule given by Eq. (4), on the top surface $A'C'D'A'$ the vertical force σ_{22} in the deformed state is transformed into a force with the following components:

$$\text{vertical: } \frac{\partial a_2}{\partial x_2}\sigma_{22} = \sigma_{22}; \quad \text{horizontal: } \frac{\partial a_1}{\partial x_2}\sigma_{22} = -k\sigma_{22}.$$

The horizontal force σ_{21} acting on the surface $ACDA$ in the deformed state is transformed into a force acting on the surface $A'C'D'A'$ in the undeformed state with the following components:

$$\text{vertical: } \frac{\partial a_2}{\partial x_1}\sigma_{21} = 0, \quad \text{horizontal: } \frac{\partial a_1}{\partial x_1}\sigma_{21} = \sigma_{21}.$$

Hence, on the surface $A'C'D'A'$, we have the vertical and horizontal stresses:

$$S_{22} = \sigma_{22}, \qquad S_{21} = \sigma_{21} - k\sigma_{22}.$$

Next, consider the wedge shown in Fig. 10.4:8(c), which remains applicable. The condition of equilibrium yields, in the horizontal direction;

$$S_{11} + kS_{21} + k\sigma_{12} = 0,$$

or

$$S_{11} = -k(\sigma_{21} - k\sigma_{22}) - k\sigma_{12} = -2k\sigma_{12} + k^2\sigma_{22}.$$

In the vertical direction:

$$S_{12} + kS_{22} - \sigma_{12} = 0, \quad \text{i.e. } S_{12} = \sigma_{12} - k\sigma_{22}.$$

It is seen that $S_{12} = S_{21}$, i.e. the Kirchhoff stress tensor remains symmetric.

These results can be obtained rapidly by substitution into the general formulas given below, see Eqs. (17), (18).

Example 4. The general case (Fig. 10.4:4). The vector $d\mathbf{T}$ denotes a force acting on a surface element in the deformed body with a unit outer normal $\mathbf{v}$: whereas $d\mathbf{T}_0$ denotes the corresponding force assigned to the corresponding surface in the original configuration with area dS_0 and unit outer normal $\mathbf{v}_0$.

If σ_{ij} is the stress tensor that refers to the strained state, we have Cauchy's relation:

$$dT_i = \sigma_{ji}v_j\,dS. \tag{14}$$

We now define stress components that refer to the undeformed state by a law similar to Eq. (14). If Eq. (3) is used, we write

$$dT_{0_i}^{(L)} = T_{ji}v_{0_j}\,dS_0 = dT_i \quad \text{(Lagrange)}. \tag{15}$$

If Eq. (4) is used, we write

$$dT_{0_i}^{(K)} = S_{ji}v_{0_j}\,dS_0 = \frac{\partial a_i}{\partial x_\alpha}dT_\alpha \quad \text{(Kirchhoff)}. \tag{16}$$

These equations define the stresses σ_{ij}, T_{ij}, and S_{ij}, which are called the *Cauchy*, *Lagrangian*, and *Kirchhoff's stress tensors*, respectively.

The following relationships between σ_{ij}, T_{ij}, and S_{ij} are given in standard textbooks:

$$T_{ji} = \frac{\rho_0}{\rho}\frac{\partial a_j}{\partial x_m}\sigma_{mi}. \tag{17}$$

$$S_{ji} = \frac{\rho_0}{\rho}\frac{\partial a_i}{\partial x_\alpha}\frac{\partial a_j}{\partial x_\beta}\sigma_{\beta\alpha}. \tag{18}$$

The Eulerian stress tensor τ_{ij} is symmetric. The Lagrangian stress tensor T_{ji} is not necessarily symmetric. The Kirchhoff's stress tensor S_{ji} is symmetric. The Lagrangian tensor is inconvenient to use in a stress-strain law in which the strain tensor is always symmetric; the Kirchhoff stress tensor is more suitable.

From Eq. (18), we have

$$S_{ji} = \frac{\partial a_i}{\partial x_\alpha}T_{j\alpha}. \tag{19}$$

$$\sigma_{ji} = \frac{\rho}{\rho_0}\frac{\partial x_i}{\partial a_p}T_{pj} = \frac{\rho}{\rho_0}\frac{\partial x_i}{\partial a_\alpha}\frac{\partial x_j}{\partial a_\beta}S_{\beta\alpha}, \tag{20}$$

$$T_{ij} = S_{ip} \frac{\partial x_j}{\partial a_p}. \tag{21}$$

Example 5. The calculation can be speeded up if one recognizes that $\partial a_i/\partial x_\alpha$ is the element in the ith row and αth column of the square matrix $\{\partial a_i/\partial x_\beta\}$; σ_{ij} is the element of the ith row and jth column of the square matrix $\{\sigma_{ij}\}$. Thus, for the calculation of Example 3, we note that

$$\left\{\frac{\partial a_j}{\partial x_m}\right\} = \begin{pmatrix} 1 & -k & 0 \\ 0 & 1 & 0 \\ 0 & 0 & 1 \end{pmatrix}, \quad \{\sigma_{ij}\} = \begin{pmatrix} 0 & \sigma_{12} & 0 \\ \sigma_{21} & \sigma_{22} & 0 \\ 0 & 0 & 0 \end{pmatrix}.$$

ρ/ρ_0 is equal to the determinant of the $\{\partial a_j/\partial x_m\}$ matrix which is equal to 1. Hence according to Eq. (17), T_{11} is equal to the scalar product of the 1st row of $\{\partial a_j/\partial x_m\}$ and the 1st column of $\{\sigma_{ij}\}$:

$$T_{11} = \frac{\partial a_1}{\partial x_1}\sigma_{11} + \frac{\partial a_1}{\partial x_2}\sigma_{21} + \frac{\partial a_1}{\partial x_3}\sigma_{31} = \sigma_{11} - k\sigma_{21} + 0.$$

Similarly,

$$T_{12} = \frac{\partial a_1}{\partial x_1}\sigma_{12} + \frac{\partial a_1}{\partial x_2}\sigma_{22} + \frac{\partial a_1}{\partial x_3}\sigma_{32} = \sigma_{12} - k\sigma_{22} + 0$$

and so on.

10.5 Equation of Motion in Lagrangian Description

Consider a continuum occupying a region V with a boundary surface S in the deformed state, which corresponds to a region V_0 with a boundary surface S_0 in the original state (Fig. 10.5:1). The body is subjected to external loads. In Eulerian description the external loads consist of a body force $\mathbf{F}$ per unit mass,

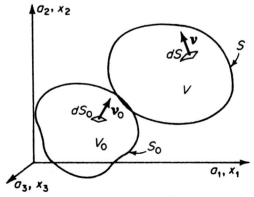

FIGURE 10.5:1 Notations.

and a surface traction $\overset{\nu}{T}$ per unit area acting on a surface element dS whose unit outer normal vector is ν. In Lagrangian description, we shall write the body force per unit mass as F_0. The original density ρ_0 corresponds to the density ρ in the deformed state. We shall specify that

$$\rho \, dv = \rho_0 \, dv_0, \qquad F_{0_i} = F_i. \tag{1}$$

We shall consider static equilibrium of the body.* The resulting body force acting on the region V or V_0 is

$$\int_V F_i \rho \, dv = \int_{V_0} F_{0_i} \rho_0 \, dv_0. \tag{2}$$

The result of the traction acting on the surface S or S_0 is

$$\int_S \overset{\nu}{T_i} \, ds = \int_S \sigma_{ji} \nu_j \, ds. \tag{3}$$

According to Eq. (10.4:3), this is equal to $\int_{S_0} T_{ji} \nu_{0_j} \, dS_0$, which can be transformed by Gauss's theorem into $\int_{V_0} (\partial T_{ji}/\partial a_j) \, dv_0$. Hence

$$\int_S \overset{\nu}{T_i} \, dS = \int_{V_0} \frac{\partial T_{ji}}{\partial a_j} \, dv_0. \tag{4}$$

The condition of equilibrium is that the sum of the body forces and surface tractions vanishes. Hence, by Eqs. (2) and (4), we have, at equilibrium,

$$\int_{V_0} \left(\rho_0 F_{0_i} + \frac{\partial T_{ji}}{\partial a_j} \right) dv_0 = 0. \tag{5}$$

Since this equation must be valid for an arbitrary region V_0, the integrand must vanish, Hence, we obtain the equation of equilibrium

$$\frac{\partial T_{ji}}{\partial a_j} + \rho_0 F_{0_i} = 0. \tag{6}$$

Further, using Eq. (10.4:2), we obtain the equation of equilibrium expressed in terms of the Kirchhoff stress tensor:

$$\frac{\partial}{\partial a_j} \left(S_{jk} \frac{\partial x_i}{\partial a_k} \right) + \rho_0 F_{0_i} = 0. \tag{7}$$

Finally, by an analysis similar to Eqs. (2)–(6) but staying in the deformed configuration, we obtain the well-known equation of equilibrium expressed in terms of Cauchy stress:

$$\frac{\partial \sigma_{ji}}{\partial x_j} + \rho F_i = 0. \tag{8}$$

---

* If the body is in motion, we can apply D'Alembert's principle to reduce an equation of motion to an equation of equilibrium. We apply the inertial force, which is the product of the mass and the negative of the acceleration, as an external load. Hence F_i and F_{0_i} include the inertial force.

10.6 Work and Strain Energy

Force multiplied by the distance it traveled in the same direction is the *work* done by the force. Force multiplied by the velocity of its travel (scalar product) is the *power* of the force. Consider a cylindrical test specimen (of muscle, or blood vessel, or bone) clamped at the top and loaded by a force at the bottom (Fig. 10.6:1). In the deformed state the length of the specimen is L, the cross-sectional area is A, the longitudinal stress is σ, so the force is σA. If the length is extended by a small amount δL then the work done is $\sigma A \delta L$. Let us express this quantity in terms of dimensions and stresses in the original configuration of the specimen, which has a length of L_0, cross-sectional area A_0, Lagrangian stress T, Kirchhoff stress S. According to Sec. 10.5 we have

$$\lambda = \frac{L}{L_0}, \qquad \delta L = L_0 \delta \lambda, \qquad TA_0 = \sigma A, \qquad SA_0 = \frac{\sigma A}{\lambda}. \tag{1}$$

Hence the work done by the force per unit original volume of the specimen is

$$\frac{\text{Work done by force}}{A_0 L_0} = \frac{\sigma A \delta L}{A_0 L_0} = \frac{\lambda S A_0 L_0 \delta \lambda}{A_0 L_0} = \frac{T A_0 L_0 \delta \lambda}{A_0 L_0}. \tag{2}$$

Let us denote the work done by the force per unit mass of the specimen by the symbol δW, and the mass per unit volume in the original state, i.e. the initial density, by ρ_0. Then $\delta \rho_0 W$ represents the work done per unit initial volume. The left-hand side of Eq. (2) can be written as $\delta \rho_0 W$. Note further that Green's strain $E_{11} = \frac{1}{2}(\lambda^2 - 1)$ implies $\lambda \delta \lambda = \delta E_{11}$. Thus Eq. (2) becomes

$$\delta \rho_0 W = \sigma \frac{A}{A_0} \delta \lambda = S \delta E_{11} = T \delta \lambda \tag{3}$$

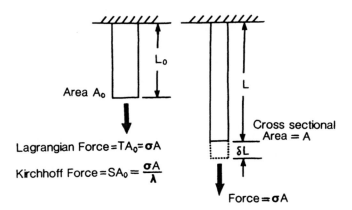

FIGURE 10.6:1 Illustration of the principle embodied in Eqs. (3) and (4).

from which we obtain:*

$$S = \frac{\partial \rho_0 W}{\partial E_{11}}, \qquad T = \frac{\partial \rho_0 W}{\partial \lambda}, \tag{4}$$

but σ does not have a correspondingly simple expression.

The work done $\delta \rho_0 W$ could be dissipated into heat, or consumed by chemical reaction, or stored in the system ready to be recalled. The last eventuality is very important. Most engineering materials and biological tissues can do that. To describe this eventuality, people invent a name: by calling the material *elastic*. Mathematically, we assume the existence of a quantity $\rho_0 W$ which is a function of the strain, and for which equations like (4) are valid. The leads to the following formal definition:

A material is said to be **elastic**, *if it possesses a strain energy function per unit mass W which is an analytic function of the strain components measured with respect to the stress-free state, with the property that the rate of change of the strain energy per unit mass is equal to the power of the stresses.*

Expressing this definition in terms of Eulerian variables, we have

$$\frac{D}{Dt} W = \frac{1}{\rho} \sigma_{ij} V_{ij}; \tag{5}$$

whereas in terms of Lagrangian variables, we have

$$\frac{D}{Dt} W = \frac{1}{\rho_0} S_{ij} \frac{D}{Dt} E_{ij} = \frac{1}{\rho_0} S_{ij} \frac{\partial}{\partial t} E_{ij}. \tag{6}$$

In these formulas, σ_{ij} is the Eulerian stress tensor, E_{ij} is the Green strain tensor, and S_{ij} is the Kirchhoff stress tensor: ρ_0 and ρ are the density in the original and deformed states, respectively. V_{ij} is the rate of deformation defined by

$$V_{ij} = \frac{1}{2}\left(\frac{\partial v_i}{\partial x_j} + \frac{\partial v_j}{\partial x_i}\right), \tag{7}$$

where v_i is the velocity field.

In Eq. (5), the material derivative DW/Dt means the rate of change of W associated with a specific group of material particles with respect to time. On the right-hand side, $\sigma_{ij} V_{ij}$ is the sum of the products of stresses and the corresponding strain rates, which is the power of the stresses per unit volume. Dividing $\sigma_{ij} V_{ij}$ by the material density ρ results in power of the stresses per unit mass. Thus Eq. (5) describes the definition of elasticity as stated.

We should like to prove that Eqs. (6) and (5) are equivalent, thus they describe the same definition. The proof is as follows: Begin with Eq. (5) of

---

* These are important formulas. On the next 4 pages a generalization of these formulas to the most general deformation of an elastic body is given, resulting in Eqs. (1), (2) of Sec. 10.7. The remark that in an incompressible material the pressure is an arbitrary quantity (Eq. (12) of Sec. 10.7), is also important.

Sec. 10.2. Taking material derivatives of terms on both sides of the equation, we obtain

$$\frac{D}{Dt}(ds^2 - ds_0^2) = \frac{D}{Dt}(dx_i\,dx_j\,\delta_{ij} - da_i\,da_j\,\delta_{ij}) = 2\left(\frac{DE_{ij}}{Dt}\right)da_i\,da_j. \tag{8}$$

Now,

$$\frac{D}{Dt}x_i = v_i, \qquad \frac{D}{Dt}dx_i = dv_i = \frac{\partial v_i}{\partial x_l}dx_l, \qquad \frac{D}{Dt}a_i = 0, \qquad \frac{D}{Dt}da_i = 0. \tag{9}$$

Hence the middle terms in Eq. (8) is

$$dv_i\,dx_j\,\delta_{ij} + dx_i\,dv_j\,\delta_{ij} = \frac{\partial v_i}{\partial x_l}dx_l\,dx_i + \frac{\partial v_j}{\partial x_l}dx_j\,dx_l$$

$$= 2V_{il}\,dx_i\,dx_l$$

$$= 2V_{il}\frac{\partial x_i}{\partial a_k}\frac{\partial x_l}{\partial a_m}da_k\,da_m. \tag{10}$$

A substitution back into Eq. (8) and changing the dummy indices several times yields the desired relation:

$$\frac{D}{Dt}E_{ij} = V_{pq}\frac{\partial x_p}{\partial a_i}\frac{\partial x_q}{\partial a_j}. \tag{11}$$

Equation (11) shows that DE_{ij}/Dt vanishes when $V_{ij} = 0$, as one should expect for a rigid-body motion. We now have, according to Eq. (10.4:20) and Eq. (11),

$$\frac{1}{\rho}\sigma_{ij}V_{ij} = \frac{1}{\rho_0}\frac{\partial x_i}{\partial a_\alpha}\frac{\partial x_j}{\partial a_\beta}S_{\beta\alpha}V_{ij} = \frac{1}{\rho_0}S_{\beta\alpha}\frac{D}{Dt}E_{\beta\alpha}. \tag{12}$$

Thus, the equivalence of Eqs. (5) and (6) is proved.

The last equality in Eq. (6) follows the definition of material derivative. E_{ij} is a function of the initial coordinates of the particles, $E_{ij} = E_{ij}(a_1, a_2, a_3, t)$. Since the labeling of each particle does not change with time, we have

$$\frac{D}{Dt}E_{ij}(a_1, a_2, a_3, t) = \frac{\partial}{\partial t}E_{ij}(a_1, a_2, a_3, t). \tag{13}$$

This is to be contrasted with the material derivative of a function $F(x_1, x_2, x_3, t)$ of the current coordinates x_1, x_2, x_3 and time. Since $x_i = x_i(a_1, a_2, a_3)$, F is an implicit function of a_1, a_2, a_3. Hence, by definition,

$$\frac{DF}{Dt} \equiv \frac{D}{Dt}F(x_1, x_2, x_3, t)\bigg|_{a_i\text{ fixed}} = \frac{\partial F}{\partial t} + \frac{\partial F}{\partial x_i}\frac{dx_i}{dt}\bigg|_{a_i\text{ fixed}}$$

$$= \frac{\partial F}{\partial t} + v_i\frac{\partial F}{\partial x_i}. \qquad \text{Q.E.D}$$

10.7 Calculation of Stresses from the Strain Energy Function

In this section we shall describe some salient properties of the strain energy function $\rho_0 W$ that characterize an elastic solid. We assume that $\rho_0 W$, as a function of the nine components of a symmetric strain tensor, is written in a form which is symmetric with respect to the symmetric strain components E_{ij} and E_{ji}, and that in forming the derivative of $\rho_0 W$ with respect to a typical component E_{ij}, the component E_{ji} will be treated as independent of E_{ij}. With this convention, we shall prove that the *Kirchhoff's stress tensor can be expressed as*

$$S_{ij} = \frac{\partial(\rho_0 W)}{\partial E_{ij}}, \tag{1}$$

where ρ_0 is the uniform mass density at the natural state, and E_{ij} are the components of Green's strain tensor. On the other hand, if W is considered as a function of the components of the *deformation gradient tensor* $\partial x_i / \partial a_j = \delta_{ij} + \partial u_i / \partial a_j$ [see Eqs. (10.2:1) and (10.2:2)],* then we shall show that *Lagrange's stress tensor* can be expressed as

$$T_{ji} = \frac{\partial(\rho_0 W)}{\partial(\partial x_i / \partial a_j)} = \frac{\partial(\rho_0 W)}{\partial(\partial u_i / \partial a_j)}. \tag{2}$$

To prove Eq. (1), we note that by the convention stated above we have

$$\frac{\partial W}{\partial E_{ij}} = \frac{\partial W}{\partial E_{ji}}, \tag{3}$$

and

$$\frac{D}{Dt} W = \frac{\partial W}{\partial E_{ij}} \frac{D E_{ij}}{Dt}. \tag{4}$$

A comparison of Eq. (4) with Eq. (10.6:6) shows that

$$\left(\frac{1}{\rho_0} S_{ij} - \frac{\partial W}{\partial E_{ij}} \right) \frac{D E_{ij}}{Dt} = 0. \tag{5}$$

Both factors of this equation are symmetric with respect to i, j. Since Eq. (5) is valid for arbitrary values of the rate of strain $D E_{ij} / Dt$, we must have

$$\frac{1}{\rho_0} S_{ij} - \frac{\partial W}{\partial E_{ij}} = 0, \tag{6}$$

which is Eq. (1).

---

* We use x_i to describe the position of a particle in the deformed configuration of the body, and a_i for that in the original, undeformed configuration. x_i is also used as the geometric variable in Eulerian description of the deformed by body, and a_i for that in Lagrangian description of the undeformed body. u_i is the displacement vector. x_i, a_i, u_i refer to the same rectangular Cartesian frame of reference.

If all the components of the rate-of-strain tensor DE_{ij}/Dt are not allowed to vary independently, we cannot claim the valididity of Eq. (6) on the basis of Eq. (5). For example, if we insist that the material is *incompressible*, then the divergence of velocity field vanishes: $\partial v_i/\partial x_i = V_{ii} = 0$. This can be expressed in terms of DE_{ij}/Dt as follows. Multiply (10.6:11) by $(\partial a_i/\partial x_r)(\partial a_j/\partial x_s)$, note that

$$\frac{\partial x_p}{\partial a_i}\frac{\partial a_i}{\partial x_r} = \delta_{pr}, \qquad \frac{\partial x_q}{\partial a_j}\frac{\partial a_j}{\partial x_s} = \delta_{qs}, \tag{7}$$

and we have

$$V_{rs} = \frac{DE_{ij}}{Dt}\frac{\partial a_i}{\partial x_r}\frac{\partial a_j}{\partial x_s}. \tag{8}$$

Hence the condition of incompressibility implies that

$$\frac{DE_{ij}}{Dt}\frac{\partial a_i}{\partial x_r}\frac{\partial a_j}{\partial x_r} = 0. \tag{9}$$

Writing

$$\frac{\partial a_i}{\partial x_r}\frac{\partial a_j}{\partial x_r} = B_{ij}, \tag{10}$$

we have

$$B_{ij}\frac{DE_{ij}}{Dt} = 0. \tag{11}$$

B_{ij} is called *Finger's strain tensor*. Since DE_{ij}/Dt are restricted by Eq. (11), we cannot say that the first factor in Eq. (5) must vanish, but only that it must be proportional to B_{ij}. Therefore, *for an incompressible elastic material, we must have*

$$S_{ij} = \frac{\partial \rho_0 W}{\partial E_{ij}} - pB_{ij}, \tag{12}$$

where p is an arbitrary scalar which can be identified with pressure.

To derive Eq. (2), we return to the general compressible material and regard W as a function of the nine components of the tensor $\partial x_i/\partial a_j$. Let us write $x_{i,j}$ for $\partial x_i/\partial a_j$. Then

$$\frac{\partial \rho_0 W}{\partial x_{i,j}} = \frac{\partial \rho_0 W}{\partial E_{kl}}\frac{\partial E_{kl}}{\partial x_{i,j}}. \tag{13}$$

From Eq. (10.2:9) we have

$$\frac{\partial E_{kl}}{\partial x_{i,j}} = \frac{1}{2}\left(\frac{\partial x_i}{\partial a_l}\delta_{kj} + \frac{\partial x_i}{\partial a_k}\delta_{lj}\right). \tag{14}$$

Hence

$$
\frac{\partial \rho_0 W}{\partial x_{i,j}} = \frac{1}{2} \left(\frac{\partial \rho_0 W}{\partial E_{jl}} \frac{\partial x_i}{\partial a_l} + \frac{\partial \rho_0 W}{\partial E_{kj}} \frac{\partial x_i}{\partial a_k} \right)
$$

$$
= \frac{\partial \rho_0 W}{\partial E_{jk}} \frac{\partial x_i}{\partial a_k} = S_{jk} \frac{\partial x_i}{\partial a_k} = T_{ji}. \tag{15}
$$

The last equality is obtained in accordance with Eq. (10.4:21). Thus Eq. (2) is verified.

If one likes to consider W as a function of the derivatives of the displacements $u_{i,j} = \partial u_i / \partial a_j$, we note that $u_{ij} = x_{i,j} - \delta_{ij}$, where δ_{ij} is the Kronecker delta. From Eq. (9) of Sec. 10.2 we obtain the following equation, which is equivalent to Eq. (14):

$$
\frac{\partial E_{kl}}{\partial u_{i,j}} = \frac{1}{2} \left(\delta_{ki} \delta_{lj} + \delta_{li} \delta_{kj} + \frac{\partial u_i}{\partial a_l} \delta_{kj} + \frac{\partial u_i}{\partial a_k} \delta_{lj} \right). \tag{16}
$$

Hence, in analog with Eq. (13) we obtain

$$
\frac{\partial \rho_0 W}{\partial u_{i,j}} = \frac{\partial \rho_0 W}{\partial E_{jk}} \left(\delta_{ik} + \frac{\partial u_i}{\partial a_k} \right) = S_{jk} \frac{\partial x_i}{\partial a_k} = T_{ji},
$$

which is the last part of Eq. (2).

Known strain energy functions of biological tissues are presented in Fung (1981).

10.8 Complementary Energy Function

If the constitutive equation

$$
S_{ij} = \frac{\partial \rho_o W}{\partial E_{ij}} \tag{1}
$$

can be inverted; i.e. if E_{ij} can be expressed as functions of $S_{11}, S_{12}, \ldots$, then there exists a function $\rho_o W_c$ which is a function of the stress components S_{ij}, and is called the *complementary energy function of the material*, with the property that

$$
\rho_o W_c = S_{ij} E_{ij} - \rho_o W, \tag{2}
$$

and

$$
E_{ij} = \frac{\partial \rho_o W_c}{\partial S_{ij}}. \tag{3}
$$

This theorem can be proved very easily, because if E_{ij} are continuous and differentiable functions of stresses then $\rho_o W$ is an implicit function of the stresses $S_{11}, S_{12}, \ldots, S_{33}$. Hence by differentiating Eq. (2) with respect to S_{ij}, we obtain

$$\frac{\partial \rho_o W_c}{\partial S_{ij}} = E_{ij} + S_{kl}\frac{\partial E_{kl}}{\partial S_{ij}} - \frac{\partial \rho_o W}{\partial E_{kl}}\frac{\partial E_{kl}}{\partial S_{ij}}$$

which reduces to Eq. (3) because the sum of the last two terms vanish on account of Eq. (1). Q.E.D.

The "If" at the beginning of the first sentence of this section is a big "If". Most nonlinear stress–strain relationships cannot be inverted. For example, a simple relation such as

$$S_{ij} = E_{ij} + E_{ik}E_{kj} + E_{ik}E_{km}E_{mj}$$

cannot be inverted explicitly. Fortunately, however, for many biological soft tissues this inversion can be done, and their complementary energy functions are known, see Fung (1979), and *Biomechanics* (Fung, 1981, pp. 253–256).

10.9 Rotation and Strain

The deformation gradient tensor given in Eq. (10.2:2) in the form of a matrix, can be resolved into the product of two matrices with special properties. Let **F** be the deformation gradient matrix of a reversible transformation (10.2:1),

$$\mathbf{F} = \left(\frac{\partial x_i}{\partial a_j}\right). \tag{1}$$

The *polar decomposition theorem* of Cauchy yields two unique expressions for **F** in terms of an *orthogonal* tensor **R** and positive symmetric tensors **U** and **V**:

$$\mathbf{F} = \mathbf{RU} = \mathbf{VR}. \tag{2}$$

R is orthogonal but need not be proper-orthogonal, so that the product of **R** and its transpose $\mathbf{R}^T$ has the value 1, and the value of the determinant of **R** is either $+1$ or -1 throughout the domain x_i and time t:

$$\mathbf{RR}^T = \mathbf{I}, \qquad \det|\mathbf{R}| = +1 \text{ or } -1. \tag{3}$$

Thus

$$\det \mathbf{U} = \det \mathbf{V} = |\det \mathbf{F}|. \tag{4}$$

R is called the *rotation tensor*. **U** and **V** are called the *right* and *left stretch tensors*, respectively. Obviously,

$$\mathbf{V} = \mathbf{RUR}^T. \tag{5}$$

The *right* and *left Cauchy–Green tensors* **C** and **B** are defined as follows:

$$\mathbf{C} \equiv \mathbf{U}^2 = \mathbf{F}^T\mathbf{F}$$

$$\mathbf{B} \equiv \mathbf{V}^2 = \mathbf{FF}^T = \mathbf{RCR}^T. \tag{6}$$

Books on continuum mechanics usually contain illustrations of the meaning of the rotation tensor **R** and the stretch tensors **C** and **B**. Note that the Green's strain tensor defined in Eq. (10.2:9) and (10.3:5) is related to **C** simply by

$$2\mathbf{E} = \mathbf{C} - \mathbf{I}. \tag{7}$$

The expression **RU** means a stretching following rotation. The expression **VR** means a rotation following stretching. The processes of stretching and rotation decomposed this way are non-commutative.

Chen (1988) established a new decomposition theorem:

$$\mathbf{F} = (\mathbf{I} + \mathbf{Y}) + \mathbf{X} \tag{8}$$

in which **I** + **Y** is an orthogonal tensor

$$(\mathbf{I} + \mathbf{Y})(\mathbf{I} + \mathbf{Y})^T = \mathbf{I}$$

$$\det |\mathbf{I} + \mathbf{Y}| = 1 \tag{9}$$

and **X** is a symmetric tensor

$$\mathbf{X}^T = \mathbf{X}. \tag{10}$$

Chen decomposes the deformation gradient tensor into the *sum* of a rotation and a strain. The processes are then commutative. Chen went on to show that his strain X defines normal strains in terms of change of *length* of linear elements, unlike those of Green's tensor which is based on the change of the *square of length* of linear elements. Chen's shear strains also have simpler interpretation than Green's shear strains. This is a significant new development which will have important applications to biomechanics especially to cases in which large local rotation is involved.

Problems

10.1 Consider a rectangular block of compressible material of unit thickness that does not change. The cross section changes from a cube to a paralleloid, with the following coordinates for their corners:

$$(a_1, a_2, a_3) \rightarrow (x_1, x_2, x_3), \qquad x_3 = a_3$$

$$(0, 0, 0) \rightarrow (0, 0, 0)$$

$$(0, 1, 0) \rightarrow (1, 2, 0)$$

$$(1, 0, 0) \rightarrow (1.5, 0, 0)$$

$$(1, 1, 0) \rightarrow (2.5, 2, 0).$$

What are the Lagrangian and Green's strain components. Using matrix algebra, determine the principal strains, stretch ratios, and the rotation.

10.2 The block named in the problem above is subject to the following stress in the deformed configuration:

$$(\sigma_{ij}) = \begin{pmatrix} 1 & 1 & 0 \\ 1 & 2 & 0 \\ 0 & 0 & 0 \end{pmatrix} \times 10^5 \ \text{N/m}^2.$$

What are the components of the Lagrangian and Kirchhoff stress tensors? N.B.: Consider a unit square in the deformed configuration. Use the mathematical transformation determined in Prob. 1 to find the shape of that element in the initial configuration. Then apply the Lagrangian and Kirchhoff method to determine the stresses.

10.3 Sketch the deformed shape of a cube defined by the following transformation.

(a) $x_1 = a_1 + a_1 a_2$
$x_2 = a_2 + a_1 a_2$
$x_3 = a_3$

(b) $x_1 = a_1 + \frac{1}{8}\sin(a_1 + 2a_2)$
$x_2 = a_2 + \frac{1}{4}\sin(a_1 + 2a_2)$
$x_3 = a_3$

10.4 Design an instrument that will enable you to measure the tension in the skin of a man in vivo. The instrument must be safe and tolerable to the subject being measured, and convenient to use by the operator. Discuss the special design considerations if the instrument is to be used by a beautician for cosmatic applications or by a surgeon in various skin cancer operations.

10.5 In an operation for hernia in the abdomen, it may be desirable to prestretch the skin so that at the end of the surgery the tension of the skin will remain normal. Too large a tension may make the operation difficult and may hinder healing. The prestretch may be accomplished by pumping air into the abdomen, so that the skin area may increase by the process of creep. Design an instrument that will measure the creep and the skin tension so that the operation can be planned scientifically.

10.6 The tension in a tout string can be measured by anchoring the string at two points, putting a lateral load perpendicular to the string, and measuring the deflection of the string as a function of the load. From this premis, design an instrument which will enable you to measure the tension in the Achilles tendon when you exercise.

10.7 A student broke one of his fingers when playing basket ball. A tendon in that finger broke. As a result he cannot straighten the finger. A surgeon can repair this trouble by taking a spare tendon from his foot and using it to reconnect the ends of the broken tendon. The problem is to decide how long the segment of the new tendon must be for him to tie and suture it to the old ends. Or, how much tension should the tendon in the finger have at the time of suturing. If the tension is too large, the finger will be bent one way. If this is too small, it will be bent the other way. How much is just right? To settle this question, a bioengineer was called in to measure the tension needed for the correct positioning of the finger, and to enable the surgeon to put in the correct tension at the time of surgery. You be the engineer. You must invent an instrument and a procedure to do it. Giving due consideration to the conditions of the operation room, present a design for this instrument.

10.8 *Blow Out of a Spherical Ballon.* Let the strain energy function of a material be

$$\phi = \frac{\mu}{k}(\lambda_1^k + \lambda_2^k + \lambda_3^k - 3), \tag{1}$$

where $\lambda_1, \lambda_2, \lambda_3$ are the principal extensions and μ and k are material constants. Let the initial (zero-pressure) thickness of the balloon be h_o, the initial radius be R_o. When the circumferential extension is λ. Show that the inflation pressure p is given by

$$p = \frac{2h_o\mu}{R_o}(\lambda^{k-3} - \lambda^{-2k-3}). \tag{2}$$

Demonstrate the possibility of blow out for a rubber balloon for which $k = 2$. Show that a ventricle will not blow out, (the value of k for myocardium is about 18).

Solution. Assuming plane stress and incompressibility, the stress strain relationship of a membrane under equibiaxial tension is

$$\sigma = \mu(\lambda^k - \lambda^{-2k}), \tag{3}$$

where σ is Eulerian stress and λ is the surface extension, $\lambda = R/R_o$. For thin-walled sphere, the equation of equilibrium is $p = 2\sigma h/R$, and the conservation of wall volume requires $4\pi R_o^2 h_o = 4\pi R^2 h$. A combination of these relations yields Eq. (2). (Ref. Ogden, R.W. Large deformation isotropic elasticity—on the correlation of theory and experiment for incompressible rubberlike solids, *Proc. Roy. Soc. London*, A., 326:565–584, 1972.)

10.9 Discuss the stability of various states of inflation of a spherical balloon from the point of view of stationary total potential energy of the system. (Ref. Alexander, H. Tensile instability of initially spherical balloons. *Int. J. Engineering Science,* 9:151–162, 1971.)

10.10 Show that the pressure p as a function of λ given by Eq. (2) has a relative maximum if $-\frac{3}{2} < k < 3$. Outside of this range of k a local pressure maximum does not exist. Hence a balloon made of a material which may be described approximately by a strain energy function given in Eq. (1) with $k \gg 3$ (such as a ventricle or a urinary bladder, or an artery, or a piece of skin) will not blow out. (Ref. Ogden, *ibid* 1972).

10.11 *Experimental Strain Analysis.* Experimental methods to reveal stress and strain in a body can be based on a wide variety of physical and chemical phenomena that are affected by stress and strain. Some examples are listed below. Expand the list.

1) Piezoelectricity:
   Electric resistance of thin wires of metals change
      with strain   – bone
   Semiconductors   – bone
   Polymer plastics   – soft tissue
   Mercury enclosed in rubber tubing   – plethysmograph
2) Optical properties, birefriengence:
   Photoelasticity   – bone model
   Frozen stress in plastics   – bone model
   Polymer fluids   – blood, body
                                   fluids

3) Thermoelasticity:
    Generation of heat by strain and blood flow   – cancer
    Revelation of temperature by color   – cancer
4) Morphological changes
    X ray of markers   – heart, lung
    Radioopaque filler   – blood vessels
5) Variation of electric property with displacement:
    Capacitance, inductance   – plethysmograph
6) Acoustic phenomena
    Ultrasound   – blood vessel
    Doppler   – blood flow
7) Electromagnetic flow meters   – blood flow
8) Chemical sensors (O_2, glucose, etc.)
    Immobilized enzyme   – diabetes
9) Nuclear magnetic resonance, metabolic imaging   – seeing proton
    contents, H,
    H_2O, P, etc.

10) CATT scan
11) Antibody marker   – cell membrane
12) Fluorescence   – cell division,
    nerve

References

Biot, M.A. (1965). *Mechanics of Incremental Deformations*. (Theory of elasticity and viscoelasticity of initially stressed solids and fluids, including thermodynamic foundations and applications to finite strain). Wiley, New York.

Chen, Zhi-da (1986). *Rational Mechanics*, (in Chinese). China Institute of Mining and Technology Press, Xuzhou, Jiangsu, China.

Eringen, A.C. (1967). *Mechanics of Continua*. Wiley, New York.

Fenton, T.R., Cherry, J.M., Klassen, G.A. (1978). Transmural myocardial deformation in the canine left ventricular wall. *Am. J. Physiol.* **235**: H523–H530.

Fung, Y.C. (1965). *Foundations of Solid Mechanics*. Prentice-Hall, Englewood Cliffs, N.J.

Fung, Y.C. (1977). *A First Course in Continuum Mechanics*. 2nd ed., Prentice-Hall, Englewood Cliffs, New Jersey.

Fung, Y.C. (1979). Inversion of a class of nonlinear stress-strain relationships of biological soft tissues. *J. Biomech. Eng.* **101**: 23–27.

Fung, Y.C. (1981). *Biomechanics: Mechanical Properties of Living Tissues*, Springer-Verlag, New York.

Green, A.E. and Zerna, W. (1954). *Theoretical Elasticity*. Oxford University Press, London.

Green, A.E. and Adkins, J.E. (1960). *Large Elastic Deformations and Nonlinear Continuum Mechanics*. Oxford University Press, London.

Malvern, L.E. (1969). *Introduction to the Mechanics of a Continuous Medium*. Prentice-Hall, N.J.

Waldman, L.K., Fung, Y.C., and Covell, J.W. (1985). Transmural myocardial deformation in the canine left ventricle. *Circ. Res.* **57**: 152–163.

CHAPTER 11

Stress, Strain, and Stability of Organs

11.1 Introduction

In this chapter we discuss the stress and strain in organs. The significance of the subject is established, the methods of approach are summarized, highlights on a few organs are surveyed, and then we focus on some topics which are important from the point of view of mechanics, such as the zero-stress state, the connection between micro- and macromechanics, the effect of surface tension, the incremental laws, the interaction between neighboring organs, the stability of some structures, and the behavior of some structures that become unstable and collapsed.

Why do we want to know the stress and strain in an organ? The primary reason is to gain an understanding of the function of the organ. Consider the lung. To analyze pulmonary gas exchange we must know the size and shape of the alveoli. The size and shape of the alveoli are related to the strain of the lung parenchyma. The gas exchange depends on blood flow, which in turn depends on the diameter of the blood vessels. Pulmonary blood vessels are embedded in an elastic parenchyma; the parenchymal stress affects the caliber of the pulmonary blood vessels. Stretching the alvelar wall narrows the capillary blood vessels, increasing the resistance to blood flow, affecting the ventilation/perfusion ratio, and modifying the interstitial fluid pressure and filtration of water from pulmonary capillaries to the interstitium, lymphatics, and aveoli.

Consider the heart, which is a muscle. The contraction of a heart muscle cell depends on its length at the time of stimulation. If the length is fixed, the maximum force developed is a function of the muscle length. If the force of contraction is fixed at any time during the course of contraction, the velocity of shortening is a function of the muscle length. What determines the length of the muscle cells in the heart? It is the stress and strain in the diastolic heart. Hence, in order to determine the performance of the heart we must know the stress and strain in the diastolic heart. In systole, the pressure of the blood in the left ventricular chamber is determined by the tension in the muscle cells. Hence the stresses in the myocardium determines the performance of the heart.

Similar considerations would show the importance of stress and strain in the smooth muscles in internal organs. This is evident in peristalsis of the intestines, stomach, ureter, Fallopian tubes, arterioles, venules, and sphincters of all kinds. Moreover, the systemic blood pressure is controlled by the vascular smooth muscle in arterioles (Ch. 5), and the smooth muscle action is controlled by stress and strain.

The autoregulation of the blood flow to vital organs such as the kidney and brain is based on the mechanical properties of vascular smooth muscles. The vascular smooth muscle of the brain can react to change of strain and stress in such a way that a ± 25 to ± 50 mmHg change in the mean arterial blood pressure would not change the rate of cerebral blood flow, neither would a ± 25 mmHg change in mean intracranial pressure affect the cerebral blood flow. The renal blood vessels have similar property. They react to changing stress and strain.

The importance of stress and strain in the musculoskeletal system is of course well known. When certain limits of stress and strain are exceeded, a tendon may be painfully sprained, a ligament may be broken, a bone may be fractured. In repair and surgery, success comes to the surgeon who understands biomechanics.

We know also that overstress or understress of a tissue may cause resorption, and change in stress level causes structural remodeling of an organ (Ch. 13). Hence stress is related to the safety, growth, and change in the organs.

Method of Approach

Analysis of stress and strain in organs requires knowledge of the anatomy of the structure, the mechanical properties of the materials, the forces that act in the organ, and the boundary conditions imposed on the organ. The method of continuum mechanics can be applied to living tissues at various levels of scale. For example, in considering the heart, one may limit one's attention to the phenomena at the scale of the whole heart. But if one also wants to know what is going on inside a single muscle cell, then the length scale of a single sarcomere is more pertinent. Further, in studying a sarcomere, one can ignore details that take place in atomic dimensions, and so on. Similarly, in considering the mechanics of the whole lung, the stresses and strains in lung parenchyma can be evaluated in regions which are large compared with an individual alveolus; whereas in considering an interalveolar septum as a continuum, the collagen fibers do not need to be individually accounted for.

At each level of consideration, the property of the continuum can be described by a constitutive equation; and each constitutive equation is written for a specific range of length scale.

At each chosen level, we choose an appropriate set of variables to describe the deformation and motion of an organ, which must satisfy the laws of conservation of mass, momentum, and energy, constitutive equations, and boundary conditions. The solution of the field equations yields stress and strain at the level of scale chosen for the investigation. On the other hand, a study of the relationship between different levels of scales (e.g., a derivation of the constitutive equation on the basis of micromechanics) would provide a deeper understanding of physiology and mechanics.

11.2 The Zero-Stress State

Any analysis of the stress and strain in an organ must begin with the zero-stress state of that organ. This state must be determind experimentally. For example, take a blood vessel. Cut a thin slice out with two plane cross-sections perpendicular to the axis of the vessel. We obtain a ring. If we cut the ring radially at one place, the ring opens up suddenly into a sector, as shown in Fig. 11.2:1. If we characterize the sector by its *opening angle*, the angle extended by two radii joining the midpoint of the inner wall to the tips of the ends, we find that the opening angle of the blood vessel can differ significantly from zero. The section opens up because certain residual strains are released.

If a body is in a zero-stress state, then any cut will not cause any deformation. One may ask whether the cut arterial ring, which has zero resultant shear

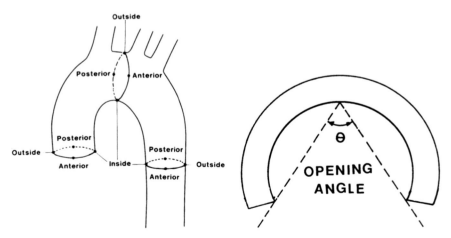

FIGURE 11.2:1 Cross section of a cut vessel at zero-stress, defining the opening angle. From Fung and Liu (1989).

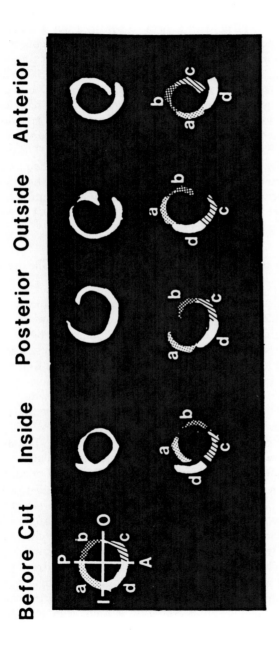

FIGURE 11.2:2 Test of the hypothesis that one cut is sufficient to reduce an arterial segment at no load to the zero-stress state. See text. From Fung and Liu (1989).

and zero bending moment, has approached the zero-stress state. To test the hypothesis that it has, one must cut the sector further into many pieces, and show that the pieces can be fitted together, without strain, into the original sector. This is obviously a difficult experiment to perform; but a somewhat reassuring example is shown in Fig. 11.2:2, which shows four consecutive segments of a thoracic aorta of a rat, each 1 mm thick. These four rings were cut at four different radial sections named "inside", "anterior", "outside", "posterior", at approximately 90° apart. The four sectors look different, implying that the aorta is not axisymmetric: its property varying circumferentially. In the lower row, four pieces cut from one of these sectors are shown; and the four pieces, named a, b, c, d, were reassembled in four different ways: abcd, bcda, cdab, dabc, beginning with a, or b, or c, or d in turn. The center line tangents of successive pieces were matched and lined up at the places of joining. A close resemblance of the reassembled configurations to the original sectors is seen. This indicates that the cut ring is a good approximation to the zero-stress state. In Sec. 11.4 we shall show an example of the heart in which this question is examined in greater detail.

Variation of Zero-Stress State with Location and Sudden Change of Blood Pressure

The opening angle varies with the location on the vascular tree, as it is demonstrated in Fig. 11.2:3. The photos in the first column show the zero-stress configurations of the sections of an aorta of a normal rat. The location of the sections is expressed by x/L, x being the distance of the section to the aortic valve, L being the length of the aorta from the aortic valve to the iliac bifurcation point. Large variation of the opening angle along the aorta is seen.

The figures in the other columns of Fig. 11.2:3 show the changes that occur in the rat aorta when the blood pressure is changed by aortic banding: the imposition of a metal clamp of 2 mm width onto the aorta at $x/L \doteq 0.60$, just above the celiac trunk, occluding the aortic lumen area locally 97%. The clamp causes the blood pressure of the upper body to rise gradually to about 25% above normal, with half of the rise achieved in 7 days. The blood pressure in the lower body decreases at first, then rises to a value somewhat below that of the upper body. The photographs in column 2 show the zero-stress configurations 2 days after banding. The third column shows the configurations 4 days after banding, etc. until 40 days after banding. Large changes in zero-stress state are seen in the aorta in response to the blood pressure change. Moreover, the peak of change is reached in 2 to 4 days. Hypertrophy and remodeling of the blood vessel occurs in this period and afterwards.

Other experiments show that the zero-stress state of arteries change also with other physical, chemical, and biological stimuli, such as hypoxia, diabetes. Similar changes are seen in small arteries and arterioles; and also in veins and in pulmonary and coronary vessels.

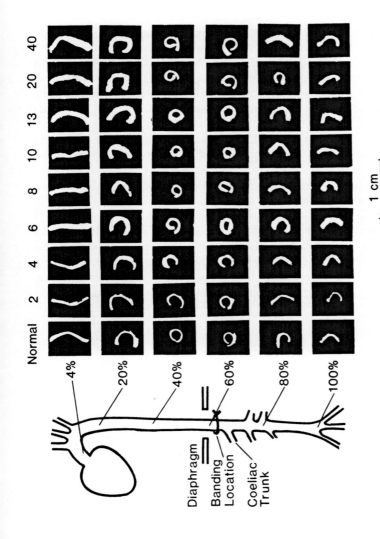

FIGURE 11.2:3 Photographs of the cross sections of the rat aorta when they were cut first transversely and then radially along the "inside" line indicated in Fig. 11.2:1. The first column shows the zero-stress state of the aorta of normal rat. The rest shows the change of zero-stress state due to vessel remodeling after a sudden onset of hypertension. The photos are arranged according to days after surgery from left to right, and according to location on the aorta from top to bottom, expressed as distance from the heart in percentage of the total length. The location of the metal clip used to induce hypertension is shown in the sketch at left. The arcs of the blood vessel wall do not appear smooth because of some tissue attached to the wall. In reading these photographs, one should mentally delete these tethered tissues. From Fung and Liu (1989).

These examples lead us to expect that in a living organ the zero-stress state is in general different from the no-load state, and that the zero-stress state can change with blood pressure or other stimuli.

11.3 Stress and Strain in Blood Vessels

Stress and strain in arteries have been analyzed by Bergel (1972), Patel and Vaishnav (1972), Chuong and Fung (1983), Fung, Fronek and Patitucci (1979), and many others. In these analyses, the assumption is made that the no-load state of an artery is its zero-stress state, and the conclusion is reached that at physiological blood pressure the circumferential stress is very nonuniform, with a high value at the inner wall and a low value at the outer wall. The ratio of the stress at the inner wall to that at the outer wall is of the order of two or three if a linear stress-strain relationship is assumed, but reaches the range of 10–20 when a nonlinear, experimentally determined constitutive equation is used.

The reason for finding this stress concentration at the inner wall under the stated hypothesis is quite simple: Arteries are thick-walled tubes at the no-load condition, with radius to thickness ratios in the range of 2–4. When they are inflated to the normal blood pressure in the range of 110–130 mmHg, the tube radius is enlarged, the wall thickness is decreased and the radius to thickness ratio is increased. Figure 11.3:1 shows that in this situation the circumferential strain at the inner wall is larger than that at the outer wall. The stress–strain relationship being exponential, the stress rises much faster than the strain

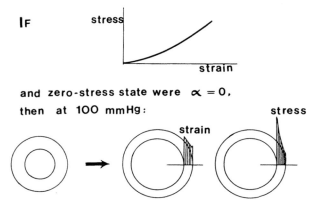

FIGURE 11.3:1 An explanation of the nonuniformity of the stress distribution in blood vessel wall. The blood vessel is a thick-walled tube at no-load state. The wall material is incompressible. When the vessel is inflated to 100 mmHg pressure, the circumferential strain at the inner wall is larger than that at the outer wall, the stress is more concentrated at the inner wall because of the nonlinear stress–strain relationship.

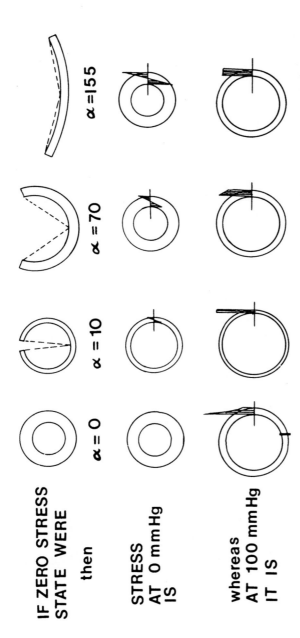

FIGURE 11.3:2 Illustration of the effect of residual strain on the homeostatic stress distribution in blood vessel wall. Successive rows refer successively to zero-stress state, state at no-load, and state at 100 mmHg transmural blood pressure. Successive columns refer successively to a thick-walled tube with no residual strain, a thin-walled tube with a small opening angle, a thick-walled tube with a moderate opening angle, and a thick-walled tube with a large opening angle. The circumferential stress distribution in the wall of each tube is shown at the right-hand side of each tube. The larger opening angle correlates with more uniformly distributed circumferential stress.

when strain increases. As a result the stress at the inner wall becomes large, and it decreases exponentially to a lower value at the outer wall.

Effect of Zero-Stress State on Stress Distribution In Vivo

Figure 11.3:2 shows qualitatively how the circumferential stress would vary throughout the vessel wall when the opening angle of the zero-stress state is changed. If the opening angle α is zero and the vessel wall is thick, the circumferential stress σ_θ is zero when blood pressure is zero, but can become very nonuniform at a blood pressure of 100 mmHg. If α is small but the vessel wall is *thin*, then the residual stress is small and the stress at 100 mmHg blood pressure is relatively uniform because of the thinness of the vessel wall. On the other hand, if the vessel wall is *thick* and the opening angle is *large*, then the residual stress would be large, but the stress distribution at 100 mmHg could be fairly uniform if α is sufficiently large. For each vessel a certain opening angle can be found which can make the circumferential stress at the inner wall equal to that at the outer wall at the normal blood pressure. This opening angle increases with increasing thickness to radius ratio. These qualitative statements can be quantitized by method presented below.

An Example of Stress Analysis

Consider a circular cylindrical artery whose zero-stress state is a circular sector (Fig. 11.3:3). Let the zero-stress state be called state 0, the unloaded

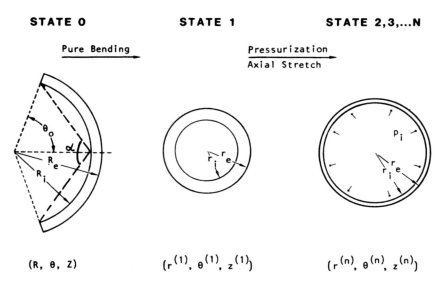

STATE 0   **STATE 1**   **STATE 2,3,...N**

Pure Bending →   Pressurization →
                                 Axial Stretch

(R, θ, z) $(r^{(1)}, \theta^{(1)}, z^{(1)})$ $(r^{(n)}, \theta^{(n)}, z^{(n)})$

FIGURE 11.3:3 The cross-sectional representations of arteries at the stress-free state, the unloaded state, and the loaded state. From Chuong and Fung (1986).

tube as state 1, and the subsequent loaded states as states 2, 3, ..., N. With cylindrical polar coordinates, a material point is denoted by (R, Θ, Z) in state 0, and (r, θ, z) in states 1, 2, ..., N. The subscripts i and e indicate the inner and external walls respectively. Θ_0 represents half of the polar angle of the zero-stress state. Note that Θ_0 is related to the opening angle α by the equation

$$\tan(\alpha/2) = \frac{\sin \Theta_0}{(1 - \cos \Theta_0)}.$$ (1)

Using the notations of Sec. 10.2, we see that the incompressibility condition of the vessel wall requires that

$$\Theta_0(R_e^2 - R_i^2) = \pi \lambda_z (r_e^2 - r_i^2),$$ (2)

where λ_z is the stretch ratio in the axial direction. The deformation of the vessel is described by the mathematical transformation:

$$r = r(R), \qquad \theta = (\pi/\Theta_0)\Theta, \qquad z = z(Z).$$ (3)

The principal stretch ratios are:

$$\lambda_r = \frac{\partial r}{\partial R}, \qquad \lambda_\theta = \left(\frac{\pi}{\Theta_0}\right)\frac{r}{R}, \qquad \lambda_z = \frac{\partial z}{\partial Z}.$$ (4)

With incompressibility condition $\lambda_r \lambda_\theta \lambda_z = 1$, Eq. (10.2:14), the first of Eqs. (3) can be written as

$$r = \sqrt{r_e^2 - \frac{\Theta_0}{\pi \lambda_z}(R_e^2 - R^2)}.$$ (5)

The pseudo-strain energy function is assumed to be (Fung, 1981)

$$\rho_0 W = \frac{c}{2} \exp Q$$ (6)

where

$$Q = b_1 E_\theta^2 + b_2 E_z^2 + b_3 E_r^2 + 2b_4 E_\theta E_z + 2b_5 E_z E_r + 2b_6 E_r E_\theta.$$ (7)

ρ_0 is mass density of the wall, $c, b_1, b_2, ..., b_6$ are material constants, E_θ, E_z, E_r are Green's strains in the circumferential, longitudinal, and radial directions, respectively.

$$E_i = \tfrac{1}{2}(\lambda_i^2 - 1), \qquad (i = r, \theta, z).$$ (8)

A Lagrangian multiplier H is introduced to impose the condition of incompressibility on the strain energy function, hence:

$$\rho W^* = \rho_0 W + \frac{H}{2}[(1 + 2E_\theta)(1 + 2E_z)(1 + 2E_r) - 1].$$ (9)

It is known that H has the significance of a hydrostatic pressure (Sec. 10.7). If a_i denotes the coordinates of a material particle at the reference state 0, and

x_i denotes that at the deformed state, then the Cauchy stress components can be obtained from

$$\sigma_{ij} = \frac{\rho}{\rho_0} \frac{\partial x_j}{\partial a_\alpha} \frac{\partial x_i}{\partial a_\beta} \frac{\partial}{\partial E_{\beta\alpha}} \rho_0 W^*, \qquad (i, j, \alpha, \beta = r, \theta, z), \qquad (10)$$

where ρ, ρ_0 denote the densities of the material in the deformed and undeformed states ($\rho = \rho_0$ in the present case). See Sec. 10.7. Introduction of Eqs. (6)–(9) into (10) yields

$$\sigma_\theta = c(1 + 2E_\theta)[b_1 E_\theta + b_4 E_z + b_6 E_r]e^Q + H$$

$$\sigma_z = c(1 + 2E_z)[b_4 E_\theta + b_2 E_z + b_5 E_r]e^Q + H \qquad (11)$$

$$\sigma_r = c(1 + 2E_r)[b_6 E_\theta + b_5 E_z + b_3 E_r]e^Q + H.$$

The problem of an artery subjected to transmural pressure and longitudinal tethering force can be solved by substituting Eq. (11) into the equation of equilibrium

$$\frac{\partial \sigma_r}{\partial r} + \frac{\sigma_r - \sigma_\theta}{r} = 0 \qquad (12)$$

and the boundary conditions:

1) $\sigma_r = 0$ on the external surface $r = r_e$,
2) $\sigma_r = p_i$ on the inner surface $r = r_i$,
3) On the ends of the blood vessel segment, there acts an external force F.

Application of the conditions 1 and 2 yields

$$p_i = \int_{r_e}^{r_i} c\{(1 + 2E_r)[b_6 E_\theta + b_5 E_z + b_3 E_r]$$

$$- (1 + 2E_\theta)[b_1 E_\theta + b_4 E_z + b_6 E_r]\}e^Q \frac{dr}{r}. \qquad (13)$$

For static equilibrium the longitudinal force $F + \pi r_i^2 p_i$ equals the integral of σ_z over the vessel wall cross section; hence

$$F + p_i \pi r_i^2 = 2\pi \int_{r_i}^{r_e} \sigma_z r \, dr. \qquad (14)$$

Use of Eqs. (11) and (13) in (14) yields

$$F = 2\pi c \int_{r_i}^{r_e} r e^Q [(1 + 2E_z)(b_4 E_\theta + b_2 E_z + b_5 E_r)$$

$$- \tfrac{1}{2}(1 + 2E_r)(b_6 E_\theta + b_5 E_z + b_3 E_r)$$

$$- \tfrac{1}{2}(1 + 2E_\theta)(b_1 E_\theta + b_4 E_z + b_6 E_r)] \, dr. \qquad (15)$$

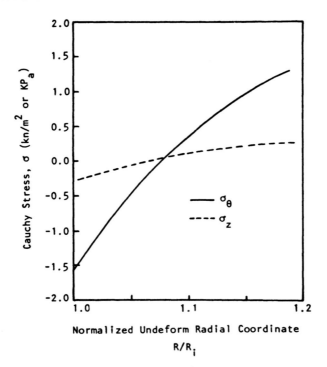

FIGURE 11.3:4 The residual Cauchy stress distribution through the thickness of a cat thoracic aorta at the no-load condition (transmural pressure and longitudinal force $=0$). $\lambda_z = 1$. For this computation, the zero-stress state of the artery is assumed to be a circular arc with a polar angle Θ_0 equal to 71.4°, outer radius $R_e = 4.52$ mm, inner radius $R_i = 3.92$ mm, as in a photograph given in Fig. 2.9:4, p. 59, Fung (1984). From Chuong and Fung (1986).

Equations (13) and (15) are two integral equations from which we can determine the material constants with the knowledge of p_i, F, r_e, r, and E_θ, E_z, E_r, i.e., intraluminal pressure, longitudinal stretch force, external radius, internal radius, and distribution of strain components, respectively. Once the material constants are determined, we can evaluate the residual stress at state 1 and the stress distribution at loaded states with residual stress and residual strain taken into consideration. Details are given in Chuong and Fung (1986). Sketches of some of the results are given in Figs. 11.3:4 and 11.3:5. Figure 11.3:4 shows the residual stress distribution. Figure 11.3:5 shows the stress distribution at a normal blood pressure of 120 mmHg. It is seen that the residual stress due to a finite opening angle of 143° has the effect of making the circumferential and longitudinal stresses σ_θ and σ_z fairly uniform throughout the blood vessel wall.

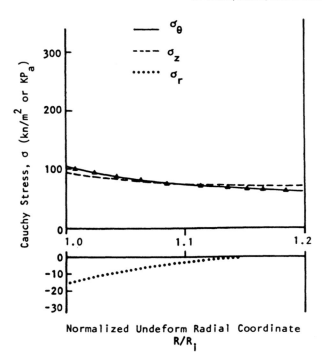

FIGURE 11.3:5 The distribution of the Cauchy stresses σ_θ. σ_z, σ_r through the thickness of a cat thoracic aorta at $p_i = 120$ mm Hg (~ 16 kP$_a$), $\lambda_z = 1.691$, under the same zero-stress state assumed in Fig. 11.3:4. From Chuong and Fung (1986).

11.4 Stress and Strain in the Heart

Stress and strain in the heart have been analyzed extensively by Grimm, Hood, Janz, Mirsky, Sandler and Dodge, Spaan, Streeter, Wong, Yin, and others under the assumption that the unloaded diastolic state is the zero-stress state. A variety of constitutive equations have been examined, as well as a variety of simplifying assumptions. See extensive reviews by Mirsky (1979) and Yin (1981).

The zero-stress state of the left ventricle of the rabbit, dog, and rat has been studied by Fung (1984) and Omens and Fung (1989). If an unloaded left ventricle is supported in such a way that the effect of its weight is minimized (e.g., floating in normal saline and resting on a soft nylon net), then a longitudinal cut would cause the left ventricle to open up as shown in Fig. 11.4:1. If an isolated, resting, diastolic left ventricle is sectioned by planes parallel to its base, Fig. 11.4:2, we obtain a ring. If the ring is cut radially, it opens up. An *opening angle* can be defined as shown in Fig. 11.4:3. Changes of strain in the myocardim from the state of unloaded whole heart to that of the ring and the

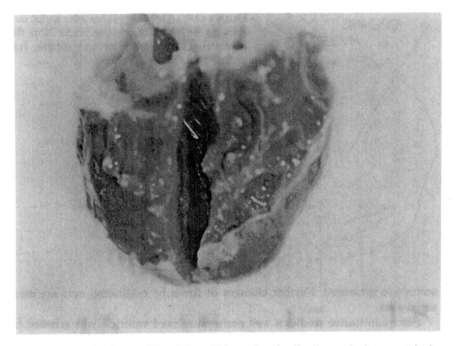

FIGURE 11.4:1 A left ventricle of the rabbit cut longitudinally on the interventricular septum. The cut line opens up. From Fung (1984).

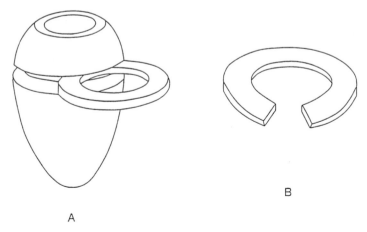

B

A

FIGURE 11.4:2 Schematic representation of a left ventricle, showing equatorial slice. This extracted slice, when placed in the holding tank, will be in the no-load condition. The slice is cut radially opposite to the right ventricle (not shown). The resulting configuration is referred to as the radially cut state. From Omens and Fung (1990).

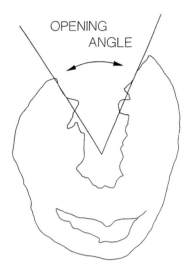

OPENING
ANGLE

FIGURE 11.4:3 Drawing of slice after radial cut, showing definition of opening angle. Note the asymmetry of the posterior and anterior free walls. From Omens and Fung (1990).

sector are measured. Further changes of strain by additional cuts are measured also.

For quantitative studies a well perfused relaxed ventricle was arrested by clamping the ascending aorta and injecting an iced, heparinized hyperkalemic Krebs-Henseleit solution directly into the left ventricle through the apex. To avoid ischemic contraction, we added a calcium channel blocker nifedipine (2×10^{-4} g/l), and a calcium chelator EGTA (10 mM/l) to the perfusate. To obtain markers for strain measurement, we sprinkled stainless steel microspheres (60–100 μm diameter) onto the ring shaped slice, and pressed them lightly into the tissue. The slice was then cut radially as indicated above. The microspheres were photographed immediately before and immediately after the cutting (Fig. 11.4:4). Computer graphics record the coordinates of light spots on selected microspheres, compute the distances between the particles, and evaluate the Green's strain tensor components E_{ij} according to the definition given in Eq. (8) of Sec. 10.2. In this computation, the distances ds, ds_0, da_i (in notations of Sec. 1.2) are measured and the strain components E_{ij} are computed from a sufficient number of simultaneous equations or by the least-squares method if the number of equations exceeds the number of unknowns. With E_{ij}, the principal strains E_1, E_2, and the principal directions are computed, and the principal stretch ratios,

$$\lambda_i = (2E_i + 1)^{1/2} \tag{1}$$

are determined.

A typical result is shown in Figure 11.4:5. If the cut ring is considered as stressfree, then what is shown here is the residual strain in the ring. The principal directions of the residual strain differ from the radial and circumferential directions within 15 ± 10 degrees (mean $\pm$ S.D.), hence the stretch

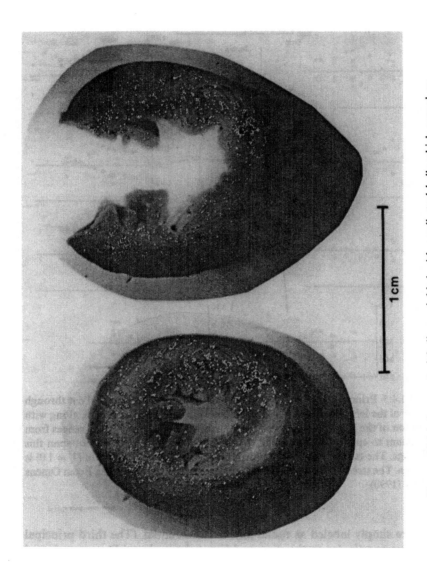

FIGURE 11.4:4 An equatorial slice sprinkled with small steel balls which served as markers of location. Local strains were computed from the displacements of the steel balls. From Omens and Fung (1990).

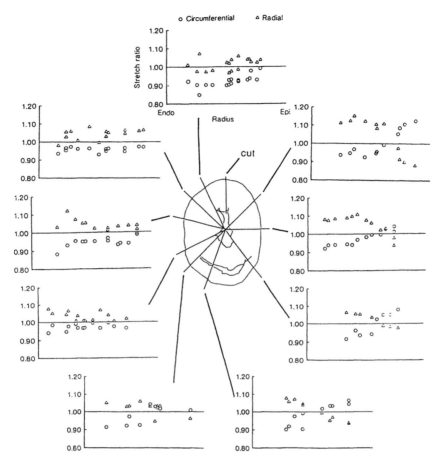

FIGURE 11.4:5 Principal stretch ratios of an equatorial slice with a radial cut through the center of the left ventricular free wall. The no-load geometry is shown, along with the location of the radial cut. The horizontal axis (radius) of each group ranges from endocardium to epicardium, although the measured stretch radios do not span this entire range. The right ventricular wall is ignored. The line of zero strain ($\lambda_i = 1.0$) is also shown. The stretch ratio values are ± 0.02 due to digitization errors. From Omens and Fung (1990).

radios are simply labeled as radial and circumferential. (The third principal strain is perpendicular to the slice, and is not shown here.) Data are plotted for points along nine radii. It is seen that the residual strain in the circumferential direction is largely compressive in the endocardial side, and tensile in the epicardial side; the magnitude being smaller in the interventricular septum. The radial strain is indeed predominant and mostly tensile. The sign and magnitude of the radial strain can be predicated by the condition of incompressibility of the material, which requires that the product of the three principal stretch ratios be exactly equal to one. If the circumferential and longitudinal stretch ratios are both less than 1, then the radial stretch ratio must be greater than 1.

One wonders whether one cut relieves all the residual stress in the slice of myocardium. The result shown in Fig. 11.4:6 partially answers this question. The figure shows the changes in strain due to a second radial cut in addition to the first one. The second cut does cause a little additional strain, implying

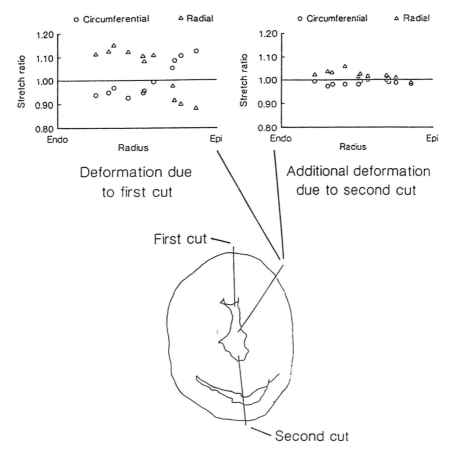

FIGURE 11.4:6 Principal stretch ratios after one radial cut, and the additional deformation due to a second radial cut. These distributions are taken from an area near the first cut (shown below plots). The stretch ratios after a second cut are referred to the configuration after the first cut. On a global scale, the second cut does not result in substantial deformations compared to those after the first cut. From Omens and Fung (1989).

that one cut is not enough; but it also shows that the changes are small. Rigorously speaking, many cuts are necessary to relieve the stress completely.

Theoretical analysis of myocardium with measured zero-stress state is best done with the finite element method. A number of good programs have been written. See an extensive review by Hunter and Smaill (1988), paper by McCulloch et al. (1987), and an ASME proceeding edited by Spilker and Simon (1989).

Accurate measurement of strain in a beating heart by Waldman (1985), Waldman et al. (1985) and others is an important new development. It yields information about the dynamics of the heart in normal and diseased conditions. The method uses embedded markers (small lead beads or stainless steel beads, or the best, gold beads) implanted into the myocardium. The three-dimensional coordinates of these beads are obtained with high-speed cine-radiography. Any four noncoplanar markers forming a small tetrahedral volume (< 0.1 cc) are used to calculate finite strains according to Eq. (10.2:8). Waldman et al (1985) showed that the directions of the principal axes of shortening vary substantially less than the variation of fiber direction across the wall ($20°$–$40°$ compared with $100°$–$140°$ for fiber direction). Hence the principal axes of the strain do not coincide with the direction of the muscle fibers. There are substantial interactions between neighboring fibers in the left ventricular wall.

11.5 The Musculoskeletal System

The spectacular development of the mechanics of the musculoskeletal system in recent years has made biomechanics an integral part of medicine and surgery. The bone has a very complex structure. Its strength, failure, growth, and resorption have been studied in detail. Replacement of damaged bone and arthritic joints by artificial bone and artificial joints has led to a rapid development of biomaterial science and technology. The soft tissues in the joints: ligaments, tendons, cartilage, and the muscles that mobilize the musculoskeletal system, have been and continue to be subjects of intensive study. There is huge literature on this topic but its review is beyond the scope of this book. Bibliographies of Chaps. 1, 2, and 12 contain references to mathematical modeling of the musculoskeletal system.

11.6 The Lung

The lung is a porous structure. In applying the concept of continuum mechanics to such a structure, it is extremely important to have the objective of the study and the size of the structure clearly stated at the beginning. For example, if one is interested in comparing the difference of strain in the upper part of the human lung from that in the lower part, then individual alveoli can be considered infinitesimal and one can speak of deformation averaged over volumes that are large compared with the volume of a single alveolus, but small compared with the whole lung. Such an approximation would be appropriate in studying the interaction of lung and chest wall, the distribution of pleural pressure, or the distribution of ventilation in the whole lung. On the other hand, if one is interested in the stress in a single alveolar wall, then even the individual collagen and elastin fibers in the wall cannot be ignored. There are interesting problems at every level; appropriate use of averaging is often the key to simplification.

In the following, several aspects of the mechanics of the lung will be discussed. Our general objective is to study the connection between the mechanical properties of the whole lung and that of the aveolar ducts and alveoli, i.e., the connection between macro- and micromechanics, with micro scale identified with alveolar dimension.

We shall take the following steps:

1) Identify the geometry of the alveoli and alveolar ducts.
2) Identify the materials of construction and the geometric configurations of the materials.
3) Determine the rheological properties of the materials of construction and the interfacial energy or surface tension.
4) Derive the constitutive equations of the alveolar walls, alveolar mouths, and the lung parenchyma.
5) Validate the derived constitutive equations and solve useful problems of the lung.

As it is stated in the Introduction (Sec. 11.1), these same steps are necessary to study any organ. The lung is used here as an example. In carrying out these steps, we shall recognize that there exist certain gaps of knowledge. These gaps will be filled in today by ad hoc hypotheses which will be replaced by fundamental data and rigorous theoretical derivations in the future.

We explain the geometry, the materials, and a very simple and very crude derivation of a constitutive equation in this section. Then in Sec. 11.7 the surface tension on alveolar wall is examined in greater detail. In Sec. 11.8 we consider small perturbations of lung parenchyma, and in Sec. 11.9 derive the incremental elastic moduli on the basis of the microstructure. The rest of the chapter deals with some applied stress problems.

Models of the Alveoli and Alveolar Ducts

Analysis of the elasticity of the lung must begin with a description of the geometric structure of the lung. Hence it is necessary to have a mathematical model of the alveolar ducts and alveoli. Looking at histological photographs such as that shown in Fig. 11.6:1, one gets the impression that alveolar ducts are bounded by curved edges, but it is very difficult to get a three-dimensional model. Hanson and Ampaya (1975) reconstructed a region of postmortem

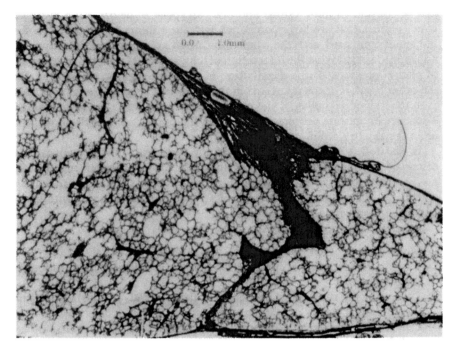

FIGURE 11.6:1 Histological photograph of alveolar ducts in a thick section (150 μm) of human lung fixed by quick-freezing technique using OsO_4 and strained by silver to reveal collagen fibers. Curvature of alveolar mouths is evident. Specimen and histology are described in Matsuda et al. (1987).

human lung wall by wall, and obtained a great deal of data, but it is still difficult to derive a mathematical model.

Pulmonary researchers have long engaged in model building. Miller (1947) reviewed the history of the development of concepts of the structure of the alveoli and alveolar ducts from Malpighi to Loeschcke. Orsos (1836), Weibel (1963), Oldmixon et al. (1987, 1988), Mercer et al. (1987) have added new results. Figure 11.6:2 shows a series of alveolar models published by famous authors. Those in the upper row visualize alveoli as bunches of grapes. Each alveolar wall has an "inside" and an "outside". On the inside there acts the alveolar gas pressure. On the outside, the pressure is presumably the pleural pressure. This model culminates in that of von Neergaard (1929), shown in the first figure of lower row, which was taken from Clements et al. (1965). This is the model that has asserted the importance of surface tension on the stability of the lung. One difficulty this model leads to is shown in the drawing (a) of Fig. 11.6:3. If two soap bubbles were blown at the ends of two tubes and these two tubes are connected and open to each other, than at equilibrium the smaller bubble

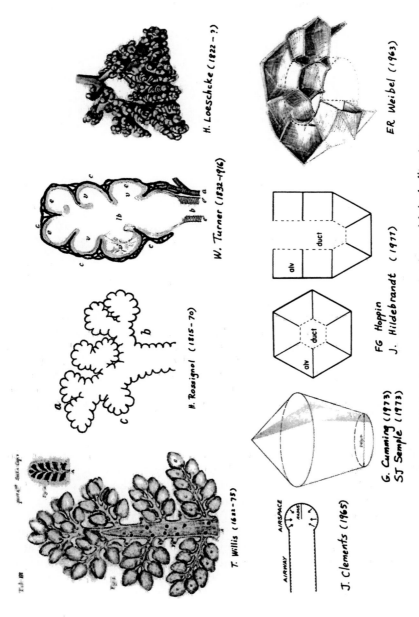

FIGURE 11.6:2 A series of well-known alveolar models in the literature.

TOPOLOGICALLY
WRONG MODEL
FOR ALVEOLAR MECHANICS

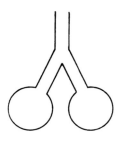

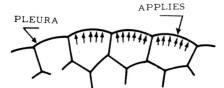

NEERGAARD MODEL
APPLIES
PLEURA

SEPT. a-a: Both faces in DUCT A.

SEPT. b-b: One face in DUCT A,
another in DUCT B.

Difference of pressures on both sides
of a-a is zero. That on the two sides
of b-b is a few μm H_2O.

FIGURE 11.6:3 Topological structure of pulmonary alveoli. *Top left*: Topologically wrong model for alveolar mechanics. *Top right*: Schematic sketch of the typical topological arrangement of interalveolar septa. Septum $a - a$: Both sides are in the same alveolar duct; the net pressure difference between the two sides is zero. Septum $b - b$: One side is in duct A, and the other side is in duct B; the difference in the pressures acting on the two sides is on the order of a few μm H_2O. *Bottom left*: Equilibrium of pulmonary (visceral) pleura, to which von Neergaard's model applies. From Fung (1975a).

would have collapsed.* Similarly, if a human lung were considered as 300 million bubbles connected in parallel to an airway, then one would conclude that at equilibrium all but one alveolus will be collapsed, unless the surface tension is zero, or the elastic force overwhelms the surface force. The lung obviously does not behave this way. The fault of this model is that it ignores the fact that an alveolar wall is an interalveolar septum. The inside of one alveolus is the outside of another alveolus. The pressure difference across each septum is negligibly small. A septum must be a minimal surface; i.e., one whose mean curvature is zero. Any calculation of septal deformation based on pressure difference of airway and pleura would be wrong.

* On account of Laplaca's formula, Eq. (11.7:1), which states that the pressure in the smaller bubble is higher; hence on opening the connecting valve, the gas in the smaller bubble will flow into the larger bubble.

Other drawings in Fig. 11.6:2 show the models of alveolar duct described by Weibel (1963) Cumming and Semple (1973) and Hoppin and Hildebrandt (1977) who did not describe the rules how these units are organized into an airway tree that fills the entire space of the lung. Karakaplan et al. (1980)'s model is similar to Hoppin and Hildebrandt's. Wilson and Bachofen (1982)'s model is based on Orsos' (1936) comments, and it seems to me an over-simplification.

To meet the space-filling requirement, the author (Fung, 1988) proposed a model and validated it against all available morphometric data. This model is based on the assumptions that all alveoli are equal and space filling before they are ventilated, that they are ventilated to ducts as uniformaly as possible, reinforced at the edges of the ventilation holes (alveolar mouths) for structural integrity, and distorted by lung weight and inflation according to the theory of elasticity.

It is well known that there exists only five regular convex polyhedra in three-dimensional space: the tetrahedron, the cube, the regular octahedron, the pentagonal dodecahedron bounded by 12 regular pentagons, and the icosahedron, bounded by 20 regular triangles. But the last two are not space filling, and the tetrahedron and octahedron are space filling only if they are used in combination. On the other hand, the cube, the nonregular octahedron (every face a triangle), the garnet-shaped rhombic dodecahedron (12 sided, each side a rhombus), and the tetrakaidekahedron (14 sided polyhedron formed by cutting the 6 corners off a regular octahedron in such a way as to leave all edges equal in length) are space filling. (See Fig. 11.6:4.) Thus only the 4- and 8-hedra with triangular faces, 6- and 12-hedra with rectangular faces, and 14-hedron with square and hexagonal faces meet the equal and space-filling requirement. Microscopic observations of the microstructure of the lung show that the alveolar walls are not all triangles or all rectangles. This leaves the 14-hedron as a more reasonable candidate.

Fung's basic unit of alveolar duct model is the *second order 14-hedron* formed by a 14-hedron surrounded by 14 identical polyhedra, Fig. 11.6:5, with all the walls of the central 14-hedron removed for ventilation. Two Order-2 14-hedra can be joined together to form a longer duct by removing a few additional walls. Figure 11.6:6 shows two ways in which two Order-2 14-hedra

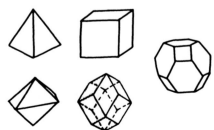

FIGURE 11.6:4 Several polyhedrons. The three on the right-hand side are space filling. From Fung (1988).

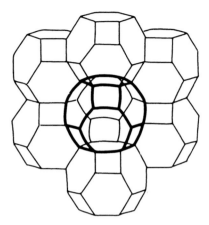

FIGURE 11.6:5 The order-2 14-hedra is the basic unit of the alveolar duct. Fourteen 14-hedra surround one central 14-hedron. Several 14-hedra in front are removed to reveal the central 14-hedron. To form a basic unit of lung structure, all faces of central 14-hedron are removed, whereas all of its edges are reinforced to take up the load and keep the structure in stable equilibrium. From Fung (1988).

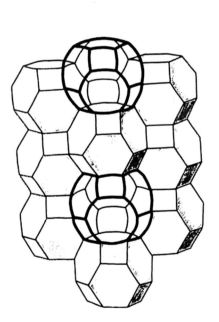

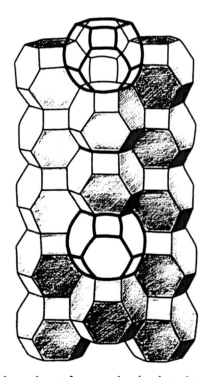

FIGURE 11.6:6 Two order-2 14-hedra are joined together to form an alveolar duct. At least one additional face must be removed in order to ventilate the alveoli. The two assemblies differ by length of one unit. The longer one can be identified as ducts of generations 4 and 5. The shorter one can be identified as ducts of generations 6 and 7. Ducts of generations 1, 2, 3 are formed by joining one more order-2 polyhedron to the longer one shown here. From Fung (1988).

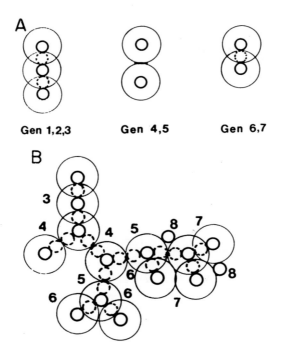

FIGURE 11.6:7 Schematic drawings of how order-2 polyhedra are assembled into alveolar ducts of generations 1, 2, ... 7 and a ductal tree. Generation 8 consists of a single 14-hedron filling in left-over space. From Fung (1988).

are joined together, differing in lengths by one basic unit (diameter of the first order 14-hedron). These units can be joined to form a ductal tree, Fig. 11.6:7. Space can be filled with a combination of first and second order 14-hedra. Ventilation of every alveolus can be assured by removing a suitable number of walls. Removal of a wall upsets the balance of forces at an edge, but the equilibrium can be restored by reinforcing the edge into a "cable" and allowing it to become curved, Fig. 11.6:8, which explains why the alveolar ducts and mouths always look rounded in Fig 11.6:1.

Comparison of morphometric predictions of this model and the experimental data published by Hanson and Ampaya (1975) and others is given in Fung (1988). The model predicts that the length of alveolar ducts of generations 1, 2, 3 is approximately 4Δ, Δ being the dimension of a single alveolus (height of a 14-hedron). The length of generations 4, 5 is 3Δ, that of generations 6, 7, 8 is 2Δ, whereas the length of the alveolar sac is $\Delta/2$. This is in good agreement with the measured data. The predicted dihedral angles between interalveolar septa, 110° and 125°, are in agreement with morphometric data. However, in order to explain the great variety of shapes of the alveoli observed by Hanson and Ampaya in human lung, it is necessary to assume that an additional

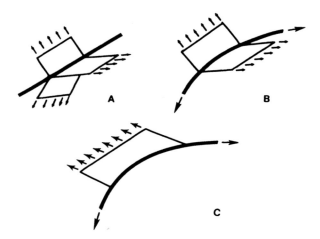

FIGURE 11.6:8 Illustration of balance of forces at an edge where 3 alveolar walls intersect. A: An edge where all 3 membranes are intact. B. One of 3 membranes is perforated. C. Two membranes are perforated. The edges with perforated membranes must carry tension and be curved in order to balance the membrane tension. From Fung (1988).

number of interalveolar septa were perforated either naturally or by disease or aging.

This model is simple enough and detailed enough to serve as the foundation for theoretical analysis of the lung elasticity and lung ventilation.

Materials of Construction

Collagen, elastin, ground substances, and cells are the principal materials of the lung parenchyma, which is composed of alveoli and alveolar ducts. Muscle exists in blood vessels and bronchi. Cartilage exists in trachea and large bronchi. The mechanical properties of these materials are discussed at length in Fung (1981). The specific information needed for the lung is their structure and distribution in lung parenchyma.

Sobin, Fung, and Tremer (1988) have measured the alveolar size and the width and curvature of collagen and elastin fibers in postmortem human pulmonary alveolar walls. Matsuda, Fung, and Sobin (1987) measured the width of these fibers in alveolar mouths. Figure 11.6:9 shows a photograph of the collagen fiber bundles in a human pulmonary alveolar wall. Figure 11.6:10 shows a photograph of the elastin fibers. Tables 11.6:1–11.6:3 list some data obtained by morphometry.

Statistical data show that the histograms of the width and curvature of collagen and elastin fibers are skewed to the right, but the square root of the width and cubic root of the curvature of the fibers in the interalveolar septa

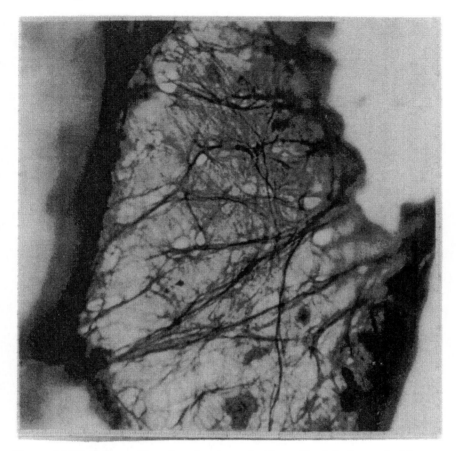

FIGURE 11.6:9 Collagen fibers in pulmonary alveolar walls of the lung of a 19 years old male, inflated to a transpulmonary pressure of 10 cm H_2O. The scale is marked on the border; 800 pixels are equal to 200 μm in the tissue. OsO_4 fixed. Silver stained. From Sobin, Fung, and Tremer (1988). Reproduced by permission.

are normally distributed, whereas the fourth root of the width of the fibers in the alveolar mouths are also normally distributed. Thus the probability frequency function is

$$f(x) = \frac{1}{\sigma\sqrt{2\pi}} e^{-(x-\mu)^2/(2\sigma^2)}, \tag{1}$$

where μ is the mean of x, σ is the standard deviation of $(x - \mu)$, and

$$x = (\text{curvature})^{1/3} \quad \text{or} \quad (\text{width})^{1/2} \text{ in wall,}$$

$$x = (\text{width})^{1/4} \text{ in mouth.} \tag{2}$$

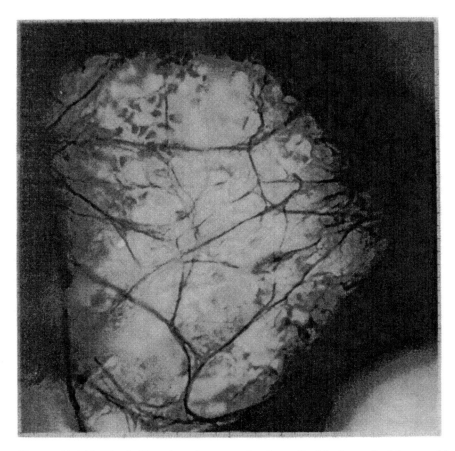

FIGURE 11.6:10 Elastin fibers in pulmonary alveolar wall of the lung of a 24 years old male inflated to a transpulmonary pressure of 10 cm H_2O. OsO_4 fixed. Orcein stained. From Sobin, Fung, and Tremer (1988). Reproduced by permission.

The data show several important features:

a) Although the fibers may show some predominant orientation in an individual septum, over all the septa examined, no predominant orientation can be identified relative to the alveolar mouth or septal edges.

b) Wider fibers are straighter, but the trend is not strong.

c) The mean value of the curvature of collagen fibers in the interalveolar septa is larger than that of the elastin fibers; but they are of the same order of magnitude. For example, in young lungs inflated at a transpulmonary pressure of 4 cm H_2O, the curvatures of collagen and elastin fibers in the septa are, respectively, 0.052 ± 0.048 (SD) vs 0.031 ± 0.030 (SD) μm^{-1}. The large standard deviations are reflections of the skewness of the distribution. The data

TABLE 11.6:1 Collagen and elastin fiber width in pulmonary interalveolar septa

Age (yrs)		Young (15–35)		Middle (36–45)		Old (65 +)	
Pressure TPP cm H_2O		4	14	4	14	4	14
Collagen							
Width D_c	Mean	0.966	0.973	0.982	1.044	1.166	1.186
μm	S.D.	0.481	0.490	0.522	0.501	0.590	0.631
$D_c^{1/2}$	Mean	0.952	0.955	0.958	0.994	1.045	1.053
$μm^{1/2}$	S.D.	0.242	0.246	0.255	0.237	0.270	0.279
Elastin							
Width D_e	Mean	0.973	0.969	1.037	1.045	1.270	1.242
μm	S.D.	0.472	0.477	0.531	0.522	0.631	0.632
$D_e^{1/2}$	Mean	0.957	0.956	0.970	0.988	1.093	1.079
$μm^{1/2}$	S.D.	0.239	0.237	0.213	0.263	0.274	0.281

Note: TPP = transpulmonary pressure = airway p − pleural pressure.

Data from Sobin, Fung, and Tremer (1988).

TABLE 11.6:2 Collagen and elastin fiber curvature in human pulmonary interalveolar septa

Age (yrs)		Young (15–35)		Middle (36–45)		Old (65 +)	
Pressure TPP cm H_2O		4	14	4	14	4	14
Collagen							
Curv	Mean	0.052	0.035	0.040	0.029	0.034	0.029
$μm^{-1}$	S.D.	0.048	0.031	0.033	0.027	0.033	0.028
$(Curv)^{1/3}$	Mean	0.349	0.305	0.319	0.286	0.297	0.280
$μm^{-1/3}$	S.D.	0.094	0.087	0.087	0.078	0.092	0.088
Elastin							
Curv	Mean	0.031	0.029	0.031	0.027	0.026	0.024
$μm^{-1}$	S.D.	0.030	0.028	0.030	0.028	0.025	0.024
$(Curv)^{1/3}$	Mean	0.288	0.285	0.288	0.270	0.273	0.267
$μm^{-1/3}$	S.D.	0.088	0.081	0.086	0.075	0.080	0.079

Note: TPP = transpulmonary pressure = airway pressure − pleural pressure.

Data from Sobin, Fung, and Tremer (1988).

TABLE 11.6.3 The width, D, in μm, and the fourth root of the width, $D^{1/4}$, in $(μm)^{1/4}$, of fibers in alveolar mouths*

Width D or $D^{1/4}$	Fiber	Lung specimen	Transpul. pressure cm H_2O	Thickness of of specimen μm	n	Mean ± S.D. for $\bar{D}$, μm for $D^{1/4}$, $μm^{1/4}$
D	Collegen	CPWA	4	150	3095	4.98 ± 2.24
	Collagen	CRQA	14	160	1405	6.37 ± 2.98
	Elastin	CGH1	10	85	403	5.72 ± 2.74
	Elastin	CIV1	4	96	466	6.19 ± 2.23
$D^{1/4}$	Collagen	CPWA	4	150	3095	1.467 ± 0.165
	Collagen	CRQA	14	160	1405	1.557 ± 0.182
	Elastin	CEH1	10	85	403	1.512 ± 0.170

*n = number of measurements.
Data from Matsuda, Fung, and Sobin (1987).

of $(curvature)^{1/3}$, $(curvature)^{1/4}$ in Table 11.6:1 show a much smaller relative SD because the distribution of these transformed variables is nearly normal.

d) The mean value of the width of collagen fibers in interalveolar septa is also of the same order of magnitude as that of the elastin fibers. In young lungs at 4 cm H_2O transpulmonary pressure, the widths of collagen and elastin fibers are 0.966 ± 0.481 (SD) and 0.973 ± 0.472 (SD) μm respectively.

e) In alveolar mouths collagen fibers run parallel, but separately, with elastin fibers. The widths of the collagen and elastin fibers in alveolar mouths vary considerably from one individual to another. For example, the widths of collagen fibers in alveolar mouths of two lungs are 4.98 ± 2.24 (SD) and 6.37 ± 2.98 (SD) μm. The smallest and largest widths of elastin fibers in nine lungs are 4.69 ± 2.22 (SD) and 7.11 ± 2.93 (SD) μm.

f) The ratio of the total volume of collagen fibers in alveolar mouths to the total volume of collagen fibers in interalveolar septa is 0.312 for young lungs, 0.460 for middle aged, 0.307 for old lungs. The corresponding ratios for elastin are 0.478, 0.422, 0.299, respectively.

g) Data from other references are discussed thoroughly in Sobin et al. (1988).

These data provide a set of fundamental information, which has to be combined with the rheological data of collagen and elastin, and information on the contributions of ground substances and cells in order to evaluate the elasticity of the interalveolar septa and alveolar mouths. Then, with the model of alveolar ducts and alveoli, we can go on to evaluate the elasticity of lung parenchyma.

Derivation of a Very Simplified Constitutive Equation

Instead of proceeding rigorously from the parenchymal model and fibrous structure outlined above, we shall make an ad hoc hypothesis and derive a

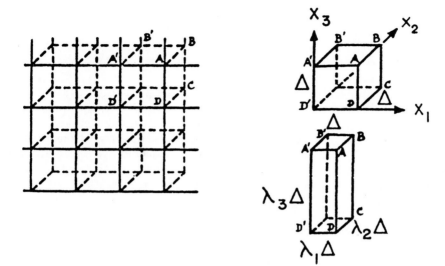

FIGURE 11.6:11 Deformation of an idealized cubic model of the alveolar structure. From Fung (1975b). Reproduced by permission.

simplified constitutive equation of the lung parenchyma.* We shall use a very crude model, representing the lung parenchyma by a cubic lattice (Fung, 1975), Fig. 11.6:11. In this model all alveoli are equal, cubic in zero-stress state, parallelopiped in deformed state. The idealized structure consists of three orthogonal sets of parallel membranes.

Let x_1, x_2, and x_3 be a rectangular Cartesian frame of reference with coordinate planes parallel to the interalveolar septa in the zero-stress state, and assume that x_1, x_2, and x_3 are the directions of principal axes of macroscopic strain. Let λ_1, λ_2, and λ_3 be the principal stretch ratios. Let Δ be the distance between the interalveolar septa in the reference state, and let 2γ, (N/m), be the surface tension (two surfaces to each membrane), and $N^{(e)}$, (N/m), be the elastic tension (stress resultant). To be specific, the tension in x_1 direction in a membrane whose normal is parallel to x_2 is denoted by $2\gamma_{21} + N_{21}^{(e)}$. Now, consider a unit square perpendicular to x_1. This section cuts $1/(\lambda_2 \Delta)$ membranes whose normals point in the x_2 direction, and $1/(\lambda_3 \Delta)$ membranes whose normals point in the x_3 direction. The force resultant in the direction of x_1 is the macroscopic stress σ_{11}. It is, therefore,

$$\sigma_{11} = [2\gamma_{21} + N_{21}^{(e)}](\lambda_2\Delta) + [2\gamma_{31} + N_{31}^{(e)}](\lambda_3\Delta) - p_A(1 - hL). \quad (3)$$

Here the first term represents the resultant of tension from all membranes whose normals point in the x_2 direction, the second term represents that from

* See Sec. 11.9 for a derivation based on the model presented in Figs. 11.6:5–11.6:7 and the rheological data of Tables 11.6:1–11.6:3 and Fung (1981).

membranes with normals in the x_3 direction, the last term is the alveolar gas pressure which acts on an area which is somewhat smaller than the cross section by an amount equal to the product of the thickness of the interalveolar septa, h, and the length of the intercepts of the interalveolar septa per unit cross-sectional area of the lung, L. Other principal stresses, σ_{22}, σ_{33}, can be obtained from Eq. (3) by cyclic permutation of the subscripts 1, 2, and 3.

Now if we can express γ and $N^{(e)}$ in terms of λ_1, λ_2, λ_3, we would have obtained the desired stress–strain relationship. The surface tension will be considered in Sec. 11.7 infra. The elastic tension can be derived from a strain energy function as discussed in Sec. 10.7. With the membranes being stressed in a state of plane stress, we can define a two-dimensional pseudo-strain–energy function for a membrane (Fung, 1981, p. 249)

$$W^{(2)} = \tfrac{1}{2} C \exp(a_1 E_1^2 + a_2 E_2^2 + 2a_4 E_1 E_2) \tag{4}$$

in which C, a_1, a_2, a_4 are constants, and E_1, E_2 are principal strains. Then

$$N_{12}^{(e)} = \frac{\lambda_1}{\lambda_2} \frac{\partial}{\partial E_1} W^{(2)} \tag{5}$$

(see Fung, 1981, pp. 246, 248, 276). The stress resultant $N_{13}^{(e)}$ can be obtained by cyclic permutation of subscripts. Substitution of (4) and (5) into (3) yields the stress-strain relationship. Expressed in terms of a pseudo-strain–energy density function of the lung parenchyma, $\rho_0 W$ per unit volume, we have

$$\rho_0 W = \frac{\rho_0}{\rho} \left\{ \frac{1}{2} \frac{C}{\Delta} \exp(a_1 E_1^2 + a_2 E_2^2 + 2a_4 E_1 E_2) \right.$$

$$+ \text{ symmetric terms by permutation of } E_1, E_2, E_3$$

$$\left. + \text{ interfacial energy} \right\} \tag{6}$$

where C, a_1, a_2, a_4, are material constants, ρ is the density of the parenchyma in deformed state, ρ_0 is that in zero-stress state, and Δ is the alveolar dimension at zero-stress state. This strain energy is called pseudo-strain–energy because loading and unloading require different sets of material constants to account for hysteresis.

These equations have been compared with experimental results by Vawter, Fung and West (1978, 1979) who tested biaxial stretching of saline-filled lung parenchyma. The elastic part of Eq. (6) compares very well with experimental results. Figure 11.6:12 shows that Eq. (6), which is derived for triaxial loading, can be used to fit biaxial loading experimental data. The three material constants, C, $a_1 = a_2$, a_4 identified for one biaxial test can be used to predict the outcome of other biaxial tests; but cannot be used to predict uniaxial test results. Reliable triaxial test data do not exist. But Eq. (6) was also used to compare with the triaxial data given by Lee and Frankus (1975), and it is shown by Fung et al. (1978) that the expression with three constants (by setting $a_1 = a_2$) fits their data as well as their polynomial with nine constants.

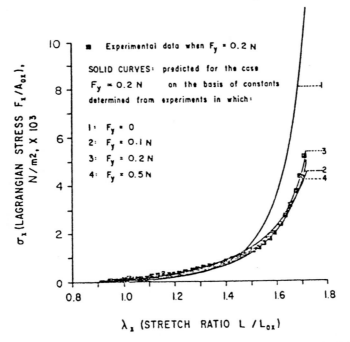

FIGURE 11.6:12 Comparison of the predicted curves for the case of a biaxial loading with a lateral load of 0.2 N with experimental data of that case (discrete squares). Curves 1, 2, 3, 4 are plots of results computed from Eq. (6) with the constants C, a_1, a_2, a_4 identified from experiments with lateral load equal to 0, 0.1, 0.2, and 0.4 N respectively. Note that curves 2, 3, 4 of the biaxial loading cases fit well; but curve 1 of the uniaxial loading case does not. From Vawter et al. (1978). Reproduced by permission.

Fung et al. (1978) derived a constitutive equation similar to the above under a different assumption. They introduced the concept of *ensemble average* of alveoli. The ensemble average alveolus is called a *mean alveolus*. They assume that the mean alveolus is a sphere in the zero-stress state, and that the mechanical properties of all the alveolar septa are the same and isotropic in their own planes. This assumption is said to be the *initially isotropic* assumption.

When the lung is deformed, the initial unit sphere is deformed into an ellipsoid with principal axes equal to the stretch ratios λ_1, λ_2, λ_3, which is the ensemble average of the deformed alveoli. A detailed calculation leads to the following strain–energy function of the lung tissue:

$$\rho_0 W^{(e)} = \frac{\rho_0}{\rho} \frac{C}{\Delta} \exp[\alpha(E_{11}^2 + E_{22}^2 + E_{33}^2) + (\beta + 2\alpha)$$

$$\times (E_{11}E_{22} + E_{22}E_{33} + E_{33}E_{11}) - \beta(E_{12}^2 + E_{23}^2 + E_{31}^2)]. \quad (7)$$

where α, β, C are constants and ρ_0, ρ, Δ have the same meaning as in Eq. (6). The last three terms in this equation show that the parameter β should be negative valued, otherwise the shear stress will not have the same sign as the shear strain because the shear stress is

$$S_{12} = \frac{\partial}{\partial E_{12}}(\rho_0 W^{(e)}) = -\beta \frac{\rho_0}{\rho} \frac{2C}{\Delta} E_{12} \exp[\quad]. \tag{8}$$

Since a positive E_{12} should correspond to a positive S_{12}, and since C is always positive, β must be negative.

Equation (7) has been used by Fung et al. (1978) to reduce the experimental data obtained by Lee and Frankus (1975), Hoppin et al. (1975), Vawter et al. (1978). It can fit any given set of data very well by proper choice of the constants.

Zeng et al. (1987) further discussed how to use the theoretical formula to fit an entire set of experimental data by a "global" method. They published a much more extensive set of experimental data. There is no doubt that the formulas (6) and (7), simple as they are, contain sufficient flexibility to accommodate the major features of the stress-strain relationship of the lung. The major weaknesses of these formulas are: (1) The cubic alveoli model is an oversimplification; hence the physical meaning of the constants C, a_1, a_2, a_4 is not clear, and cannot be related to the microstructure and microrheology of the materials in detail. (2) There is a hidden hypothesis that the microstrain is the same as the macrostrain. The alveolar mouths or ducts are not given an additional degree of freedom. (3) Experimental data used for checking are derived from biaxial loading of liquid-filled lung parenchyma. The surface tension part needs a more refined treatment, see Sec. 11.7. (4) There is a lack of a good method to perform triaxial loading experiments on lung parenchyma. The conventional pressure-volume test of the whole isolated lung does not yield precise information on stress-strain relationship because one has to evaluate the effect of pleura and to assume a state of uniform isotropic strain in the parenchyma when it is not. Hajji et al. (1979) have estimated that the pleura contributes as much as 25% to the bulk modulus of the lung. The biaxial testing method (Fung, 1981, p. 243; Vawter et al., 1978) solves the problem of edge condition, but the alveolar configuration in biaxial loading is different from that in normal lung. Hence, the elastic constants determined may not apply to normal lung. For these reasons, further improvements are necessary. In Secs. 11.8 and 11.9, we shall discuss the derivation of constitutive equation of the lung more thoroughly.

Boundary-Value Problems

Some problems can be solved by the equations of equilibrium (Eq. 1.7:1) alone:

$$\frac{\partial \sigma_{i1}}{\partial x_1} + \frac{\partial \sigma_{i2}}{\partial x_2} + \frac{\partial \sigma_{i3}}{\partial x_3} + X_i = 0, \qquad i = 1, 2, 3 \tag{9}$$

where σ_{ij} are the stress components in the parenchyma, X_i is the body force per unit volume, and x_1, x_2, x_3 are rectangular cartesian coordinates. For example, when $X_i = 0$, a solution of (9) is:

$$\sigma_{11} = \sigma_{22} = \sigma_{33} = \text{a constant, } \sigma.$$
$$\sigma_{ij} = 0 \qquad \text{if } i \neq j. \tag{10}$$

This represents a uniformly inflated weightless lung parenchyma. It can be an exact solution if the boundary condition on the pleura is satisfied. As seen from the sketch in Fig. 11.6:13, the boundary condition is

$$\sigma = (1 - hL)p_A - p_{PL} - N_{PL}^{(1)}\kappa_{PL}^{(1)} - N_{PL}^{(2)}\kappa_{PL}^{(2)}, \tag{11}$$

where σ is the tissue stress, p_A is the alveolar pressure, p_{PL} is the pressure acting on the pleura, $N_{PL}^{(1)}$, $N_{PL}^{(2)}$ are the principal membrane tensions in the pleura,

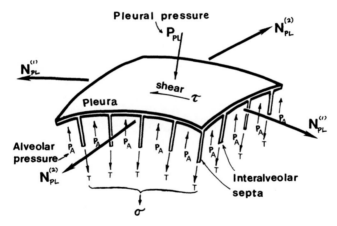

$\sigma' =$ normal component of the resultant
force per unit area of the tension
in interalveolar septa

FIGURE 11.6:13 The boundary condition on the pulmonary pleura. On the side of the pleural cavity, there acts the intrapleural pressure P_{PL} and the shear stress τ due to the motion of the intrapleural fluid. On the side of the alveoli, there acts the alveolar gas pressure P_A, and the surface traction $\overset{v}{T}_i = \sigma_{ij}\nu_j$ due to stress σ_{ij} in the lung parenchyma (see Sec. 1.7), ν_i being the normal vector of the pleura. The normal component of the surface traction is $\sigma = \overset{v}{T}_i \nu_i$. In the plane of the pleura, there acts principal stress resultants $N_{PL}^{(1)}$ and $N_{PL}^{(2)}$ in two orthogonal principal directions. The pleura has principal curvatures $\kappa_{PL}^{(1)}$ and $\kappa_{PL}^{(2)}$. We assume that the directions of the principal axes of pleural curvature coincide with those of pleural membrane stress, and place the local frame of reference with origin on the pleura and x_1, x_2 axes pointing in the direction of these principal axes. The x_3 axis is normal to the pleura. The equation of equilibrium of forces in the direction normal to the pleura is Eq. (11) or (13) in the case illustrated.

$\kappa_{PL}^{(1)}$, $\kappa_{PL}^{(2)}$ are the principal curvatures of the pleura, $(1 - hL)$ is the fraction of the pleura surface that is exposed to alveolar gas as explained in Eq. (3); the rest being occupied by the interalveolar septa.

Another solution of Eq. (9) when $X_1 = \rho g$ is

$$\sigma_{11} = \sigma_{22} = \sigma_{33} = c_h - \rho g x_1,$$
$$\sigma_{ij} = 0 \qquad \text{if } i \neq j, \tag{12}$$

where c_h is a constant, and g is the gravitational acceleration. The solution is exact if the shear on the pleura is zero, and the normal stress on the pleura is

$$c_h - \rho g x_1 = (1 - hL)p_A - p_{PL} - N_{PL}^{(1)}\kappa_{PL}^{(1)} - N_{PL}^{(2)}\kappa_{PL}^{(2)}. \tag{13}$$

Eq. (13) defines the pleural pressure that must exist for the solution given by Eq. (12) to be correct.

For a lung in the chest, the boundary conditions are imposed by the chest wall and diaphragm, and the motion of the lung relative to the chest wall. Equations (11) and (13) are not realizable. Hence Eqs. (10) and (12) are only mathematical ideals.

The boundary conditions on the visceral pleura of the lung are 1), the normal component of pleural displacement is confined by the chest wall, and 2), the pleura is subjected to the shear stress and pressure imposed by the intrapleural fluid which flows in a very narrow space. The normal amount of intrapleural fluid between the visceral and parietal pleura is small; the gap between the pleura is only a few μm. But the intrapleural fluid is viscous. The flow of a viscous fluid in the very narrow gap caused by gravity and relative motion of the lung and the chest wall can generate significant shear stress and pressure gradient. To determine the boundary condition of the lung, the problem of flow in the pleural space must be solved. Hence, a study of the mechanics of the lung in the chest must solve three problems simultaneously: the lung, the intrapleural flow, and the chest wall motion.

11.7 Surface Tension at the Interface of the Alveolar Gas and Interalveolar Septa

Energy per unit area of an interface between two materials or two phases of a material can be revealed as surface tension, which has physical units of force per unit length (e.g. N/m). Figure 11.7:1 shows a surface S containing an arbitrary line element of length L. In a "free-body" diagram of a small element of surface containing L, T represents the tension per unit length acting on the line element. In a length L, the total force is TL. If the line is displaced to the right by a distance dx, the tensile force will do work equal to $TL\,dx$. The surface area is increased by $L\,dx$. If W represents the energy per unit area of the membrane, the total increase in energy is $WL\,dx$. This increase in surface energy is equal to the work done. Hence $TL\,dx = WL\,dx$ or $T = W$, i.e. the surface tension per unit length is equal to the surface energy per unit area.

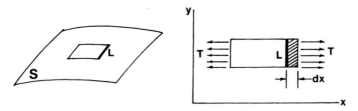

FIGURE 11.7:1 Illustration of the equivalence of surface tension and energy per unit area of interface.

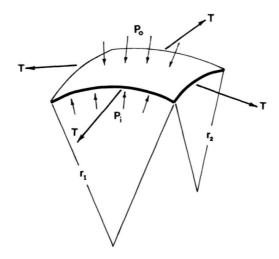

FIGURE 11.7:2 The condition of equilibrium of a membrane subjected to internal and external pressures and tensile stress in the membrane, leading to the Laplace's equation (1).

Surface tension depends, of course, on the media on both sides of the interface. Examples are interfaces between oil and water, water and air.

If the surface is curved, Fig. 11.7:2, the effect of surface tension on the balance of pressures acting on the two sides of the surface is given by Laplace's equation (see any elementary book on continuum mechanics, e.g., Fung, 1977):

$$T\left(\frac{1}{r_1} + \frac{1}{r_2}\right) = p_i - p_o, \tag{1}$$

where r_1, r_2 are the principal radii of curvature of the surface, and p_i and p_o are the internal and external pressures, respectively. If r_1 and r_2 are small, $p_i - p_o$ can be large. For example, the surface tension between pure water and air being 7.8×10^{-4} N/cm, if $r_1 = r_2 = 1$ μm, the equilibrium pressure difference is 15.6 N/cm^2 (1.54 atm).

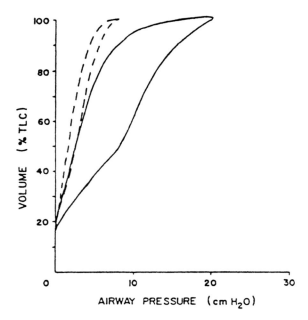

FIGURE 11.7:3 Pressure-volume curves of a cat's lung. (a) Solid curves: lung inflated by air. p = airway pressure-pleural pressure. (b) Dotted curves: lung inflated by filling the airway with saline. The liquid-air interface was thus eliminated. From Bachofen et al. (1979). Reproduced by permission.

Because the interalveolar septa are moist and are exposed to alveolar gas, surface tension is significant. Its effect can be seen in Fig. 11.7:3 which shows two pressure–volume relationships of a dog lung, one inflated by air, the other inflated by filling the airway with saline. Filling the lung with saline eliminates the gas-liquid interface, decreases the surface tension, and dramatically alters the pressure–volume relationship.

When a newborn baby takes the first breath to inflate its lungs, it has to create the needed interface. If the surface tension is too high, inflation can be difficult (see Sec. 11.12). Hence the fetus must reduce the surface tension of the amniotic fluid to an acceptable level. The fetus accomplishes this in the last few weeks of pregnancy, see Fig. 11.7:4. In this period, certain cells on the alveolar wall begin to secrete *surfactants* such as lecithin, sphingomyelin, etc. which reduce surface tension. Birth will be safer if a sufficient concentration of surfactants exist in the amniotic fluid. Gluck et al. (1971) first based clinical decisions on whether a baby is ready to come into this world or not by the lecithin/sphingomyelin ratio (Fig. 11.7:4) of the amniotic fluid, which is easily measured by chromatography. Since then childbirth has become much safer.

Methods of measuring surface tension of a fluid-gas interface are described by Shaw (1970) and others. Clements and Tierney (1965) used a Wilhelmy-

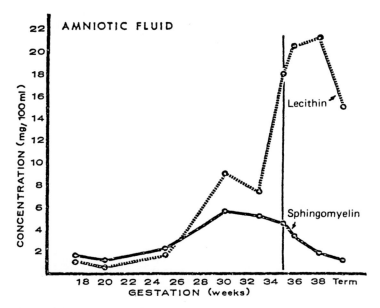

FIGURE 11.7:4 Mean concentrations in amniotic fluid of sphingomyelin and decithin during gestation. The acute rise in lecithin at 35 weeks marks pulmonary maturity. From Gluck et al. (1971). Reproduced by permission.

Langmuir trough to measure the surface tension of pulmonary surfactants. A surfactant is spread on the surface of water in the trough in a layer of about one molecule thick. The interface is then compressed or extended by a sliding bar while all the surfactant is confined in the surface and the surface tension is measured. Figure 11.7:5 shows a typical result. It is seen that the surface tension varies with the surface area. Increasing area decreases surfactant concentration and increases surface tension. In a cyclic variation of area there is a large hysteresis loop, because as the area changes the configurations of the surfactant molecules are changed, stretched, tumbled, overlapped, piled up, etc. in an irreversible way, which affects their effectiveness to reduce surface tension.

Reifenrath and Zimmermann (1973) collected fluid directly from the alveolar walls of a dog with a microneedle, injected the fluid into a small cubicle of water, then measured the surface tension of a small gas bubble formed at the tip of a pipette. They showed that the hysteresis loop of the surface tension-area relationship can be stabilized only after hundreds of cycles. This method is discussed by Enhorning (1977).

Schürch, Goerke, and Clements (1976) attempted to measure the surface tension of an interface on an interalveolar septum directly by depositing a metal microsphere on a septum in vivo and observing the change of curvature of the septum under the weight of the sphere. Flicker and Lee (1974) measured

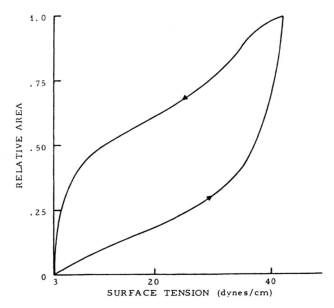

FIGURE 11.7:5 Surface area-surface tension relationship of lung extract from an experiment using Wilhelmy trough. From Clements and Tierney (1965). Reproduced by permission.

the change of dimensions of the pulmonary alveoli just beneath the pleural surface of a dog's lung, combined the data with the pressure–volume curves of air-filled lungs, and computed the surface tension-area relationship under the hypothesis that the alveoli far from the surface behave the same way as the subpleural alveoli.

To describe the surface tension-area relationship in cyclic changes of area mathematically, Fung et al. (1978) wrote γ_{min}, γ, γ_{max} for the surface tension at the minimum area (A_{min}), intermediate area (A), and maximum area (A_{max}) of the interface, respectively, after the hysteresis loop is stabilized by repetition of periodic changes of area. Define a dimensionless variable ξ:

$$\xi = (A - A_{min})/(A_{max} - A_{min}).$$ (2)

Then the curves as shown in Fig. 11.7:5 can be represented by a straight line plus a Fourier sine series:

$$\gamma = \gamma_{min} + (\gamma_{max} - \gamma_{min})\left(\xi + \sum_{n=1}^{\infty} c_n \sin n\pi\xi\right).$$ (3)

One set of the coefficients c_n applies when the area is increasing, another set applies when the area is decreasing. Zupkas (1977) has evaluated the coefficient c_n from published results of surfactants tested in Wilhelmy-Langmuir trough type of experiment. Some of his results are listed in Table 11.7:1.

TABLE 11.7:1 Fourier coefficients of the surface tension-surface area relationship given by Eq. (3). Data by Clements (1962). Computation by Zupkas (1977)

Harmonic	Inflation		Deflation	
	$C_n \times 10^3$	C_n/C_1	$C_n \times 10^3$	$C_n/|C_1|$
1	366	1.000	−185	−1.000
2	−174	−0.475	−133	−0.719
3	94.1	0.257	−72.6	−0.398
4	−61.0	−0.167	−46.4	−0.251
5	50.0	0.139	−40.3	−0.218
6	−32.6	−0.089	−24.4	−0.132
7	27.1	0.074	−21.3	−0.115
8	−22.2	−0.061	−14.3	−0.077
9	18.9	0.052	−15.7	−0.085
10	−15.6	−0.043	−10.6	−0.057
11	15.2	0.042	−9.42	−0.051
12	−12.1	−0.040	−5.90	−0.032
13	13.8	0.038	−6.27	−0.034
14	−9.73	−0.027	−4.31	−0.023
15	11.3	0.031	−4.72	−0.026
16	−10.8	−0.030	−2.16	−0.012

On the other hand, Flicker and Lee (1974) expressed their results in the form

$$\gamma = c_1(A/A_{min})^{c_2} \tag{4}$$

whereas Vawter and Shields (1982) wrote

$$\gamma = c_1'[1 - c_2' \exp(-c_3' A/A_{min})]. \tag{5}$$

Some typical values are: $c_1/D_A = 0.1966$ cm H_2O, D_A = alveolar diameter = 60.1 μm for the dog lung, $c_2 = 2.71$; $c_1'/D_A = 6.78$ cm H_2O, $c_2' = 4.83$, $c_3' = 2.35$.

In 1982, Wilson introduced a method to evaluate the surface tension on the alveolar walls of a lung. Their method is based on energy consideration. They assume that the total energy of an air-filled lung, E, is the sum of the elastic energy stored in the tissue and the surface energy. The tissue energy is the sum of two parts, the energy of a saline-filled lung (in which the surface tension can be ignored), U_s, and the additional strain energy associated with the distortion of the alveolar wall caused by the imposition of the surface tension when the lung became air-filled, $\Delta U(V, S)$. S represents the total area of the alveolar walls. V represents the lung volume. It is assumed that U_s is a function of V alone, and ΔU is a function of two independent variables V and S. The surface tension γ is a function of S. Thus

$$E = U_s(V) + \Delta U(V, S) + \int_0^S \gamma dS. \tag{6}$$

At a fixed lung volume, the equilibrium state of the lung structure minimizes the total energy. Hence the partial derivative of E with respect to S must be zero at the equilibrium state:

$$\frac{\partial \Delta U}{\partial S} + \gamma = 0. \tag{7}$$

If the lung volume is increased by dV, the work done by the transpulmonary pressure $p_A - p_{PL}$ is $(p_A - p_{PL})dV$, whereas the energy is increased by dE. These must be equal. Hence

$$p_A - p_{PL} = \frac{dE}{dV} = \frac{dU_s}{dV} + \frac{\partial \Delta U}{\partial V} + \frac{\partial \Delta U}{\partial S}\frac{dS}{dV} + \gamma\frac{dS}{dV}. \tag{8}$$

The sum of the last two terms vanishes on account of Eq. (7). The first term on the right-hand side, dU_s/dV, can be identified as the recoil pressure of the saline-filled lung P_s. Therefore, Eq. (8) is reduced to the following form:

$$p_A - p_{PL} - p_S = \frac{\partial \Delta U}{\partial V}. \tag{9}$$

Differentiating Eq. (7) with respect to V and Eq. (9) with respect to S and eliminating $\partial^2 \Delta U/\partial V\,\partial S$, one obtains

$$\frac{\partial \gamma}{\partial V} = -\frac{\partial(p_A - p_{PL} - p_s)}{\partial S}. \tag{10}$$

Then an integration yields Wilson's equation:

$$\gamma = -\int_{V_s}^{V} \frac{\partial(p_A - p_{PL} - p_s)}{\partial S}\,dV, \tag{11}$$

where V_s is the volume of the saline-filled lung at which the surface tension vanishes. The integrand in Eq. (11) is a function of the alveolar area A and lung volume V; and can be determined experimentally. Hence Eq. (11) can be used to calculate $\gamma(V, S)$.

Bachofen et al. (1979) and Gil et al. (1979) fixed rabbit lungs by perfusing fixatives through pulmonary blood vessels, then prepared histological slides from which the surface area of the alveolar walls was measured by stereological methods. They obtained the surface area of alveolar walls in air-filled, saline-filled, and detergent-rinsed and then air-filled rabbit lungs inflated to various percentages of total lung capacity. From these data, Wilson and Bachofen (1982) derived $p_A - p_{PL} - p_s$ and V_s as functions of V and S and calculated γ as a function of S. Their results are shown in Fig. 11.7:6. It is seen that the calculated values of surface tension decrease to less than 2 dyn/cm as surface area decreases along the deflation limb of the pressure–volume curve. Surface tension increases very steeply with surface area on the inflation limbs, reaching a limiting value of just under 30 dynes/cm.

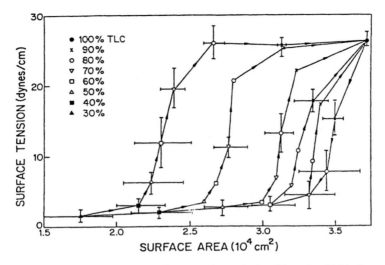

FIGURE 11.7:6 The surface area-surface tension relationship of rabbit's lung as a function of lung size. TLC = total lung capacity. From Wilson (1982). Reproduced by permission.

11.8 Small Perturbations Superposed on Large Deformation

Limiting considerations to small perturbations superposed on a known large deformation of a body is one of the ways to linearize the nonlinear equations of biomechanics. For example, we may superimpose a small perturbation on a uniformly inflated lung, so that the transformation law for the location of a particle a_i in the zero-stress state to the location of the same particle in the deformed state x_i is

$$x_i = \lambda_0 a_i + u_i, \qquad (i = 1, 2, 3). \tag{1}$$

Here a_i and x_i are referred to the same rectangular cartesian frame of reference, λ_0 is a constant and u_i is a function of a_1, a_2, a_3. If u_i is so small that the derivatives $|\partial u_i/\partial a_j|$ are infinitesimals of the first order, then on substituting Eq. (1) into the definition of the strain Eq. (10.2:9) and neglecting the second order terms, we obtain the green's strain

$$E_{ij} \equiv \frac{1}{2}\left(\frac{\partial x_k}{\partial a_i}\frac{\partial x_k}{\partial a_j} - \delta_{ij}\right)$$
$$= \frac{(\lambda_0^2 - 1)}{2}\delta_{ij} + \lambda_0 \frac{1}{2}\left(\frac{\partial u_i}{\partial a_j} + \frac{\partial u_j}{\partial a_i}\right). \tag{2}$$

On defining the Cauchy infinitesimal strain e'_{ij} in the usual way,

$$e'_{ij} \equiv \frac{1}{2}\left(\frac{\partial u_i}{\partial x_j} + \frac{\partial u_j}{\partial x_i}\right) \tag{3}$$

and noting that $\partial u_i / \partial a_j$ is equal to $(\partial u_i / \partial x_k)(\partial x_k / \partial a_j) = \lambda_0(\partial u_i / \partial x_k)$ according to Eq. (1) and to the infinitesimals of the first order, we can write Eq. (2) as

$$E_{ij} = \frac{1}{2}(\lambda_0^2 - 1)\delta_{ij} + \lambda_0^2 e'_{ij}. \tag{4}$$

If the stress σ_{ij} is an analytic function of E_{ij}, then on substituting Eq. (4) into the analytic function we can expand it into a power series of e'_{ij}, and on keeping only linear terms of e'_{ij}, σ_{ij} can be written in the form

$$\sigma_{ij} = \sigma_{ij}^{(0)} + \sigma'_{ij}, \tag{5}$$

where $\sigma_{ij}^{(0)}$ depends on λ_0, but is independent of u_i, and σ'_{ij} is a linear function of e'_{ij}. The terms e'_{ij}, σ'_{ij} are called the *incremental strains* and *incremental stresses*, respectively.

If the relationship between σ'_{ij} and e'_{ij} is the same irrespective of how the frame of reference is oriented, then the relationship is said to be *isotropic*. It is important to note that isotropy in stress-strain relationship does not require that the microstructure have a spherical symmetry. Only the stress-strain relationship needs to be. If a stress-strain relationship is derived from a microstructure presented in Secs. 11.6 and 11.9, and it is shown that the resulting equation is the same in three different orientations of the frame of reference, then by tensor transformation rules it can be shown that the equation will be the same under arbitrary rotation of coordinates, and the isotropy is established.

An isotropic nonlinear constitutive relation will lead to an isotropic incremental stress-strain relationship in the case of small perturbations of a uniformly inflated lung, because in this case the stress $\sigma_{ij}^{(0)}$ is isotropic. If the large deformation is anisotropic, then $\sigma_{ij}^{(0)}$ is anisotropic and the incremental stress-strain relationship will be anisotropic.

A basic theorem in continuum mechanics (see, e.g., Fung, 1977, p. 168) states that a linear isotropic relation between two symmetric tensors of the second order can be defined by two constants. For stress and strain tensors the linear isotropic relationship is the Hooke's law,

$$\sigma'_{ij} = \lambda e'_{\alpha\alpha}\delta_{ij} + 2\mu e'_{ij} \tag{6}$$

or

$$e'_{ij} = \frac{1 + v}{E}\sigma'_{ij} - \frac{v}{E}\sigma'_{\alpha\alpha}\delta_{ij}; \quad \left(E = \frac{\mu(3\lambda + 2\mu)}{\lambda + \mu}, v = \frac{\lambda}{2(\lambda + \mu)}\right), \tag{7}$$

where λ and μ are the Lamé constants, E is the Young's modulus, v is the *Poisson's ratio*. In engineering literature μ is usually denoted by G, and is called the *shear modulus*. If such a material is subjected to a uniform pressure p under which the volume v changes by an amount Δv, then the ratio of $-p$ divided by $\Delta v / v$ is the *bulk modulus*, K, which is related to other constants by the formula

$$K = \frac{E}{3(1 - 2v)} = \frac{2G(1 + v)}{3(1 - 2v)} = \lambda + \frac{2}{3}G, \tag{8a}$$

$$E = \frac{9KG}{3K + G}, \qquad G = \frac{E}{2(1 + v)}, \qquad v = \frac{3K - 2G}{2(3K + G)}. \tag{8b}$$

See Fung (1977), pp. 195, 217. These equations are applicable to the perturbation of a uniformly inflated lung. Many solutions of the classical theory of elasticity can be applied to the lung. A few examples follow:

Example 1. Cylindrical hole in an initially uniformly expanded parenchyma (*Fig.* 11.8:1). Assume isotropic incremental stress-strain relationship. In cylindrical polar coordinates, a solution that satisfies the equations of equilibrium is

$$\xi_r = \frac{A}{r}, \qquad \xi_\theta = 0, \qquad \xi_z = 0, \tag{9}$$

where ξ_r, ξ_θ, ξ_z are displacements in the radial, circumferential, and axial directions respectively; A being a constant, and r the radial coordinate. See *Foundations of Solid Mechanics* (Fung, 1965, pp. 243, 244). This solution can satisfy the boundary conditions of a specified displacement or radial stress at

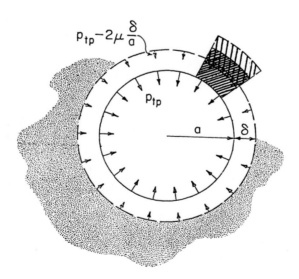

FIGURE 11.8:1 A circular cylindrical hole in a uniformly expanded lung requires a uniform tension to act on the surface of the hole in order to maintain the uniform stress field. If the hole expands the parenchyma will be distorted with no change in volume. The radial stress at the wall is reduced by 2G times the fractional change in hole radius, G being the shear modulus. This disturbance in displacement dies away at a rate proportional to r^{-1}. From Wilson (1986), reproduced by permission.

the inner wall of a cylindrical hole; and zero stress and zero displacement at $r \to \infty$. This solution can be used to estimate the interaction between the lung parenchyma and an embedded blood vessel or bronchus, see Sec. 11.10. The deformation described by Eq. (9) is such that a cross-hatched element of the parenchyma shown at the one o'clock position in Fig. 11.8:1 will deform into a shape shown by the dotted line. The element expands in the circumferential direction and shrinks in the radial direction, with no change of volume. It is a "pure shear". If the boundary condition is that $\xi_r = \delta$ at $r = a$, then the constant A is equal to $a\delta$. The stresses that accompany the deformation depend on the shear modulus of the parenchyma alone, not on the bulk modulus. The decrease in normal stress at the boundary is $2G\delta/a$. Hence when a vessel expands the perivascular pressure increases by $2G$ times the fractional change in vessel radius. Since the volume of the parenchyma does not change with this deformation, ventilation per unit mass of parenchyma is not affected by the expansion of the vessel.

Example 2. Gravitational deformation of a cylinder of parenchyma in a rigid container (Fig. 11.8:2). Consider a linear elastic solid in a rigid circular cylindrical container of radius R and height L. Assume that in the initial state the solid is subjected to a uniform isotropic tensile stress σ_0. Now impose the gravitational force in the direction of the z-axis. The material is then subjected to a body force in the z-direction, of magnitude ρg per unit volume, ρ being the density of the material and g the gravitational acceleration. Let the body be free to slide but remain in contact with the container. Then the equations

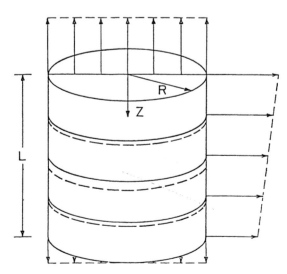

FIGURE 11.8:2 Distortion of a linearly elastic solid in a rigid circular cylindrical container due to gravitational force acting in the axial direction. The distortion produces local changes in volume and shape. From Wilson (1986), reproduced by permission.

of equilibrium and boundary conditions are satisfied by a solution in which the displacements ξ_r, ξ_θ, ξ_z referred to a set of cylindrical polar coordinates r, θ, z are:

$$\xi_r = 0, \qquad \xi_\theta = 0, \qquad \xi_z = \frac{\rho g}{2} \frac{1}{K + \frac{4}{3}G} z(L - z). \tag{10}$$

The displacement is zero at the top and bottom as required by the boundary conditions and is maximum at midheight. Horizontal planes shown by the solid circles in Fig. 11.8:2 are displaced to the positions shown by the dashed lines. The top half of the solid is expanded relative to the initial state of uniform expansion, whereas the bottom half is compressed. The corresponding stress perturbation is:

$$\sigma'_{rr} = \rho g \frac{K - \frac{2}{3}G}{K + \frac{4}{3}G}\left(\frac{L}{2} - z\right), \qquad \sigma'_{\theta\theta} = 0, \qquad \sigma'_{zz} = \rho g\left(\frac{L}{2} - z\right). \tag{11}$$

Thus the normal stress in the z-direction is tensile at the top and compressive at the bottom. The radial normal stress is linear in z as shown by the dashed lines in Fig. 11.8:2. The vertical gradient of the pressure acting on the cylindrical wall (analog of the pleural pressure in the lung) is smaller than the hydrostatic gradient in a fluid with specific weight ρg by the factor $[1 - (2G/3K)]/[1 + (4G/3K)]$.

This simple example cannot model a lung, but it leads one to expect an intrapleural pressure gradient different from ρg.

Example 3. Half space loaded on the surface. A classical solution for a semi-infinite space bounded by a plane loaded by a concentrated force has been used by Hajji et al. (1979) and Lai-Fook et al. (1978) to determine the incremental modulus of a uniformly inflated lung. The classical solution is associated with the names of Boussinesq, Kelvin, Hertz, Mindlin, and others, and can be found in books on elasticity, e.g., Fung (1965).

Example 4. Incremental modulus of lung parenchyma. Experimentally determined values of the bulk modulus (K) and shear modulus (G) of a dog's lung for small perturbations of a uniformly inflated state are shown in Figs. 11.8:3 and 11.8:4, from Lai Fook et al. (1978), and Lai-Fook and Toperoff (1980). As functions of the inflation pressure and volume, we have, roughly

$$K \approx 4(p_A - p_{PL}), \qquad G \approx 0.7(p_A - p_{PL}), \quad \text{(dog)}. \tag{12}$$

If we compute the Young's modulus E and Poisson's ratio v from the formulas (8b), we obtain, from Eqs. (12),

$$E \approx 2(p_A - p_{PL}), \qquad v \approx 0.42. \tag{13}$$

Thus, the Poisson's ratio is about 0.42, not far from 0.5.

It is well known that in small deformation, a Poisson's ratio of 0.5 means that the material is incompressible. For finite deformation this conclusion is generally untrue even for a material that is isotropic and obeys Hooke's law.

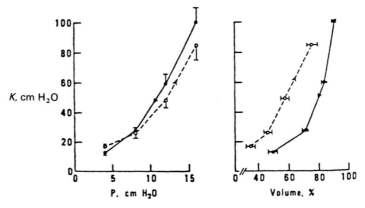

FIGURE 11.8:3 Bulk modulus of lung parenchyma of excised dog lungs obtained by small volume perturbations is shown as a function of transpulmonary pressure on the left and lung volume on the right. Inflation (open circles and dashed line) and deflation (closed circles and solid) histories are shown. Reprinted with permission from *J. Biomechanics*, **12**: 757–764, "Elastic properties of lung parenchyma: The effect of pressure-volume hysteresis on the behavior of large blood vessels," copyright © 1979, Pergamon Press plc.

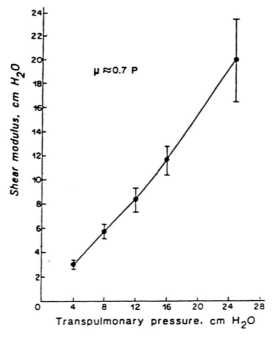

FIGURE 11.8:4 The shear modulus of excised dog lungs determined by punch indentation tests is shown as a function of transpulmonary pressure. From Lai-Fook (1979a). Reproduced by permission.

Vawter (1983) demonstrated this as follows: The condition of incompressibility, expressed in terms of the principal stresses $\sigma_1, \sigma_2, \sigma_3$ for a material obeying Eq. (7), is

$$\left\{1 + \frac{2}{E}[\sigma_1 - v(\sigma_2 + \sigma_3)]\right\}\left\{1 + \frac{2}{E}[\sigma_2 - v(\sigma_3 + \sigma_1)]\right\}$$

$$\times \left\{1 + \frac{2}{E}[\sigma_3 - v(\sigma_1 + \sigma_2)]\right\} = 1. \tag{14}$$

This equation is satisfied by $v = 0.5$ if the strains $\sigma_1/E, \sigma_2/E, \sigma_3/E$ are infinitesimal or if the stress state is isotropic ($\sigma_1 = \sigma_2 = \sigma_3$). Otherwise Eq. (14) cannot be satisfied for arbitrary loading whether $v = 0.5$ or not. Hence $v = 0.5$ does not guarantee the absence of volume change in finite strain if the stress is not isotropic.

11.9 Derivation of Constitutive Equation on the Basis of Microstructure: Connection Between Micro and Macro Mechanics

Constitutive equations are phenomenological. They are regarded as empirical by experimenters, and axiomatic by mathematicians. In biomechanics, we often try to derive them on the basis of microstructure, e.g., Eq. (11.6:6), in order to gain a better understanding, or to get some guidance to the mathematical form. Our approach is discussed more fully below.

Consider a lung. If we want to derive a constitutive equation of the whole lung, we have to solve the problem of stress and strain distribution in the alveoli and alveolar ducts, for which we need the constitutive equations of the interalveolar septa and the alveolar mouths. In order to derive the constitutive equations of the interalveolar septa and alveolar mouths, we have to solve the problem of stress and strain distribution in the collagen and elastin fibers in the interalveolar septa and mouths, for which we need the constitutive equations of the collagen fibers, elastin fibers, ground substances, and cells. Conversely, if the constitutive equations of the fibers and ground substances are known, then we can derive, in succession, the constitutive equations of the interalveolar septa and alveolar mouths, and that of the whole lung.

The hierarchy of the system is illustrated in Fig. 11.9:1. To derive the constitutive equation of the whole lung (the first column), we need to know the zero-stress state of the whole lung (the upper left-hand box). A macrostrain of the parenchyma can then be defined relative to the zero-stress state. In response to this macrostrain, the microstructures of the alveoli and alveolar ducts deform and produce stresses. The problem can be solved according to steps listed in the second column. In the process, we need to solve the sub-problem of the interalveolar septa and alveolar mouths by steps listed in the third column. The solution of the problems in the third column is then used to complete the solutions of the problems listed in the second column,

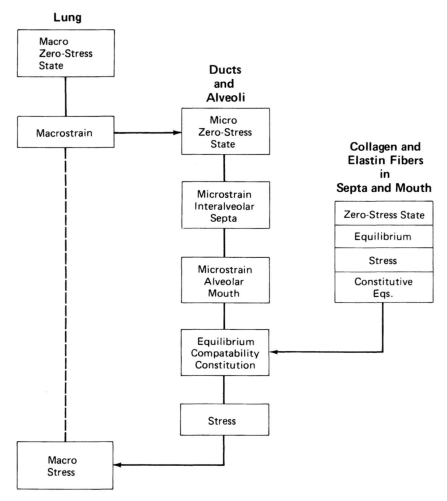

FIGURE 11.9:1 Steps involved in the derivation of a macroscopic stress-strain relationship of the lung.

which in turn gives answer to the box at the bottom of the first column. The relation between the top and bottom boxes of the first column yields the desired constitutive relationship.

Example: Incremental Elastic Moduli of Lung

We take the first order 14-hedron as the basic model of the alveoli, and the second order 14-hedron as the basic unit of the alveolar ducts. As explained in Sec. 11.6, an assemblage of order 1 and order 2 14-hedra is the model of the lung parenchyma.

Suppose that we wish to study the incremental elasticity of a uniformly inflated lung. We choose a rectangular cartesian frame of reference x, y, z with the origin located at the center of a second-order 14-hedron as shown in Fig. 11.9:2. Impose an incremental macroscopic strain described by the tensor

$$\begin{pmatrix} \varepsilon_{xx} & 0 & 0 \\ 0 & \varepsilon_{yy} & 0 \\ 0 & 0 & \varepsilon_{zz} \end{pmatrix} \tag{1}$$

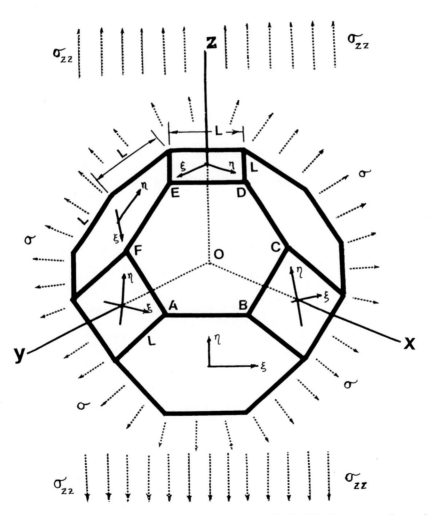

FIGURE 11.9:2 Coordinates for strain and stress analysis. The lung parenchyma is represented by a continuum of 14-hedra. The incremental loads shown here are: a uniform tensile stress σ, and a uniaxial tension σ_{zz}. The length of every edge is L at the homeostatic condition.

on this continuum of 14-hedra. Consider first the case in which $\varepsilon_{xx} = \varepsilon_{yy} = \varepsilon_{zz} = \varepsilon$. We expect the incremental macroscopic stress tensor to be

$$
\begin{pmatrix}
\sigma & 0 & 0 \\
0 & \sigma & 0 \\
0 & 0 & \sigma
\end{pmatrix}. \tag{2}
$$

If σ can be found, then $\sigma/(3\varepsilon)$ is the *bulk modulus* of the lung. To find σ, we note that since the 14-hedra fill the entire space, the condition of compatibility is satisfied if the strains specified by Eq. (1) apply to the outer shell of all order-2 14-hedra as illustrated in Figs. 11.6:5–11.6:7. However, internally, the strain in the walls of the 14-hedra may be different from Eq. (1) because the duct has the freedom to deform differently from the macroscopic average. The strain of the central duct (shown by heavy lines in Figs. 11.6:5–11.6:7) can be specified by a microscopic strain tensor

$$
\begin{pmatrix}
\varepsilon'_{xx} & 0 & 0 \\
0 & \varepsilon'_{yy} & 0 \\
0 & 0 & \varepsilon'_{zz}
\end{pmatrix}. \tag{3}
$$

By specifying tensors (1) and (3) on the outer and inner borders of the second order 14-hedra, the interalveolar septum and alveolar mouths are strained and stressed in specific ways. The mathematical problem is to determine ε'_{ij}

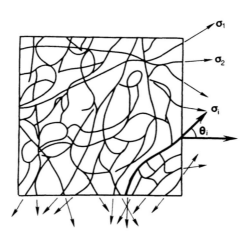

FIGURE 11.9:3 Schematic drawing of an infinitesimal element of a pulmonary alveolar wall membrane with collagen and elastin fibers embedded in it. Stresses exist in the fibers, ground substances, cell, interfaces. The forces that act on an edge perpendicular to the x-axis on the right-hand side are depicted here: $\sigma_1, \sigma_2, \ldots, \sigma_i$ acting in the fibers in directions tangential to the axes of the fibers. The normal force f_{xx} representing forces derived from the surface tension, ground substances, cells, and bending and transverse shear of the fibers; and the shear force tangential to the edge are not shown. The dimensions Δx, Δy, may be considered as unit length.

and the stresses in the interalveolar septa and mouths according to the theory of elasticity.

In the process of the solution we need the constitutive equation of the interalveolar septa. Figure 11.9:3 shows such a septum with collagen and elastin fiber bundles in them. When displacements are imposed on the borders of the septum, the fibers are stressed and strained. Analysis of the detail is a sub-problem which requires the constitutive equations of the fibers.

The principle of the analysis is thus very clear. Generalization to other types of macroscopic and microscopic strains can be done.

The Mathematical Problem

The simplest method of solution is the method of minimum potential energy. First, the strain distribution in every interalveolar septum and alveolar mouth is written according to the method of finite elements consistent with the strains given in Eq. (1) on the outer shell, and Eq. (3) on the inner shell. Second, the stress-strain laws of the interalveolar septa and the alveolar mouths are derived according to the morphometric and rheologic data of the collagen and elastin fibers and ground substances. Third, the strain energy of every interalveolar septum and every edge of the duct is derived. Fourth, the strain energy of the whole parenchyma is obtained by summing together the contributions from all elements. Adding to the strain energy the potential energy of the external load, we obtain the total potential energy. By minimizing the total potential energy with respect to the unknown variables, we obtain as many linear equations as the number of unknowns. The solution of these equations as functions of the specified boundary conditions yields the answer.

Lack of information on the zero-stress states of the collagen and elastin fiber bundles in the lung is the major hurdle at present. Quantitative determination of the constitutive equations of these fibers must be pursued. Many types of collagen have been identified in the lung parenchyma. The constitutive equation of every type needs to be identified. All fibers may not have the same zero-stress state in the interalveolar septa. Statistics on the zero-stress states of the collagen and elastin fibers are of fundamental importance.

Experiments on the zero-stress states of the collagen and elastin fibers and ground substances in the interalveolar septa can be done by puncturing slits and circular holes in the septa. If there were residual stresses in these fibers or ground substance, they will be revealed by the changes of the shape and size of the slits and circles. The holes and slits are stress indicators which can be interpreted on the basis of known solutions in the theory of elasticity.

The Influence of Tissue Growth and Change on the Homeostatic State

In Chapter 13, the relationship between stress and growth is studied. We shall see that the homeostatic state is, among other things, a result of the stress state. When the stress state changes, for example, by immobilization or vigorous exertion, the tissue may change its shape, size, composition, and zero-stress state.

In Sec. 13.10, especially in Fig. 13.10:4, a basic hypothesis is proposed which states that the rate of growth of a tissue varies with stress acting in the tissue. There exists several stress levels at which the rate of growth is zero, i.e., homeostatic. A deviation from these stresses will cause either growth or resorption. If this hypothesis is verified, then it follows that at a homeostatic condition there is a unique stress. If this rule applies to the collagen fibers in the interalveolar septa, then we may assume that all collagen fibers have the same stress. This would greatly simplify the analysis. In particular, the derivation of the incremental stress-strain law, which describes the relationship between small incremental strains and stresses superposed on a homeostatic condition (Sec. 11.8), will be simplified greatly.

Incremental Stress-Strain Relationship

Using the conclusion named at the end of the preceeding paragraph, we assume that all collagen fibers at the homeostatic state have the same (collagen) tensile stress; all elastin fibers also have a uniform (elastin) stress at the state of homeostatis. Then it can be shown that the stress-strain relationship of the interalveolar septa can be put into the following form:

$$N_{11} = (3C + B)e'_{11} + (C + B)e'_{22} + (N_{11})_h,$$
$$N_{22} = (C + B)e'_{11} + (3C + B)e'_{22} + (N_{22})_h, \tag{4}$$
$$N_{12} = 2C\, e'_{12}.$$

Here e'_{11}, e'_{12}, e'_{22} are the microstrains in each membrane relative to a set of rectangular Cartesian coordinates (ξ, η) introduced in each membrane, and N_{11}, N_{12}, N_{22} are stress resultants. $(N_{11})_h$, $(N_{22})_h$ are the stress resultants at homeostatic condition. B and C are constants. B, C are functions of the fiber statistics, collagen and elastin elasticity, and $(N_{11})_h$, $(N_{22})_h$. For fiber statistics, we use the data given in Tables 11.6:1–11.6:3 for human lung. For collagen elasticity we use the formula given in Fung (1981):

$$T = (T^* + \beta)e^{\alpha(\lambda - \lambda^*)} - \beta, \tag{5}$$
$$\beta = T^* e^{-\alpha(\lambda^* - 1)}[1 - e^{-\alpha(\lambda^* - 1)}]^{-1}. \tag{6}$$

Here T is the Lagrangian stress, λ is the stretch ratio. α and T^* are two material constants defined by Eq. (5); they are determined by fitting experiment data with Eq. (5). Physically, α is the ratio of the incremental modulus of elasticity to the stress, i.e., $(dT/d\lambda)/T$. For a linear material obeying Hooke's law, $dT/d\lambda$ is a constant. Collagen is nonlinear: its $dT/d\lambda$ is proportional to T, and the constant of proportionality is α. T^* is the *homeostatic tensile stress* at a *homeostatic stretch ratio of* λ^*. β is a constant computed from Eq. (6). Accord-

TABLE 11.9:1 The coefficients C and B defined in Eq. (4), in units of dynes/cm

C, B	Age	Transpulm. pressure cm H_2O	Hexagonal Membrane at surface tension γ (dyn/cm)		Square membrane at surface tension γ (dyn/cm)	
			10	20	10	20
C	Young	4	91.0	56.1	111.5	75.2
	Young	14	517.9	475.2	612.2	567.0
	Middle	4	106.2	70.3	129.8	92.7
	Middle	14	684.9	641.3	808.3	764.2
	Old	4	147.8	112.4	178.7	142.1
	Old	14	905.2	861.6	1066.7	1022.5
B	Young	4	1.5	0.9	1.8	1.2
	Young	14	4.5	4.1	5.3	5.0
	Middle	4	0.4	0.3	0.6	0.4
	Middle	14	1.8	1.7	2.2	2.1
	Old	4	1.0	0.7	1.2	0.9
	Old	14	4.7	4.5	5.5	5.3

ing to the assumption named at the beginning of this paragraph, the stress T^* is the same for all collagen fibers (hence the fiber force is proportional to the fiber cross-sectional area). The values of α have been determined for tendon and for arteries; their values differ a great deal. We use α of the arteries, and T^* computed from equilibrium. For elastin, we used a linear stress-strain curve (Fung, 1981). For ground substances it was found that their contribution is negligible. The computed values of B and C depend largely on the values of α and surface tension. Examples are given in Table 11.9:1.

The elasticity of alveolar mouth is computed similarly. The strains in the interalveolar septa and alveolar mouth are then computed from the assumed macroscopic strain, Eq. (1), applied to the outer border of the 2nd order 14-hedra and microscopic strain, Eq. (3), applied to the central ducts, by finite element method. The stress is calculated by Eq. (4). Two cases shown in Fig. 11.9:2 are considered: incremental expansion and incremental uniaxial tension superposed on a uniformly inflated lung. The macroscopic incremental stress is computed for each case. The bulk modulus K, Young's modulus E, and shear modulus G, and Poisson's ratio v are computed. The results are listed in Table 11.9:2.

Note that the principal axes of strain and stress coincide. In fact, the x, y, z axes in Fig. 11.9:2 are axis of material symmetry of the 14-hedra. Hence the incremental stress-strain relationship of the lung parenchyma is isotropic, i.e., it is the Hooke's law (Eq. 11.8:7).

The computed results compare reasonably well with the experimental results on postmortem human lung obtained by Yen et al. (1987). Further calculations show that the bulk modulus and the shear modulus are nearly

TABLE 11.9:2 The effect of the material constant α of collagen on the Young's modulus E, bulk modulus K, shear modulus G, and Poisson's ratio v of human lung parenchyma under the assumptions of $T^* = 6.56 \times 10^6$ dyn/cm^2 when $\lambda^* = 1.3$ for collagen, $E_h = 4 \times 10^6$ dyn/cm^2 for elastin, surface tension $= 10$ dyn/cm, morphometric data of middle aged human

Transpulmonary pressure P	α	7.5	10.0	12.5	15.0
	E (cm H$_2$O)	11.77	14.97	18.08	21.13
	E/P	2.94	3.74	4.52	5.28
4 cm H$_2$O	K/P	2.40	3.05	3.67	4.28
	G/P	1.14	1.45	1.75	2.05
	v	0.292	0.291	0.289	0.288
	E (cm H$_2$O)	53.88	70.68	87.48	104.26
	E/P	3.85	5.05	6.25	7.45
14 cm H$_2$O	K/P	3.13	4.11	5.09	6.07
	G/P	1.50	1.96	2.42	2.89
	v	0.287	0.287	0.287	0.287

proportional to the transpulmonary pressure, agreeing with dog lung result shown in Fig. 11.8:3. But reasonable agreement with experimental results does not guarantee the correctness of the basic hypotheses underlying the analysis. This example should encourage us to really carefully study the zero-stress state of collagen and elastin fibers in the interalveolar septa, and determine the constitutive equations of these fibers.

11.10 Interdependence of Mechanical Properties of Neighboring Organs

When different organs are put next to each other, they will of course interact mechanically. The mechanical property of an organ in isolation is not the same as that organ in contact with neighbors. We have illustrated this in earlier chapters. For example, Fung et al. (1966), Fung (1966) have shown that the capillary blood vessels in the mesentery are well supported by surrounding gel and are rather rigid. The capillary blood vessels in a bat's wing, which is a very thin membrane of approximately 30–40 μm thickness, is fairly distensible (Bouskela and Wiederhielm, 1979) because the surrounding tissue is relatively small. The capillaries in the pulmonary interalveolar septa are distensible in the direction perpendicular to the septa because they are unsupported in that direction, but are indistensible in the planes of the septa with respect to blood pressure because there are no free space to move (Sec. 6.6). For comparison, the distensibility (change of diameter per cm H$_2$O divided by diameter) of the mesenteric capillaries of the frog given by Baldwin and Gore (1989) is of the order of 0.1% per cm H_2O, that of the pulmonary capillaries of the cat is about 5% per cm H_2O.

The pulmonary arteries and veins are embedded in the lung parenchyma. When the lung is inflated the parenchyma behaves as an elastic body. The pulmonary arterial wall is internally subjected to blood pressure and shear stress, externally to alveolar gas pressure and parenchymal stress, and longitudinally to stretch due to lung inflation. The dimensions of the vessel are influenced by these loads, which are usually coupled together. For example, when the blood pressure is increased the diameter of the blood vessel will increase. The increase of diameter induces incremental stresses in the parenchyma which resists the expansion of the vessel. Conversely, in a reduction of blood pressure the lung parenchyma participates in resisting the reduction of vessel diameter. This is undoubtedly the reason why the pressure–diameter relationship of the pulmonary blood vessels shown in Fig. 6.6:4 is so different from that of the aorta shown in figures of Ch. 8 of Fung (1981), the former being linear whereas the latter being exponential.

Lai-Fook (1979a, b) and Lai-Fook et al. (1978, 1980, 1982) analyzed the situation by considering a circular cylindrical hole in the parenchyma into which is fitted a blood vessel. By matching the boundary conditions at the interface, the radial stress at the interface can be determined, and the "interdependence" of the vessel and parenchyma clarified.

11.11 Instability of Structures

In this and the next sections instability and atelectasis of the lung are discussed. The word *instability*, like the word *disease*, has no unique meaning. Here I define stability as the tendency of a system toward returning to the initial state after an arbitrary infinitesimal perturbation: A system is *stable* if it would, *unstable* if the returning is not guaranteed.

The term *atelectasis* may also mean different things to different people. Here I mean the existence of some groups of completely collapsed alveoli in which there is no ventilation. In contrast, Wilson (1982) considered the pressure-volume curves (PV) of the lung and used a positiveness of the bulk modulus of the parenchyma as a criterion for stability. Stamenovic (1986) generalized this concept and defined atelectasis as a coexistence of several phases of expansion, with each phase having a uniform volume expansion ratio. While their investigations throw light on the phenomenon, their definitions are different from ours.

Thus defined, atelectasis and instability are not the same thing. A lung which is stable with respect to small perturbations may be changed into an atelectatic state by a "large" deformation. On the other hand, an atelectatic lung may be quite stable in the atelectatic state. However, it is most likely that atelectasis follows instability. Hence we investigate them both: the initial tendency toward instability, and the persistent atelectatic plaque.

Atelectasis is often seen in surgery, trauma, disease, airway obstruction, high oxygen breathing, high acceleration, etc. To a patient or a physician, the most important question is how to reinflate the collpased region. A related question is the inflation of a new born baby's liquid filled lung. The answers are discussed below.

Physical Principles of Stability Analysis

Figure 11.11:1 shows the common concept of equilibrium and stability. A ball in a concave dish will rest at the bottom and is stable there. Turn the dish over, the ball will be in equilibrium at the top but unstable there. This concept is expressed mathematically by saying that the potential energy of the ball is a relative minimum at the stable equilibrium position, and a relative maximum at the unstable equilibrium position. The potential energy of the ball, derived from gravitation, is equal to the product of the weight of the ball and the height. The relative maximum or minimum is examined against variation of the radial distance from the equilibrium position.

This concept can be generalized to an elastic system by the method of calculus of variation (see, for example, Fung, 1965, Chap. 10). One examines the variation of the potential energy of the system against all possible displacements. If the first variation of the potential energy function is zero at a certain state of strain, then that state is in equilibrium. If the second variation of the potential energy is positive at the equilibrium state, then that state is stable. If the second variation of the potential energy were negative, then that state of equilibrium is unstable. This is illustrated in the lower half of Fig. 11.11:1. In this generalization, the slope of the dish is replaced by the first variation,

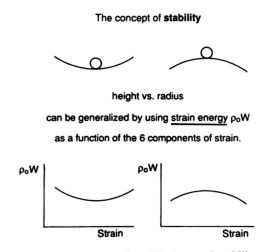

The concept of **stability**

height vs. radius

can be generalized by using <u>strain energy $\rho_0 W$</u>

as a function of the 6 components of strain.

FIGURE 11.11:1 The common concept of equilibrium and stability and its generalization to a complex elastic structure.

the curvature of the dish is replaced by the second variation of the potential energy.

From the general theory of thermodynamics we recognize the existence of internal energy. For the lung, the internal energy per unit volume is designated as $\rho_0 W$, and it consists of two parts: 1) the strain energy in the tissue, and 2) the surface energy of the liquid-gas interfaces of alveolar walls. As to the external forces acting on the lung, some, like gravity, is conservative and has a potential function; others, like aerodynamic forces in airway and viscous shear force in blood vessels, are nonconservative. All external forces multiplied by the corresponding displacements is equal to the work done, which has the same dimension as energy. When the lung deforms a little in an arbitrary way, the displacement of the lung can be described by a continuous vector field δu. This displacement causes a change of strain δE_{ij}, a change of stress δS_{ij}, a change of internal energy per unit volume $\delta(\rho_0 W)$, a work done by body force X_i equal to $X_i \, \delta u_i$, and work done by surface force T_i equal to $T_i \, \delta u_i$. Then a rigorous analysis yields the result (see, e.g., Fung, 1965, p. 450)

$$S_{ij} = \frac{\partial \rho_0 W}{\partial E_{ij}}, \tag{1}$$

$$\delta^2(\rho_0 W) > 0: \text{stable}, \tag{2}$$

$$\delta^2(\rho_0 W) < 0: \text{unstable}. \tag{3}$$

The critical condition of instability is

$$\delta^2(\rho_0 W) = 0. \tag{4}$$

Here $\delta^2(\rho_0 W)$ is the second variation of $\rho_0 W$, i.e. the second order terms of the change of $\rho_0 W$ due to the small variations of displacements δu_i.

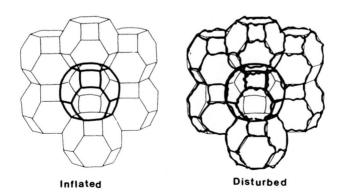

Inflated   Disturbed

FIGURE 11.11:2 Illustration of the idea that a general, arbitrary deformation is considered in the stability study. The *left-hand side* is a mathematical model of a basic unit of pulmonary alveolar duct. The *right-hand side* shows a perturbation.

Applying this theorem to the lung, we must allow δu_i to be completely arbitrary (subjected only to the restrictions at the boundary of pleura). This is illustrated in Fig. 11.11:2, in which a model of an alveolar duct is shown in the left, with each alveolus represented by a 14-sided polyhedron, ventilated to the duct bounded by heavy lines at the center. A disturbed configuration is shown on the right hand side, in which every wall and every edge is deformed in an arbitrary manner. In the theoretical examination of the critical condition of stability, the tendency to collapse is examined against any possible (accidental or otherwise) small perturbations. If the lung is stable, then any small transient disturbance will die away. If it is unstable, then a small disturbance may cause the lung to collapse.

The principle of stability analysis is embodied in the four equations given above. The steps to be taken for a quantitative evaluation of lung stability are then very clear: One must evaluate the potential energy, i.e. the sum of strain energy, surface energy and the potential of external load, as a function of arbitrary deformation of the lung.

The Potential Energy Function

We have discussed the strain energy function in Secs. 10.6, 10.7, 11.6 (Eqs. 11.6:6, 11.6:7), and 11.7 (Eq. 11.7:6). Since the stress tensor σ_{ij} can be obtained by differentiating the strain energy function with respect to the strain components E_{ij}, an integration of the stress with respect to strain yields the strain energy per unit volume, $\rho_0 W$:

$$\rho_0 W = \int S_{ij} \, dE_{ij}. \tag{5}$$

In the lung, the surface energy of the gas-alveolar wall interface is part of the strain energy $\rho_0 W$.

The potential energy is the sum of $\rho_0 W$ and the potential of the external loading, which has been enumerated earlier.

The Stability Criterion

Consider arbitrary small perturbations of an inflated lung. Let the stretch ratios at equilibrium be λ_{10}, λ_{20}, and λ_{30}. Let the stretch ratios of the perturbed lung be $\lambda_1 = \lambda_{10} + \delta\lambda_1$, $\lambda_2 = \lambda_{20} + \delta\lambda_2$, and $\lambda_3 = \lambda_{30} + \delta\lambda_3$. The $\delta\lambda$ values are arbitrary, but infinitesimal. The strain energy $\rho_0 W$ is changed to $\rho_0 W + \delta(\rho_0 W)$ due to the perturbation. The stability is determined by the second variation of the strain energy, $\delta^2(\rho_0 W)$. The system is stable if $\delta^2(\rho_0 W) \geq 0$; otherwise it is unstable.

Substituting $\lambda_i = \lambda_{i0} + \delta\lambda_i$ into $\rho_0 W$ and retaining only the second-order terms results in

$$\delta^2(\rho_0 W) = \sum_{i,j=1}^{3} k_{ij} \delta\lambda_i \delta\lambda_j, \tag{6}$$

where k_{ij} are the values of the second derivatives evaluated at $\lambda_i = \lambda_{i0}$:

$$k_{ij} = \frac{\partial^2 \rho_0 W}{\partial \lambda_i \partial \lambda_j}. \qquad (i, j = i, 2, 3). \qquad (7)$$

The right side of Eq. (6) is a quadratic form. If the equilibrium is to be stable, the quadratic form must be positive definite, i.e., >0 for whatever values of $\delta\lambda_i$ and $\delta\lambda_j$, and 0 only when $\delta\lambda_i = \delta\lambda_j = 0$. The conditions for the positive definiteness are (see Fung, 1965, pp. 29–30):

$$k_{11} + k_{22} + k_{33} > 0, \qquad (8)$$

$$\begin{vmatrix} k_{11} & k_{12} \\ k_{21} & k_{22} \end{vmatrix} + \begin{vmatrix} k_{22} & k_{23} \\ k_{32} & k_{33} \end{vmatrix} + \begin{vmatrix} k_{33} & k_{31} \\ k_{13} & k_{11} \end{vmatrix} > 0, \qquad (9)$$

$$\begin{vmatrix} k_{11} & k_{12} & k_{13} \\ k_{21} & k_{22} & k_{23} \\ k_{31} & k_{32} & k_{33} \end{vmatrix} > 0. \qquad (10)$$

The quantities in Eqs. (9) and (10) are determinants. Thus the problem of stability is reduced to the checking of the three equations, (8), (9), and (10).

The simplest and most important case to check is the uniformly inflated state. Detailed examination shows (Fung, 1975) that the inflated state is stable above a critical transpulmonary pressure. On the other hand, an unstable state is obtained if the surface tension is a constant independent of area, and the elastic stress is zero (in which case $k_{11} = 0$).

11.12 Collapsed Structure. Example of Atelectatic Lung

When the critical condition for instability is met, the lung will have a tendency toward collapse. The following three types of collapsed lung are of special interest: (1) atelectasis of the *focal type*, in which the alveoli collapse toward a central focus, (2) atelectasis of the *axial type*, in which the alveoli collapse toward a line, and (3) atelectasis of the *planar type*, in which the alveoli collapse toward a plane. These are illustrated in Figs. 11.12:1 and 11.12:2.

In each case it is assumed that at the core a number of the interalveolar septa are coalesced. When two interalveolar septa touch each other, their liquid coverings will fuse, thus eliminating surface tension. Pressure, tension, and shear stress can be transmitted through the coalesced alveolar septa. Immediately next to these coalesced septa are the open alveoli, whose walls are moist, on which surface tension acts. The dimensions of the alveoli next to the coalesced region may conceivably be reduced because of the necessary continuity of the membranes (septa). Farther away from the atelectatic core, the alveoli are less and less influenced by the localized perturbation; they become those of the normally inflated lung.

There are, however, considerable differences in the behavior of the three types of atelectasis. For the planar type, the transition from the coalesced

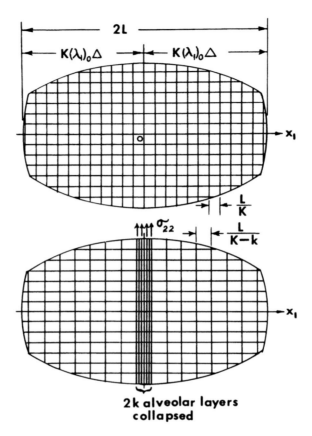

FIGURE 11.12:1 Conceptual description of a planar type of atelectasis. 2k layers of alveoli are collapsed into a plaque. From Fung (1975b).

region to that of the normally inflated alveoli can be immediate. In other words, once created, a planar atelectasis can exist in an inflated lung. Since an inflated lung is stable with respect to all infinitesimal disturbances, an atelectic plaque can be introduced only by a large disturbance, such as an obstruction of the airway, a compression by a tumor, a pressure by a surgeon's finger, or a transient local reduction of stress to a level below the critical stress of instability. Because planar atelectasis is stable, it can be very persistent.

For atelectasis of the axial type, we ask two questions: 1) How large is the transition zone between the coalesced core and the normal alveoli? 2) Can we pull the coalesced alveolar walls out by inflating the lung? The answer to 1) is zero; that to 2) is that we can. These answers are based on exact solutions of the equations of equilibrium, compatibility, and boundary conditions (Fung, 1975). In the transition zone it is shown that the displacement must be proportional to r, resulting in a constant radial strain similar to a pattern

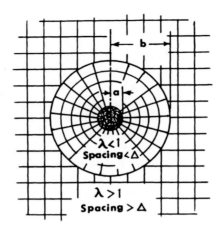

FIGURE 11.12:2 An *axial* or a *focal* atelectasis. The alveoli in the central region are collapsed. A transition zone between circles of radius a and radius b is assumed. Beyond radius b is the normally inflated zone. A rigorous mathematical analysis shows that $b = a$, i.e., the transition zone does not exist. *Lower figures*: Illustration that (1) the direction of the idealized alveoli is immaterial in the stability argument and (2) the edge of a planar atelectatic plaque provides a region in which the interalveolar septa can be pulled out to reinflate the alveoli. From Fung (1975b).

given by Eq. (11.8:9). The boundary condition with the normal alveoli then leads to the first conclusion. For the answer to the second question, we consider a small perturbation of the uniformly inflated region of the lung surrounding the core. For small perturbations a linearized theory suffices. An exact linear solution is that the radial displacement is inversely proportional to r. Consequently, the radial strain decreases with r, the circumferential strain increases with r, the areal strain is zero, i.e., the cross-sectional area of any small element in r, θ plane is preserved. This means that the tissue can be pulled out of the core and the alveoli reinflated by an outward radial movement, or coalesced into the core by a radial movement in the opposite direction.

The focal type of atelectasis behave similarly. A transition zone does not exist. Small centripetal radial displacement may push the parenchyma into the core, outward radial displacement may pull the alveoli coalesced in the core out into the inflated lung.

Reinflation of Atelectatic Lung

We have explained above that from a mechanical point of view, focal and axial atelectasis can be pulled out by reinflating the lung to a larger size. But how can a planar atelectasis be pulled out? It is not effective to pull the planar coalesced region which can transmit tensile stress. But the edges of a plaque of planar atelectasis must behave somewhat like half of an axial atelectatic core. Since there is no axial symmetry, the condition there is quite complex, but it is clear that to remove a planar atelectasis one should work on the edges of the plaque. Overinflation will pull out the alveolar septa at the edges. One needs to overcome, in this process, not only the surface tension of the newly created interfaces but also the viscous friction between the septa because of their relative motion. Therefore, the duration of overinflation should not be too brief. But excessive positive airway pressure compresses the alveolar capillaries and decreases the blood flow; hence, the duration of overinflation should not be too long. Intermittently applied positive-pressure breathing at a suitable frequency, or negative pressure applied to the chest or abdomen, are local procedures. When a surgeon massages a lung after an operation to remove signs of atelectasis, he is applying this principle. The same principle tells why it is difficult to open an atelectatic plaque which borders on a pleura where the peeling action is ineffective.

Fleischner (1936) first described platelike atelectasis in roentgenograms of the lung. The linear shadows in the lung, variously called Fleischner's lines, platter atelectasis, or discoid atelectasis, have been discussed in detail by Fraser and Paré (1970, p. 301). These lines are undoubtedly shadows of planar atelectasis. Fraser and Paré (pp. 196–239) presented detailed patterns of lobar and total pulmonary collapse. They showed that when atelectasis is approached a collapsed lobe or segment tends to look like a curved plate whose edges tend not to retract from the chest wall and the mediastinum. Thus, an atelectatic lobe is also planar.

Problems

11.1 Develop a finite element program to analyze the stress distribution in arterial wall by taking into account the open zero-stress configurations as shown in Fig. 11.2:3.

11.2 Continuing the development suggested in Problem 11.1, develop a finite element program to analyze the stress distribution in the vessel wall in the aortic arch region, where the vessel is toroidal.

11.3 Develop further a program for computing the stress distribution in the left ventricle, taking into account the zero-stress configurations shown in Fig. 11.4:4.

11.4 In order to measure the mechanical properties of vascular smooth muscles, small arteries are used because smooth muscle cells occupy a larger portion of the vessel wall in smaller vessels. In vascular smooth muscle research, arterioles at the periphery are favored because smooth muscle cells occupy 80–90% of the walls of these vessels (the rest is the intima, collagen, elastin, and adventitia). Propose a constitutive equation for the smooth muscle, containing several unknown material constants. In order to identify these unknown constants, it is expedient to test the vessel in unusual deformation modes in addition to the usual inflation by internal pressure. Fung used a pair of micropipettes with rectangular mouths to deform the vessel wall. The pipettes can push the vessel into elliptical shape. Using a vacuum pump, the pipette can suck the wall into the pipette mouth to an extent depending on the transmural pressure. The deformation is measured. Develop a computing program to identify the material constants.

11.5 Similar to the scheme outlined in Prob. 11.4, micropipette aspiration method can be used to study the mechanical properties of cell membranes. Develop a computing program to evaluate the material constants of the membrane constitutive equation from cell aspiration data. (Cf.: Chien et al., 1978.)

11.6 One of the simplest ways to evaluate residual stress in a membrane is to poke holes in an unloaded membrane and observe the change in geometry of the holes. A circular punch creating a circular hole, or a knife creating a slit would provide very useful data. To understand the results quantitatively, it is necessary to assume a constitutive equation of the membrane material. Develop a computing program to relate the deformation of the hole to the residual stress and constitutive equation. Include the identification of material constants in your program.

References

Ardila, R., Horie, T., and Hildebrandt, J. (1974). Macroscopic isotropy of lung expansion. *Resp. Physiol.* **20**: 105–115.

Bachofen, H., Gehr, P., and Weibel, E.R. (1979). Alterations of mechanical properties and morphology in excised rabbit lungs rinsed with a detergent. *J. Appl. Physiol.: Resp. Environ. Exercise Physiol.* **47**: 1002–1010.

Bachofen, H., Hildebrandt, J., and Bachofen, M. (1970). Pressure-volume curves of air and saline-filled excised lungs: Surface tension in situ. *J. Appl. Physiol.* **29**: 422–431.

Baldwin, A.L. and Gore, R.W. (1989). Simultaneous measurement of capillary distensibility and hydraulic conductance. *Microv. Res.* **36**: 1–22.

Bergel, D.H. (1972). The properties of blood vessels. In: *Biomechanics: Its Foundations and Objectives*, (Y.C. Fung and M. Anliker, eds.). Englewood Cliffs, N.J. Prentice-Hall, Ch. 5, pp. 105–140.

Bergel, D.H. (1961). The static elastic properties of the arterial wall. The dynamic elastic properties of the arterial wall. *J. Physiol.* (London) **156**: 445–469.

Bouskela, E. and Wiederhielm, C.A. (1979). Microvascular myogenic reaction in the wing of the intact unanesthetized bat. *Amer. J. Physiol.* **237**(1): H59–H65.

Chien, S., Sung, P., Skalak, R., Usami, S., and Tözeren, A. (1978). Theoretical and experimental studies on viscoelastic properties of erythrocytes membrane. *Biophy. J.* **24**: 463–487.

Chuong, C.J. and Fung, Y.C. (1983). Three-dimensional stress distribution in arteries. *J. Biomech. Eng.* **105**: 268–274.

Chuong, C.J. and Fung, Y.C. (1986). Residual stress in arteries. In: *Frontiers in Biomechanics* (G.W. Schmid-Schönbein, S.L.-Y Woo, and B.W. Zweifach, eds.), Springer-Verlag, New York, pp. 117–129.

Clements, J.A. (1962). Surface phenomena in relation to pulmonary function. *Physiologist* **5**: 11–28.

Clements, J.A. and Tierney, D.F. (1965). Alveolar instability associated with altered surface tension. In *Handbook of Physiology, Sec.* 3, *Respiration*, Vol. 2 (W.O. Fenn and H. Rahn, eds.). Amer. Physiol. Soc., Washington, D.C. pp. 1565–1583.

Cumming, G. and Semple, S.J. (1973). *Disorders of the Respiratory System.* Blackwell, Oxford.

Dale, P.J., Mathews, F.L., and Schrotter, R.C. (1980). Finite element analysis of lung alveolus. *J. Biomech.* **13**: 865–873.

Doyle, J.M. and Dobrin, P.B. (1973). Stress gradients in the walls of large arteries. *J. Biomech.* **6**: 631–639.

Enhorning, G. (1977). Pulsating bubble techniques for evaluating pulmonary surfactant. *J. Appl. Physiol.: Resp. Environ. Exercise Physiol.* **43**: 198–203.

Fleischner, F.G. (1936). Über das Wessen der baselan horizontalen Schattenstreifen im Lungenfeld. *Wien Arch. Inn. Med.* **28**: 461–480.

Flicker, E. and Lee, J.S. (1974). Equilibrium of force of subpleural alveoli: implications to lung mechanics. *J. Appl. Physiol.* **36**: 366–374.

Frankus, A. and Lee, G.C. (1974). A theory of distortion of lung parenchyma based on alveolar membrane properties. *J. Biomech.* **7**: 101–107.

Fraser, R.G. and Paré, J.A.P. (1970, 1977, 1979). *Diagnosis of Diseases of the Chest.* 1st ed. 1970, 2nd ed., Vol. 1, 2, 1977, Vol. 3, 4, 1979. Saunders, Philadelphia.

Fung, Y.C. (1965). *Foundations of Solid Mechanics*, Prentice-Hall, Englewood Cliffs, New Jersey.

Fung, Y.C. (1966). Theoretical considerations of the elasticity of red cells and small blood vessels. *Fed. Proc.* **25**: 1761–1772.

Fung, Y.C. (1974). A theory of elasticity of the lung. *J. Appl. Mech.* **41**: 8–14.

Fung, Y.C. (1975a). Does the surface tension make the lung inherently unstable? *Cir. Res.* **37**: 497–502.

Fung, Y.C. (1975b). Stress, deformation, and atelectasis of the lung. *Circ. Res.* **37**: 481–496.

Fung, Y.C. (1977). *A First Course in Continuum Mechanics*, 2nd Edn, Prentice-Hall, Englewood Cliffs, N.J.

Fung, Y.C. (1981). *Biomechanics: Mechanical Properties of Living Tissues.* Springer-Verlag, N.Y.

Fung, Y.C. (1984). *Biodynamics: Circulation.* Springer-Verlag, New York, pp. 53–66.

Fung, Y.C. (1985). What principle governs the stress distribution in living organs? In: *Biomechanics in China, Japan, and U.S.A.* (Y.C. Fung, E. Fukada, and J.J. Wang, eds.). Science Press, Beijing, China, pp. 1–13.

Fung, Y.C. (1988). A model of the alveolar ducts of lung and its validation. *J. Apl. Physiol.* **64**: 2132–2141.

Fung, Y.C., Zweifach, B.W., Intaglietta, M. (1966). Elastic environment of the capillary bed. *Circ. Res.* **19**: 441–461.

Fung, Y.C., Tong, P., and Patitucci, P. (1978). Stress and strain in the lung. *J. Eng. Mech. Div., ASCE* **104**: 201–223.

Fung, Y.C., Fronek, K., and Patitucci, P. (1979). Pseudoelasticity of arteries and the choice of its mathematical expression. *Amer J. Physiol. (Heart, Circ.)* **237**: H620–H631.

Fung, Y.C. and Sobin, S.S. (1982). On the constitutive equation of the lung tissue. In *1982 Advances in Bioengineering.* Amer. Soc. Mech. Engineers, New York, pp. 84–87.

Fung, Y.C., Sobin, S.S., Tremer, H., Yen, R.T. and Ho, H.H. (1983). Patency and compliance of pulmonary veins when airway pressure exceeds blood pressure. *J. Appl. Physiol.: Respirat. Environ. Exer. Physiol.* **54**(6): 1538–1549.

Fung, Y.C., Yen, R.T., Tao, Z.L., and Liu, S.Q. (1988). A hypothesis on the mechanism of trauma of lung tissue subjected to impact load. *J. Biomech. Eng.* **110**: 50–56.

Fung, Y.C. and Liu, S.Q. (1989). Change of residual stress in arteries due to hypertrophy caused by aortic constriction. *Circ. Res.* **65**: 1340–1349.

Gil, J., Bachofen, H., Gehr, P., and Weibel, E.R. (1979). The alveolar volume-to-surface-area relationship in air-and-saline-filled lungs fixed by vascular perfusion. *J. Appl. Physiol.: Resp. Environ. Exercise Physiol.* **47**: 990–1001.

Gluck, L., Kulovich, M.V., Borer, R.C., Jr., Brenner, P.H., Anderson, G.G., and Sepllacy, W.N. (1971). Diagnosis of the respiratory stress syndrome by amniocentesis. *Am. J. Obstet. Gynecol.* **109**: 440–445.

Hajji, M.A., Wilson, T.A., and Lai-Fook. S.J. (1979). Improved measurements of the shear modulus and pleural membrane tension of the lung. *J. Appl. Physiol.: Resp. Environ. Exercise Physiol.* **47**(1): 175–181.

Hansen, J.E. and Ampaya, E.P. (1975). Human air space shapes, sizes, areas, and volumes. *J. Appl. Physiol.* **38**: 990–995.

Hayashi, K., Handa, H., Mori, K., and Moritake, K. (1971). Mechanical behavior of vascular walls. *J. Soc. Material Science Japan* **20**: 1001–1011.

Hildebrandt, J. (1969). Dynamic properties of air-filled excised cat lungs determined by fluid plethysmograph. *J. Appl. Physiol.* **27**: 246–250.

Hoppin, F.G., Jr., Lee, G.C., and Dawson, S.V. (1975). Properties of lung parenchyma in distortion. *J. Appl. Physiol.* **39**: 742–751.

Hoppin, F.G., Jr. and Hildebrandt, J. (1977). Mechanical properties of the lung. In *Bioengineering Aspects of the Lung*, (J.B. West, ed.). Marcel Dekker, New York, pp. 83–162.

Hunter, P.J. and Smail, B.H. (1988). The analysis of cardiac function: a continuum approach. *Prog. Biophys. Molec. Biol.* **52**: 101–164.

Karakaplan, A.D., Bieniek, M.P., and Skalak, R. (1980). A mathematical model of lung parenchyma. *J. Biomech. Eng.* **102**: 124–136.

Lai-Fook, S.J. (1979a). A continuum mechanics analysis of pulmonary vascular interdependence in isolated dog lobes. *J. Appl. Physiol.: Resp. Environ. Exercise Physiol.* **46**: 419–429.

Lai-Fook, S.J. (1979b). Elastic properties of lung parenchyma: the effect of pressure-volume hysteresis on the behaviour of large blood vessels. *J. Biomech.* **12**: 757–764.

Lai-Fook, S.J., Wilson, T.A., Hyatt, R.E., and Rodarte, J.R. (1976). The elastic constants of inflated lobes of dog lungs. *J. Appl. Physiol.* **40**: 408–513.

Lai-Fook, S.J., Hyatt, R.E., and Rodarte, J.R. (1978). The effect of parenchymal shear modulus and lung volume on bronchial pressure–diameter behavior. *J. Appl. Physiol.: Resp. Environ. Exercise Physiol.* **44**: 859–868.

Lai-Fook, S.J. and Toperoff, B. (1980). Pressure-volume behavior of perivascular interstitium measured directly in isolated dog lung. *J. Appl. Physiol.: Resp. Environ. Exercise Physiol.* **48**: 939–946.

Lai-Fook, S.J. and Kallok, M.J. (1982). Bronchial–arterial interdependence in isolated dog lung. *J. Appl. Physiol.: Resp. Environ. Exercise Physiol.* **52**: 1000–1007.

Lambert, R.K. and Wilson, T.A. (1973). A model for the elastic properties of the lung and their effect on expiratory flow. *J. Appl. Physiol.* **34**: 34–48.

Lanir, Y. (1983). Constitutive equation for lung tissue. *J. Biomech. Eng.* **105**: 374–380.

Lee, G.C. (1978). Solid mechanics of lungs. *J. Eng. Mech. Trans. ASCE*, **104**: 177–200.

Lee, G.C. and Frankus, A. (1975). Elasticity properties of lung parenchyma derived from experimental distortion data. *Biophys. J.* **15**: 481–493.

Lee, G.C., Frankus, A., and Chen, P.D. (1976). Small distortion properties of lung parenchyma as a compressible continuum. *J. Biomech.* **9**: 641–648.

Liu, J.T. and Lee, G.C. (1978). Static finite deformation analysis of the lung. *J. Eng. Mech. Div., ASCE* **104**: 225–238.

Liu, S.Q. and Fung, Y.C. (1988). Zero-stress states of arteries. *J. Biomech. Eng.* **110**: 82–84.

Liu, S.Q. and Fung, Y.C. (1989). Relationship between hypertension, hypertrophy, and opening angle of zero-stress state of arteries following aortic constriction. *J. Biomech. Eng.* **111**: 325–335.

Love, A.E.H. *The Mathematical Theory of Elasticity*. Cambridge University Press, Cambridge, 1st ed., 1892, 1893; 4th ed., 1927. Reprinted Dover Publication, New York, 1944. Note especially the "Historical Introduction", pp. 1–31.

Matsuda, M., Fung, Y.C., and Sobin, S.S. (1987) Collagen and elastin fibers in human pulmonary alveolar mouths and ducts. *J. Appl. Physiol.* **63**: 1185–1194.

McCulloch, A.D., Smaill, B.H. and Hunter, P.J. (1987). Left ventricular epicardial deformation in isolated arrested dog heart. *Am. J. Physiol.* **252**: H233–H241.

Mead, J., Takishima, T., and Leith, D. (1970). Stress distribution in lungs: a model of pulmonary elasticity. *J. Appl. Physiol.* **28**: 596–608.

Mercer, R.R. and Crapo, J.D. (1987). Three-dimensional reconstruction of the rat acinus. *J. Appl. Physiol.* **63**: 785–794.

Miller, W.S. (1947) *The Lung*, 2nd ed. Springfield, IL., C.C. Thomas.

Mirsky, I. (1979). Elastic properties of the myocardium: a quantitative approach with physiological and clinical applications. In: *Handbook of Physiology*, Sec. 2, *The Cardiovascular System*, Vol. 1, *The Heart* (R.M. Berne and N. Sperelakis, eds.). Amer. Physiol. Soci., Bethesda, MD. pp. 497–531.

Oldmixon, E.H. and F.G. Hoppin. (1987) Lengths and topology of septal borders (Abstract) *Federation Proc.* **46**: 820, 1987.

Oldmixon, E.H., Butler, J.P., and Hoppin, F.G., Jr. (1988) Dihedral angles between alveolar septa. *J. Appl. Physiol.* **64**: 299–307.

Omens, J. and Fung, Y.C. (1989). Residual strain in the rat left ventricle. *Circ. Res.* **66**: 37–45.

Orsos, F. (1936). The frameworks of the lung and their physiological and pathological significance. *Beiträge zur Klinik der Tuberkulose und speziefischen Tuberkuloseforschung.* **87**: 568–609.

Patel, D.J. and Vaishnav, R.N. (1972). The rheology of large blood vessels. In: *Cardiovascular Fluid Dynamics*, (D.H. Bergel, ed.). Academic Press, New York, Vol. 2: Ch. 11, pp. 2–65.

Reifenrath, R. and Zimmermann, I. (1973). Surface tension properties of lung alveolar surfactant obtained by alveolar micropuncture. *Resp. Physiol.* **19**: 369–393.

Rodarte, J. and Fung, Y.C. (1986). Distribution of stresses within the lung. In: *Handbook of Physiology*, Sec. 3, *The Respiration System* Vol. 3, Part 1. (A.P. Fishman, P.T. Macklem, and J. Mead, eds.). American Physiological Society, Williams and Wilkins, Baltimore, MD. pp. 233–246.

Rosenquist, T.H., Bernick, S., Sobin, S.S., and Fung, Y.C. (1973). The structure of the pulmonary interalveolar microvascular sheet. *Microvasc. Res.* **5**: 199–212.

Schürch, S., Goerke, J., and Clements, J.A. (1976). Direct determination of surface tension in the lung. *Proc. Natl. Acad. Sci. USA*, **73**: 4693–4702.

Seguchi, Y., Fung, Y.C., and Ishida, T. (1986). Respiratory dynamics—computer simulation. In: *Frontiers of Biomechanics.* (G. Schmid-Schönbein, S. Woo, and B.W. Zweifach, eds.), Springer-Verlag, New York, pp. 377–391.

Shaw, D.J. (1970). *Introduction to Colloid and Surface Chemistry*, 2nd. ed. Butterworths, London.

Sobin, S.S., Fung, Y.C., and Tremer, H.M. (1982). The effect of incomplete fixation of elastin on the appearance of pulmonary alveoli. *J. Biomech. Eng.* **104**: 68–71.

Sobin, S.S., Fung, Y.C., and Tremer, H.M. (1988). Collagen and elastin fibers in human pulmonary alveolar walls. *J. Appl. Physiol.* **64**(4): 1659–1675.

Spilker, R.L. and Simon, B.R. (eds.) (1989). *Computational Methods in Bioengineering.* ASME BED-Vol. 9 Am. Soc. Mech. Eng., New York.

Stamenovic, D. (1986) The mixture of phases and elastic stability of lungs with constant surface forces. *Mathematical Modeling* **7**: 1071–1082.

Stamenovic, D. and Wilson, T.A. (1985). A strain energy function for lung parenchyma. *J. Biomech. Eng.* **107**: 81–86.

Takamizawa, K. and Hayashi, K. (1987). Strain energy density function and uniform strain hypothesis for arterial mechanics. *J. Biomech.* **20**: 7–17.

Takishima, T. and Mead, J. (1972). Tests of a model of pulmonary elasticity. *J. Appl. Physiol.* **28**: 596–608.

Tao, Z.L. and Fung, Y.C. (1987). Lungs under cyclic compression and expansion. *J. Biomech. Eng.* **109**: 160–162.

Vaishnav, R.N. and Vossoughi, J. (1987). Residual stress and strain in aortic segments. *J. Biomech.* **20**: 235–239.

Valberg, P.A. and Brain, J.D. (1977). Lung surface tension and air space dimensions from multiple pressure-volume curves. *J. Appl. Physiol.: Resp. Environ. Exercises Physiol.* **43**: 730–738.

Vawter, D.L. (1983). Poisson's ratio and incompressibility. *J. Biomech. Eng.* **105**: 194–195.

Vawter, D.L., Fung, Y.C. and West, J.B. (1978). Elasticity of excised dog lung parenchyma. *J. Appl. Physiol.* **45**: 261–269.

Vawter, D.L., Fung, Y.C., and West, J.B. (1979). Constitutive equation of lung tissue elasticity. *J. Biomech. Eng.*, Trans. of Amer. Soc. Mech. Engineers **101**(1): 38–45.

Vawter, D.L. and Shields, W.H. (1982). Deformation of the lung: the role of interfacial forces. In *Finite Elements in Biomechanics* (R.H. Gallagher, et al., eds.). Wiley, New York, pp. 83–110.

von Neergaard, K. (1929). Neue Auffassungen über einen Grundbegriff der Atemmechanik: Die Retraktionskraft der Lunge, abhängig von der Oberflächenspannung in den Alveolen. *Z. Ges. Exp. Med.* **66**: 373–394.

Waldman, L.K. (1985). In-vivo measurement of regional strains in myocardium. In: *Frontiers in Biomechanics.* (G.W. Schmid-Schönbein, S.L. Woo, and B.W. Zweifach, eds.). Springer-Verlag, New York, pp. 99–116.

Waldman, L.K., Fung, Y.C., and Covell, J.W. (1985). Transmural myocardial deformation in the canine left ventricle. *Circ. Res.* **57**: 152–163.

Weibel, E.G. (1963) *Morphometry of the Human Lung.* Academic Press, New York.

West, J.B. and Matthews, F.L. (1972). Stresses, strains, and surface pressures in the lung caused by its weight. *J. Appl. Physiol.* **32**: 332–345.

Weyl, H. (1952). *Symmetry.* Princeton Univ. Press, Princeton, N.J.

Wilson, T.A. (1972). A continuum analysis of a two-dimensional mechanical model of the lung parenchyma. *J. Appl. Physiol.* **33**: 472–478.

Wilson, T.A. (1981). Relations among recoil pressure, surface area and surface tension in the lung. *J. Appl. Physiol.: Resp. Environ. Exercise Physiol.* **50**: 921–926.

Wilson, T.A. (1982). Surface-tension-surface-area curves calculated from pressure-volume loops. *J. Appl. Physiol: Resp. Environ. Exercise Physiol.* **53**: 1512–1520.

Wilson, T.A. (1983). Nonuniform lung deformation. *J. Appl. Physiol.: Resp. Environ. Exercise Physiol.* **54**(6): 1443–1450.

Wilson, T.A. (1986). Solid Mechanics. In *Handbook of Physiology*, Sec. 3. *The Respiratory System.* III. *Mechanics of Breathing*, Part I (A.P. Fishman, P.T. Macklem, J. Mead, eds.). American Physiol. Soc. Bethesda, MD.

Wilson, T.A. and Bachofen, H. (1982). A model for the mechanical structure of the alveolar duct. *J. Appl. Physiol.: Resp. Environ. Exercise. Physiol.* **52**: 1064–1070.

Yen, M.R.T., Fung, Y.C., and Artaud, C. (1987). The incremental elastic moduli of the lungs of rabbit, cat and man. *The 1987 Adv. in Biomech.* Amer. Soc. Mech. Engineers, New York, pp. 39–40.

Yen, M.R.T., Fung, Y.C., Ho, H.H., and Butterman, G. (1986). Speed of stress wave propagation in lung. *J. Appl. Physiol.* **61**: 701–705.

Yen, R.T., Fung, Y.C., and Bingham, N. (1980). Elasticity of small pulmonary arteries in the cat. *J. Biomech. Eng.* **102**: 170–177.

Yin, F.C.P. (1981). Ventricular wall stress. *Circ. Res.* **49**: 829–842.

Zeng, Y.J., Yager, D., and Fung, Y.C. (1987). Measurement of the mechanical properties of the human lung tissue. *J. Biomech. Eng.* **109**: 169–174.

Zupkas, P. (1977). Mathematical analysis of surface tension diagrams of mammalian lung components. M.S. Thesis, Univ. of Calif., San Diego.

CHAPTER 12

Strength, Trauma, and Tolerance

12.1 Introduction

There are many reasons why the study of the strength of biological tissues and organs is important. In the first place any living organism must be strong enough to withstand the loads imposed on it by its environment and its activities. The history of evolution is a history of cells forming more efficient organizations for competition and survival. The shapes of plants and animals depend largely on the structural materials these organisms can manufacture and organize into structures of adequate strength. See Currey (1970) and Wainwright et al. (1976).

Engineering science is concerned with the description and measurement of the strength of natural or artificial biological materials, and the determination of their significance. The determination of the *failure characteristics* of living tissues and organs is especially complex, because there are many ways a material can "fail" in biological sense. Besides yielding, plastic deformation, creep, rupture, fatigue, corrosion, wear, and impact fracture, one has to consider other kinds of failure. An impact causing severe edema in the lung can be fatal. A sprain of the ankle can be very painful. A concussion of the brain has neurological effects. To study the strength of biological materials one has to correlate clinical observations and pathological lesions with stress and strain in the tissues.

In this chapter we shall discuss the strength of organs from the point of view of trauma research. Trauma is a Greek work for wound, meaning an injury to a living body caused by the application of external force or violence. Next to heart disease and cancer, trauma is the third greatest killer in the United States. For people between 15 and 45 years of age, it is the number 1 killer. Minimizing trauma is on everybody's mind. To the medical community the problem is critical care and management. To an automotive engineer the problem is design for safety and protection. To police and paramedics the problem is quick transportation. To government the problem is legislation. To parents and teachers the problem is safety education. Everybody has a role to play. Everybody has something at stake. From the point of view of reducing the possibility of getting into a traumatic situation in the first place, it is a problem of culture, of life style, of war and peace, of law and order. When it gets to the stage when bioengineers, physicians, surgeons, and nurses can do something about it, it is already toward the end of the line.

Biomechanics is involved in many of these stages. Specifically, it is involved in process causing traumas, in trauma management, in surgical treatment, in pathophysiology, in recovery, in physical therapy, in rehabilitation, in the design of vehicles, and in personal protection. A clear understanding of biomechanics in all these stages will be helpful to minimize trauma as a national and personal problem.

Trauma is a vast subject. In this chapter we can only provide a survey of basic principles in trauma research. We shall discuss the types of loading that may cause trauma, the failure modes of materials, the strength and tolerance of organs, and biomechanics in trauma management, healing, recovery, and rehabilitation. We shall discuss engineering for trauma prevention, taking passenger aircraft design as a concrete example; and end the chapter with suggestions for future research.

A tool for trauma research is modeling. What this chapter offers is a sketch of the basic features that detailed calculations are expected to reveal.

12.2 Failure Modes of Materials

The mechanical properties of biological tissues have some features similar to those of the familiar engineering materials and other features very different from them. To study this very complex subject let us consider the metals first. If rods of metals are pulled in a testing machine at room temperature, typical load-elongation relationships shown in Fig. 12.2:1 are obtained. The initial region, appearing as a straight line, is the region in which the law of linear elasticity holds. The maximum load M is called the *ultimate load*. At the point C the specimen breaks.

For a given material, the stress–strain relationships in tension, shear, compression, bending, and torsion are somewhat different, and the loading condition should be mentioned in stating the values of the elastic limit, yield stress, and ultimate stress.

The stress–strain relationship of bone is somewhat similar to that shown in Fig. 12.2:1c. Typical examples are shown in Fig. 12.2:2. See Yamada (1970) for a comprehensive set of data. Saha (1982) for dynamic strength of bone.

Soft tissues have very different stress–strain relationships. See some typical examples in Fig. 12.2:3. In the physiological range the stress generally in-

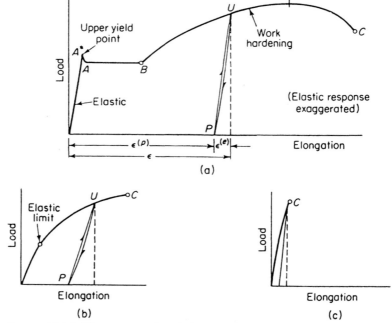

FIGURE 12.2:1 The stress-strain relationship of several engineering materials.

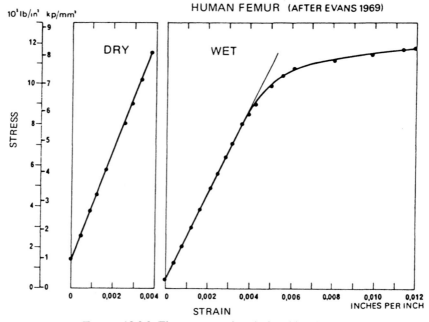

FIGURE 12.2:2 The stress-strain relationship of bone.

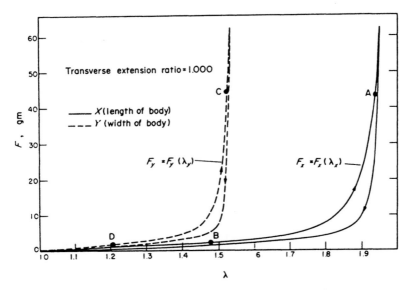

FIGURE 12.2:3 The stress-strain relationship of soft tissues.

creases exponentially with increasing strain. At certain strain higher than physiological the tissue yields and breaks. Hysteresis, creep, and relaxation exist in both hard and soft tissues. See Fung (1981) for a detailed discussion of the mechanical properties of soft tissues.

For most tissues the strength depends on the strain rate. The degree of dependence varies with the tissue: strain rate effect is significant for the bone, less so for the ligament.

Experimental data on the strength of soft tissues are difficult to obtain. Test specimens of suitable sizes are often unavailable, or are difficult to keep in in vivo condition. Clamping of the specimen for strength testing can be a difficult problem. Furthermore, it is important to realize that the damage that a force can do to an organ depends not only on the magnitude of that force, but also on a number of other factors. Consider the following simple experiments (Fig. 12.2:4).

(1) A twine is to be cut by a pair of dull scissors. I have difficulty cutting it when the twine is relaxed. But if I pull it tight and then cut it, it breaks easily. Why?
(2) A stalk of fresh celery breaks very easily in bending. An old, dehydrated one does not. Practice on carrots also!
(3) A balloon is inflated. Another is not inflated but is stretched to a great length. Prick them with a needle. One explodes. The other does not. Why?
(4) A thin-walled metal tube is filled with a liquid. Strike it on one side. Sometimes the shell fails on the other side. This is known as "contre coup". How can this happen?

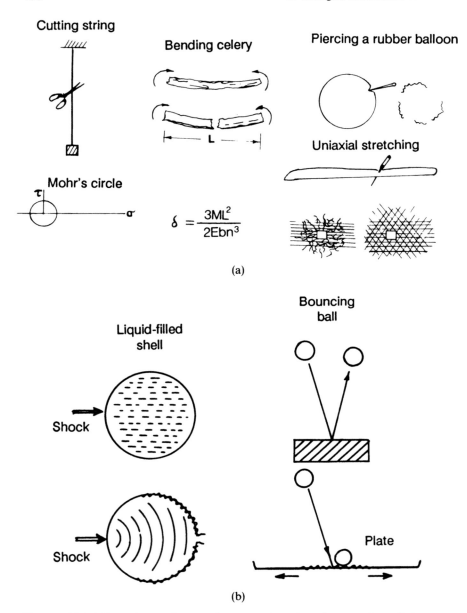

Cutting string

Bending celery

Piercing a rubber balloon

Uniaxial stretching

Mohr's circle

$$\delta = \frac{3ML^2}{2Ebn^3}$$

(a)

Liquid-filled shell

Bouncing ball

Shock

Shock

Plate

(b)

FIGURE 12.2:4 Several experiments demonstrating that the meaning of the term "strength" depends on the condition the specimen is in: whether it is subjected to large initial strain, internal fluid pressure, uniaxial versus biaxial or triaxial loading condition, or focused elastic waves. See text.

(5) Take a small nylon ball, or a pearl, or a ball bearing, throw it onto a hard surface. It bounces. Throw it onto a thin metal plate such as that used in the kitchen for baking, it won't bounce.

Think of biological analogs of these experiments. The twine is similar to a blood vessel, a tendon, or a muscle. The celery is similar to an erectile organ. The balloon experiment shows the difference between the behavior of a material under a uniaxial tension and one subjected to biaxial tension. Many organs of our body are subjected to biaxial tension: pericardium, pulmonary pleura, interalveolar septa of the lung, a taut skin, a diaphragm, a filled bladder, etc. A human head impacting on a windshield of a car is not unlike a tube filled with liquid in the contre coup experiment, or the ball against a hard surface in the bouncing experiment. The same head impacting on a thin sheet of metal may have its kinetic energy transferred to the sheet without bouncing. The kinetic energy is absorbed and the danger of head injury lessened. These examples show that it is instructive to understand what is going on in these experiments.

In the first example, the twine can be understood if we postulate that the fibers in the twine break when the maximum principal tensile stress exceeds the ultimate stress. If shear is applied to the twine when it is slack, the principal tensile stress in the twine is numerically equal to the shear stress imposed. See the Mohr's circle on the lower left part of Fig. 12.2:4(a) which is a graphical method for determining the principal stresses. On the other hand, if the twine is pulled taut and then the shear is applied, the principal tensile stress is numerically equal to the initial tensile stress plus the shear stress (draw a Mohr's circle for this case) and is, therefore, larger than that in the slack case. Therefore it is easier to cut the twine when it is taut.

In the second example, the specimen fails by bending. When the specimen bends, however, half of the specimen is stretched, whereas the other half is compressed. The fibers in the stretched side are expected to be taut in the fresh and plump celery, whereas those in the dehydrated specimen are likely to be slack. Upon bending, the taut fibers in the fresh celery will be stressed more, the slack fibers in the dehydrated celery will still be slack. The highly stressed fibers break when the ultimate stress is reached; the slack fibers will not break. Thus, the difference in the failure characteristics of these two celeries can be explained in the same way as in the first example. This is an interesting example. The contrast between a fresh celery and a dehydrated one is not so different from that of some tissues in vivo and in vitro, with blood perfusion and without, edematous or normal. The expected difference in the mechanical properties of these specimens is worth remembering.

The third example is also an interesting one to remember. It shows that some materials are ductile under uniaxial loading; but become brittle under biaxial and triaxial tension. The reason for the difference is illustrated in the sketches given in the lower right corner of the figure. Rubber is a high polymer of long chain molecules. These molecules are bent and twisted in a complex and random fashion. When the rubber membrane is stretched uni-axially, some molecules in the direction of stretching become straightened and take up the load. If a hole is now made in the membrane (as by the needle), some of those tauted molecules in the direction of stretching will be broken, but the molecules in other directions remain bent and twisted, the hole remains a hole, and nothing dramatic happens. Consider, on the other hand, the situation of the inflated balloon. In this case the membrane is stretched in every direction. The long-chain molecules in every direction are stretched straight and taut. If a hole is made in the middle, the chains in every taut molecule intersecting the hole will be broken, and an explosion results!

Biological soft tissues are composed of collagen and elastin fibers and other long chain molecules embedded in ground substances. The fibers and chains can be stretched when the tissue is under strain. The relevance of the example is evident.

The fourth example shows what focusing of stress waves can do. The compression wave in the fluid initiated by the impact moves to the right. The flexural wave of the metal shell also moves to the right along the curved surface of the tube wall. If the flexural wave and the compression wave arrive at the other side simultaneously, a concentration of stress may occur which may exceed the ultimate stress of the materials and cause fracture on the far side. This is a mechanism that may occur in head injury.

The fifth experiment illustrates the possibility of transferring energy be-tween two bodies in impact. When a ball impacts a hard surface, it bounces; its kinetic energy is transformed first into elastic strain energy in the ball when the ball is stopped; then the elastic strain energy in the ball is transformed back into kinetic energy propelling the ball upward. Against a good hard surface the ball loses little energy in the collision. On the other hand, when the ball hits a thin plate which is more flexible than the ball, the plate deforms and sets off elastic waves which propagate away from the point of impact, carrying kinetic energy with the motion and storing elastic energy in the plate as it deforms. By this process only a small part of the original kinetic energy of the ball is transformed into strain energy in the ball. At the time of rebound the strain energy in the ball is small, therefore the rebound is small.

The design principle for using the sheet metal (or plastic) as shock absorber is its flexural compliance against a concentrated load. The plate should be more flexible than the ball (or head).

In examining the questions of strength and tolerance of man to impact loading, it may be beneficial to remember these simple examples. The magni-tude of initial stresses, the difference between uniaxial, biaxial, and triaxial states of stress, and the dynamic effects are important factors determining the strength of the organs.

There are other factors that may have great effect. The existence or absence of stress raisers such as notches or microscopic cracks is important. Cracks, though small, may induce large stress concentration at their ends. A sharp notch has the same effect. So notches and cracks should be avoided in load bearing members. This is an important principle to remember in surgery and in manufacturing or installing prosthesis.

The art of predicting the strength of biological tissues from their structure and ultrastructure is still in the developing phase. Yamada (1970) has presented collections of strength data on bones, tendons, cartilage, ligaments, skin, muscles, and some other tissues of man and other animals of different ages and sexes under static tension, compression, torsion, and bending, dynamic impact loads, or fatigue oscillations. Standard engineering testing machines are used to obtain data on bone specimens or soft tissues in simple elongation. Special testing methods have been developed to test soft tissues in biaxial loading condition to take care of finite deformation and large strain (Lanir and Fung, 1974; Fung, 1981, p. 242). A convenient reference to testing methods is the CRC Handbook edited by Feinberg and Fleming (1978).

To find the strength of tissues such as the skin, a new *tearing test* was developed by Schneider (1982). The method was originally used by Rivlin and Thomas (1953) for testing rubber membrane and adopted by Sharma (1979) for the testing of blood vessels. A rectangular test specimen is cut as shown in Fig. 12.2:5. The middle flap is folded over and pulled. The critical value of the tension at which the specimen is torn at constant velocity is measured. This method has the advantage that the configuration of the tear remains

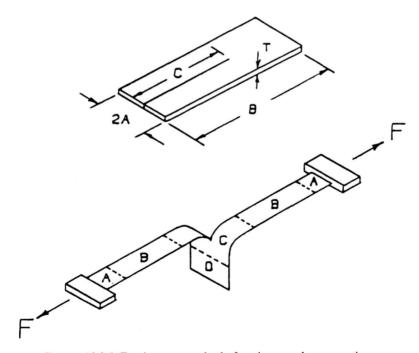

FIGURE 12.2:5 Tearing as a method of testing membraneous tissue.

self-similar during failure, yielding meaningful data with little interference from the method of clamping of the specimen in the testing machine.

12.3 Injury and Repair of Organs

For man, the level of injury that one is willing to tolerate may be much lower than the breaking failure discussed in Sec. 12.2. Often a minor pain is intolerable to some. The physician, surgeon, physiologist, bioengineer, patient, and lawyer are interested in evidences of injury and repair at the cellular level. Injury is deviation from the normal. All the knowledge and tools of science, technology, and medicine must be brought to bear to develop this subject. A recent publication edited by Woo and Buckwalter (1988) presents a detailed discussion of injury and repair of the musculoskeletal soft tissues, including tendon, ligament, bone-tendon and myotendinous junctions, skeletal muscle, peripheral nerve, peripheral blood vessel, articular cartilage, and meniscus. Future directions of research are pointed out. The biomechanics of injury and repair, however, are largely untouched. Evidently, this is a field for future cultivation.

12.4 Shock Loading and Structural Response

Penetration by bullets and bomb fragments is a well-known cause of trauma. Strong electromagnetic and heat radiations from nuclear weapons are, of course, traumatic to the extreme and we hope that these weapons will never be used again. Control of these things require intelligence of man beyond technology.

Blunt impacts occur daily and are more controllable. They are our main concern in the following discussion. Blunt impacts do not penetrate the skin. They occur in automobile crashes, survivable airplane crashes, falling, colliding, boxing, diving, football, or shock waves of explosives. How much impact a man can tolerate depends on many factors: the strength of the shock, the magnitude of the peak force, the duration of the pulse, and how rapidly the load is applied. Is it slow enough for the organs to respond quasistatically? Or fast enough to induce vibration? Or even faster so that the major feature of the response consists of stress waves in the organs? It depends also on the initial stress in the organ. For example, if one tenses up, one can induce large stresses in muscles, bones and joints. This initial stress is superposed on the response to the shock loading. The sum is what the body has to bear.

In an automobile crash accident, if a passenger did not fasten his seat belt, then he will fly off as a projectile, and be decelerated when he hits the car. If he has his seat belt fastened, then he will acquire certain deceleration. Each part of his body has a different deceleration history. To analyze the stress in any of his organs, a free-body diagram of that organ can be constructed. According to D'Alembert's principle, the body can be considered as in a static equilibrium with the inertial force (mass × deceleration) applied as an external load. To find the stress in the organ is to find the response of the organ to the dynamic load. Although the full analysis usually require elaborate mathematical modeling, major features can be understood through simple examples.

As a simple example, consider an elastic wire of uniform cross-sectional area A and length L, made of a uniform linear elastic material with a Young's modulus E (see Fig. 12.4:1). One end of the wire is attached to a rigid ceiling, the other end has a stopper and hangs free. Now let a load W be applied, infinitely slowly, to the free end. The wire extends, infinitely slowly, so that equilibrium is maintained at all times. When the load is fully applied, the free end extends a distance δ. The strain in the wire must be δ/L. The longitudinal stress σ is, therefore, $E\delta/L$. A multiplication with the cross-sectional area A gives the total tension in the wire, $AE\delta/L$. Since this must be equal to W in an equilibrium condition, we see that

$$AE\delta/L = W, \qquad \text{hence} \qquad \sigma = E\frac{\delta}{L} = \frac{W}{A}. \tag{1}$$

Now let us consider a different situation. Let us lift the weight W slightly above the stopper. At $t = 0$ we suddenly drop the weight W on the stopper. The wire extends, but equilibrium cannot be maintained. The free end accelerates downward at first, then it is pulled back by the tension in the wire, and finally at a certain instant of time $t = t_0$ the free end becomes stationary, with a zero velocity and a finite displacement δ'. At this instant, $t = t_0$, the wire and the weight are not in equilibrium; an upward acceleration exists and the weight starts an upward motion. Eventually the weight vibrates about an equilibrium position. If we make the assumptions that the strain in the wire

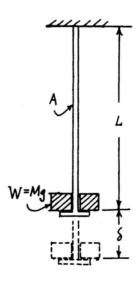

FIGURE 12.4:1 A suddenly applied load on a wire.

is uniform throughout the length of the wire at all times, and that the weight W and the stopper are perfectly rigid, then the maximum strain and stress in the wire are reached at $t = t_0$, and can be calculated as follows. The strain at $t = t_0$ is δ'/L, the stress is $E\delta'/L$, and the total strain energy in the wire is $\frac{1}{2}E\left(\dfrac{\delta'}{L}\right)^2 AL$. This strain energy must be equal to the work done by the external load, $W\delta'$. The balance of energy, therefore, yields the equation

$$\frac{1}{2}E\left(\frac{\delta'}{L}\right)^2 AL = W\delta'.$$

Hence

$$\delta' = 2\frac{WL}{AE}, \qquad \sigma' = E\frac{\delta'}{L} = 2\frac{W}{A}. \tag{2}$$

Comparing Eqs. (1) and (2), we see that

$$\delta' = 2\delta, \qquad \sigma' = 2\sigma. \tag{3}$$

Thus the dynamic stress and strain are twice the corresponding static values.

The difference between the two cases examined above arises from the difference in the work done by the external load. In the first case the equilibrium is maintained so that the load experienced by the wire is linearly proportional to the elongation, as shown in Fig. 12.4:2 and the work done

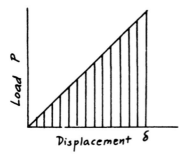

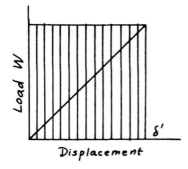

FIGURE 12.4:2 The load-deflection relationship of a spring when the load is applied slowly. How slow the loading must be to qualify for such a "static" response is discussed in the text.

by the external load is equal to the area of the triangle. In the second case the full load W acts on the wire, and the work performed is equal to the area of the shaded rectangle in Fig. 12.4:2.

The reasoning given above is so simple that it cannot fail to impress us. But is the numerical factor 2 infallible? Under what conditions is it valid? Under what conditions would it be in gross error?

The condition of validity is contained in the statements "we suddenly drop the weight W on the stopper", and "we assume that the strain in the wire is uniform throughout the length of the wire". If these assumptions are not valid, the ratio between the maximum dynamic stress and the static stress will be different from 2. In fact, it may be much smaller or larger than 2. This will become clear in the following sections.

12.5 Vibration and the Amplification Spectrum of Dynamic Structural Response

The problem discussed in Sec. 12.4 and sketched in Fig. 12.4:1 can be presented as follows. A mass M is attached to a spring (wire). Under the assumption of uniform stress distribution in the spring, let the length of the spring be L and the deflection be u. The tension in the spring is then EAu/L. At time $t = 0$ a gravitational acceleration g is suddenly imposed on the system. The equation of motion of the mass is then

$$M\frac{d^2u}{dt^2} = -\frac{EA}{L}u + Mg. \tag{1}$$

The initial condition is

$$u = \frac{du}{dt} = 0 \quad \text{when } t = 0. \tag{2}$$

The solution of Eq. (1) is

$$u = C_1 \sin \omega t + c_2 \cos \omega t + \frac{LM}{EA}g, \tag{3}$$

where ω is the circular frequency of free vibration

$$\omega = \sqrt{\frac{EA}{LM}}. \tag{4}$$

The initial conditions Eq. (2) are satisfied if

$$c_1 = 0, \qquad c_2 = -\frac{LM}{EA}g. \tag{5}$$

Hence

$$u = \frac{LM}{EA}g\,(1 - \cos \omega t). \tag{6}$$

Thus the deflection history is sinusoidal. The maximum deflection is reached when $\omega t = n\pi$ (n being odd integers), at which the deflection is $2LMg/(EA)$, corresponding to a stress $2Mg/A$ in the spring. This proves that the maximum dynamic stress in the spring is twice the static value Mg/A.   Q.E.D.

This simple example can be easily extended to give many applications. We know that shock problems facing an engineer do not always arise from solid-to-solid impact. In most cases the blow reaches an object after being softened by some intermediate structures. Such are the cases of a passenger in a train, an instrument in a package, a machine soft-mounted on the ground, a building subjected to an earthquake.

The ground shock problem can be illustrated by a simple case shown in Fig. 12.5:1. A mass M is attached to a massless spring, which in turn is built-in to a "ground". When the ground moves horizontally with a displacement history $s(t)$, the mass M moves to $x(t)$. If the spring constant (force per unit displacement) is K, and the damping constant (force per unit velocity) is c, then the equation of motion of the mass is

$$M\ddot{x} + c(\dot{x} - \dot{s}) + K(x - s) = 0, \tag{7}$$

where x represents the horizontal displacement of the mass, and a dot over x indicates a differentiation with respect to time. If we let

$$y = x - s \tag{8}$$

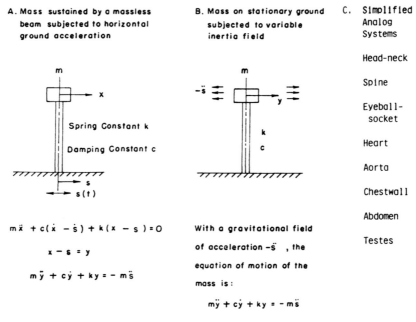

FIGURE 12.5:1 Ground shock of a cantilevered structure.

represent the displacement of the mass relative to the ground, then Eq. (7) may be written as

$$M\ddot{y} + c\dot{y} + Ky = -M\ddot{s}.$$ (9)

It is convenient to write this equation in the form

$$\ddot{y} + 2\varepsilon\omega\dot{y} + \omega^2 y = -\ddot{s},$$ (10)

$$\omega^2 = \frac{K}{M}, \qquad \varepsilon = \frac{c}{2\sqrt{KM}},$$ (11)

where ω is the natural frequency of the system, and ε is the ratio of actual damping to the critical damping of the system.

The solution of Eq. (10) for an arbitrary ground acceleration $\ddot{s}(t)$ and arbitrary initial conditions

$$y(0) = y_0, \qquad \dot{y}(0) = \dot{y}_0 \quad \text{when } t = 0$$ (12)

can be written in a closed form:

$$y(t) = y_0 e^{-\varepsilon\omega t}\cos\omega\sqrt{1-\varepsilon^2}\,t + \frac{1}{\omega}(\dot{y}_0 + y_0\varepsilon\omega)e^{-\varepsilon\omega t}\sin\omega\sqrt{1-\varepsilon^2}\,t$$

$$- \frac{1}{\omega}\int_0^t \ddot{s}(\xi)e^{-\varepsilon\omega t}\sin\sqrt{1-\varepsilon^2}\,\omega(t-\xi)\,d\xi,$$ (13)

and mathematically the problem is solved.

The motion expressed by Eq. (13) represents a forced motion during the interval in which $\ddot{s}(t) \neq 0$, and a residual oscillation after $\ddot{s}(t)$ transpires. An engineer is interested in $y(t)$ because it gives information on the dependence of the response on the structural parameters ω, the natural frequency, ε, the damping factor, and $\ddot{s}(t)$, the pulse history. In particular, the maxima and minima of $y(t)$ as functions of these parameters are of interest. The maximum response as a function of the natural frequency of the structure is known as the *response spectrum*.

Real problems of structural response to ground acceleration differ from the idealized example considered above by having many (or an infinite number of) degrees of freedom. But there exists a classical method of generalized coordinates which enables us to write down the Lagrangian equations of motion. See Chap. 2. For certain types of damping and forcing functions these equations can be decoupled by introducing appropriate normal modes of free vibration as generalized coordinates (Sec. 2.10). In that case for each normal mode the mathematical problem is the same as that of the single-degree-of-freedom case.

Example 1. Let the pulse $\ddot{s}(t)$ be a trapezoid as shown in the first panel of Fig. 12.5:2 (rise from 0 to peak in time t_m, constant for a period t_m, then decay in another t_m). The response $y(t)$, calculated from Eq. (13), is shown in the figure for several values of the parameters $2ft_m$ and the damping factor ε. Here

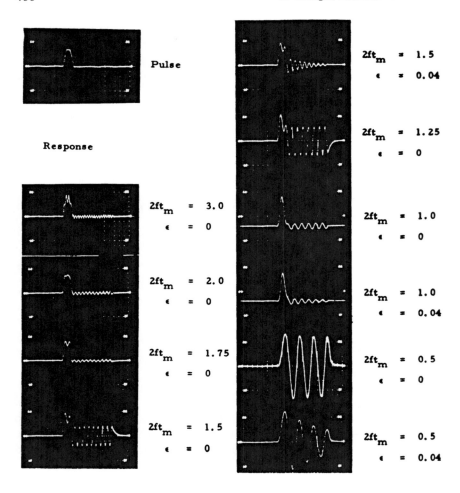

f = Frequency of oscillator, cycles per sec.

t_m = Rise time to peak of pulse, sec.

ϵ = Percent critical damping

FIGURE 12.5:2 The dynamic response of a single degree of freedom elastic oscillator to a trapezoidal pulse.

t_m is the rise time of the pulse, f is the frequency of natural vibration of the system. The parameter $2ft_m$ is called the *rise-time parameter*:

$$2ft_m = \frac{\text{rise time}}{\text{half period of vibration}}. \tag{14}$$

Example 2. *Amplification spectrum.* For an impact force, the maximum static displacement under the maximum absolute value of the force is

$$\delta = \frac{\max|F(t)|}{K} = \frac{\max|F(t)|}{m\omega^2}. \tag{15}$$

The ratio of the maximum deflection of the spring in a mass-spring system to δ is defined as the *dynamic amplification factor* or *amplification spectrum*:

$$\text{Amplification spectrum} = \frac{|y|_{\max}}{\delta} = \frac{\text{max. dynamic displacement}}{\text{max static displacement}}. \tag{16}$$

If an acceleration pulse is considered, we define

$$\delta = \frac{\max|\ddot{s}(t)|}{\omega^2} \tag{17}$$

and the amplification spectrum as

$$\frac{|y|_{\max}}{\delta} = \frac{|\ddot{y} + \ddot{s}|_{\max}}{|\ddot{s}|_{\max}} = \frac{\text{max acc. of mass}}{\text{max ground acc.}}. \tag{18}$$

The amplification spectrum for the trapezoidal pulse illustrated in Fig. 12.5:2 shows that the maximum response is reached when $2ft_m$ is in the neighborhood of 0.5. When $2ft_m$ is small, the amplification factor increases linearly with $2ft_m$. When $2ft_m$ is greater than 2, the amplification factor tends to 1.

The force shock amplification spectrum is the same as the ground acceleration spectrum for $\ddot{s}(t) = F(t)/m$. The t_m for the pulses are the time at which the peak impact force is reached in each impact.

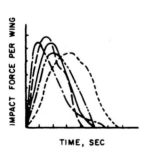 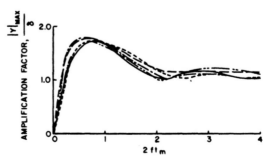

FIGURE 12.5:3 The *amplification spectrum* of a single-degree-of-freedom oscillator to a set of impact loads recorded from air-plane landings by a variety of airplanes and airports. ε is the damping factor, $\varepsilon = 1$ corresponds to critical damping.

Example 3. *Amplification spectrum of aircraft landing pulses.* Figure 12.5:3 shows a set of records of the landing impact force of an airplane and the amplification spectrum computed for each impact. It is seen that although the impact force histories vary considerably from one landing to another, the amplification spectra look much alike. For the purpose of engineering design and analysis of injury potential, it often suffices to know the mean amplification spectrum.

12.6 Impact and Elastic Waves

John Hopkinson (1872) performed an interesting experiment of which the explanation would help us understand the nature of elastic wave propagation and its significance in shock and trauma. He tried to measure the strength of a steel wire by suddenly stretching the wire. As is shown schematically in Fig. 12.6:1, a ball-shaped weight pierced by a hole was threaded on the wire and dropped down from a known height so that it struck a clamp attached to the bottom of the wire. For a given weight we expect the existence of a critical height beyond which the falling weight would break the wire. Using different weights dropped from different heights, however, Hopkinson obtained the remarkable result that the minimum height from which a weight had to be dropped to break the wire was, within certain ranges, almost independent of the size of the weight!

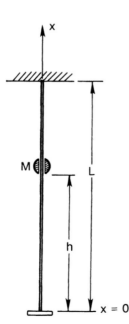

FIGURE 12.6:1 John Hopkinson's experiment on a weight falling on a stopper supported by a metal wire. The experiment was designed to test the strength of the metal.

Now, when different weights are dropped from a given height, the velocity reached at a given level is independent of the magnitude of the weight. Hopkinson's result suggests that in breaking the wire it is the velocity of the loading that counts. Following this lead, Hopkinson explained his result on the basis of elastic wave propagation. He knew that in a plane progressive wave propagating in a homogeneous isotropic elastic medium (see Sec. 5.7), the stress σ is proportional to the particle velocity v (Eq. 5.7:12):

$$\sigma = \rho c v, \tag{1}$$

where $c = \sqrt{E/\rho}$ is the speed of longitudinal waves in the wire, ρ is the density of the material, E is Young's modulus of elasticity. The stress in the wire, however, is not largest at the instant of impact at the lower end. The largest value is reached sometime later when the elastic wave had propagated up and down the wire a few times. When this largest stress equals the ultimate stress of the wire, the wire breaks.

When the weight hits the clamp in Hopkinson's test, the end of the wire acquires a particle velocity V_0 which is equal to that of the weight. A steep-fronted tension wave is generated and propagated up the wire. In the meantime, the weight is slowed down by the tension of the wire. The elastic wave, on reaching the fixed end at the top, is reflected as a tension wave of twice the intensity of the incident wave. The reflected wave is reflected again at the lower end, and so on. John Hopkinson used a 27-foot long wire and weights ranging from 7 lbs to 41 lbs; and the absolute maximum tensile stress was reached near the top after a number of reflections. Bertram Hopkinson (1914), in repeating his father's experiment, used a smaller weight (1 lb) so that the weight was slowed down faster. G.I. Taylor (1946) showed theoretically that the maximum tensile stress in B. Hopkinson's experiment occurred at the third reflection, i.e. the second reflection at the top of the wire when the tensile stress reached $2.15 \, \rho c V_0$.

The impact wave analysis is very similar to the arterial pulse wave analysis presented in Secs. 5.7–5.9. Consider a wire as shown in Fig. 12.6:1. Choose an axis x along the length of the wire, with the origin o located at the lower end. When the wire is loaded, each particle in the wire is displaced longitudinally from its original position by a small amount u, which is positive in the direction of x. Assuming that plane cross sections remain plane, then u is a function only of x and time t. The strain in the wire is

$$e = \frac{\partial u}{\partial x}. \tag{2}$$

If the wire material obeys Hooke's law, then the axial stress is

$$\sigma = Ee = E \frac{\partial u}{\partial x}. \tag{3}$$

The equation of motion, Eq. (1) of Sec. 1.7, becomes, in the present case

$$\rho \frac{\partial^2 u}{\partial t^2} = \frac{\partial \sigma}{\partial x}. \tag{4}$$

A substitution of Eq. (3) yields the wave equation

$$\frac{\partial^2 u}{\partial x^2} - \frac{1}{c^2} \frac{\partial^2 u}{\partial t^2} = 0 \tag{5}$$

with the wave speed

$$c = \sqrt{\frac{E}{\rho}}. \tag{6}$$

The general solution of Eq. (5) is

$$u = f(x - ct) - g(x + ct), \tag{7}$$

where f and g are two arbitrary functions. The function $u = f(x - ct)$ represents a wave propagating in the positive x direction; whereas $u = g(x + ct)$ represents a wave propagating in the negative x direction. In either case we have

$$\frac{\partial u}{\partial x} = \pm \frac{1}{c} \frac{\partial u}{\partial t} = \pm \frac{v}{c} \tag{8}$$

with $v = \partial u / \partial t$ denoting the particle velocity. A substitution of Eq. (8) into Eq. (3) yields the formula

$$\sigma = \pm \frac{E}{c} v = \pm \rho c v \tag{9}$$

where the $-$ sign applies to a wave propagating in the positive x-direction, and the $+$ sign applies to a wave in the other direction. This is the equation quoted at the beginning of this section.

The Hopkinson problem (Fig. 12.6:1) is specified by the initial and boundary conditions

$$u = 0 \quad \text{at } x = L \text{ for all } t, \text{ and when } t \leqslant 0 \text{ for all } x, \tag{10}$$

$$M \frac{dV}{dt} = A\sigma - Mg \quad \text{at } x = 0, \tag{11}$$

$$V = -V_0 \quad \text{when } t = 0. \tag{12}$$

Here V is the value of $\dfrac{\partial u(x, t)}{\partial t}$ at $x = 0$. The mathematical problem is reduced to finding the arbitrary functions $f(x - ct)$ and $g(x + ct)$ so that $u(x, t)$ given by Eq. (7) satisfies Eqs. (10)–(12).

G.I. Taylor's calculated results about the elastic stress history at the two ends of the wire in Hopkinson's experiments are shown in Figs. 12.6:2 and 12.6:3. Figure 12.6:2 refers to one of J. Hopkinson's experiments, with the mass

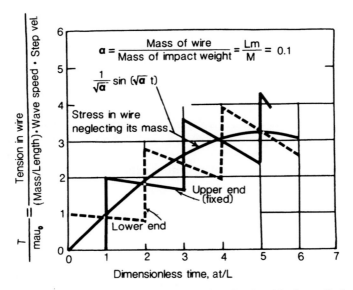

FIGURE 12.6:2 Elastic waves in a spring induced by a load suddenly applied to a mass attached to the end of the spring. Solution by G.I. Taylor (1946) when the mass of the wire is equal to 10% of the mass of the impact weight.

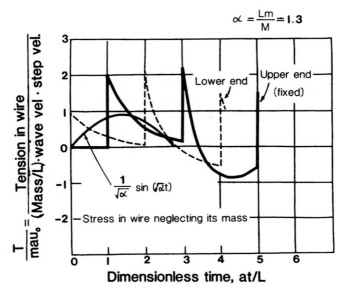

FIGURE 12.6:3 Taylor's solution when the mass of the impact weight is comparable to the mass of the wire.

of the wire equal to 10 percent of the mass of the impact weight. The clamp at the top of the wire and the impact weight itself were assumed to be rigid. The ordinate represents $\sigma/\rho c V_0$, i.e., the ratio of the stress at any time to the maximum stress in the wave before the first reflection. The abscissa represents ct/L so that reflections occur at the top when $(ct/L) = 1, 3$, and 5, and at the bottom when $(ct/L) = 2, 4$, and 6. Figure 12.6:3 refers to B. Hopkinson's experiments with $\rho AL/M = 1.3$.

If the density of the wire tends to zero, then the wave speed tends to infinity, and the tension in the wire becomes uniform and varies harmonically in the manner of the simple spring-mass system discussed in Sec. 12.4. As a spring-mass system the stress in the spring can be expressed by the formula

$$\frac{\sigma}{\rho c V_0} = \frac{1}{\sqrt{\alpha}} \sin\left(\sqrt{\alpha}\frac{ct}{L}\right) \tag{13}$$

where, in Taylor's notation,

$$\alpha = \frac{\rho AL}{M} = \frac{\text{length of wire}}{\text{characteristic length } M/(\rho A)}. \tag{14}$$

In Fig. 12.6:2 it is seen that when $\alpha = 0.1$ the stress at each end varies above and below the stress represented by Eq. (13) by an amount approximately equal to $\rho c V_0$, the overall stress history is a disturbed sine curve. On the other hand, for $\alpha = 1.3$, the stress history has lost all resemblance of a sine wave, as shown in Fig. 12.6:3.

This simple analysis calls attention to the importance of elastic waves, sound speed, and particle velocity in the creation of trauma.

The speed of sound in several human organs is listed in Table 12.6:1. It is seen that the cortical bone has a sound speed of 3500 m/sec. This may be compared with the sound speed of 4800 m/sec in steel, aluminum, copper, etc. The speed of elastic waves in the lung is of the order of 30–45 m/sec (Yen et al., 1986). The lung has a complex structure. There are many pathways along which forces can be transmitted; hence there are many types of sound waves and sound speeds. Table 12.6:1 lists several sound speeds in the lung. The speeds given by Yen et al. (1986) were measured from lungs of man, cat, and rabbit under impact pressure and wall velocity comparable with those induced by shock waves in an air blast; and the values depend on the transpulmonary pressure and animal species. The sound speed given by Rice (1983) was measured by a microphone listening to a sound made by an electric spark. The speed given by Dunn and Fry (1961) was measured by ultrasound waves. The ultrasound waves are apparently very different from the waves measured by Yen et al. and Rice. Note that the speed of sound in the lung is much lower than both the speed of sound in the gas and in the tissue, in analogy with the fact that the sound in water containing gas bubbles is much slower than the sound speeds in pure water and in gas alone.

TABLE 12.6:1 Velocity of sound in various tissues, air, and water

Tissue	Density (g/cm^3)	TPP* (kPa)	Sound speed mean ± S.D. (m/sec)	Reference
Muscle	1		1580	Ludwig (1950), Frucht (1953), von Gierke (1964)
Fat	1		1450	Ludwig (1950), Frucht (1953)
Bone	2.0		3500	Clemedson and Jönsson (1962)
Collapsed lung	0.4		650 (ultrasound)	Dunn and Fry (1961)
Collapsed lung pneumonitis	0.8		320 (ultrasound)	Dunn and Fry (1961)
Lung, horse	0.6		25	Rice (1983)
Lung, horse	0.125		70	Rice (1983)
Lung, calf			24–30	Clemedson and Jönsson (1962)
Lung, goat		0	31.4 ± 0.4	Yen et al. (1986)
		0.5	33.9 ± 2.3	
		1.0	36.1 ± 1.9	
		1.5	46.8 ± 1.8	
		2.0	64.7 ± 3.9	
Lung, rabbit		0	16.5 ± 2.4	Yen et al. (1986)
		0.4	28.9 ± 3.3	
		0.8	31.3 ± 0.9	
		1.2	35.3 ± 0.8	
		1.6	36.9 ± 1.7	
Air			340	Dunn and Fry (1961)
Water, distilled, 0°C			1407	Kaye and Labby (1960)
Air bubbles (45% by vol.) in glycerol and H$_2$O			20	Campbell and Pitcher (1958)

* TPP = Transpulmonary pressure = airway pressure − pleural pressure.
  1 kPa = 10^3 N/m^2 ∼ 10.2 cm H$_2$O.

12.7 Wave Focusing and Stress Concentration

In the preceding three sections, we have analyzed a mass on an elastic wire three ways. In Sec. 12.4, we obtain a dynamic amplification factor of 2 under the assumption of uniform stress in the wire. In Sec. 12.5, we obtain a dynamic amplification spectrum when the stress in the wire is assumed to be uniform but the loading is not instant but consists of a pulse of specified shape, and the system is allowed to vibrate. In Sec. 12.6, the assumption of uniform stress throughout the wire is removed so that elastic waves are revealed. The third analysis is of course the most general. If the second analysis is generalized to include all the normal modes of the wire (Chap. 2), the wave features will be revealed. If the third analysis is applied to a loading pulse of specified shape and continued long enough, the vibration aspect will be seen. Thus, both the second and third analysis are valuable, and in special situations simplified versions reveal important aspects of dynamic response. To appreciate the special situations, note that the loading pulse and the structure have the following characteristic times:

> Loading pulse: rise time t_m.
> Vibration mode: period, inverse of frequency, f^{-1}.
> Elastic wave: $L/c \equiv$ body length/wave speed.

When $t_m \ll L/c$, the simplified analysis of Sec. 12.5 reveals almost the whole story. When $t_m \gg L/c$, the vibration analysis, generalized to include all relevant normal modes, solves the problem of maximum stress faster. In the latter case, the dimensionless variable $2 f t_m$ is an important parameter with regard to the prediction of maximum response, as demonstrated in Fig. 12.5:3.

The same comment applies to the three-dimensional elastic bodies of more complex geometry.

In an infinite domain of a linear, homogeneous, isotropic, elastic medium, there can be *longitudinal waves* in which the material particles of the medium move in the direction of propagation of the waves, and *transverse waves* in which the material particles move in a direction perpendicular to the direction of propagation of the waves. The longitudinal waves are also said to be *irrotational* waves, or *dilatational* waves, or *compression* waves. In seismology, it is called a *P-wave*, P signifying "push". The transverse waves are also said to be *rotational waves*, or *shear waves*. In seismology, it is called an *S-wave*, S signifying "shake".

The wave speed for a longitudinal wave is denoted by C_L; that for a transverse wave is denoted by C_T. C_L and C_T are material constants of the medium. For a linear isotropic elastic medium, they are related to the Young's modulus E, shear modulus G, Poisson's ratio v, and the density of the material ρ, by the equations

$$C_L = \sqrt{\frac{E(1 - v)}{\rho(1 + v)(1 - 2v)}}, \qquad C_t = \sqrt{\frac{G}{\rho}}. \tag{1}$$

See e.g., Fung (1977), p. 310.

In a semi-infinite medium bounded by a plane which is free from surface traction, a plane *S*-wave incident upon the free surface will be reflected as a combination of a *P*-wave and an *S*-wave. Similarly, an incident *P*-wave will be reflected as both *P*- and *S*-waves.

If two elastic media which have different wave speeds are joined together in a plane, then the diffracted waves also consist of both *P*- and *S*-waves.

In a semi-infinite body there exists a type of surface wave called the *Rayleigh wave*, which propagates at a slower speed then C_T, and for which the elastic displacement decreases exponentially as the distance from the free surface increases. For a layered medium there exists a *Love wave*, which is caused by the interface between the media. These surface waves are dispersive. More complex wave patterns arise when the interface is curved.

In a finite elastic body, the wave patterns are more complex. In a sphere there exists spherical waves which converge at the center with increasing intensity. In a cylinder there exists similarly cylindrical waves which converge at the center. These are examples of focusing. Focusing is not limited to spheres and cylinders. A curved surface or interface in an elastic body can cause focusing and increased intensity of wave motion.

Extending these concepts to man and animals subjected to impact load, one can visualize the complexity of stress distribution in internal organs.

The stress–strain relationships of the soft tissues of man and animals are nonlinear. Their elastic moduli increase with increasing stress. This kind of strain-stiffening material property can lead to shock waves analogous to the strong shock of supersonic airplane and the surf breakers in water waves along the sea shore. The basic cause of these shock waves is that particles that have higher stress also have faster speeds of propagation. Hence the strong stress is pushed to the wave front, causing a jump condition.

Although a detailed analysis is beyond the scope of this book, the general features of impact and trauma are not beyond comprehension.

Focusing by curved surfaces is an effective way to concentrate energy into small regions in a material. If the energy of an impact on a body can be concentrated into a small region by a mechanism of focusing, the small region may be first to become endangered. For example, if a man is subjected to a pressure wave due to explosion, the gas-containing organs, ear, lung and intestine, are the most susceptible to damage (Secs. 12.8, 12.9). The convergence of stress waves and the reflection of waves on the heart and spine undoubtedly play a role in this case.

12.8 Trauma of the Lung Due to Impact Load

In the next two sections, we analyze the trauma of the lung in order to illustrate the application of the wave theory and to discuss the meaning of injury and tolerance levels.

It is known that when a man or animal is subjected to a shock wave due to a bomb or industrial explosion, the ear and lung are most prone to damage (Bowen et al., 1965; Clemedson, 1956; Clemedson and Jönsson, 1962). Lung injury is revealed by edema and hemorrhage. Trauma patients of automobile crash accidents may also suffer lung injury and pulmonary edema.

The following sequence of facts may be noted: (1) In a plane progressive wave, the stress is equal to the product of the tissue velocity, sound velocity, and the mass density of the tissue, Eq. (12.6:9). (2) If the impact load is applied rapidly, the induced velocity of the lung tissue can be high. (3) The velocity of sound in the lung is singularly low among all organs, Table 12.6:1. (4) The Mach number of impact of the lung, i.e., the ratio of the velocity of lung tissue to the sound velocity of lung, can approach 1 or exceed 1, i.e. can be transonic or supersonic. A supersonic shock wave concentrates disturbance and energy in a small region behind the wave front. (5) Trauma occurs when stress or strain exceed certain critical values. (6) High speed impact injury is therefore expected to be localized. (7) Focusing by a curved wave front may cause further concentration of damage. Evidences of hemorrhagic injury in the lung are usually localized and are usually most severe next to the spine, heart, ribs. These facts suggest the importance of the wave feature of the phenomena.

Clemedson (1956) was the first to state that the initial surface velocity of the lung is the key parameter with respect to lung injury. Jönsson et al. (1979) constructed a drop tower to test lung injury based on this concept. Clemedson and Jönsson (1962) stated that the lethal injury level for rabbit lung lies at a velocity of around 15 ms^{-1}.

Yen et al. (1988) experimented on a greatly simplified model: They excised rabbit lung, supported it on a fine soft net (Fig. 12.8:1), perfused it with saline at isogravimetric condition (i.e. a condition in which the lung weight does not change with time), impacted it with a shock wave or a light weight pellet, measured the initial velocity and the maximum displacement at the point of impact, and continuously monitored the weight of the lung following impact for many hours. The pulmonary venous pressure was fixed at 3 cm H_2O, the airway pressure was held at 10 cm H_2O, and the pulmonary arterial pressure ranged from 12.5–15 cm H_2O.

Two kinds of impactors were used: A shock tube which sends air shock

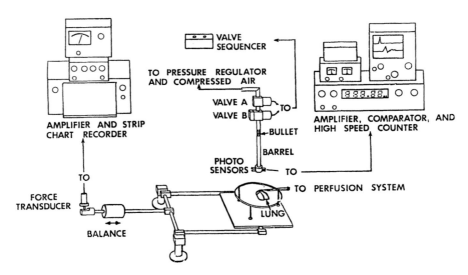

FIGURE 12.8:1 Test set-up used by Yen et al. (1988) to study lung edema due to impact load. The lung was perfused. Isogravimetric condition must be established before testing. Change of lung weight after impact was measured.

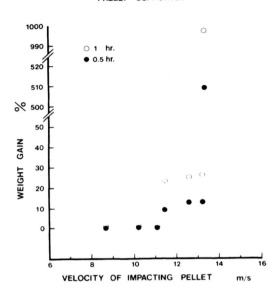

FIGURE 12.8:2 Rate of the gain of lung weight (as a percentage of the weight of the lung at isogravimetric condition before impact) as a function of impact velocity of the pellet. Pellet weighed 1.49 g with a cross-sectional area of 2 cm^2. From Yen et al. (1988).

waves of prescribed Mach number and overpressure, and a compressed air gun which shoots a pellet of 1.49g with a cross section of 2 cm^2. The surface deflection was measured by a SP 2000 Motion Analyzer System (Spin Physics, Inc.) at 2000 frames per sec.

Pellet test results are shown in Fig. 12.8:2. The lungs were "freely" supported on a nylon net. No edema was found if the impacting pellet velocity was less than 11 ms^{-1}. But edema reached 20% of initial lung weight 1 h after impact when the impacting pellet velocity was 11.5 ms^{-1}. At an impact velocity of 13.5 ms^{-1} and above, the lung weight increased at a rapid rate. Then edema was massive, and the shock loading has done a great damage.

The advantage of the excised-lung approach is a better defined measurement. The disadvantage is that the effects of many other factors are omitted from consideration, hence it cannot directly yield a tolerance level of the whole animal.

Lau and Viano (1981), Viano and Lau (1988) found that impulsive injury primarily of the alveolar region is prevalent at impact velocities above 15 ms^{-1}. In studying anesthetized intact rabbits, Yen et al. (1988) found that the critical impact velocity for the initiation of edema is lower for the whole animal than that for the isolated lung. This is probably due to the more complex geometric structure of the whole animal, resulting in more complex wave patterns and stronger stress concentration.

Edema is not the only feature of lung trauma. Hemorrhage is another. There are evidences that hemorrhage is not necessarily bad; it leads to blood coagulation and then stoppage of flow, cutting off edema. In post-mortem examination of victims of impact accidents, it was often noted that there were red marks on the surface of the lung. They are named "rib markings". Since the rib markings are of red color, they are often considered a sign of hemorrhage of the delicate pulmonary capillary blood vessels (Clemedson and Jönsson, 1962, Frisoli and Cassen, 1950). Yen and Fung (1985) showed, however, that rib markings often are not marks of hemorrhage; Often they are marks of atelectasis, i.e., collapsed alveoli, removable by rebreathing or reinflation.

12.9 Cause of Pulmonary Edema in Trauma

Since soft tissues have good strength in compression, why does a compression wave cause edema? The hypothesis of Fung et al. (1988) is that tensile and shear stresses are induced in the alveolar wall on rebound from compression, and that the maximum principal stress (tensile) or maximum shear stress in the lung during the dynamic process may exceed critical values for increased permeability of endothelium and epithelium to small solutes, or even fracture. Furthermore, small airways may collapse and trap gas in alveoli at a critical strain, causing traumatic atelectasis. The collapsed airways reopen at a higher strain after the wave passes, during which the expansion of the trapped gas will induce additional tension in the alveolar wall.

Increased permeability of the epithelium of the interalveolar septa due to traumatic stretching is considered to be a factor of great importance in trauma. If fluid movement obeys Starling's law, Eq. (8.6:14), then the rate of movement of fluid per unit area of a membrane is proportional to the difference of static pressures on the two sides of the membrane minus the osmotic pressure difference. Edema can be caused by a change of the distributions of the concentrations of the solutes or the static pressures in such a way that fluid will move from the interstitium to the alveolar space. For example, if the epithelium becomes permeable to a certain small solute, then that solute in the interstitium will cross the epithelium into the alveolar side, increase the osmotic pressure there, and pull fluid from the interstitium into the alveolus. The movement of fluid and solutes can be studied by the isogravimetric method, indicator dilution method, electron microscopy, lymph measurement, etc. (Chap. 9). See Crone and Lassen (1970), Fishman and Hecht (1969), Giuntini (1971), and Staub (1978). Brigham (1978), Effros et al. (1982), Egan et al. (1976), Nicolaysen and Hauge (1982) have shown that the epithelium is less permeable to small solutes than the endothelium, and that the change of permeability of the epithelium to small solutes is probably the reason for alveolar edema.

Our hypothesis stated above assumes that it is the tensile stress that does the damage. There are two ways to induce tensile strains in alveolar walls. One is by macroscopic dynamic response of the lung–chest system to the impact load. The other is the micromechanical response of the alveoli to stress waves. The dynamics of the chest and lung was analyzed by Bowen et al. (1965) and White et al. (1971) as a single degree-of-freedom elastic shell enclosing a gas which has a uniform pressure. Chuong (1985) analyzed the stress waves with the finite-element method and showed that the pressure distribution is very nonuniform when the lung is subjected to a traveling shock wave, and that in some locations tensile strain with a magnitude comparable with the absolute value of the maximum initial compressive strain is induced. Superposed on this macroscopic dynamic response are the micromechanical response of the airways and alveoli, as well as the normal tensile stress due to inflation. The bronchioles are compressed when the shock wave arrives. If a critical level of strain is exceeded, the bronchioles will collapse, trapping gas in the alveoli. After the shock wave passes, the trapped and compressed gas in the alveoli rebounds, creating tension in the interalveolar septa. The tension causes damage. This mechanism needs an experimental verification. Our validating experiments are described below.

Experimental Test of the Hypothesis

To test this hypothesis, Fung et al. (1988) used isolated rabbit lung perfused with saline at the isogravimetric condition and measured the rate of lung weight increase due to a transient increase of stretch of the alveolar membrane. The stretch lasted for 1 min., then the lung was returned to the normal condition to measure the rate of weight change. The weighing platform is similar to that shown in Fig. 12.8:1. At the 5 minutes resting period, the airway pressure was maintained at 10 cm H_2O, and the lung weight was monitored.

The results for four rabbits are given in Fig. 12.9:1. The rate of increase of lung weight in the resting period, dw/dt, is expressed as a percent of the initial lung weight per hour, and is plotted against the 1 minute stretching pressure preceding that period. It is seen that as the lung was increasingly stretched, the rate of increase of lung weight rises, i.e. edema increases. Hence stretching the lung promotes edema.

The second experiment demonstrates gas trapping in compressed lung and the re-expansion of the gas when the compression wave passes.

Tao and Fung (1987) compressed isolated rabbit lungs with trachea open to the atmosphere and recorded the relationship between the lung volume and pleural pressure. The lung was hung in a Lucite box, which can be pressurized to a variable pressure acting on the pleura, P_{PL}. The airway was open to the atmosphere so that the alveolar gas pressure, P_A, is zero. The transpulmonary pressure, $P_t = P_A - P_{PL}$, was 6.4 cm H_2O at the outset. With increasing pleural pressure they measured the lung volume. They found that for the rabbit lung a range of pleural pressure exists in which gas will be trapped in the alveoli, see Fig. 12.9:2. At a critical closing pleural pressure, P_{cl}, a limiting lung volume

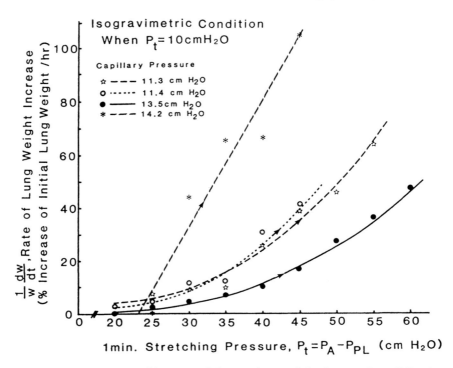

FIGURE 12.9:1 The rate of increase of the wet lung weight due to edema following successive steps of increased stretching of rabbit lung. Each stretch lasted 1 min. Each resting period was 5 min.; during which the transpulmonary pressure was 10 cm H_2O and the rate of weight increase was measured. Four rabbits; each symbol represents a rabbit. The initial isogravimetric condition is stated in the inset at top left. The rate of increase of lung weight is expressed as percent of the initial lung weight per hour. The regression lines and the initial weight are, from top down,

(1) $y = 4.58x - 104.30$, $(r = 0.9151)$, $w = 13.30g$.

(2) $y = 0.06x^2 - 2.04x + 25.75$, $(r = 0.9882)$, $w = 26.55g$.

(3) $y = 0.05x^2 - 2.25x + 29.14$, $(r = 0.9884)$, $w = 23.40g$.

(4) $y = 0.03x^2 - 1.33x + 16.04$, $(r = 0.9961)$, $w = 41.50g$.

is reached at which the lung behaves like a closed balloon obeying Boyle's law on further compression. In the rabbit this limiting volume was roughly one-quarter to one-half of the initial lung volume which was about 60% of the total lung capacity. On inflating the lung again, there exists another critical pressure, P_{re-op}, at which the lung volume begins to increase again. This critical pressure for reopening is higher than that for closing, and varies with the initial lung volume, the rate of strain, and the maximum compression imposed on the lung. When Boyle's law applies, gas is trapped in the alveoli. In the case

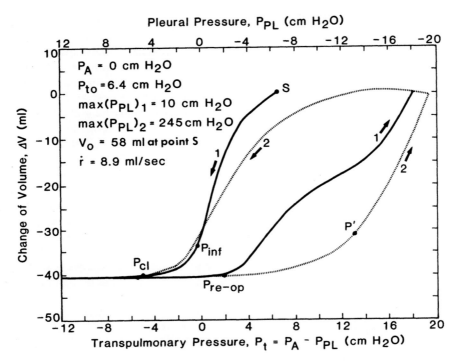

FIGURE 12.9:2 The pressure-volume relationship of rabbit lung with the regime of small and negative transpulmonary pressure emphasized. The airway pressure was atmospheric (zero). At the starting point, S, the transpulmonary pressure was 6.4 cm H_2O, and volume was 58 ml. The rate of volume reduction was 8.9 ml/sec. In application, these curves may be read as a stress-strain relationship of the lung because the tissue stress is equal to the transpulmonary pressure (1 cm H_2O = 98 N/m^2), and the change of volume divided by initial volume is the volumetric strain. From Tao and Fung (1987).

shown in Fig. 12.9:2, the lung was compressed to $P_t = -10$ cm H_2O in the first cycle (solid curve), but to a maximum of $P_t = -245$ cm H_2O in the second cycle (dotted curve). At the point P' the slope of the PV curve in expansion is equal to that in compression at the same volume.

If the strain rate changes, the pressure at the inflection point, P_{inf}, and the closing pressure, P_{cl}, remain relatively unchanged, but the reopening pressure, P_{re-op}, and the characteristic pressure P', where the PV curve bends upward sharped on reinflation, will change to the extent of several cm H_2O. If the lung is compressed to a much higher pressure, the reopening pressure is increased. If the transpulmonary pressure remains lower than the reopening pressure, then the collapsed airways will not be reopened and traumatic atelectasis results. These results show that gas trapping in the compressed lung is a reality.

One should note that the gas trapping mechanism is related to the lung strain only. P_A or P_{PL} alone does not cause the alveoli or small airways to collapse. It is the transpulmonary pressure $P_A - P_{PL}$ that is relevant.

Fung et al. (1988) also presented a theoretical analysis of the response of a group of alveoli with trapped gas to a tension or compression wave. It was shown that when a traveling stress wave arrives and passes, the alveoli respond and moves past the equilibrium condition by the inertial force; and an oscillation follows. In the oscillation the alveoli contracts and expands. In expansion tensile stress is generated in the wall of the alveoli. These facts support the hypothesis that lung trauma is caused by overstretching the alveolar membrane.

Clinical and experimental studies of gas trapping in the lung have been reviewed by Anthonisen (1977), Bates et al., (1971), and Hoppin and Hildebrandt (1977) in relation to the maximal expiratory flow phenomena (Sec. 7.6). Kooyman (1981) has shown that diving animals such as Weddell seal have cartilage in their bronchioles all the way to the alveoli, presumably to prevent the closure of the bronchioles before the closure of the alveoli. Therefore in deep sea diving nitrogen will not be trapped in the seal lung. This may explain why seal do not get bends or decompression sickness whereas human divers do in deep sea diving. In man the trapped nitrogen will be dissolved in blood under high pressure, and evolve as bubbles when the pressure returns to normal. These observations are not directly related to trauma. But according to our reasoning, we anticipate that a Weddell seal has a better tolerance to impact trauma than man.

These theories and experiments, taken together, show the reasonableness of the expectation that pulmonary edema due to traumatic impact is related to the velocity of the surface of the lung induced by the impact.

12.10 Tolerance of Organs to Impact Loads

In the 1950's and 60's an audacious group of people in the United States undertook a remarkable research program to determine the tolerance of man to body acceleration. They built rocket sleds on long tracks in Western deserts. Col. John Paul Stapp and others tested themselves to the limits of their physical tolerance. They recorded their observations and collected data on some primates and animals. The results were extremely helpful to the development of aeronautics and astronautics. In the meantime, people concerned with highway safety and automobile design began their research on the tolerance of various organs to impact load. They realized that only through such quantitative research can progress be made. In the following, we present some data on human tolerance to whole-body acceleration, head impact, and spinal injury. Data on other organs are not summarized, but some leading references are given.

Human tolerance data are difficult to obtain. Data from accident investigations are often clouded. Use of volunteers is expensive and difficult. Use of cadavers or animal models requires extrapolation and has questions in interpretation. Every piece of information in the literature came from some special circumstances and cannot be used indiscriminately.

Whole-Body Acceleration Tolerance

Approximate tolerance limits for well-restrained sitting human subjects have been summarized by Eiband (1959), Roth et al. (1968), and Stapp (1957, 1961, 1970). Unfortunately, evaluated test data are available only in terms of idealized trapezoidal forcing functions with peak acceleration, duration, and rate of onset as the only parameters evaluated.

Tolerance to spineward acceleration (eyeballs-out, $-G_x$) as a function of magnitude and duration of impulse is illustrated in Fig. 12.10:1. For sternumward acceleration (eyeballs-in, $+G_x$), the tolerance limit is similar, but can be higher or lower depending on how well the head is supported. In the longitudinal axis, tolerance to headward acceleration $(+G_z)$ (Fig. 12.10:2) exceeds tolerance to footward acceleration (eyeballs-up, $-G_z$) by a small margin. But note the great difference between human tolerances to G_x and G_z: some volunteers have withstood 45 G with duration less than 0.044 sec or

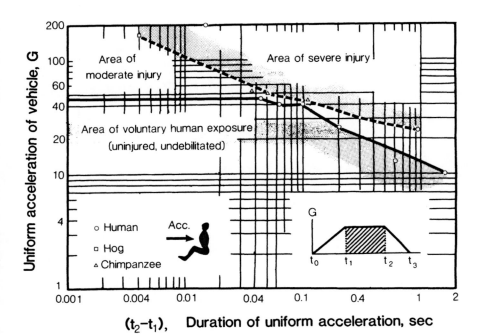

FIGURE 12.10:1 Duration and magnitude of spineward acceleration endured by various subjects. Survivable exposures. Maximal body support used in all cases. From Eiband (1959), by permission.

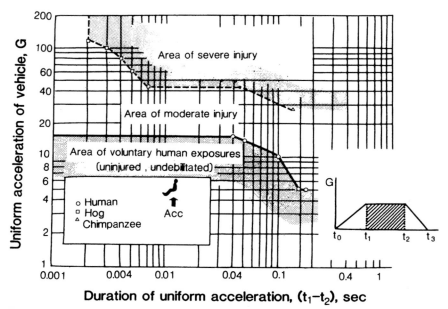

FIGURE 12.10:2 Duration and magnitude of headward acceleration endured by various subjects. From Eiband (1959), by permission.

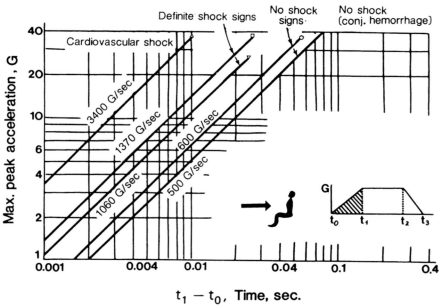

FIGURE 12.10:3 Initial rate of change of spineward acceleration endured by various subjects. The 500–600 G/s curve were obtained by human volunteers fully strapped. The 1060 G/s curve was obtained from Chimpanzee, 1370 G/s curve was from human, 3400 G/s curve was from Chimpanzee. From Eiband (1959), by permission.

25 G at duration 0.2 sec in eyeballs-out accelerations; but in headward, eyeballs-up accelerations the tolerance limit is only 15 G for a duration of 0.1 sec. All the data quoted in these figures are for well-restrained young male adult subjects in top physical condition, with a variety of full-torso and head restraints, not just a lap belt. It is expected that the tolerance limits would be much lower if the subjects were not so well restrained.

The rate of onset of the applied force also has a definite effect on human tolerance. Under some impact conditions, the rate of onset appears to be a determining factor, as indicated in Fig. 12.10:3. The rate effect can be understood from the point of view of the dynamic amplification factor and elastic waves, as was discussed earlier in Secs. 12.5 and 12.6.

Tolerance to lateral (G_y) accelerations seems lower (9 G) for a duration of 0.1 sec. Other data are summarized by Stapp (1970).

A variety of factors are involved for human volunteers in determining where the tolerance limits lie. A variety of trauma are quoted in the area of "severe injury" in these figures. There exists a wide corridor of uncertainty.

Tolerance Level of Individual Organs

More definitive tolerance information can be obtained from experiments designed to study individual organs. Then the injury level can be correlated with the input acceleration pulse.

A certain amount of data exists with respect to the fracture of the cranium, deformation and fracture of facial bones, injuries to the neck, thorax, abdomen, and arms and legs. A general reference is the book edited by Nahum and Melvin (1985). Additional data on tolerance level can be found in the following, which contain further references. For *cranium*, Hodgson and Thomas (1972). For *thorax*, Lau and Viano (1986), Neathery (1974), Stalnaker et al. (1973), Viano and Lau (1987). For *abdomen*, Mertz and Kroell (1970), Stalnaker et al. (1972). For *liver* and *kidney*, Melvin et al. (1973).

In the following we discuss two topics: head and spinal injuries.

Head Injury

Head injury is the most serious frequent trauma of automobile accidents. It is also the most frequent cause of death of passengers in military aircraft in survivable crashes. Table 12.10:1 lists some data from the U.S. Army Safety Center. It is seen that next to the head, chest injuries are the major cause of death. The frequency of serious vertebral injuries is lower for light fixed-wing aircraft than it is for helicopters, implying that the vertical component of impact load experienced by the occupants of helicopters during impact is larger than that in light, fixed-wing aircraft. Rollover of helicopter is a major cause of fatalities. The British Royal Air Force experience was similar (Hill, 1978).

The brain can be injured by fracture, impingement, excessive acceleration,

TABLE 12.10:1 *Frequency of injuries to each body part* as percentages of total injuries (U.S. Army Aircraft, 1971 through 1976, total injuries in parentheses). Data from Laananen (1980)

	Major and fatal injuries combined		Fatal injuries only	
	Helicopters (1,114)	Light fixed-wing (104)	Helicopters (403)	Light fixed-wing (53)
Head	19.7	19.2	31.5	30.2
Face	9.4	14.4	5.0	5.7
Neck	2.6	0.0	2.7	0.0
Arms, hands	12.1	11.5	7.7	5.7
Thorax	12.5	19.2	21.6	28.3
Abdomen	7.1	5.8	11.7	9.4
Pelvis	3.0	1.9	1.0	0.0
Spine	16.5	12.5	6.5	9.4
Legs, feet	17.1	15.4	12.4	11.3

high localized pressure or tensile stress, high localized shear stress and strain, and cavitation in high-tension regions. The regions where the maximum normal stress occurs are usually different from where the maximum shear stress occurs; and they are affected significantly by the flow through foramen magnum (opening at the base of the skull) during impact. The brain tissue can be contused and blood vessels ruptured.

A trauma most widely studied in *brain concussion*, which is defined as a clinical syndrome characterized by immediate transient impairment of neural function, such as loss of consciousness, and disturbances of vision and equilibrium due to mechanical forces. Normally, concussion does not cause permanent damage. It is the first functional impairment of the brain to occur as the severity of head impact increases. It is reproducible in experimental animals.

Concussion has been studied with respect to rotational acceleration (Holbourne, 1943, Gurdjian et al., 1955, Gennarelli et al., 1971, and Hirsch et al., 1970), translational acceleration (Lissner et al., 1960), and flexion-extension of the upper cervical cord during motion of the head-neck junction. Hirsch et al. (1970) have attempted to establish injury criteria and tolerance levels for rotational acceleration on the basis of experiments on monkeys. Lissner et al. (1960), Gurdjian et al. (1953, 1955) proposed a tolerance specification for translational acceleration on the basis of experiments on cadavers and animals. Their proposal is known as the Wayne State University Concussion Tolerance Curve. As it is shown in Fig. 12.10:4, the ordinate is the "effective" acceleration (which is an average front-to-back acceleration of the skull measured at the occipital bone over a "duration" T) for impacts of the

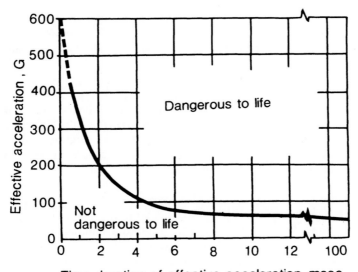

FIGURE 12.10:4 The "Wayne State" tolerance curve for the human brain in forehead impacts against plane, unyielding surfaces.

forehead against a plane, unyielding surface; the abscissa is the duration of the effective part of the pulse. The curve was derived from clinical observations of skull fracture, cadaver experiments with the duration of pulse in the 2–6 msec range, animal experiments with pulse duration in the 6–10 msec range, and long-duration human volunteer experiments by Stapp (1957), Figs. 12.10:1 and 12.10:3. Considerable exploration of observed data was involved.

Gadd (1966) showed that the curve shown in Fig. 2.10:4 can be represented fairly well in the 2.5 to 50 msec range by the equation

$$\bar{a}^{2.5}T = \text{const.} \tag{1}$$

(but badly elsewhere). Here $\bar{a}$ is the "effective" or "average" acceleration in the "duration" T. Determination of the duration T was troublesome, so Gadd (1966) replaced Eq. (1) with the following:

$$\int_0^\infty [a(t)]^{2.5}\,dt = \text{a constant called the } Severity\ Index. \tag{2}$$

If the acceleration is given in units of the earth's gravitational acceleration, G, and the time in seconds, then a severity index of 1000 is recommended by the Society of Automotive Engineers as a criterion of head injury for car design with respect to frontal impacts (Gadd, 1966).

Versace (1971) pointed out that the upper limit of the integral in Eq. (2)

causes difficulty if the acceleration pulse has a long tail which is hard to measure accurately and is probably unimportant. To remedy the situation, an arbitrary measure called *Head Injury Criterion* (HIC) was devised:

$$\text{HIC} = \text{Max}\left\{\frac{1}{t_2 - t_1}\int_{t_1}^{t_2}[a(t)]^{2.5}\,dt\right\}(t_2 - t_1) < 1000, \qquad (3)$$

t_1, t_2 are time in seconds, $(t_1 < t_2)$. The quantity on the left-hand side of the inequality sign is a function of t_1 and t_2. The criterion calls for varying t_1, t_2 to obtain a maximum of the quantity in the braces. Hodgson and Thomas (1972) have shown that $t_2 - t_1$ must be less than 15 msec in order to pose a concussion hazard, even if the HIC value exceeds 1000.

Similar investigation has been done with regard to lateral impact of the head. Nahum et al. (1980) measured the pressure in cadaver skulls when they were subjected to lateral loads. A contracoup phenomenon was observed as in frontal blows. Stalnaker et al. (1973) reported a side-impact threshhold of head injury at a peak translational head acceleration of 76 *G* with a pulse duration of 20 msec. On the other hand, in a rocket sled test, Stapp sustained without injury an acceleration pulse of Severity Index 1500 at the head (without impact with a solid surface) while 45 *G* was measured on the seat.

Ommaya (1968), Hirsch et al. (1970) have shown that head injury occurs much more readily if the head is allowed to rotate. The brain in the skull can tolerate considerable frontal or side impact if rotation of the brain relative to the skull does not occur. On the other hand, if rotation does occur, the same impact that was tolerable for translational motion may cause severe injury.

Numerous other criteria have been proposed (see King's review, 1975). Some are approximations of an approximation. As experimental data accumulate, they appear to be less and less justifiable. In practical applications, their apparent simplicity is superficial, because an engineer has to predict the acceleration pulse $a(t)$ in the first place. This is often done by mathematical modeling. There is a trend toward mathematical modeling of the brain (see Goldsmith, 1972; Ward et al., 1980) A future task of trauma research is to correlate the clinical syndromes and pathological lesions with the stress and strain in the brain, not merely with the deceleration of the head.

Spinal Injury

As it is shown in Table 12.10:2, spinal injuries are the most prevalent injuries among the crew and passengers of modern civil air transport accidents. These injuries may range from fatalities to minor complaints. Often the complaints are not due to broken bones or burst discs. A disturbance sufficient to pinch the nerves in the spine may elicit pain in the lower back, limbs, or other parts of the body.

The load acting at any point of the backbone during an impact does not simply arise from the inertial force of the body, but also from the reaction of the muscles and ligaments attached to the spine, and the variable pressure in

TABLE 12.10:2 Impact injuries to crew and passengers in civil air transport accidents in the 1970–78 period in the United States mentioned in Chandler et al.'s (1980) report*

	No. of individuals	Percentage**
Spine, vertebra, neck	80	48%
Lower extremities	49	30%
Head	11	6.6%
Ribs	11	6.6%
Other bone	5	3%
Shoulder, clavicle, hand	5	3%
Hip	2	1.2%
Kidney	1	0.6%
Spleen	1	0.6%
Intestine	1	0.6%
Unclassified (contusions, head injury, abrasions, lacerations, ribs, limbs, vertebra, pelvis, arms legs)	232	

* Data are from 30 major "survivable" accidents, involving Douglas DC8, DC9, DC10, Boeing 707, 727, 737, Martin 404, Convair 340, 440, Fairchild FH227, F27B, Lockheed L1011, 188A. Some persons who suffered multiple injuries are listed in more than one category.

** Percentages are calculated under the assumption that the injuries included in the "unclassified" category are distributed in the same ratio as those itemized above.

the abdomen caused by the reaction of the abdominal muscles to the impact. The loads in the vertebrae and discs are usually much larger than those required by static equilibrium against external load alone, because there exists a system of redundant self-equilibrating forces in the muscles which act on the spine. In Chapter 1, Table 1.4:1, we have shown the results obtained by Nachemson and Elfström (1970) on the pressure in the lumbar disc (the *nucleus pulposus* of L3) of the spine. Note how large the load in the lumbar spine is when one lifts weight the "wrong way". If one lifts a 20 kg weight with back straight and knees bent, the load acting in the lumbar disc (L3) is 185 kg. If, however, one lifts the same 20 kg weight with knees straight and back bent, the load on the lumbar disc becomes as high as 390 kg. A similar difference can be expected in the crash impact situation. If a vertical impact load is applied on a spine that is kept straight, the stress in the spine would be much lower than that acting on a spine that is bent. In other words, a bent spine is in no position to resist large loads.

The spine is a very complex structure. Spinal injury is a very complex problem. Tolerance data is woefully lacking. Many people complain of low back pain in normal life, suggesting that people's tolerance level to spinal load may be quite low.

In the case of forced landing of an air transport, some general observation can be made from a biomechanical point of view. For a passenger sitting in an airplane subjected to verticle deceleration, a straight spine is estimated to be 2 to 5 times stronger than a bent spine (Kazarian and Graves, 1977; Chandler and Trout, 1979). The reason is as follows: If the spine is straight and the inertia force is parallel to the spine, then the bending moment which tends to bend the spine forward is relatively small. This moment is resisted by the back muscles, and in the lower thoracic and lumbar region, also by the abdominal pressure (Fung, 1977, p. 29). In contrast, if the person sits with a bent spine, then the vertical inertial force will create a larger bending moment on the spine because of the larger moment arm. If the person bends forward with arms hugging the knee as some airlines recommend, then the bending moment in the waist region tends to bend the spine backward. In this position the abdominal pressure and the back muscle cannot help, and the bending moment would have to be resisted by compression and shear between the superior and inferior articular processes of successive lumbar vertebrae (see, e.g., Jacob et al. (1982, p. 133) and Fig. 1.4:2, 1.4:3). This may cause slip, undue deformation, or dislocation, which endangers the integrity of the spinal cord. Hence a straight spine is stronger.

As the proverb says: An ounce of prevention is worth a pound of cure. A good design should put a person in the strongest configuration. In this respect, the rearward-facing seating in an airplane looks very attractive (Fung, 1982). Support of spine can be obtained easily in a forward crash if the seats face rearward, and are designed to be strong enough to survive crashes without collapsing. The rearward facing seats should be high enough to support the head to avoid whiplash in a plane crash. In a forward crash the deceleration of the plane will keep the passenger's spine straight against a rearward facing seat, support his head naturally and firmly, keep his arms and legs in position, not to fly off and injure themselves. Seated rearward, children and pregnant women can be supported much more easily compared with wearing conventional seat belts in the forward facing arrangement. The only requirement is that the seats must be designed to take the load.

Since today's airplanes are not so designed, and since spinal injuries incur in many other human activities, it is necessary to study the problem closely. For reviews of current state-of-the-art, see Jacobs and Ghista (1982), Liu (1980), Panjabi and White (1982), Saha (1982), and Sonnerup (1982).

12.11 Biomechanical Modeling

In assessing human tolerance to impact loads, and in the design of vehicles for crashworthiness, it is necessary to calculate the stress and strain at specific points in various organs, and this is best done by mathematical modeling. Through mathematical modeling, one can connect pieces of information on anatomy, physiology, and clinical observations with people, vehicle, and accident. A validated model can then become a foundation of engineering.

Biomechanical modeling is in its infancy. Vigorous development is sure to come. The trend is toward dynamics of three-dimensional bodies capable of revealing the vibrational amplification and wave propagation features discussed in Sections 12.4–12.6. Much closer study of the physiological, pathological, and clinical aspects will be done.

A sampling of the huge literature is given below: A general review is given by King (1984). Head injury is reviewed by Goldsmith (1972), Ward et al. (1980). Spinal injury is reviewed by Huston and Perrone (1980). The effect of wearing helmets was studied by Huston and Sears (1979). Intervertebral discs were modeled by Kulak et al. (1976). Spinal load was analyzed by Schultz and Anderson (1981), Koogle et al. (1979), Privitzer and Belytschko (1980), and validated by Kazarian and Graves (1977), Engin (1979). Bones have been modeled by Brown et al. (1980), Piziali et al. (1976), Hayes et al. (1978), Orne and Young (1976), Viano et al. (1976).

Chest models were developed by Roberts and Chen (1970), Sundaram and Geng (1977). Chaffin (1969), Kane and Scher (1970), Passerello and Huston (1971) modeled the astronauts, swimming and kicking, and human attitude control. Muskian and Nash (1974), Fleck (1975) modeled a seated man.

And then there are models of heart, lung, blood vessels, muscles (Chapters 6, 11) and single cells (see References in Ch. 4), as well as astronauts in spaceship (Thomson and Fung, 1965).

12.12 Engineering for Trauma Prevention

A bioengineer can help to make this world a safer place to live. Understanding trauma is an important step. Improving the design of automobiles, aircraft, factories, work places, and sports equipment, etc. to make them intrinsically better for people is the next step. Search, develop, and promote safer and better work habits, sport techniques, general culture is our responsibility. Manufacturing and marketing safer and better products, and teaching the public about healthier ways of life is a further step.

In trauma research, the application of biomechanics is not limited to the engineering aspects. It must be extended to all aspects of emergency handling, patient evaluation, testing, treatment, healing, recovery, and rehabilitation. A deeper understanding of tissue engineering, of growth and resorption is necessary. A broad horizon is in front of us.

Trauma is a world problem. Trauma research is expensive. Its objective is to benefit all mankind. It is a fitting topic for international cooperation.

Problems

12.1 In testing a rectangular specimen of cortical bone in bending, it was found that when the bending moment reached a certain limit, microcracks were formed on the tensile side. At a somewhat greater bending moment, a few cracks appeared on the compressive side while many more microcracks formed on the tensile side. The cracks on the tensile side were perpendicular to the beam axis whereas the cracks on the compressive side were inclined at about 45°. Based on a consideration of principal stresses and maximum shear (use Mohr's circle if you wish, see Fung, 1977, Chap. 4), explain what do these experimental results reveal about the relative magnitude of fracture strength of the bone in tension, compression, and shear?

Given the dimensions of the beam, derive formulas expressing the strength of the bone in tension, shear and compression.

12.2 In a torsion test of a long bone, the bone broke along a line at approximately 45° to the torsion axis. What does this tell about the strength of the bone with respect to shear, tension, and compression.

Given the dimensions of the shaft, derive formulas expressing the strength of the bone.

12.3 A bone subjected to a loading applied very slowly breaks into two pieces. A load of similar magnitude applied impulsively breaks the bone into many small pieces. How do you explain this?

12.4 Consider Hopkinson's experiment, Fig. 12.4:1. The mass M has a velocity v_0 when it hits the stopper. As the elastic wave goes up the wire, a force $A\sigma = A\rho c v$ acts on the mass. The equation of motion of the mass is, therefore,

$$M\frac{dv}{dt} = A\rho c v.$$

Solve this equation with the initial condition $v = v_0$ at $t = 0$. With this result, discuss the waves in the wire.

12.5 Show that the equations of motion of an elastic oscillator subjected to a force acting on the mass and one subjected to a ground acceleration are similar.

12.6 Derive wave equation in spherical polar coordinates. Find a solution symmetric with respect to the origin. Discuss the phenomenon of focusing (Cf.: S. Temkin, *Elements of Acoustics*, Wiley, New York, 1981, Ch. 4).

12.7 Design a program to study the tolerance levels of the brain in response to impact load. List possible impairment of functions, clinical symptoms, and the corresponding lesions. Discuss how the lesions can be observed and correlated to the impact load; what should be measured in your proposed experiment; and how.

12.8 Design a program to study the tolerance levels of the spine to various kinds of injuries that may arise in people of different sex, age, and occupation.

12.9 Verify the statements made in Sec. 12.2 with regard to experiments 1 and 2 through a calculation of the principal stresses. Set up equations to analyze experiments 4 and 5.

12.10 Assessment of gait may be helped by the use of electromyograms of major muscles during walking. Design an equipment for use on patients. (Cf.: R. Shiavi et al., *J. Rehabilitation Res. and Dev.* 24: 13–30, 1987.)

12.11 *High pressures in the human hip joint.* W.A. Hodge et al. (*Proc. National Acad. Sci.* 83: 2879–2883, 1986) reported that the local pressure measured on the opposing layers of cartilage of the human joint can be as high as 18 MPa ($\sim$ 180 atm), which was obtained when a 73-year old woman (68 kg) was rising from the sitting to the standing position. In normal level walking, the maximum pressure was 4 MPa. Records show also that the local joint pressures were only 5% greater with the use of a cane than with a crutch. Formulate a mathematical model to analyze these movements.

12.12 In a head-on car crash from a speed of 60 mph, compute the impulse on the heart, aortic arch, lung, head, neck, and thoracic and lumbar spines. Explain clearly all the assumptions used in your calculation.

References

Anthonisen, N.R. (1977). Closing volume. In: *Regional Differences in the Lung*, (J.B. West, ed.), Academic Press, New York, pp. 451–482.

Bates, D.V., Macklem, P.T., and Christie, R.V. (1971). *Respiratory Function in Disease*, W.B. Saunders, Philadelphia.

Belytschko, T., Kulak, R.F., and Schultz, A.B. (1974). Finite element stress analysis of an intervertebral disc. *J. Biochem.* **7**: 277–285.

Bowen, I.G., Holladay, A., Fletcher, E.R., Richmond, D.R., and White, C.S. (1965). A fluid-mechanical model of the thoraco-abdominal system with applications to blast biology. Report No. DASA-1675, Lovelace Foundation for Medical Education and Research, Albuquerque, New Mexico.

Brigham, K.L. (1978). Lung edema due to increased vascular permeability. In: *Lung Water and Solute Exchange* (N. Staub, ed.), Marcel-Dekker, New York, pp. 235–276.

Brown, T.D., Way, M.E., and Ferguson, Jr., A.B. (1980). Stress transmission anomalies in femoral heads altered by asectic necrosis. *J. Biochem.* **13**: 687–699.

Campbell, I.J. and Pitcher, A.S. (1958). Shock wave in a liquid containing gas bubbles. *Proc. Roy. Soc.* (London), A, **243**, 534–545.

Chaffin, D.B. (1969). A computerized biomechanical model—development of and use in studying gross body actions. *J. Biomech.* **2**: 429–441.

Chandler, R.F., Neri, L.M., Pollard, D.W., and Caiafa, C.A. (1980). Crash injury protection in survivable air transport accidents—U.S. Civil Aircraft Experience from 1970 to 1978. FAA Report No. FAA-CT-80-34.

Chandler, R.F. and Trout, E.M. (1979). Evaluation of seating and restraint systems conducted during fiscal year 1978. FAA Report No. AM-79-17, ADA074881/4.

Chuong, C.J. (1985). Biomechanical model of thorax response to blast loading. Report on Contract No. DAMD 17-82-C-2062. Jaycor, San Diego, CA, 92138.

Clemedson, C-J. (1956). Blast injury. *Physiol. Rev.* **36**: 336–354.

Clemedson, C-J. and Jönsson, A. (1962, 1964) Distribution of extra-and intra-thoracic pressure variations in rabbits exposed to air shock waves. *Acta Physiol. Scand.* **54**: 18–29, 1962. See also, ibid, **62**: Suppl. **233**: 3–31, 1964.

Courant, R. and Friedrichs, K.O. (1948). *Supersonic Flow and Shock Waves*. Interscience, New York.

Crone, C. and Lassen, N.A. (eds.) (1970). *Capillary Permeability*, Proc. of a Sym., Academic Press, New York.

Currey, J.D. (1970). *Animal Skeletons*. Edward Arnold, London.

Dunn, F. and Fry, W.J. (1961). Ultrasonic absorption and reflection by lung tissue. *Phys. Med. Biol.* **5**: 401–410.

Effros, R.M., Mason, G., Uszler, J.M., and Chang, R.S.Y. (1982). Exchange of small molecules in the pulmonary microcirculation. In *Mechanisms of Lung Microvascular Injury. Annals of N.Y. Acad. of Sci.* **384**: 235–245.

Egan, E.A., Nelson, R.M., and Oliver, R.E. (1976). Lung inflation and alveolar permeability to non-electrolytes in the adult sheep in vivo. *J. Physiol. (London)* **260**: 409–424.

Eiband, A.M. (1959). Human tolerance to rapidly applied accelerations: A summary of the literature. NASA Memorandum No. 5-19-59E.

Engin, A.E. (1979). Measurement of resistive torques in major human joints. Report No. AMRL-TR-79-4. Air Force Aerospace Medicine Res. Lab., Wright-Patterson Air Force Base, Ohio.

Feinberg, B.N. and Fleming, D. G. (eds) (1978). *CRC Handbook of Engineering in Medicine and Biology. Sec. B. Instruments and Measurements* Vol. I (A. Burstein and E. Bahniuk, eds.). CRC Press, Boca Raton, Florida.

Fishman, A.P. and Hecht, H.H. (eds.) (1969). *The Pulmonary Circulation and Interstitial Space.* The Univ. of Chicago Press, Chicago.

Fleck, J.T. (1975). Calspan three-dimensional crash victim simulation program. In *"Aircraft Crashworthiness"*. (K. Saczalski, G.T. Singley, W.D. Pilkey, and R.L. Huston, eds.), Univ. of Virginia Press, Charlottesville.

Frisoli, A. and Cassen, B. (1950). A study of hemorrhagic rib markings produced in rats by air blast. *J. Aviation Med.* **21**: 510–513.

Frucht, A.-H. (1953). Die Schallgeschwindigkeit in menschlichen und tierischen Geweben. *Ges. Exp. Med.* **120**: 526–557.

Fung, Y.C. (1957). Some general properties of the dynamic amplification spectra. *J. Aeronautical Sci.* **24**: 547–548.

Fung, Y.C. (1962). On the response spectrum of low-frequency mass-spring systems subjected to ground shock. *J. Aeronautical Sci.* **29**: 100–101.

Fung, Y.C. (1965). *Foundations of Solid Mechanics.* Prentice-Hall, Englewood Cliffs, New Jersey.

Fung, Y.C. (1977). *A First Course in Continuum Mechanics.* 2nd ed., Prentice-Hall, Englewood Cliffs, New Jersey.

Fung, Y.C. (1981). *Biomechanics: Mechanical Properties of Living Tissues.* Springer-Verlag, New York.

Fung, Y.C. (1982). On human tolerance to impact loads. Report to U.S. Department of Transportation, Transportation Center, Cambridge, Massachusetts.

Fung, Y.C. (1984). The application of biomechanics to the understanding and analysis of trauma. In: *The Biomechanics of Trauma* (A.M. Naham, ed.). Appleton-Century-Crofts, Norwalk, Connecticut, pp. 1–16.

Fung, Y.C. (1985). The application of biomechanics to the understanding and analysis of trauma. In *The Mechanics of Trauma*, (A.M. Nahum and J. Melvin, eds.), Appleton-Century, Crofts, Norwalk, Conn., pp. 1–16.

Fung, Y.C. and Barton, M.V. (1958). Some shock spectra characteristics and uses. *J. Appl. Mech.* **25**: 365–372.

Fung, Y.C. and Barton, M.V. (1962). Shock response of a nonlinear system. *J. Appl. Mech.* **29**: 465–476.

Fung, Y.C., Yen, R.T., Tao, Z.L., and Liu, S.Q. (1988): A hypothesis on the mechanism of trauma of lung tissue subjected to impact load. *J. Biomech. Eng.* **110**: 50–56.

Gadd, C.W. (1966). Use of a weighted impulse criterion for estimating injury hazard. *10th Stapp Car Crash Conf.*, Soc. of Automotive Engineers. pp. 95–100.

Gadd, C.W., Culver, C.C., and Nahum, A.M. (1971). A study of responses and tolerances of the neck. *15th Stapp Car Crash Conf.*, Soc. of Automotive Engineers.

Gennarelli, T.A., Ommaya, A.K., and Thibault, L.E. (1971). Comparison of translational and rotational head motions in experimental cerebral concussions. *15th Stapp Car Crash Conf.*, Soc. of Automotive Engineers.

Giuntini, C. (ed.) (1971). *Central Hemodynamics and Gas Exchange.* Minerva Medica, Torino, Italy.

Goldsmith, W. (1972). Biomechanics of head injury. In *"Biomechanics: Its Foundations and Objectives"*. (Y.C. Fung, ed.), Prentice-Hall, Englewood Cliffs, N.J., pp. 585–634.

Griffith, A.A. (1920). The phenomena of rupture and flow in solids. *Phil. Trans. Roy. Soc. London,* **A221**: 163–198.

Gurdjian, E.S., Lissner, H.R., Latimer, F.R., Haddad, B.F., and Webster, J.E. (1953). Quantitative determination of acceleration and intracranial pressure in experimental head injury. *Neurology* 3: 417–423.

Gurdjian, E.S., Webster, J.E., and Lissner, H.R. (1955). Observations on the mechanism of brain concussion, contusion and laceration. *Surgery, Gynecology and Obstetrics* 101: 688–890.

Hayes, W.C., Swenson, Jr., L.W., and Schurman, D.J. (1978). Axisymmetric finite element analysis of the lateral tibial plateau. *J. Biochem.* 11: 21–33.

Hill, I.R. (1978). Injury mechanisms analysis in aircraft accidents (of Royal Air Force, U.K.) In AGARD Conf. Proc. No. 253: *Models and Analogues for the Evaluation of Human Biodynamic Response, Performance and Protection.* NATO, AGARD, Paris. Article A12.

Hirsch, A.E., Ommaya, A.K., and Mahone, R.H. (1970). Tolerance of sub-human primate brain to cerebral concussion. In *Impact Injury and Crash Protection*, Charles C. Thomas, Springfield, IL., pp. 352–369.

Hodgson, V.R. and Thomas, L.M. (1972). Effect of long-duration impact on head. *16th Stapp Car Crash Conf.*, Soc. of Automotive Engineers.

Holbourne, A.H.S. (1943). Mechanics of head injury. *Lancet* 245: 438–441.

Hopkinson, J. (1872). *Collected Scientific Papers*, Vol. II, p. 316.

Hopkinson, B. (1914). A method of measuring the pressure produced in the detonation of high explosives or by the impact of bullets. *Proc. Royal Soc. (London)* 213: 437–456.

Hopping, F.G. and Hildebrandt, J. (1977). Mechanical properties of the lung. In: *Bioengineering Aspects of the Lung*, (J.B. West, ed.), Marcel Dekker, New York.

Huston, R.L. and Perrone, N. (1980). Dynamic response and protection of the human body and skull in impact situations. In *"Perspective in Biomechanics* (H. Reul, D.N. Ghista, and G. Rau, eds.), Harwood Academic Publishers, New York, pp. 531–572.

Huston, R.L. and Sears, J. (1979). Effect of protective helmet mass on head/neck dynamics, *1979 Biomechanics Symposium*, ASME, (W.C. Van Buskirk, ed.), pp. 227–229.

Jacob, S.W., Francone, C.A., and Lossow, W.J. (1982): *Structure and Function in Man.* Saunders, Philadelphia.

Jacobs, R.R. and Ghista, D.N. (1982). A biomechanical basis for treatment of injuries of the dorsolumbar spine. In *Osteoarthromechanics*, (D.N. Ghista, ed.), McGraw-Hill, New York, pp. 435–472.

Jönsson, A., Clemedson, C-J., Sundqvist, A-B., and Arvebo, E. (1979): Dynamic factors influencing the production of lung injury in rabbits subjected to blunt chest wall impact. *Aviation, Space & Envir. Medicine* 50: 325–337.

Kane, T.P. and Scher, M.P. (1970). Human self-rotation by means of limb movements. *J. Biomech.* 3: 39–49.

Kaye, G.W.C. and Labby, T.H. (1960). *Tables of Physical and Chemical Constants*, 12th Ed., Longmans, Green, New York.

Kazarian, L. and Graves, G.A., Jr. (1977). Compression strength characteristics of the human vertebracolumn. *Spine* 2 (No. 1).

King, A.I. (1975). Survey of the state of the art of human biodynamic response. In *Aircraft Crashworthiness*, (K. Saczalski, G.T. Singley III, W.D. Pilkey and R.L. Huston, eds.), Univ. Press of Virginia, Charlottesville, VA., pp. 83–120.

King, A. (1984). A review of biomechanical models. *J. Biomech. Eng.* 106: 97–104.

Koogle, T.A., Swenson, Jr., L.W., and Piziali, R.L. (1979). Dynamic three-dimensional modelling of the human lumbar spine. In *"Advances in Bioengineering"* (M.K. Wells, ed.) Amer. Soc. Mech. Engineers, pp. 65–68.

Kooyman, G.L. (1981). *Weddell Seal: Consummate Diver.* Cambridge Univ. Press, London.

Kraman, S.S. (1983). Speed of low-frequency sound through lungs of normal man. *J. Appl. Physiol.* 55: 1862–1867.

Kulak, R.F., Belytschko, T.B., Schultz, A.B. and Galante, J.O. (1976). Nonlinear behavior of human intevertebral disc under axial load. *J. Biomech.* 9: 377–386.

Laananen, D.H. (1980). Aircraft crash survival design guide Vol. 2. Aircraft Crash Environment and Human Tolerance. Report prepared by Simula Inc., Tempe, Arizona for Applied Tech.

Lab., U.S. Army Res. and Tech. Labs., Fort Eustis, Virginia. Report No. USARTL-TR-79-22B.

Lanir, Y. and Fung, Y.C. (1974): Two-dimensional mechanical properties of the rabbit skin. I. Experimental system. *J. Biomech.* 7: 29–34. II. Experimental results. *J. Biomech.* 7: 171–182.

Lau, V.K. and Viano, D.C. (1981). Influence of impact velocity and chest compression on experimental pulmonary injury severity in rabbits. *The Journal of Trauma,* 21: No. 12.

Lissner, H.R., Lebow, M., and Evans, F.G. (1960). Experimental studies on the relation between acceleration and intracranial pressure changes in man. *Surgery Gynecology, and Obstetrics* 3: 329–338.

Liu, Y.K. (1980). Mechanisms, evaluation, prognosis, and management of head and neck injury. In *Perspectives in Biomechanics,* (H. Reul, D.N. Ghista, and G. Rau, eds.), Harwood Academic Publishers, New York. pp. 537–581.

Ludwig, G.D. (1950). The velocity of sound through tissues and the acoustic impedance of tissues. *J. Acoustic Soc. Am.* 22: 862–866.

Melvin, J.W., Stalnaker, R.L., Roberts, V.L., and Trollope, M.L. (1973). Impact injury mechanisms in abdominal organs. *17th Stapp Car Crash Conf.,* Soc. of Automotive Engineers.

Mertz, Jr., H.J. and Kroell, C.K. (1970). Tolerance of the thorax and abdomen. In *Impact Injury and Crash Protection,* Charles C. Thomas, Springfield, Ill.

Muskian, R. and Nash, C.D. (1974). A model for the response of seated humans to sinusoidal displacements of the seat. *J. Biomech.* 7: 209–215.

Nachemson, A., and Elfstrom, G. (1970). Intravital dynamic pressure measurements in lumbar discs. A study of common movements, manoeuvres, and exercises. *Scand. J. Rehab. Med.,* *Suppl* 1. See also author's paper in *Perspectives in Biomedical Engineering,* (R.M. Kenedi, ed.), Univ. Park Press, New York, 1973, pp. 111–119.

Nahum, A.M. and Melvin, J. (eds) (1985). *The Biomechanics of Trauma.* Appleton-Century-Crofts, Norwalk, Connecticut.

Nahum, A., Ward, C., Raasch, E., Adams, S., and Schneider, D. (1980). Experimental studies of side impact to the human head. *24th Stapp Car Crash Conf.,* Soc. of Automotive Engineers.

Neathery, R.F. (1974). Analysis of chest impact response data and scaled performance recommendations. *18th Stapp Car Crash Conf.,* Soc. of Automotive Engineers, pp. 459–493.

Nicolaysen, G. and Hauge, A. (1982). Fluid exchange in the isolated perfused lung. In *Mechanisms of Lung Microvascular Injury, Annals of N.Y. Acad. of Sci.* 384: 115–125.

Ommaya, A.K. (1968). Mechanical properties of tissues of the nervous system. *J. Biomech.* 1: 127–138.

Ommaya, A.K., Fass, F., and Yarnell, P. (1968). Whiplash injury and brain damage. *J. Amer. Med. Assoc.* 204: 285–289.

Orne, D. and Young, D.R. (1976). The effects of variable mass and geometry, pretwist, shear deformation and rotary inertia on the resonant frequencies of intact long bone: A finite element model analysis. *J. Biomech.* 9: 763–770.

Panjabi, M.M. and White, A.A. III (1982). Spinal mechanics: Kinematics, kinetics, and mathematical models, procedures and devices for correction of deformaties and fixation of spinal fractures. In *Perspectives in Biomechanics,* (H. Reul, D.N. Ghista and R. Raul, Eds.), pp. 617–682. Harwood Academic Pub. New York.

Park, J.B. (1984). *Biomaterial Science and Engineering.* Plenum Press, New York.

Passerello, C.E. and Huston, R.L. (1971). Human attitude control. *J. Biomech.* 4: 95–102.

Perrone, N. (1972). Biomechanical problems related to vehicle impact. In: *Biomechanics: Its Foundations and Objectives.* (Y.C. Fung, ed.), Prentice-Hall, Englewood Cliffs, New Jersey, pp. 567–584.

Piziali, R.L., Hight, T.K., and Nagel, D.A. (1976). An extended structural analysis of long bones—Application to the human tibia. *J. Biomech.* 9: 695–701.

Privitzer, E. and Belytschko, T. (1980). Impedance of a three dimensional head-spine model. *Intern. J. of Math. Modeling* (No. 2)

Rice, D.A. (1983). Sound speed in pulmonary parenchyma. *J. Appl. Physiol.* 54(1): 304–308, 1983.

Richmond, D.R., Damon, E.G., Fletcher, E.R., Bowen, I.G., and White, C.S. (1968). The relationship between selected blast-wave parameters and the response of mammals exposed to air blast. *Annals of N.Y. Acad. Sciences*, Vol. 152, pp. 103–121.

Rivlin, R.S. and Thomas, A.G. (1953). Rupture of rubber. I. Characteristic energy for tearing. *J. Polymer Sci.* **10**: 291–318.

Roberts, S.B. and Chen, P.H. (1970). Elastostatic analysis of the human thoracic skeleton, *J. Biomech.* **3**: 527–545.

Roth, E.M., Teichner, W.G., and Craig, R.L. (1968). *Compendium of human response to the aerospace environment*, (E.M. Roth Ed.), Vol. II, Sec. 7. *Acceleration. NASA Contractor Report* No. CR-1205 (II), Prepared by Lovelace Foundation, Albuquerque, N.M.

Saha, S. (1982). The dynamic strength of bone and its relevance. In *Osteoarthromechanics*, (D.N. Ghista, ed.), McGraw-Hill, New York, pp. 1–44.

Schneider, D.C. (1982). Viscoelasticity and tearing strength of the human skin. Ph.D. Thesis, University of California, San Diego, Department of AMES/Bioengineering.

Schultz, A.B. and Andersson, G.B.J. (1981). Analysis of loads on the lumbar spine. *Spine* **6**: 76–82.

Sharma, M.G. (1979). Development of a tear criterion for rupture of thoracic aortas in vitro. In *1979 Adv. in Biomech.* 119–122. Amer. Soc. Mech. Eng. New York.

Sonnerup, L. (1982). Stress and strain in the intervertebral disk in relation to spinal disorders. In *Osteoarthromech.* (D.N. Ghista, ed.), McGraw-Hill, New York, pp. 315–352.

Stalnaker, R.L., Roberts, V.L., and McElhaney, J.H. (1973). Side impact tolerance to blunt trauma. *17th Stapp Car Crash Conf.*, Soc. of Automotive Engineers.

Stapp, J.P. (1957). Human tolerance to deceleration. *Am. J. of Surgery* **93**: 734–740.

Stapp, J.P. (1961). Human tolerance to severe, abrupt deceleration. In *Gravitational Stress in Aerospace Medicine*, (O.H. Gauer and G.D. Zuidema, eds.), Little, Brown, Boston, pp. 165–88.

Stapp, J.P. (1970). Voluntary human tolerance levels. In *Impact Injury and Crash Protection*, (E.S. Gurdjian, W.A. Lange, L.M. Partrick, and L.M. Thomas, eds.), Charles C. Thomas, Springfield, Ill.

Staub, N. (1978). Lung fluid and solute exchange. In *"Lung Water and Solute Exchange"* (N. Staub, ed.). Marcel Dekker, New York, pp. 3–16.

Sundaram, S.H. and Geng, C.C. (1977). Finite element analysis of the human thorax. *J. Biomech.* **10**: 505–516.

Tao, Z.L. and Fung, Y.C. (1987). Lungs under cyclic compression and expansion. *J. Biomech. Eng.* **109**: 160–162.

Taylor, G.I. (1946). The testing of materials at high rates of loading. *J. Inst. Civil Engineers* **26**: 486–519.

Thomas, A.G. (1960). Rupture of rubber. VI. Further experiments on the tear criterion. *J. Appl. Polymer Sci.* **3**: 168–174.

Thomson, W.T. and Fung, Y.C. (1965). Instability of spinning space stations due to crew motion. *AIAA J.* **3**: 1082–1087.

Versace, J. (1971). A review of the severity index. *15th Stapp Car Crash Conf.*, Soc. of Automotive Engineers.

Viano, D.C. Helfenstein, U., Anliker, M., and Ruegsegger, P. (1976). Elastic properties of cortical bone in female human femurs. *J. Biomech.* **9**: 703–710.

Viano, D.C. and Lau, V. (1988). A viscous tolerance criterion for soft tissue injury assessment. *J. Biomech.* **21**: 387–399.

von Gierke, H.E. (1964). Biodynamic response of human body. *Appl. Mechanics Review* **17**: 951–958.

von Gierke, H.E. (1971). Biodynamic models and their applications. *J. Acoustic Soc. of America* **50**: 1397–1412.

Wainwright, S.A., Biggs, W.D., Currey, J.D. and Gosline, J.M. (1976). *Mechanical Design in Organisms*. Wiley, New York.

Ward, C., Chan, M., and Nahum, A. (1980). Intracranial pressure—A brain injury criterion. *24th Stapp Car Crash Conf.*, Soc. of Automotive Engineers, pp. 161–185.

White, C.S., Jones, R.K., Damon, E.G., Fletcher, E.R., and Richmond, D.R. (1971). The biodynamics of airblast. Report No. DNA 2738T, Lovelace Foundation for Medical Education and Research, Albuquerque, New Mexico.

Woo, S.L.Y. and Buckwalter, J.A. (eds.) (1988). *Injury and Repair of the Musculoskeletal Soft Tissues.* Park Ridge, IL. American Academy of Orthopedic Surgeons.

Yamada, H. (1970). *Strength of Biological Materials.* (F.G. Evans, ed.), Williams and Wilkins, Baltimore.

Yen, R.T. and Fung, Y.C. (1985). Thoracic trauma study: Rib markings on the lung due to impact are marks of collapsed alveoli; but not hemorrhage. *J. Biomech. Eng.* **107**: 291–292.

Yen, R.T., Fung, Y.C., Ho, H.H. and Butterman, G. (1986). Speed of stress wave propagation in the lung. *J. Appl. Physiol.* **61**(2): 701–705.

Yen, R.T., Ho, H.H., Tao, Z.L. and Fung, Y.C. (1988). Trauma of the lung due to impact load. *J. Biomech.* **21**: 745–753, 1988.

The flower of water lily is loved for its poetic purity. But it revealed to me a certain aspect of growth. Last summer I returned home after a three-month trip and found my lily pond almost dry and that the lilies were shriveled and had turned black. Each plant was reduced to the size of a hand. I drained and refilled the pond. The next day some tiny leaves sprouted. They grew rapidly; their slender stems lengthened at a rate of about a foot a day. In three days new round green leaves were floating on the surface of water, and before long I got the picture shown here. What tenacity! What is their secret?

CHAPTER 13

Biomechanical Aspects of Growth and Tissue Engineering

13.1 Introduction

All parents want healthy children. Everyone wants a strong and handsome body. We all believe that a proper level of exercise, i.e. a proper level of stress and strain, is necessary for health. This is common sense. If we can turn this common sense into a precise knowledge about stress and growth, then it will enlighten biology in general, help surgeons to engineer healing, throw light on physical education, sports techniques, health care, rehabilitation.

In our bodies, cells either live in steady state (homeostasis), reproduce, move, or die. Tissues hypertrophy or resorb. In this chapter we are concerned with tissues and organs, and not with cell culture. The distinction is important because in cell culture, cells may proliferate at a great rate, and are active individually. When cells form a confluent layer, they become quiescent and then change slowly. In a tissue, every cell has neighbors on all sides, and the rate of turnover is usually very low.

There is no doubt that cells and tissues change and grow molecule by molecule and must have a biochemical foundation. But it must have a biophysical foundation also. This chapter is concerned about the latter, with a focus on the effects of stress and strain.

In the following, we shall first present a historical review of the concept of growth and change in bone, then present information on the growth and change of soft tissues, especially the heart, lung, and blood vessels. We then offer the idea that growth and change in an organ can be revealed simply by examining the changes of its zero-stress state. A theoretical examination of the meaning of the residual stresses in organs then leads to an experimental hypothesis on growth presented in Sec. 13.10.

The concepts of growth and change lead to the proposal of *tissue engineer-*

ing as a new discipline. Tissue engineering is defined as the application of the principles of engineering and biology toward a fundamental understanding of the structure-function relationship of tissues, and the development of biological substitutes to restore, maintain, or improve tissue functions. Examples are artificial blood vessels covered with the patient's own endothelial cells; skin substitute made of the patient's own keratinocytes, etc. The present status and future prospects are discussed in Secs. 13.11–13.13.

13.2 Wolff's Law and Roux's Functional Adaptation Concept

On the subject of the relationship between structure and stress, Wolff's law is the most famous, although its meaning and foundation are hazy. In 1981, Roesler presented a thorough account of the history of Wolff's law. The story is interesting and instructive, hence we present an outline of Roesler's findings below.

In 1866, G.H. Meyer presented a paper on the structure of cancellous bone at a meeting of the Züricher Naturforschende Gesellschaft. Meyer demonstrated that "the spongiosa showed a well-motivated architecture which is closely connected with the statics and mechanics of bone." C. Culmann, a mathematician who had published a book on graphic statics in the same year was in the audience. Culmann remarked that the lines in Meyer's drawings resembled the principal stress trajectories in cantilever beams. Meyer, in his paper published the following year (1867), stated that Culmann, stimulated by the drawings of bone structure, asked a student of his to construct the principal stress trajectories in a crane-like curved bar loaded in a fashion similar to the human femur (see Fig. 13.2:1). This was later to become the famous "Culmann's crane."

Meyer (1867) formulated his ideas in three questions:

1) Is it possible that structures like the observed ones are formed by static (force-equilibrium) conditions?
2) What is the internal metamorphosis that makes these structures so "fit for service?"
3) Can these structures be understood if one adds to the external loads the mechanical influence of the traction of muscles and ligaments?

Two years later, Wolff (1869) claimed that he could prove the following:

1) There is a perfect mathematical correspondence between the structure of cancellous bone in the proximal end of the femur and the trajectories in Culmann's crane.
2) There is a statical importance and necessity of the trajectorial structure of the bone.
3) Bone growth can occur only in the interstitial space.
4) The compact bone is nothing but a compressed cancellous bone.

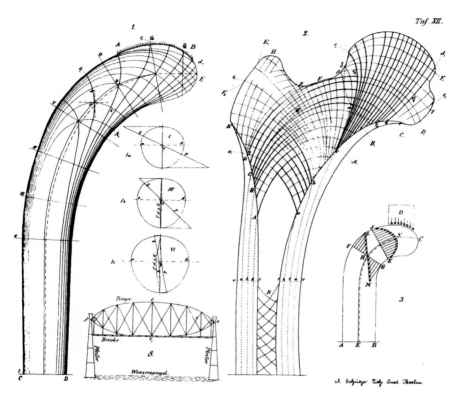

FIGURE 13.2:1 Culmann's crane presented by Wolff in his 1870 paper in Virchow's Archiv.

Wolff then published a sequence of papers in 1870, 1872, 1874, 1884, 1891, and a book in 1892. In his 1884 paper he called this the "law of bone transformation". His "proof", besides relying on the rough similarity between the trabecular pattern in the cancellous bone of the femur to Culmann's crane, (Fig. 13.2:1), does contain a valid detail: just as the principal directions of a stress tensor at any given point are orthogonal, the trabeculae in the bone, when they intersect, do seem to intersect at right angles (Fig. 13.2:2). Otherwise there are many questions: why should the bone be molded according to the stress trajectories of one particular loading? What happens to other loading conditions? Why is it sufficient to consider a curved beam of homogeneous, isotropic, elastic material, when the bone is not? Furthermore, Wolff's ideas about interstitial growth, and medullary cavity, turned out to be wrong.

The "crane" constructed by Culmann's student was also questionable. The figure obviously does not agree with the known results of the theory of elasticity of curved beams, which have stress concentration on the concave side. In Wolff's 1870 paper, a figure of the stress distribution in the crane

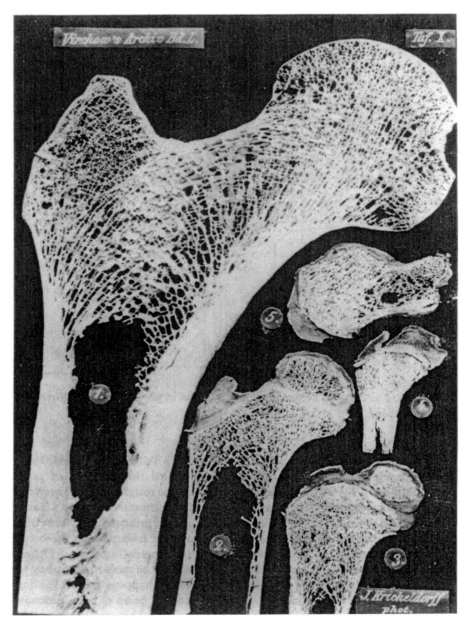

FIGURE 13.2:2 Photograph of bone structure presented in Wolff's paper of 1870 in Virchow's Archiv.

showed a linear distribution of the normal stress and a parabolic distribution of the shear stress in the beam cross section, which are true only for straight beams. Roesler (1981), after a detailed investigation, concluded that Culmann (or his student) just used the St. Venant theory of a straight cantilever rectangular beam loaded by shear at the free end, plotted the intercepts of the trajectories on 8 normal sections of the crane, and faired in with smooth curves. Yet this drawing became a historical landmark!

The three questions asked by Meyer, however, remain meaningful and need to be answered.

Roux's Functional Adaptation Concept

In 1880 and 1881 Wilhelm Roux introduced the idea of "functional adaptation", which means "adaptation to a function by making use of it". He called the process of functional adaptation a "quantitative self-regulating mechanism" controlled by a "functional stimulus". Being a contemporary of Charles Darwin (1809–1882) and Alfred Wallace (1823–1913), Roux (1850–1924) formulated his idea in a general way to deal with all aspects of evolution. One of his important examples is the trajectory hypothesis of cancellous bone structure. He said: "There exists a group of very fine appropriate structural formations which could not result from selection alone. ... This is the structure of cancellous bone corresponding to the static pressure lines, which enables bone to resist external forces with a minimum material consumption". "Many cells are influenced by functional stimuli, the nerve, muscle, gland cells by the related (electric) pulse, bone cells and connective tissue cells by pressure and tension. ... It is plausible that, if this influence is disadvantageous for some cells of a tissue or some parts of it, then these cells will disappear during the process of physiological regeneration. If, however, variations appear for which the functional stimulus is favorable, then in the course of time these cells will replace all the other cells, which are indifferent to the functional stimulus" (Roux, 1881).

In this way, Roux put the question of the structure of cancellous bone in a more fundamental form. By focusing on functional stimulus, the question can be examined in detail and answered quantitatively. But his own analysis of the cancellous bone seems to be faulty, as was pointed out later by Pauwels (1980), see the following section, and Roesler (1981).

13.3 Healing of Bone Fracture

Bone is a solid material which may fail in modes discussed in Sec. 12.2. If a crack exists in a material, it may lead to fracture because the ends of the crack are stress raisers. The critical state of stress at which a crack will enlarge itself is reached when the energy needed for the creation of new free surface of an advancing crack is balanced or exceeded by the sum of the energies released from the strain in the body due to the formation of new surface, the work done by the loading caused by the advancing crack, and the energy dissipation due to plastic deformation in regions near the ends of the crack. This basic concept by Griffith (1920) has lead to important advances in fracture mechanics. See reviews in Erdogen (1983), and Sih (1981). To the bone, additional considerations are needed to account for the composite structure of the bone (being an organized mixture of collagen and crystalline hydroxyapatite), the unit structure of the osteon, and the existence of voids in cancellous bone.

Pauwels (1935, see his book, 1980, pp. 1–105) has shown that fractures of the femoral neck can be classified into three types on the basis of stress acting on the surface of fracture imposed by the muscles and the weight of the body. If the normal stress is denoted by σ_n and the shear stress is denoted by τ, then Pauwel's three types of fractures are:

$$\text{Type 1:} \quad \sigma_n < 0 \quad \text{(in compression),} \quad \tau < f$$
$$\text{Type 2:} \quad \sigma_n < 0 \quad \text{(in compression),} \quad \tau > f$$
$$\text{Type 3:} \quad \sigma_n > 0 \quad \text{(in tension).}$$

Here f is a certain critical value of shear stress. Pauwels said that the Type 1 fracture usually heals by itself into a bony union without treatment. Type 3 fractures usually will not form a bony union. Fractures of this type that appear to have been healed often progressively evolve into a pseudoarthrosis (false joint) after being loaded. The Type 2 fracture generally has a pessimistic prognosis. He argues that the newly regenerated bone tissue cannot resist a shear stress τ appreciably higher than f.

Figure 13.3:1 shows Pauwels examples of the three types of fracture. In this figure, R is the resultant of the forces in the muscles and the body weight. K_s is the "effective" shearing force. Z is tension, D is compression. The figure

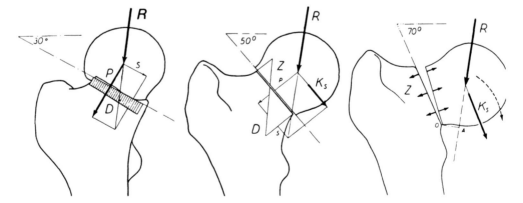

FIGURE 13.3:1 Pauwels' illustration of three types of fracture. From Pauwels (1980). Reproduced by permission.

shows that the angle between the normal to the plane of fracture of the femoral neck and the verticle axis is a deciding factor. If the angle is less than 30°, the fracture is Type 1 and healing can be assured. If it is greater than 60°, the fracture is Type 3 and pseudoarthrosis will result. In between, much more carefully analysis is needed, and some ingenious methods of treatment have been devised.

The healing of a fracture in the femoral neck is very different from that of a long bone (such as a diaphyseal fracture). In the latter, a collagenous periosteal callus will develop first to take up any tensile stress if it exists. In the femoral neck, no such periosteal callus will form. The healing of the femoral neck relies on the medullary healing tissue.

How does the medullary healing tissue develop? It is known that the new tissue can develop into either a connective tissue, or a cartilage, or a bone. If it develops into a connective tissue or cartilage, then the fracture will become a false joint. If it develops into a bone, then the fracture will be healed.

Roux and Pauwels' Hypotheses

Roux and Pauwels recognized the importance of stress in healing. Roux (1985) formulated the following hypotheses: tensile and shear stresses provoke the formation of connective tissue. Friction or shearing movement leads to the formation of cartilage. 'Functional' compression (defined as a periodically varying compressive stress) provokes bone formation.

Pauwels (1980, pp. 1–105) at first believed in Roux's hypotheses. Then he expressed doubts (1980, pp. 106–137), and finally presented (1980, pp. 375–407) the following hypothesis: *elongation* and *hydrostatic 'pressure'* are the two specific stimuli for the differentiation of the mesenchymal tissue. Elongation (i.e. unequal principal strains) provokes the formation of collagen fibrils, and therefore of connective tissue. Hydrostatic "pressure" (here defined as a state of strain in which the three principal strains are equal) stimulates the formation of cartilage tissue. There is no specific mechanical stimulus for the formation of bone tissue.

Pauwels quotes many observations (e.g., Krompecker, 1937) that contradict Roux's hypotheses and support his own. Among these, the most important is the following: In bone remodeling of the femoral neck which is cancellous, the regions in which the largest principal stress is in tension are calcified along the tensile trajectories, showing that tensile stress stimulates bone growth as well as the compressive stress.

Carter's Hypothesis

The question was studied recently by Carter et al. (1987), Carter (1988), Carter and Wong (1988). They examined the growth, maturation, and aging of the skeleton in relation to the proliferation, maturation, degeneration, and ossification of cartilage and the local stress or strain histories of the skeletal tissue. Using a strain energy function as a probe, and finite element analysis as a tool, they reached a qualitative conclusion as shown in Fig. 13.3:2. The figure delineates the tissue development under cyclic loading of an undifferentiated mesenchymal tissue. The left figure is drawn for a tissue with good blood supply and adequate tissue oxygen tension. The right figure is for a tissue with poor blood supply and low oxygen tension. The abscissa refers to the amplitude of cyclically applied dilatational (mean) stress, the ordinate refers to the amplitude of the cyclic octahedral shear stress. For each stress state defined by the coordinates, the end product is labeled in the figure. The coordinates

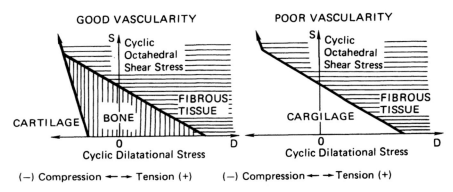

FIGURE 13.3:2 Carter's hypothesis.

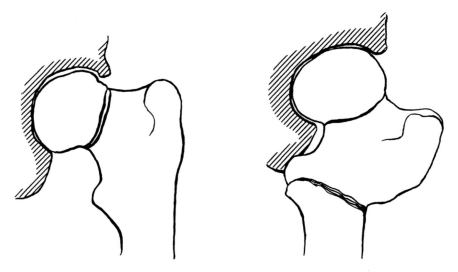

FIGURE 13.3:3 "Reorientation" surgery of a femeral head. Drawn from photographs given by Pauwels (1980).

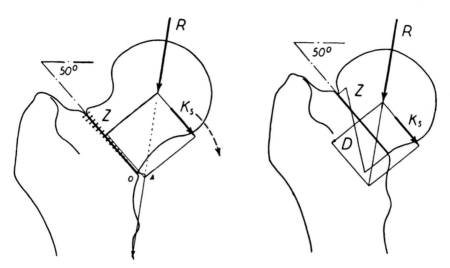

FIGURE 13.3:4 Resorption process turns a bad type 3 fracture into a healable type 2 fracture. From Pauwels (1980). Reproduced by permission.

are supposed to represent some kind of weighted average of the entire history of loading imposed on the bone over a period of time. A few "bad" cycles with high shear or tensile hydrostatic stress may be sufficient to promote fibrous tissue formation at the expense of osteogenesis. The details are not entirely clear.

While the cellular dynamics remain to be clarified, it is clear that consideration of mechanical stress acting on the wound should guide any surgical procedures for the treatment of bone fracture.

Example of Application: Beneficial use of Resorption. Figure 13.3:3 shows an operation of "reorientation", which turns a pseudoarthrosis into a Type 1 fracture which heals. Figure 13.3:4 shows a Type 3 fracture turned into a Type 2 fracture after resorption and controlled later by nailing. Both of these examples are from Pauwels (1980, pp. 36, 37).

Piezoelectric Effect

Bone is piezoelectric. In fact, all collagenous tissues, soft and hard, are piezoelectric. Fukada (1974), Bassett (1978), Yasuda (1974), and others have proposed that the piezoelectric property can be used clinically to heal fractures and nonunions in bone. The electromagnetic waves can be imposed on the organ noninvasively. Bassett stresses the importance of the wave form because he thinks that the electromagnetic waves must provoke interactions between electrically charged cell membranes and charged hormones, antibodies, and ions as normally achieved in nature. Impressive clinical results have been reported.

Bioelectric Effect on the Growth of Whole Organ

There is no doubt that electromagnetic waves induce certain changes in cells. Bioelectricity was brought to public attention in the mid 1700's by Luigi Galvani. In 1903, A.P. Matthews discovered that living hydroids (small relatives of sea anemones) are electrically polarized—that is, the 'head' end of the animal, with the flowerlike hydranth, has a different electric potential than the 'foot' end, which is fastened to rock. Subsequently, E.J. Lund (1921) observed the growth of eggs on the seaweed *Fucus* in sea water in the presence of an applied current created by placing electrodes at the opposite sides of a petri dish, and found that the eggs always produced their first outgrowths directed toward the negative electrode. He then showed that he could control which end of a piece of hydroid would become the head and which the foot by applying an electric field. In 1920, Ingvar showed that chick nerve cells cultured in vitro would survive and send out nerves toward a cathode when a small DC current was applied to the culture dish. This was confirmed by Sisken and Smith (1975) who found also that the growth rate in the electric field was two or three times faster than normal. More recent efforts on electric stimulation of nerve growth include the use of implantation of a bimetal (platinum and silver) electrode (insulated except at the tip) across severed ends of a nerve (with platinum in the proximal segment and silver in the distal segment) and electromagnetic waves. Faster and better organized growth has been reported.

How about limbs? Robert Becker (1961) discovered that the electric behavior of an amputated frog's limb was different from that of a newt's limb. Now, frog's limbs are no better than human's so far as regeneration is concerned, but a newt can grow new ones in a month or two when its limbs are cut off. Smith (1974) experimented on frog's ability to regenerate limbs by imposing the electric pattern seen in an amputated newt's limb. He implanted a bimetallic couple of the sort mentioned in the preceding paragraph in the limb stump of an amputated adult frog's forearm. When the cathode was at the wound surface, he found that the frog grew regenerates which contained recognizable organized limb structures.

13.4 Mathematical Formulations of Wolff's Law

Continued efforts have been made to clarify Wolff's Law. On the side of morphology, quantitative stereology is used to determine the directions and areal density of the trabeculae in any cross section. On the side of mechanics, finite element method has been used to determine the stresses and strain in the structure. Photoelasticity has served for many years as a means to determine the principal stresses in two-dimensional lucite models of bone. The use of "frozen stress" in transparent plastics has extended photoelasticity to three-dimensional models.

In the meantime, bone cells are studied with electron microscopes and biochemical means. Menton et al. (1984) noted that all periosteal and endosteal bone surfaces are covered with bone lining cells. Among the bone lining cells are *osteoblastic precursor cells* (Gr. *blastos*, germ). As the precursor cell becomes an osteoblast and matures, it develops numerous pseudopods that contact the pseudopods of maturing and mature osteocytes. As the mineralization front approaches the osteoblast, the osteoblast appears to round up and begins to function as an osteocyte.

Chambers (1985) describes *osteoclastic cells* (Gr. *Klastos*, broken) which can reach the bone via blood circulation. Whether these are mononuclear phagocytes or not is not clear. They participate in calcium homeostasis of the blood and bone. Chambers (1985) presents a hypothesis which assumes that strain stimulates the osteocytes which in turn induces osteoblasts to produce osteokinetic agents that attract the osteoclasts. Thus the osteoblasts have the capacity either to suppress (through prostaglandin production) or to stimulate (through mineral exposure) osteoclastic resorption.

Lanyon (1984) has shown that bone resorption will occur for strains less than about 0.001 and bone deposition will occur for strains greater than about 0.003. How osteocytes can sense strains so small is not known. Gross and Williams (1982) and Salzstein and Pollack (1987) proposed that the mechanism may be the ion streaming of fluid in cavities in bone tissue, including lacunae, canaliculi, Haversian, and Volkmann canals. El Haj et al. (1988) showed that the orientation of the proteoglycan molecules within bone tissue is related to strain rate. If there were no further loading, the proteoglycan orientations persist for over a day. Further data are provided by Yeh and Rodan (1984), Nulend (1987), and Gross et al. (1988).

Surface Remodeling

Cowin and associates (1979, 81, 84, 85) have studied the mathematical form of Wolff's law of bone remodeling. Their basic idea is to describe the osteoblastic and osteoclastic activities on the surface of bone tissue, including trabecula, Haversian canal, and the periosteal or endosteal surface of a whole bone. Bone cell activity is assumed to depend upon genetic, metabolic, and hormonal factors as well as the strain history experienced by the bone cell. The osteoblastic and osteoclastic function are written as a_b and a_c respectively. The rate of surface deposition or resorption is expressed in terms of U, with the dimensions of velocity (typically μm/day or mm/year). Then, following the ideas of Martin (1972) and Hart (1983), Cowin and Van Buskirk (1979) wrote

$$U = n_b A_b a_b - n_c A_c a_c \tag{1}$$

in which n_b and n_c denote the number of osteoblasts and osteoclasts, respectively, per unit area, and A_b, A_c are the surface areas available to the osteoblasts and osteoclasts, respectively. Further, for simplicity, they assumed

$$a_b = H_b \varepsilon + G_b, \qquad a_c = H_c \varepsilon + G_c \tag{2}$$

$$U = C(\varepsilon - \varepsilon_0) \tag{3}$$

in which ε is the strain, H_b, H_c, G_b, G_c, C, ε_0 are constants. ε_0 represents a strain at which no remodeling occurs. C is the remodeling rate constant. Cowin et al. (1981, 1985) have applied these equations to the remodeling of long bones with the finite element method.

Mathematical Description of the Cancellous Bone Architecture

A cross section of a cancellous bone specimen is shown in Fig. 13.2:2. The calcified bone tissue, called trabeculae, is surrounded by bone marrow. Whitehouse (1974) and Whitehouse and Dyson (1974), used stereological methods to measure the mean intercept length of the trabeculae as a function of the direction of the test lines. Let θ be the angle between a test line and an arbitrarily chosen x-axis. Let L be the average distance between two bone/marrow interfaces measured along all test lines inclined at an angle θ. Whitehouse showed that the following equation fits the measured results very well

$$\frac{1}{L^2(\theta)} = M_{11} \cos^2\theta + M_{22} \sin^2\theta + 2M_{12} \sin\theta \cos\theta, \tag{4}$$

where M_{11}, M_{22}, M_{12} are constants. Harrigan and Mann (1984) then showed that in three dimensions their experimental results can be fitted by an equation which is a generalization of Eq. (4):

$$\frac{1}{L^2(\mathbf{n})} = M_{ij} n_i n_j, \qquad (M_{ij} = M_{ji}) \tag{5}$$

in which $\mathbf{n}$, with components (n_i, n_2, n_3) referred to a rectangular cartesian frame of reference, denotes a unit vector in the direction of the test line. Equations (4) and (5) represent an ellipse and an ellipsoid, respectively, in the space of n_1, n_2, n_3. M_{ij} can be considered the components of a tensor of rank 2, $\mathbf{M}$, with M_{ij} as components. In matrix form,

$$(\mathbf{M}) = \begin{pmatrix} M_{11} & M_{12} & M_{13} \\ M_{21} & M_{22} & M_{23} \\ M_{31} & M_{32} & M_{33} \end{pmatrix}. \tag{6}$$

The quadratic form defined by the right-hand sides of Eqs. (4) and (5) are positive definite. Hence an inverse of the matrix $\mathbf{M}$ exists. The validity of Eq. (5) was confirmed by Cassidy and Davy (1985).

Cowin (1986) then introduced a tensor H defined as the inverse of the square root of $\mathbf{M}$:

$$\mathbf{H} = \mathbf{M}^{-1/2} \tag{7}$$

and called it the *fabric tensor* of cancellous bone. He showed that the Young's modulus of cancellous bone is larger if the H_{ij} values are larger. Hence it is preferable to correlate H_{ij} with the mechanical properties of the cancellous bone.

Mathematical Statement of Wolff's Law

When a cancellous bone is subjected to a loading, resulting in a stress represented by a tensor T_{ij}, a remodeling process will go on so that the fabric tensor H_{ij} will change with time. If T_{ij} were a constant, then after the elapse of a certain period of time the remodeling process will cease when a new homeostatic condition (remodeling equilibrium) is reached. Concerning this homeostatic condition, Cowin (1986) states Wolff's law in the following form:

"At a homeostatic condition, the principal axes of the stress tensor T_{ij} and those of the fabric tensor H_{ij} coincide."

Cowin shows that this coincidence of principal axes is assured if the matrix multiplication of (T_{ij}) and (H_{ij}) is commutative, i.e., if

$$(T_{ij})(H_{ij}) = (H_{ij})(T_{ij}). \tag{8}$$

The proof of condition (8) is the same as that described in Sec. 2.10.

Wolff's law must be part of a more general law defining the rate of change of the fabric tensor as a functional of the stress T_{ij}, strain E_{ij}, tensor H_{ij}, blood flow $\dot{Q}$, biochemical factors c_k, age, and time t:

$$dH_{ij}/dt = f(T_{ij}, E_{ij}, H_{ij}, \dot{Q}, c_k, t). \tag{9}$$

By saying that it is a functional, we mean that it is a function of the entire histories of these variables. Finding the functional and validating it are objectives of future research.

One aspect of bone remodeling is the change of the density and strength of the bone, or more restrictively, the solid volume fraction of trabecular structure, denoted by v and defined as the volume of solid trabecular struts per unit bulk volume of the tissue. At homeostatic condition of the patella bone, Hayes and Snyder (1981) found a strong correlation of v with the von Mises effective stress component of T_{ij} in two-dimensions. A subsequent three-dimensional study by Stone et al. (1984) found, however, that v correlates not with von Mises stress, but with principal stresses. Fyhrie and Carter (1985) suggested that v correlates with the strain energy density. So again future research has to focus on validation.

13.5 Remodeling of Soft Tissues in Response to Stress Changes

Heart, lung, blood vessels, and muscle also remodel in response to stress and strain. Their features are described below.

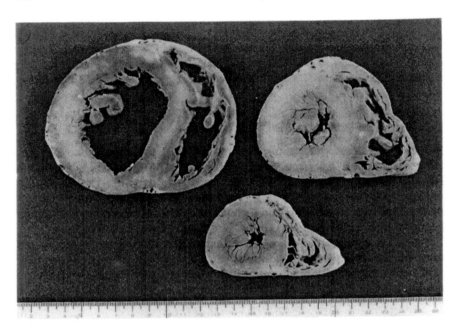

FIGURE 13.5:1 A comparison of three dog's hearts: (a) *Top left*: Volume-overloaded heart. (b) *Top right*: Pressure-overloaded heart. (c) *Bottom*: Normal. The normal heart was photographed 4 hours post mortem. Photo by courtesy of Dr. Toshio Nakamura, Tokai University, Isehara, Kanagawa, Japan.

Hypertrophy of the Heart

It is well known that when the heart is overloaded its muscle cells increase in size. Figure 13.5:1 shows a comparison of the cross sections of (c) a normal heart, (a) a heart chronically overloaded by an unusually large blood volume (b) a heart chronically overloaded by an unusually large diastolic and systolic left ventricular pressure, all from autopsies of man. Figure 13.5:2 shows a comparison of the histology of a normal and a pressure-overloaded heart. It is seen that the muscles in hypertrophic heart are much bigger in diameter than those of the normal heart. Figure 13.5:3 shows the changes in myocytes in hypertrophy.

Note the difference of these two types of hypertrophy. The volume-overloaded heart increases its ventricular volume. The pressure-overloaded heart increases its wall thickness both inward and outward, resulted in a decrease in the volume of the ventricular chamber. Ultrastructural studies of the muscle fibers showed (Anversa et al., 1971) that the structure of the muscle cells of the pressure-overloaded hypertrophic heart are pretty uniform throughout the ventricular wall, as expressed in terms of the volumetric density of the contractile mass, mitochondria, and T system, and the surface-

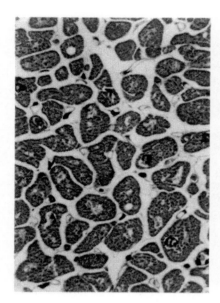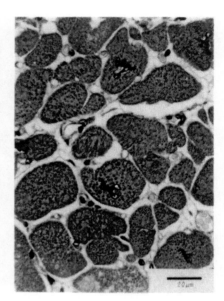

FIGURE 13.5:2 A comparison of the histology of a normal (left) and a pressure-over-loaded heart (right). Photographed at the same magnification. Courtesy of Dr. Toshio Nakamura.

to-volume ratio of these components in the outer, middle, and inner layers of the ventricular wall. The fiber diameter of the hypertrophic muscle cells of the left ventricle was said (Lund and Tomanek, 1978) to be larger in the sub-endocardium than in the subepicardium, consistent with the concept that the tensile stress is higher in the endocardial region; whereas the muscle fiber diameter of the right ventricle was the same as the control. The muscle fiber orientation was found (Tezuka, 1975) to be somewhat rearranged in the inner myocardial layer: the average angle between the muscle fibers at the endo-cardial surface and the basal plane was 63° for the normal heart, 56° for pressure overloaded heart, 50° for the volume overloaded heart.

Nakamura et al. (1982), using data on the pressure-volume relationship of the left ventricle to deduce the stress–strain relationship of the myocardium, found that the stress–strain relationship of the hypertrophic myocardium is not different from that of the normal. Arai et al. (1968) measured the growth of the coronary artery and the capillary blood vessels in hypertrophic hearts and found that the mean blood flow per unit volume of the heart muscle estimated by the anatomical radii of the three major branches of coronary arteries was practically the same as that in the normal heart.* The total

---

* Based on Suwa et al.'s (1968) result that the mean blood flow of an arterial branch, $\dot{Q}$, can be related to the radius of that branch, r, by the equation $\dot{Q} = qr^n$, where q and n are constants. The value $n = 2.7$ was found to hold for all the arterial systems examined.

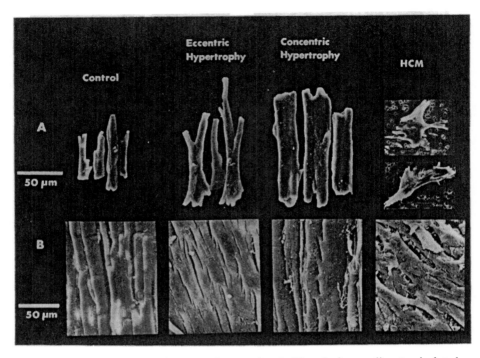

FIGURE 13.5:3 Scanning electron micrographs. A: Ventricular cardiocytes isolated from autopsied human hearts. HCM = hypertrophic cardiomyopathy with asymmetrical hypertrophy. B: En face views of cellular architecture in situ in ventricular myocardium. From K. Kawamura (1982). Reproduced by permission.

capillary length was found to increase in hypertrophic hearts in parallel with the increase in the total surface area of the muscle fibers. The capillary length per unit volume of the heart muscle, was found to be somewhat decreased in hypertrophic heart.

The change in the behaviour of the hypertrophic heart seems to be due more to the changed geometry of the ventricular chamber than to the changed properties of the muscle fibers or blood vessels. Nakamura et al. (1982) showed that at an end-diastolic pressure of 20 mmHg the end-diastolic volume of hypertrophic heart was smaller than the normal by 47% ($p < 0.01$), and the peak systolic pressure in isovolumic beats was higher than the control by 24% ($p < 0.01$). When ejection pressure was the same, the difference between the stroke volumes of the hypertrophic and control hearts was insignificant. The isovolumic pressure line (see *Biodynamics: Circulation*, Fung (1984), p. 39) changed, its slope and volume-axis intercept decreased in proportion to the change in chamber geometry.

The question arises why is hypertrophic heart prone to failure? (Braunwald, 1980). The results quoted above suggest no simple answer. It is clear, however, that the process of hypertrophy, its initiation and its equilibrium state, must be stress related.

Volume and pressure overload are not the only causes for hypertrophy of the heart. Frolich (1983) and Alpert (1983) reviewed other hemodynamic and nonhemodynamic factors such as the increased heart rate, the augmented ventricular contractility, the increased total peripheral resistance, aging, collagen deposition, coexistant diseases (such as atherosclerosis, myocarditis, diabetes mellitus), as well as race, sex, obesity, hormones (catecholamines, angiotension II, growth hormone), and therapy. Limas (1983) showed that the microtubules are involved in isoproterenol-induced hypertrophy of the heart.

Remodeling of the Lung

Cowan and Crystal (1975) showed that when one lung of a rabbit was excised, the remaining lung expanded to fill the thoracic cavity, and it grew until it weighed approximately the weight of both lungs. What are the factors responsible for this growth response? A study of the effects of growth hormone on lung growth after unilateral pneumonectomy suggests that hormone is not directly involved (Brody and Buhain, 1973). Increasing the blood flow to one lung does not result in a growth response of the lung (Romanova et al., 1970). The hypoxemia seen immediately after pneumonectomy resolves shortly after surgery and is probably not a major factor (Nattie et al., 1974). It remains to consider the mechanical factor of increased stress and strain (Cohn, 1939 and Gaensler and Strieder, 1951).

For the purpose of quantitative investigation, collagen synthesis before and after surgery was studied by Crystal (1974), Cowan and Crystal (1975), using New Zealand white rabbits. They determined the course of collagen synthesis in the lungs, performed surgery at 70 days of age, and continued monitoring until 180 days after birth. At a specified time the lungs were excised and weighed. Tissue slices (0.5 mm) were incubated from 1 to 4 hours in a medium containing 0.012 mM [^{14}C]proline. After incubation, the slices were rinsed, homogenized, and measured for dry weight, DNA, protein, hydroxyproline, [^{14}C]proline in noncollagen protein, [^{14}C]hydroxyproline, and the specific activity of free [^{14}C]proline as described by Bradley et al. (1974). The rate of collagen synthesis (nmols [^{14}C]proline incorporated into [^{14}C]hydroxyproline per mg of DNA per hour) was calculated. This reflected the synthesis of collagen chains (proline is hydroxylated only after it is incorporated into collagen), because it has been verified that the number of hydroxylated prolines per collagen molecule remained constant.

Cowin and Crystal's (1975) results are shown in Fig. 13.5:4. The ordinate shows the total content of collagen in the right lung. Normal growth, marked by open circles, shows a rapid increase of collagen in the 1st month of life. A similar rapid growth is seen after left pneumonectomy. The rate of collagen synthesis as a percentage of the total protein synthesis of the right lung is high in the first month after birth and 1st month after left pneumonectomy.

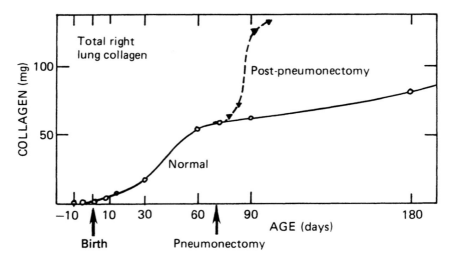

FIGURE 13.5:4 Cowan and Crystal's experiment showing lung collagen content in the right lung after left pneumonectomy (●--●) compared with the normal right lung at several ages (o——o). From Cowan and Crystal (1975). Reproduced by permission.

Cowan and Crystal saw similar results if the left lung was tied off at the hilum and left in the chest. On the other hand, if the left lung was excised and the cavity left by the excised lung was filled with liquid paraffin embedding wax after chest closure, then they found little change in the right lung from normal.

These experiments show that there is a direct relationship between the changed strain level and collagen synthesis in the lung.

Remodeling of Pulmonary Blood Vessels in Hypertension

When the oxygen supply to the lung is reduced suddenly, pulmonary blood pressure increases and changes in blood vessels occur. The thicknesses of the adventitia and the smooth muscle layer in the pulmonary arteries increase, the lumens of small arteries deccrease, and new muscle growth occurs on smaller and more peripheral arteries than normal. To identify cell division activities in these vessels, Meyrick and Reid (1979) studied the incorporation of ^{3}H-thymidine into the DNA of the new cells in rats exposed to hypobaric hypoxia (exposure to 380 torr) after 1, 2, 5, 7, 10, and 14 days. Using autoradiographs of 1-μm sections, they measured and expressed the ^{3}H-thymidine incorporation by a "labeling index". Figure 13.5:5 shows one of their results. It is seen that the remodeling occurs within days. In a separate study, Meyrick and Reid (1982) showed that pulmonary arterial endothelial cell mitosis

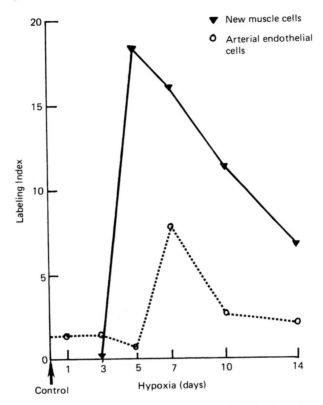

FIGURE 13.5:5 The cell division activities reflected by the "labeling index" of incorporating ^{3}H-thymidine in the pulmonary peripheral intra-acinar arteries (external diameter between 15 and 100 μm) after exposure to hypoxia. Standard deviations too small to be shown. From Meyrick and Reid (1979). Reproduced by permission.

occurred also in rats fed with *Crotalaria spectabilis* seeds which caused pulmonary artery hypertension; but in this case not only did the changes take weeks to occur, but also there was a lack of evidence of cell division of small muscle cells.

Further, Sobin et al. (1983) have shown that when rabbits were put into a chamber low in partial pressure of oxygen, changes in the structure of their pulmonary arterioles occurred within a few hours. Bevan (1976) induced hypertension in dogs by tying off a renal artery, and showed that mitotic activity began in the otolic artery soon afterwards, reaching a maximum in about 2 weeks, and then subsiding gradually. Such mitotic activity was associated with an increase in blood pressure, and hence is stress related.

Further discussion of the remodeling of blood vessels due to physical, chemical, and biological stimuli is given in Sec. 13.9.

Changes in Blood Vessel Diameter Caused by Increased Blood Flow

Kamiya and Togawa (1980) presented results of a very interesting experiment on the adaptive change of canine carotid artery diameter in response to changes in blood flow. They constructed surgically an arteriovenous shunt from the common carotid artery to the external jugular vein. As a consequence, blood flow was increased in one segment of the artery while decreased in another. They showed that in 6 to 8 months after the operation, the segment with increased flow dilated while the other segment with decreased flow atrophied to a smaller diameter. The diameter of the artery increased or decreased in such a way that the wall shear rate* $\dot{\gamma}$ remained almost constant (within 15%) if the change of flow was within 4 times of the control. The transendothelial protein permeability, evaluated at the Evans-blue-dye-T-1824-stained surface by a reflectometric method, also showed a close correlation with wall shear ($r = 0.934$). They therefore suggested a local regulatory mechanism of wall shear stress controlling the vessel growth involving protein turnover in the vascular wall.

According to Liebow (1963), Thoma (1893) first observed in chicken embryos that the pathways of the fastest blood velocity became the main arteries while those with slower velocity atrophied. Liebow (1963) and others, on studying arteriovenous fistulas and collateral circulation, have also showed that increased blood flow induces blood vessel dilatation. Rodbard (1975) collected clinical evidences of the same. More recent work on the relationship between shear stress and changes in vascular endothelial cells initiated by Fry (1968) in search of atherogenesis has also indicated this trend.

Kamiya et al. (1984) collected data from the literature and obtained the following estimates of the wall shear stresses (dynes/cm^2) in the dog:

aorta	large arteries	small arteries	arterioles	venule	veins	vena cava
12	10–16	19	14	3	1.5	6.3

For the capillaries, data from rat and cat show that the wall shear stress lies in the range of 10–26 dyn/cm^2. The narrowness of the range of the wall shear stress in the entire vascular tree is remarkable.

If the shear stress on the endothelial wall due to the viscosity of the flowing blood really controls the blood vessel caliber, then the feat is all the more remarkable in view of the smallness of the shear stress. We know that the state of stress at any point in a continuum can be stated either in terms of three principal stresses (tensile or compressive) or in terms of two maximum shear stresses and a mean stress. The two maximum shear stresses are numerically equal to one half of the differences of the principal stresses, and act on planes inclined at 45° to the principal planes. The mean stress is equal to one third of the sum of the principal stresses. When blood flows in a blood vessel, the blood pressure induces stresses in the vessel wall. At a normal blood pressure

* $\dot{\gamma}$ is calculated from $\dot{\gamma} = 4\dot{Q}/(\pi r^3)$ under the assumption of laminar flow. $\dot{Q}$ is the flow rate.

the circumferential stress is of the order of 10^6 dyn/cm^2, the longitudinal stress is smaller but of the same order of magnitude, the radial stress is an order of magnitude smaller. The maximum shear stresses in the arterial wall due to blood pressure are, therefore, 5 orders of magnitude larger than the shear stress imposed by the blood on the vessel endothelium. Yet we know that pressure overloading induces thickening of the vessel wall (usually with decreased vessel lumen), whereas the flow overloading induces enlargement of the vessel diameter. This suggests that there is somethig very special about the endothelial cells, and that the biochemical-biomechanical pathway must be very carefully studied.

Healing of Surgical Wounds

Healing of wounds and development of new strength reveal structural and mechanical changes in a tissue. The author and his associates (Lee et al., 1985) have measured the strength of surgically anastomosed arteries of the rat sutured with a polyglycolic acid surgical thread (Dexon). The abdominal aortas and the carotid arteries were severed, sutured, and then the wounds were closed and the animal healed. After a specific period of time, the vessels were taken out and tested in uniaxial loading condition. The stress–strain relationship of the vessels was measured, and then the vessels were pulled to failure. It was found that the strength of the anastomosis was the lowest in about 4 months. Figure 13.5:6. In the first day, the force at failure was about the same as that of the control. Then the strength decreased with time, until a minimum was reached in 4 to 6 months; at which time the tensile force to failure was about 25% of the control for the carotid artery and 49% of the control for the abdominal aorta. The corresponding values of the tensile stress at failure were 17 and 11%, respectively. The differences between ultimate forces and stresses were caused by the thickening of the vessel wall in the neighborhood of the suture line in the healing process. After 4 to 6 months, the strength gained again. At 13 months, the strength of the anastomosis was about the same as that of the control. The stretch ratio at failure was constant through all periods.

Thus the healing of the artery after surgery appears to be quite slow: return of the maximum tensile force to normal takes 12 months. In contrast, wound healing in the gastrointestinal tract is much faster. Gottrup (1980) has shown that when incisions were made in rat stomach and duodenum and immediately resutured with polypropylene thread 6–0, the wounded tissue can regain the breaking strength of the intact tissue in 5 to 10 days.

The development of the strength of an artery after surgery depends on the process of resorption of the suture, the regeneration of the smooth muscle and the connective tissues, and the changes in structure in the process of resorption and regeneration. It is believed, but not proven, that adequate stress is necessary. Gottrup (1981) has attempted to show a correlation between the collagen content and the gain in strength in wound healing in gastrointestinal tract.

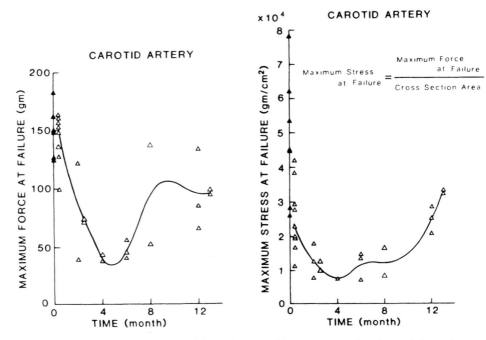

FIGURE 13.5:6 *Left*: Force at failure for carotid artery as a function of time after surgery. Failure loads of the controls at day zero are shown in solid black triangles. The curve is a smooth cubic spline curve. *Right*: Stress at failure of carotid artery as a function of time after operation. Black triangles are for controls at day zero. From Lee et al. (1985). Reproduced by permission.

Goldin et al. (1980) have examined the change of various mechanical and biochemical factors in the process of wound healing of digital tendon of the rabbit. Theoretical analysis of tissue strength based on morphological, rheological, and structural data needs to be developed.

Muscle Atrophy During Space Flight or Immobilization

Animals exposed to the weightless condition of space flight have demonstrated skeletal muscle atrophy. Leg volumes of astronauts are diminished upon return to earth. In flight vigorous daily exercise is necessary to keep astronauts in good physical fitness over a longer period of time. Herbison and Talbot (1985) reviewed the literature of muscle growth and noted the following: The rate of muscle protein synthesis decreases within 6 h of immobilization of rat limbs (Booth, 1982); it returns to normal 6 hrs after remobilization (Booth et al., 1982). Insulin appears to be an important hormonal factor for short-term regulation and maintenance of protein balance in skeletal muscle (Goldberg, 1979). Immobilized muscle atrophies; but there is a marked difference between stretched immobilized muscle versus muscles immobilized in the resting or shortened position (Booth, 1982; Simrad et al., 1982, Spector et al., 1982). When muscle is denervated, strong electric stimulation may retard atrophy (Karpovich, 1968). During limb immobilization and space flight, the blood cortisone level is elevated (Leach and Rambaut, 1977). Passive tension or repetitive stimulation retards protein degradation in isolated muscle preparation (Goldberg et al., 1975). Muscle protein degradation may be induced by thyroid hormones (Goldberg et al., 1980). Inactivity of skeletal muscle induced by joint immobilization or by pharmacologic or surgical denervation is associated with an increase in extrajunctional acetylcholine receptors (Fischbach and Robbins, 1971; Lavoie et al., 1976; Pestronk et al., 1976).

Fundamentally, growth is a cell biological phenomenon at molecular level. Stress and strain keep the cells in a certain specific configuration. Since growth depends on this configuration as well as other factors, it depends on the stress and strain (Coan and Tomanek, 1979; Faulkner et al., 1983.)

13.6 Stress Field Created by Fibroblast Cells and Collagen Synthesis

In a continuum, any local change in stress and strain affects the entire body. In fact, it can be proven mathematically that if a simply connected body obeys the linear law of elasticity and is free from body force, then if there exists a finite domain in the body, no matter how small, in which all strains vanish, then the strain must vanish in the entire body, which is thus stress free. Similarly, in a field of flow of a Newtonian fluid in a simply connected domain free from body force, one can show that if the velocity is identically zero in a small finite region, then the flow is zero everywhere. Mathematically, this is a property of the partial differential equations of the 'elliptic' type which governs these continua. Conversely, if strain or flow is changed in one region, no matter how small, then the whole field will be changed. Applying this mathematical result to a living body, we infer that if one cell changes its volume or shape, then every cell in the body is stressed and strained. Thus, in principle, the stress field of each cell is influenced by other cells in the body no matter how remote they are. The intensity of the signal, of course, decreases with increasing distance between the cells.

A biological experiment by Harris et al. (1981) shows this beautifully. They prepared a thick gel of reprecipitated collagen and planted two groups of chicken heart fibroblasts in it at a distance of about 1–2 cm apart. When the growing cells spread and cluster, they create a field of tension between the two explants. The tensile stress stretches the collagen fibers into alignment. Figure 13.6:1 shows a photograph of the straight bundles of collagen that form between the adjacent explants.

The phenomenon illustrated by Fig. 13.6:1 suggests that tissues such as tendons, ligaments, organ capsules and dermal tessellations do not need to be

FIGURE 13.6:1 Harris et al.'s experiment on stress field created by growing cell colonies. Two groups of cells separated by a distance of 1.5 cm were grown in a collagen gel. In the process of polymerization, the newly formed collagen fibers become aligned into long axially oriented tracts interconnecting two centers of traction. Heart explants from 8-day chick embryos after 96 h in culture. Scale bar, 1 mm. From Harris et al. (1981), *Nature* Vol. *290*, p. 251. Reproduced by permission.

laid down by direct secretion of collagen fibres into their eventual arrangement. Harris et al. (1981) believe that fibroblast traction can rearrange and repack collagen into these patterns, even beginning from a totally random meshwork.

An earlier experiment by Harris et al. (1980) may help visualize what is going on. They prepared very thin sheets of silicone rubber membrane by briefly exposing a silicone fluid to a flame. The outer layer of the fluid was cross-linked into a membrane of about 1 μm thick. This membrane was then lifted up and put in a tissue culture fluid and a fragment of embryonic chick heart was placed on the membrane. Figure 13.6:2 shows what was seen after 48 hours. Under dark-field illumination, the bright radiating lines show the wrinkles in the membrane, wrinkles similar to those in cloth fabric under tension. In engineering this is known as 'tension field' in thin-walled structures. These lines represent the lines of tension (a principal stress trajectory) in the membrane. The membrane is compressed and wrinkled ('buckled' in engineering terminology) in the direction perpendicular to the tension line; and since the membrane is very thin the compressive stress (the so-called critical buckling stress) is very small and can be treated as zero. Figure 13.6:2 is a graphic demonstration of the tension field created by the fibroblasts. The rays are the principal tensile stress trajectories.

When the area of contact between the fibroblasts and the membrane substrate was examined under the microscope, it was seen that the cells adhere to the membrane along the outer border of the area of contact. Behind the border (of width 5 to 25 μm) the membrane was wrinkled, with the direction of the wrinkles perpendicular to the rays outside, indicating that the membrane was under compression beneath the cells. These facts are interpreted as follows (Fig. 13.6:3): The outgrowing cells are rather flat (like pancakes). When they thicken and round up (into balls), the edge pulls the membrane substrate in, creating compression wrinkles under the cells and tension rays radiating away from the cells.

Thus the fibroblasts first grow out thin and/or elongated and then round up and contract into a ball. In so doing, depending on the circumstance, it

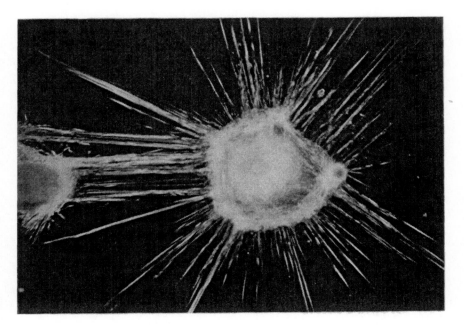

FIGURE 13.6:2 Harris et al.'s experiment on cells grown on very thin membranes of silicone rubber. The contraction of the cells caused wrinkles in the rubber membrane. Photograph shows a low magnification, dark-field illumination view of a chick heart explant that had been spreading on the membrane for 48 hours. The bright radiating lines are the tension wrinkles. The bar is 1 mm long. From Harris et al. (1980), *Science*, Vol. *208*, p. 178. Reproduced by permission.

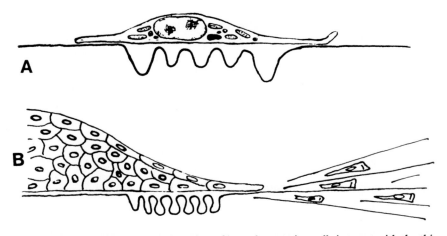

FIGURE 13.6:3 Harris et al.'s explanation of how the growing cells interact with the thin membrane on which they rest. (A) Diagrammatic side view of an individual cultured fibroblast distorting and wrinkling the elastic silicone substratum upon which it has spread and is crawling. (B) Diagrammatic side view of the margin of an explant whose cells are spreading outword on a silicone rubber substratum. The traction forces exerted by the outgrowing cells compress the rubber sheet beneath the explant and stretch it into long radial wrinkles in the surrounding area. From Harris et al. (1980). *Science*, Vol. *208*, p. 177. Reproduced by permission.

may stretch the surrounding medium and induce tension in a certain direction. The tension may serve to line up the procollagen and then form collagen fibers. If there were two colonies of fibroblasts, each anchored in the medium, then their contraction may induce tension between these colonies and eventually create collagen fibers linking these colonies. This is believed to be the cause of the condition shown in Fig. 13.6:1. It resembles Paul Weiss's (1955) "center effects" theory of nerve growth.

The apparent contraction of collagen networks around deep wounds, burns, and surgically implanted prostheses may be caused by the same process. The idea is very suggestive and may have many applications.

The concept that an internal stress field is created by the growing cells themselves adds an exciting dimension to the stress modulation of growth. In engineering language it adds a feedback control channel. It suggests that with a proper design, tissue growth can be engineered. See Secs. 13.8–13.10.

13.7 Growth Factors

Biological growth is a cellular activity. At the foundation, its study belongs properly to molecular biology, genetics and biochemistry. From molecules to organism, mechanics plays a role because there are many steps that are influenced by flow and stress.

To study growth from the point of view of physiology is to begin at the macroscopic end. One accepts the existence of a homeostatic state of an individual organism, and asks how does the homeostatic state change when the environment and the internal stress are changed.

The trend of research is to bring the microscopic and macroscopic views together. In cellular biomechanics, biochemistry, cell biology and biomechanics are brought together. In biomechanics of growth, growth hormone, and growth factors are essential players. But the subject is vast, and it is beyond the scope of this book to describe it in any detail. In the following, a bare outline of the historical highlights is given to serve as an introduction.

Growth Hormone

The term hormone was chosen by Bayliss and Starling in 1902 from a Greek work *hormaein* meaning "I arouse to activity". Earlier, in 1825, Leuret and Lassaigne showed that in the dog, acidification of the upper intestine led to secretion of pancreatic juice. In 1849, Berthold observed that the atrophy of the comb and loss of male behavior in cockerels following castration could be prevented by grafting a testis into the abdominal cavity. These and other experiments lead to the idea that endocrine glands regularly release substances into the blood stream that are necessary for the normal development and function of other parts of the body. In 1902, Bayliss and Starling demonstrated that an extract of dog intestinal mucosa could, on intravenous injection into the dog, cause intense pancreatic secretion. They gave the name *secretin* to the agent in the extract that caused the pancreas to secrete, and the term *hormone* to describe similar messengers.

The following are endocrine glands: hypothalamus, hypophysis (pituitary gland), thyroid, parathyroid, adrenal, islet of Langerhans of the pancreas, ovary, testis, pineal gland, placenta. *Growth hormone* (GH) was first found in pituitary gland, *growth hormone-releasing factor* (GHRF) and *growth hormone release-inhibiting hormone* (GHRIH, also called *somatostatin*) exist in hypothalamus, but was first found from pancreatic tumors.

The great stimulus to an understanding of the pituitary was a burst of interest in acromegaly toward the end of the nineteenth century. The interest was stimulated by Pierre Marie in 1886 who described and named *acromegaly* (enlargement of the extremitas of the skeleton—the nose, jaws, fingers, and toes) as a disease (see Raben, 1959). In 1921, Evans and Long showed that a pituitary extract could induce gigantism in rats. Cho-Hao Li and associates (1948) then produced highly purified crystalline growth hormone in 1940's, and Li and associates determined the complete structure of the molecule in 1971.

The growth hormone accelerates growth, increasing the size of all organs and promoting the growth of bone before closure of the epiphyses. It increases protein formation, decreases carbohydrate utilization, and increases the mobilization of fat for energy use (Knobil and Greep, 1959).

In growth hormone experiments, the rat has been favored because it is exceptional in two ways. The epiphyses of almost all of its long bones remain open throughout life, and growth continues for as long as good nutrition and good health prevail. It is also peculiar in responding indiscriminately to the growth hormone of all other mammals thus far explored.

Growth in human dwarfs in response to human growth hormone is normally proportioned, but there is no sexual maturation. No features of gigantism or acromegaly have yet been described, but prolonged treatment with large doses has not yet been attempted.

Growth Hormone Releasing Factor

The discovery of a peptide with specific and unusually potent stimulatory effect on growth hormone secretion was achieved in 1982. The source of this peptide was not the hypothalamus, but from pancreactic tumors obtained from two patients. In both cases, the patient had symptoms of acromegaly. In one patient, pituitary enlargement was diagnosed and transsphenoidal surgery undertaken; the pituitary showed hyperplasia of the GH-secreting cells, the symptoms of acromegaly persisted postoperatively, and a pancreatic tumor was subsequently found. The second patient had a normal pituitary and two pancreatic tumors. In both cases extracts of the pancreatic tumors revealed a peptide of similar structure. Within a few weeks of each other, Guillemin et al. (1982) and Vale et al. (Rivier et al., 1982) revealed the structure of the human pancreas growth hormone releasing factor.

Guillemin's Language of Polypeptides

With modern techniques of cell biology, many growth factors have been identified. Peptides can be recognized by radioimmunoassay techniques and visualized in cells that produce them by immunocytochemical methods. Neurons can be stained for various hormone-releasing factors. The morphologist can work out a complete map for the distribution of each kind of peptide-containing neuron in the brain and spinal cord. Palkovits (1984) has shown that particular peptides are not confined to specific regions but are instead widely distributed in patterns which have no known or obvious functional correlations. Guillemin (1985) likened the peptides in the brain as a language. He concludes that the central nervous system is a gigantic multi-endocrine organ secreting within itself or to very closely attached structures (the pituitary gland, the periventricular organs) multiplicity of peptide molecules, most of them are also known to be secreted by nonneuronal endocrine cells or tissues at the periphery (and thus not specifically neuropeptides). He said that we do not really know the role(s) of these peptides in the brain. But we know what many of these same peptides do at the periphery. So he sums it up by saying that we know a local pituitary "dialect" for some peptides, a local pancreatic "dialect", but we do not know the brain "dialect(s)".

NGF, EGF

Then 1986 Nobel Prize in Medicine was awarded to Rita Levi-Montalcini and Stanley Cohen for their discoveries of *nerve growth factor* (NGF), and *epidermal growth factor* (EGF), respectively.

The story began when Levi-Montalcini was invited to St. Louis by Viktor Hamburger. In St. Louis they repeated an experiment by Elmer Bueker in which he had noted that mouse sarcomas transplanted to chick embryos induced vivid nerve growth from the chick embryo nervous system into the tumor. Levi-Montalcini noted that sensory and sympathetic ganglia that remote from the tumor also become enlarged. So she suggested and proved that a diffusible substance (NGF) exists that stimulates nerve growth. Joined by Cohen, they began to characterize this material enzymatically, and discovered that the addition of snake venom greatly enhanced the nerve growth—stimulating activity of the tumor extract. They soon realize that the snake venom itself was able to stimulate nerve growth. They reasoned that if venom glands in snakes contained NGF-like activity, perhaps salivary glands of mammals might also be active. Thus it was discovered that the submandibular gland of adult male mice is an incredibly rich source of NGF. Cohen then isolated NGF. Levi-Montalcini developed an NGF bioassay.

Stanley Cohen injected salivary gland extracts into newborn mice. He found precocious opening of the eyes and precocious tooth eruption. He soon realized that the salivary gland extract contained a second biologically active factor different from NGF. He isolated this factor, named it epidermal growth factor (EGF), and demonstrated its effect on the epithelial cells of the skin, mucous membrane, cornea, mammary gland; and several other types of cells such as connective tissue, cartilage, liver, smooth muscle, and hormone-producing cells. Typical references are Cohen (1959), Levi-Montalcini and Angeletti (1966). See Schreiber et al. (1986).

Angiogenesis: ECGF, FGF

Folkman et al. started the study of angiogenesis in the 1960's by observing that transplanted tumors in mouse may or may not grow depending upon whether the transplant stimulates the growth of new capillaries or not (Folkman, 1985, Thomas et al., 1985). Using cell culture, rabbit cornea, biopolymers, and chick embryo chorioallantoic membranes as media to observe capillary growth, they saw new capillaries arise as offshoots of established vessels. In the presence of an angiogenic stimulus, endothelial cells begin to extrude themselves from their parent vessels by releasing specific collagenases that cause local lysis of the vascular basement membrane. The endothelial cells elongate, undergo directional locomotion toward the angiogenic stimulus, and form a spout in which endothelial cells proliferate. A lumen is then formed. Next, two sprouts join to form a loop through which blood begins to flow. New sprouts branch from the loop and a capillary network is gradually constructed.

Folkman et al. found that the tissue mast cells are associated with increased capillary growth. Since mast cells secrete heparin, they were led to the discovery that heparin can potentiate angiogenesis: it has a high binding affinity for the endothelial growth factors (ECGF). This discovery led to the isolation and purification of ECGF by heparin-affinity chromatography in 1986. Now an entire class of ECGF has been identified. They are peptides, and are related either to the basic fibroblast growth factor (FGF) or to acidic FGF. Genes for basic FGF from the brain and the ECGF, a precursor of acidic FGF, were cloned. Three other angiogenic factors were reported: angiogenin, transforming growth factor alpha, and transforming growth factor beta, all purified from human tumor cells, but were later found in a variety of normal cells.

Steroids that inhibit angiogenesis in the presence of heparin have been found. The tetrahydro derivatives of cortisone were found to be the most potent angiogenesis inhibitors.

13.8 Significance of Zero-Stress State: Changes Reveal Nonuniform Remodeling of Tissue

Blood vessel remodeling can be revealed in many ways, such as:

(1) Overall changes in vessel lumen diameter, wall thickness/lumen diameter ratio, connective tissue/smooth muscle ratio.

(2) Nonuniform growth in different parts of the vessel wall. Change of opening angle at zero-stress state.

(3) Mechanical property changes in stress-strain relationship, smooth muscle tone, viscoelasticity, and constitutive equation.

(4) Cellular and molecular changes. Cell morphology. Ultrastructure. Biochemical changes.

Among these ways the change of zero-stress state has a unique significance, because the zero-stress state is the only state at which every cell and extracellular material is at natural shape. When the tissue is remodeled because of growth or resorption of the cells and extracellular matter, we would like to see how do the shape and size of these matters change. For this purpose it is best to compare everything at zero-stress state. To compare them in any other state is to bring in the complication of deformation due to internal stress.

In Fig. 11.2:3 we have seen that the zero-stress state of arteries change a great deal due to a step change of blood pressure. The change in opening angle is large and easy to observe.

In Fig. 13.8:1 a simple illustration of several types of remodeling is shown. Uniform changes lead to elongation, widening, shortening, shrinking. But bending must be caused by a nonuniform growth. The opening angle changes shown in Fig. 11.2:3 are due to bending, hence they exhibit nonuniform remodeling of the blood vessel wall caused by hypertension.

Conceptually, tissue growth is expected to be nonuniform because cells die or divide one by one. Death of a cell leaves a vacancy which will be filled by

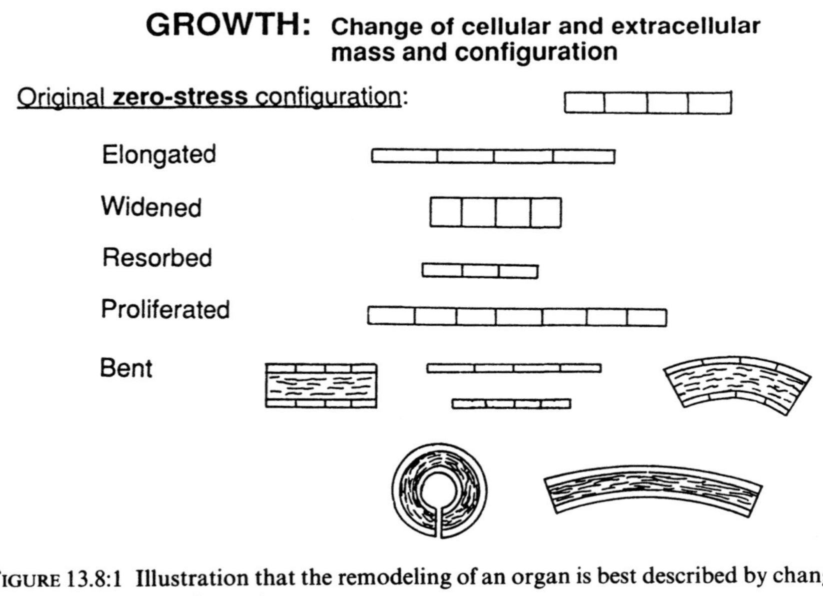

FIGURE 13.8:1 Illustration that the remodeling of an organ is best described by change of its zero-stress configuration.

neighboring cells, in the process these cells deform and create residual stresses in themselves. To see how much these neighboring cells have changed, we must make suitable number of cuts to render the tissue stress-free.

In Sec. 11.3 we have explained that a consequence of the residual stress coming from the opening angle of the blood vessel is to reduce the stress concentration in the inner wall region at a normal physiological blood pressure. Calculations of the blood vessel wall stress distribution by Fung (1984, 1985), Chuong and Fung (1986), Vaishnav and Vossoughi (1986), Takamizawa and Hayashi (1987) (see ref. in Ch. 11) have shown the consistency of the opening angle and the hypothesis of uniform circumferential stress distribution in arterial wall at homeostatic condition at normal blood pressure. When the blood pressure is raised above the normal value, the tensile stress at the inner wall is increased more than that at the outer wall. The inner wall hypertrophies more than the outer wall, and the opening angle at zero-stress increases. As remodeling proceeds, the opening angle continues to change until a new homeostatic condition is established, at which the circumferential stress distribution is again quite uniform at the new steady blood pressure.

Conversely, if one assumes that tensile stress is uniformly distributed throughout the blood vessel wall at the homeostatic (physiological) condition, then one can calculate the needed residual strain, and the opening angle of the zero-stress state. The results have some resemblance with experimental data. Fung (1984, 1985) examined the dense body distribution in the vascular smooth muscles of the arteries of the rabbit mesentery and showed that they are quite uniform at physiological blood pressure. Similarly, the sarcomere length of the myocardium in the left ventricle has been found by Hort (1960) and others (see Fung, 1984b, p. 61) to be fairly uniform across the ventricular wall. But the scatter of the data is large. The ratio of the standard deviation to the mean are so large that any attempt to prove uniformity is difficult. Furthermore, a sarcomere length distribution that is spatially uniform at diastole may become nonuniform in systolic period. Hence in an organ in which stress and strain vary cyclically there is a fundamental question of how to define homeostasis. Could it be better defined in terms of work done per unit volume of tissue per unit time? Or in terms of oxygen demand? Or microcirculatory blood flow? These questions are topics for further study.

13.9 A Hypothesis on Growth

To help theoretical and experimental studies, I would like to propose a hypothesis. The hypothesis is that the remodeling of the blood vessel involving growth or resorption of cells and extracellular materials is linked to the stress in the vessel. We recognize that (1) the transport of matter through cell membranes by active or passive mechanism depends on strain in the cell membranes, (2) the granular-to-fibrous transformation of actin molecules and

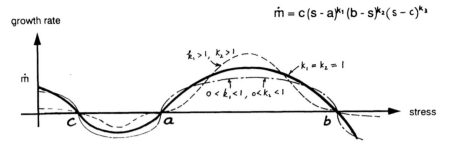

FIGURE 13.9:1 The author's proposed stress-growth law.

the function of actin-myosin cross-bridges are strain dependent, (3) chemical reaction rate depends on pressure, stress, and strain. Hence, specifically, we propose that growth depends on stress and strain in a manner shown in Fig. 13.9:1. There exist equilibrium states of stress, a, b, c, ... At a, an increase of stress causes growth, a decrease of stress causes resorption. At b or c an increase of stress causes resorption, a decrease of stress causes growth. Joining the trends at a, b, c together by a smooth curve, I shall express the growth rate by an equation

$$\dot{m} = C(s - a)^{k_1}(b - s)^{k_2}(s - c)^{k_3}. \tag{1}$$

Here $\dot{m}$ is the rate of growth, s is stress, C, k_1, k_2, k_3, a, b and c are constants. If k_1, k_2, k_3 are less than one, the curve is very steep at a, b, c. If k_1, k_2, k_3 are greater than one, the curve is very flat at a, b, c. Taking this as an experimental hypothesis, then the problem is to determine the constants as functions of growth factors and other physical, chemical, and biological stimuli, and whether s should be dilatational, or deviatoric, or principal stresses or their invariants.

Some explanations are due. The existence of homeostatic states is axiomatic in physiology. The normal physiological condition of blood vessel behaves like point a. Hypertrophy due to increased blood pressure agrees with the condition to the right of a. Astronauts at zero G seems to be in a condition to the left of a (Mack et al., 1967). A metal screw installed in a bone may reach the state b: an over-tightening leads to resorption. Tissue culture in stress-free state is possible; hence I give $\dot{m}$ a positive value near the $s = 0$ axis. Joining that to the point a requires a zero-crossing homeostatic state c. A step reduction of blood pressure so that the stress is reduced to a value below c will induce growth.

Some consequences of this hypothesis are:

1) At homeostasis, all collagen fibers have the same stress, all elastin fibers have another uniform stress, all sarcomeres in skeletal or cardiac muscle cells have the same length, all dense bodies in smooth muscle cells have the same spacing.

2) If the homeostatic condition is described by the equilibrium state *a*, then in hypertension, growth starts at the inner wall, and the opening angle at zero-stress would be increased.

The first conclusion follows our hypothesis if the hypothesis holds for the collagen and elastin and muscle fibers. If this conclusion is true, then it is tremendously important to the derivation of constitutive equations. We have used it in Sec. 11.9.

The experimental results on the sarcomere length and dense body spacing mentioned in Sec. 13.8 support this conclusion.

The second conclusion is supported by the features illustrated in Fig. 11.2:3.

A feature deducible from the results of hypertension experiment presented in Fig. 11.2:3 also lends support to our hypothesis: If we plot the opening angle of various segments of the rat aorta against the number of days post surgery, we obtain the result shown in Fig. 13.9:2. If we compute the hoop stress in the artery by the formula

$$\sigma = \frac{pr}{h} \tag{2}$$

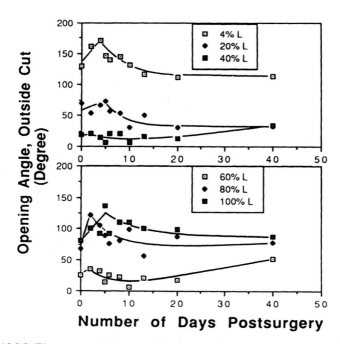

FIGURE 13.9:2 The course of change of the state of zero-stress of rat aorta following aortic banding described in Sec. 11.2 and shown in Fig. 11.2:2. The opening angle (inside cut) is plotted against the number of days following surgery. %L marks the location of a section along the aorta, see Sec. 11.2, Fig. 11.2:3.

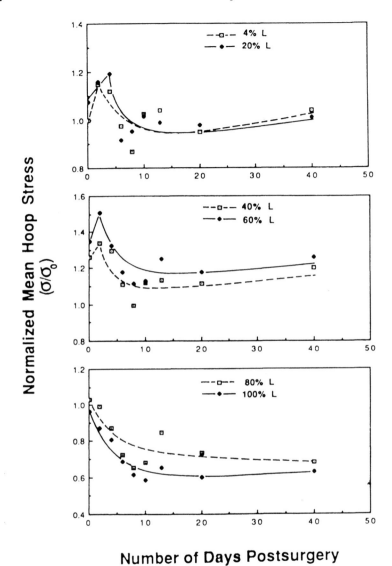

Number of Days Postsurgery

FIGURE 13.9:3 Change of hoop stress in the aorta following the occlusion of aorta mentioned in Fig. 13.9:2.

where p is the blood pressure, r is the lumen radius, h is the wall thickness, σ is the mean circumferential stress, and plot it against the number of days post surgery, we obtain the result shown in Fig. 13.9:3. Now the hoop stress is the mean circumferential stress in the vessel wall. If we assume that the maximum circumferential stress in the vessel wall is proportional to the mean, then the stress that drives the hypertrophy and change of opening angle is proportional to the hoop stress. If, in addition, we assume that in the neighborhood of the equilibrium point a, $\dot{m}$ is proportional to $s - a$, then the curve of $\dot{m}$ vs days will look like the hoops stress vs days curve, Fig. 13.9:3. The opening angle change is due to a cummulative effect of $\dot{m}$, i.e., α is proportional to $\int \dot{m}\, dt$. Thus the curve in Fig. 13.9:2 should be the result of integrating the curve in Fig. 13.9:3, as it seems to be.

The choice of the quantity s for the abscissa is still debatable. In the discussion of Wolff's law of the bone, Sec. 13.4, we have seen advocates for principal stress in tension or compression, von Mises stress, mean stress, strain energy, and full stress tensor. The corresponding $\dot{m}$ could be the rate of increase of tissue mass, calcified tissue, or fabric tensor.

Validation of the hypothesis is a task for the future.

13.10 Engineering of Blood Vessels

Many vascular reconstruction surgeries involve the use of some type of graft as a conduit to bypass an occluded arterial segment. The graft may be natural, such as saphenous vein or internal mammary artery, or synthetic, such as dacron. Use of saphenous vein has been very successful with 90% long-term patency rates. Large synthetic vessels have been quite successful, but smaller synthetic vessels (say, less than 6 mm in diameter) have difficulty remaining patent over a period of several years. Among many reasons for this difficulty, two principal ones are (1), cellular hyperplasia at or around the junction between the vascular graft and the native artery, leading to progressive stenosis (Clowes et al., 1985), and (2), failure to grow a layer of endothelial cells to cover the graft surface to protect it against blood clotting. Greater understanding of these phenomena must be obtained.

Endothelial Cell Seeding

Stanley et al. (1985) and Zamora et al. (1986) have shown that pre-seeding of synthetic vascular graft improves graft patency. Davies et al. (1986) showed that turbulence enhances cell proliferation. Jarrell et al. (1986) showed that microvascular endothelial cells can be isolated from fat tissue, and deposited on synthetic polymer such as polystyrene and dacron by gravity. Adhesion develops within a few minutes of incubation. Adhesion is improved by coating the surface with fibronectin. The rate of spreading varies greatly with the nature of the surface. With suitable optimization, it is possible to cover the surface with a confluent monolayer of endothelial cells in less than an hour following incubation.

Litwak et al. (1988) developed a small diameter, compliant vascular prosthesis using polyurethaneurea as basic material. A surface active component polydimethylsiloxane-based polyurethane is added to obtain a composite material. Grafts are extruded as multilayered tubes. The inner, blood contacting layer presents an open foam face. A solid impervious layer, adherent to the inner foam, provides the graft with strength and prevents fluid transfer. An outer layer of textured foam serves as a tissue anchor. Optimal bulk and biocompatibility properties for vascular implants have been obtained.

Autologous Blood Vessel Substitute

Bowald et al. (1980), Greisler et al. (1985) have formulated a research hypothesis that a normal arterial wall can be formed through natural repair processes if a scaffold of a vascular graft is made of a slowly disappearing, nontoxic mesh. Galletti et al. (1988), Gogolewski et al. (1987), and Yue et al. (1988) have studied transient biocompatible scaffolds for regeneration of arterial wall. The challenge is to design bioresorbable tubular fabrics which at the time of implantation has suitable mechanical strength and anticlotting property, but lead, after the disappearance of the synthetic polymers, to a regenerated vessel of adequate strength and hemocompatibility, without rupturing at any time. The development of the new blood vessel substitute involves simultaneous decay of the polymer and growth of cellular constituents.

Galletti et al.'s experiments have yielded encouraging results. They used homopolymers such as polydioxanone or copolymers such as composites of polyurethanes and polylactides or fibrin, or fast resorbing polyglycolide yarns coated with "retardant" polymers, e.g., a mixture of polylactide and poly 2, 3 butylene succinate. They implanted 8 mm internal diameter, 8–9 cm long aortic graft in dogs and found that the graft functioned effectively for six months when the animal was sacrificed. The polymers have largely decayed at three months. No infection was found with fully bioresorbable prosthesis. Galletti and other authors state that the compliance of the original prosthesis is a critical factor in obtaining the development of a circumferentially oriented smooth muscle layer and the deposition of elastin.

Initial success has to be bolstered by true understanding. On the chemical side, the use of peptide factors that promote cell recruitment and proliferation to build tissue of host origin must be studied. Types of cells recruited into the implant must be identified. Angiogenesis and development of microcirculation in the implant must be clarified. The reaction of the implant and cells to a variety of physical, chemical, and biological stimuli such as activated white blood cells and bacteria must be understood. The design and manufacturing of the bioresorbable polymer scaffold must be optimized. Mechanical forces involved must be understood. But the road to success is discernible.

13.11 Tissue Engineering of Skin

Each year more than 100,000 people are hospitalized and about 10,000 die from burn injuries in the United States. Conventional treatment for severe burns includes early excision of the eschar, coverage of wounds with cadaver allograft, or topical anti-microbial agents and dressings. Definitive wound closure is achieved by grafting skin from an unburned area on the patient. These established procedures for victims of severe burns result in long hospitalizations and multiple surgery. Greater availability of skin substitutes for wound closure offers the prospects for reduction or elimination of the donor site for skin autograft, fewer surgical procedures, shorter hospitalization time, improved functional and cosmetic results, and substantial economic benefits in reduced health care costs. Skin substitutes can also be used for testing the safety of consumer products that contact the body, of foods and drugs, and of occupational and environmental hazards; including skin irritancy test, skin corrosion, and percutaneous absorption, and skin inflammation.

Skin Substitute Made of Patient's Own Cells

A permanent skin substitute can be created by growing the patient's own skin cells in a collagen scaffold. The purpose is to heal a deep skin wound of a patient with his own cells rather than his own skin, thus minimizing the trauma of an accident to a much smaller degree. The collagen scaffold is biodegradable. When the skin substitute is grafted surgically, and becomes well vascularized so that microcirculation is reestablished, then a new permanent skin is obtained. Such a substitute for skin autograft has been developed which presently consists of cultured human epidermal keratinocytes attached to a porous and resorbable collagen and chondroitin sulfate membrane that is populated with human fibroblasts (HF). Initial application of the substitute to athymic mouse has shown advantages such as a large saving of donor's skin, and a reduction of the contraction of the healing wound.

A variety of materials that are obtained from either in vitro or ex vivo preparations have been proposed. The use of tissue culture techniques for normal human epidermal keratinocytes can accomplish expansion ratios of area coverage that can exceed 1000-fold in a period of 3–4 weeks. Cultures of these kinds will form multilayered sheets and have been shown to provide wound closure after application to excised full thickness burns. See Bell et al. (1983), Boyce and Hansbrough (1988), Burke et al. (1981), Gallico et al. (1984), Ham (1981), Pittlekow and Scott (1986), Yannas and his associates (1980a, b, 1982), Hefton et al. (1987).

Biodegradable Scaffold

Synthesis of collagen-GAG membranes was first accomplished by Yannas et al. (1980a, b, 1982). Comminuted bovine collagen is partially solubilized in $0.05M$ acetic acid and coprecipitated with chondroitin-6-sulfate in a refrigerated homogenizer. The coprecipitate is cast into sheets and frozen. The frozen coprecipitate is lyophilized overnight and cross-linked in a vacuum oven at $105°C$ and 10^{-4} torr for 24 hours. Dry membranes may be stored for extended periods of time.

Before tissue culture or grafting, dry membranes are rehydrated in $0.05M$

acetic acid, cross linked with 0.25% glutaraldehyde, washed exhaustively, and stored in 70% isopropanol.

Pore size can be controlled by controlling the freezing temperature and the concentration of starting materials. By controlling the freezing temperature gradient, a layer of the substitute with smaller pores can be created by quick freezing to a lower temperature.

Wound Contraction

Wound contraction is a principal problem with skin graft and wound healing. The contraction creates tension in the neighboring healthy skin and distorts the skin and the limb. A wound caused by burn and healed without graft may contract to 15–20% of the original wound size. Skin substitute must be designed to reduce this wound contraction. A thorough understanding of growth and stress would have tremendous applications here. Cf. Doillon et al. (1987), Kennedy and Cliff (1979), McGrath and Simon (1983), Snowden and Cliff (1985).

Problems to Be Solved

Tissue engineering of the skin needs the solution of the following problems:

1) Creation of an optimal collagen scaffold with optimal pore size in relation to cellular mechanics, strength, contraction, and adhesion, and optimal compliance with regard to the controlling of contraction. If *angiogenic factors* were needed, they can be immobilized in the scaffold.

2) Understand the contraction of healing wound. Pharmaceutical, chemical (e.g. nerve growth factor) and mechanical means have been sought to control wound contraction, but the degree of success has been rather limited. The study of skin substitute provides a unique opportunity to understand and to control wound contraction. Molecular mechanics of the actin fibers and cell mechanics will play a central role. On the other hand, contraction is related to growth and resorption. Since growth is modulated by stress (Sec. 13.9), control of stress during the healing process is believed to be a key to the solution of the contraction problem.

3) Vascularization of the skin substitute.

4) Cellular adhesion and the mechanical properties of the skin substitute. It is known that the adhesive stress between cells is very sensitive to the environmental parameters such as pH, temperature, and chemicals or pharmaceutical agents in the bathing fluid (Evans, 1985). Hence it is feasible to design a fluid so that a desired strength of adhesion is obtained.

Future Prospects

One can think of many other tissues for engineering: cartilage, blood, nerves, etc. To make living substitutes for them is an exciting objective. Although nobody can say when such an objective will be achived, we are sure that a sound engineering knowledge is a prerequisite.

Conclusion

Here we bring our discussion of biomechanics to conclusion. Biomechanics helps us understand nature. It sensitizes us to observe, and to appreciate. It is a gentle tool: sharp, precise, natural, and unavoidable. It is the foundation of engineering for prosthesis, rehabilitation, tissue substitutes, trauma prevention, and health.

Problems

13.1 To search for evidence of residual strains in bone, the place to look is around the growth plate. To find any changes of the zero-stress state with respect to time, one should consider the period of growth in young animals, or period of healing of a fracture. Design a research plan to study this question.

13.2 The stress in living organisms is influenced by earth's gravitation. To study the effect of stress on growth and change, one of the best ways is to study space flight. Design experiments to study astronaut's bone demineralization and cardiovascular deconditioning during space flight. (Cf. Anderson, S.A., Cohn, S.H.: Bone Demineralization During Space Flight. *The Physiologist*, **28**: 213–217, 1985; Levy, M.N., Talbot, J.M.: Cardiovascular Deconditioning of Space Flight. *The Physiologist*, **26**: 297–303, 1983. Consult the IUPS and NASA publications on Proc. Annual Mtg. of IUPS Commission on Gravitational Physiology, as *Suppl. to Physiologist*.)

13.3 Aortic arch and the first couple of generations of pulmonary arteries are highly curved. They have large opening angles at zero-stress state (often in the range of 180° to over 360°). Locally, the geometry of these vessels may be approximated by a torus, a curved shell with a planar circular centerline of radius ρ_0 and a circular cross section of radius a. At the center of curvature of the centerline, erect an axis perpendicular to the plane of the centerline, and use it as the polar axis of a cylindrical-polar frame of reference, with coordinates (ρ, θ, z) to describe the location of points on he shell. The nature of the shell depends on the radial coordinate ρ. For the *outer* shell, $\rho > \rho_0$, the two principal curvatures of the midsurface of the shell have the same sign. For the *inner* shell, $\rho < \rho_0$, the signs of the two principal curvatures are different. In differential geometry, the outer shell is said to be elliptic, the inner shell is said to be hyperbolic. The stress distribution in the shell in response to the pressure in the tube is different in these two parts. Set up equations to analyze the stress distribution in the vessel wall. For simplicity, assume that the vessel wall obeys Hooke's law of elasticity.

13.4 Using the equations derived above, make analysis to answer the following questions: (1) If we assume that at the equilibrium condition subjected to certain internal pressure, the circumferential stresses are uniformly distributed, what would the incremental membrane stresses be when the internal pressure is changed by an infinitesimal amount? (2) By reducing the internal pressure step by step until it is zero, what residual stresses remain in the shell? (3) What are the residual strains, and the opening angle at zero-stress state if the shell is cut by the method described in Sec. 11.2?

13.5 A description of cancer growth. Let $x(t)$ be the number of living cancer cells in an organ at time t. Let $q(t)$ be the fraction of cancer cells eliminated in the growth period. Then in an infinitesimal period of time Δt the number of cancer cells increased may be assumed to be proportional to $[1 - q(t)]x(t)$. Hence

$$\Delta x = K(t)[1 - q(t)]x(t)\Delta t$$

when $K(t)$ is the constant of proportionality. If the number of cancer cells at $t = 0$ is $x(0)$, then an integration of the equation above yields

$$x(t) = x(0)\exp\left\{\int_0^t K(t)[1 - q(t)]\,dt\right\}.$$

Experiments suggest that the integrand is of the form $e^{-\alpha t^n}$. In chemotherapy, α is proportional to the concentration of drug above the minimal effective concentration. Propose a way to evaluate the effectiveness of drugs through the evaluation of α and n.

13.6 The mechanics of cellular growth is beautifully discussed by Ingber et al. (1985, 1989). Follow up the literature and offer a critical discussion. Suggest a theory of your own if you feel there is a need.

13.7 Consider further the aortic arch studied in Problems 13.3 and 13.4. Based on the effects of centrifugal forces, curvature, Bernoulli's equation, and boundary conditions, give a qualitative description of the flow field. Estimate the pressure distribution, and wall shear stress. These variables change rapidly, of course, with the activity of the left ventricle and the aortic valve. Describe the course of variation of these variables thorough a cycle of heart beat.

13.8 Problems 13.3 and 13.4 are concerned with stress distribution in the vessel wall. The endothelial cells seem to be very sensitive to the shear stress. What is known about the effect of shear stress on the morphology of the endothelial cells, and on the orientation and shape of the cell nuclei? What is known about the effect of the shear stress on the mass transport of cholestrol: LDL, HDL?

13.9 The stress in the smooth muscle cells is important to the muscle tone and contraction force and shortening. Describe the mechanical behavior of the vascular smooth muscle in the aorta. The connective tissues, especially collagen and elastin fibers, are stressed. How can these stresses be calculated?

13.10 Stress is a tensor. If one knows the principal stresses, one can easily compute the shear stress acting on any plane. Consider a blood vessel in physiological condition. Compare the maximum shear stress in the vessel wall with the surface shear stress acting on the endothelium due to the flowing blood. How different are they?

13.11 In response to stresses, how do the various components of the blood vessel wall change instantaneously and in long term? If the stresses become abnormal, how do the various components in the vessel wall remodel themselves? How does the remodeling of one component interfere with that of another neighboring component? How can the whole system remodel? Perhaps you cannot find answers to some of your questions in the library. Then it is necessary to experiment or theorize. Formulate a research proposal to settle some of your questions.

13.12 Blood rheology has a major effect on the phenomenon of flow separation, i.e., the failure of streamlines near a solid wall to follow the boundary of the solid wall. Where in the circulation system of man are the sites of possible flow separation? What characteristics of blood rheology affect separation? How can these characteristics be modified in chemical condition?

13.13 What are the effects of flow separation on the pressure and shear stress in the blood vessel? What are the effects of the stress distribution in the blood vessel wall on the tendency of the blood vessel to remodel itself? If the flow separation condition changes, how would the blood vessel remodel itself in the long run?

References

Alpert, N.R. (ed.) (1983). *Myocardial Hypertrophy and Failure.* Raven Press, New York.

Anversa, P., Vitali-Mazza, L., Visioli, O. and Marchetti, G. (1971). Experimental cardiac hypertrophy: a quantitative ultrastructural study in the compensatory stage. *J. Molecular and Cellular Cardiology* 3: 213–227.

Arai, S., Machida, A. and Nakamura, T. (1968). Myocardial structure and vascularization of hypertrophied hearts. *Tohoku J. Exp. Medicine* 95: 35–54.

Arai, S., Nakamura, I. and Suwa, N. (1976). Quantitative analysis of cardiac hypertrophy due to pressure load in reference to the relations of blood pressure, left ventricular weight, and left ventricular capacity. *Tohoku J. Exp. Medicine* 118: 299–309.

Bassett, C.A.L. (1978). Pulsing electromagnetic fields: A new approach to surgical problems. In *Metabolic Surgery*, (H. Buchwald and R.L. Varco, eds), Grune and Stratton, New York, pp. 255–306.

Bayliss, W.M. and Starling, E.H. (1902). Proc. Roy. Soc. (London) 69: 352. *J. Physiology* 28: 325.

Becker, R.O. (1961). The bioelectric factors in amphibian limb regeneration. *J. Bone Joint Surg.* 43A: 6431.

Bell, E., Ivarsson, B. and Merrill, C. (1979). Production of a tissue-like structure by contraction of collagen lattices by human fibroblasts of different proliferation potential in vitro. *Proc. Natl. Acad. Sci. U.S.A.* 76: 1274–1278.

Bell, E., Sher, S., Hull, B., Merrill, C., Rosen, S., Chamson, S., Asselineau, D., Dubertret, L., Coulomb, B., Lapiere, C., Nusgens, B. and Neveux, Y. (1983). The reconstitution of living Skin. *J. Invest. Dermatol.* 81(1), Supp. 2S–10S.

Bevan, R.D. (1976). An autoradiographic and pathological study of cellular proliferation in rabbit arteries correlated with an increase in arterial pressure. *Blood Vessels* 13: 100–128.

Booth, F.W. (1982). Effect of limb immobilization on skeletal muscle. *J. Appl. Physiol.* 52: 1113–1118.

Booth, F.W., Nicholson, W.F. and Watson, P.A. (1982). Influence of muscle use on protein synthesis and degradation. *Exercise, Sport Sci. Rev.* 10: 27–48.

Bowald, S., Busch, C. and Erikson, I. (1980). Absorbable meterial in vascular prosthesis: A new device. *Acta Chir. Scand.* 146: 391.

Boyce, S.T. and Hansbrough, J.F. (1988). Biologic attachment, growth and differentiation of humen epiderman keratinocytes onto a graftable collagen and chondroitin-6-sulfate membrane. *Surgery* 103(4): 421–431.

Bradley, K.H., McConnell, S.D. and Crystal, R.G. (1974). Lung collagen composition and synthesis. *J. Biol. Chem.* **249**: 2674–2683.

Braunwald, E. (1980). Pathophysiology of heart failure. In *Heart Disease: A Textbook of Cardiovascular Medicine.* Saunders, Philadelphia, pp. 453–471.

Brody, J.S. and Buhain, W.J. (1973). Hormonal influence on post-pneumonectomy lung growth in the rat. *Respir. Physiol.* **19**: 344–355.

Burke, J.F., Yannas, I.V., Quinby, W.C., Bondoc, C.C. and Jung, W.K. (1981). Successful use of a physiologically acceptable artificial skin in the treatment of extensive burn injury. *Ann. Surgery* **194**: 413–428.

Carter, D.R. (1987). Mechanical loading history and skeletal biology. *J. Biomech.* **20**: 1095–1109.

Carter, D.R. (1988). The regulation of skeletal biology by mechanical stress histories. In *Tissue Eng.* (R. Skalak and C.F. Fox, eds.), Alan R. Liss, New York, pp. 173–178.

Carter, D.R., Fyhrie, D.P. and Whalen, R.T. (1987). Trabecular bone density and loading history: Regulation of connective tissue biology by mechanical energy. *J. Biomech.* **20**: 785–794.

Carter, D.R. and Wong, M. (1988). Mechanical stresses and endochondral ossification in the chondroepiphysis. *J. Orthop. Res.* **6**: 148–154.

Cassidy, J.J. and Davy, D.T. (1985). Mechanical and Architectural Properties in Bovine Cancellous Bone, *Trans. Orthop. Res. Soc.* **31**: 354.

Chambers, T.J. (1985). The pathobiology of the Osteoclast. *J. Clin. Pathol.* **38**: 241.

Chuong, C.J. and Fung, Y.C. (1986). Residual stress in arteries. In *Frontiers in Biomechanics* (G.W. Schmid-Schönbein, S.L.-Y. Woo, B.W. Zweifach, eds.), Springer-Verlag, New York, pp. 117–129.

Cliff, W.J. (1963). Observations on healing tissue: A combined light and electron-microscopic investigation. *Roy. Soc. of London, Phil. Trans.* B, **233**: 305

Clowes, A.W., Gown, A.M., Hanson, S.R. and Reidy, M.A. (1985). Mechanisms of arterial graft failure. I. Role of cellular proliferation in early healing of PTFE prostheses. *Am. J. Path.* **118**: 43.

Coan, M.R. and Tomanek, R.J. (1979). Regeneration of transplanted muscle with reference to overload. In A. Mauro, (ed.). *Muscle Regeneration,* Raven Press, New York, pp. 509–522.

Cohen, S. (1959). Purification and metabolic effects of a nerve growth-promoting protein from snake venom. *J. Biol. Chem.* **234**: 1129–1137.

Cohn, R. (1939). Factors affecting the postnatal growth of the lung. *Anat. Rec.* **75**: 195–205.

Cowan, M.J. and Crystal, R.G. (1975). Lung growth after unilateral pneumonectomy: Quantitation of collagen synthesis and content. *Am. Rev. Respir. Disease* **111**: 267–276.

Cowin, S.C. (1984). Modeling of the Stress Adaptation Process in Bone, *Cal. Tissue Int.* **36**, (Supp. S99–S104.

Cowin, S.C. (1986). Wolff's law of trabecular architecture at remodeling equilibrium. *J. Biomech. Eng.* **108**: 83–88.

Cowin, S.C. and Van Buskirk, W.C. (1979). Surface bone remodeling induced by a medullary pin. *J. Biomech.* **12**: 269.

Cowin, S.C. and Firoozbakhsh, K. (1981). Bone remodeling of diaphyseal surfaces under constant load: theoretical predictions. *J. Biomech.* **14**: 471.

Cowin, S.C., Hart, R.T., Balser, J.R. and Kohn, D.H. (1985). Functional adaptation in long bones: establishing in vivo values for surface remodeling rate coefficients. *J. Biomech.* **12**: 269.

Crystal, R.G. (1974). Lung Collagen: Definition, diversity, and development. *Fed. Proc.* **33**: 2243–2255.

Culmann, C. (1866) *Die graphische Statik.* Meyer und Zeller, Zurich.

Davies, P.F., Remuzzi, S., Gordon, E.F., Dewey, C.F., Jr., Gimbrone, M.A., Jr., (1986). Turbulent fluid shear stress induces vascular endothelial cell turnover *in vitro. Proc. Natl. Acad. Sci.* **83**: 2114–2117.

Doillon, C.J., Hembry, R.M., Ehrlich, H.P. and Burke, J.F. (1987). Actin Filaments in Normal Dermis and During Wound Healing. *Am. J. Path.* **126**(1): 164–170.

El Haj, A.J., Skerry, T.M., Caterson, B. and Lanyon, L.E. (1988). Proteoglycans in bone tissue:

identification and possible function in strain related bone remodeling. *Trans. Orth. Res. Soc.* **13**: 538.

Erdogen, F. (1983). Stress intensity factors. *J. Appl. Mech.* **50**(4b), 992–1002.

Evans, E.A. (1985). Membrane Mechanics and Cell Adhesion. In: *Frontiers in Biomechanics* (G. Schmid-Schönbein, S. Woo and B.W. Zweifach, eds.), Springer-Verlag, New York, pp. 1–17.

Fischbach, G.D. and Robbins, N. (1971). Effect of chronic disuse of rat soleus neuromuscular junctions on postsynaptic membrane. *J. Neurophysiol.* **34**: 562–569.

Faulkner, J.A., Weiss, S.W., and McGeachie, J.K. (1983). Revascularization of skeletal muscle transplanted into the hamster cheek pouch: intravital and light microscopy. *Microvasc. Res.* **26**: 49–64.

Folkman, J. (1985). Tumor angiogenesis. *Adv. Cancer Res.* **43**: 175–203.

Frolich, E.D. (1983). Hemodynamics and other determinants in development of left ventricular hypertrophy. *Fed. Proc.* **42**: 2709–2715.

Fry, D.L. (1968. Acute vascular endothelial changes associated with increased blood velocity gradients. *Circ. Res.* **22**: 165–197.

Fry, D.L. (1977). Aortic Evans blue dye accumulation: Its measurement and interpretation. *Am. J. Physiol.* (*Heart, Circ. Physiol.* 1) **232**: H204–H222.

Fry, D.L., Mahley, R.W., Weisgraber, K.H. and Oh, S.K. (1977). Simultaneous accumulation of Evans blue dye and albumin in the canine aortic wall. *Am. J. Physiol.* **233** (*Heart Circ. Physiol.*) **2**: H66–H79.

Fukada, E. (1974). Piezoelectric properties of biological macromolecules. *Adv. Biophys.* **6**: 121.

Fung, Y.C. (1981). *Biomechanics: Mechanical Properties of Living Tissues.* Springer-Verlag, New York.

Fung, Y.C. (1984). *Biodynamics: Circulation*, Springer-Verlag, New York.

Fung, Y.C. (1985). What principle governs the stress distribution in living organs? In *Biomechanics in China, Japan, and USA.* (Y.C. Fung, E. Fukada, and J.J. Wang, eds.) Science Press, Beijing, China, pp. 1–13.

Fung, Y.C. (1988). Cellular growth in soft tissues affected by the stress level in service. In: *Tissue Engineering* (R. Skalak and C.F. Fox, eds.), Alan Liss, New York, pp. 45–50.

Fyhrie, D.P. and Carter, D.R. (1985). A Unifying Principle Relating Stress State to Trabecular Bone Morphology, *Trans. Orthop. Res. Soc.* **31**: 337.

Gaensler, E.A. and Strieder, J.W. (1951). Progressive changes in pulmonary function after pneumonectomy: The influence of thoracoplasty, pneumothorax, oleothorax, and plastic sponge plombage on the side of pneumonectomy. *J. Thorac. Surgery* **22**: 1.

Galletti, P.M., Aebischer, P., Sasken, H.F., Goddard, M.B. and Chiu, T.H. (1988). Experience with fully bioresorbable aortic grafts in the dog. *Surgery* **103**: 231.

Gallico, G.G., O'Conner, N.E., Compton, C.C., Kehinde, O. and Green, H. (1984). Permanent coverage of large burn wounds with autologous cultured human epithelium. *New Eng. J. Medicine* **311**(7): 448–451.

Gogolewski, S., Galletti, G. and Ussia, G. (1987). Polyurethane vascular prostheses in pigs. *Colloid Polymer Sci.* **265**: 774.

Goldberg, A.L. (1979). Influence of insulin and contractile activity on muscle size and protein balance. *Diabetes* **28**: 18–24.

Goldberg, A.L., Etlinger, J.D., Goldspink, D.F. and Jablecki, C. (1975). Mechanism of work-induced hypertrophy of skeletal muscle. *Med. Sci. Sports* **7**: 148–261.

Goldberg, A.L., Tischler, L.M., DeMartino, G. and Griffin, A. (1980). Hormonal regulation of protein degradation and synthesis in skeletal muscle. *Federation Proc.* **39**: 31–36.

Goldin, B., Block, W.D. and Pearson, J.R. (1980). Wound healing of tendon. I. Physical, Mechanical, and metabolic changes. II. A mathematical model. *J. Biomech.* **13**: 241–256; 257–264.

Gospodarowicz, D., Bialecki, H. and Greenburg, G. (1978). Purification of the fibroblast growth factor activity from bovine brain. *J. Biol. Chem.* **253**: 3736–3743.

Gottrup, F. (1980, 1981). Healing of incisional wounds in stomach and duodenum. *Am. J. Surg.*

140: 296–301; **141**: 222–277.

Greisler, H.P., Kim, D.V., Price, J.B. and Voorhees, A.B. (1985). Arterial regeneration activity after prosthetic implantation. *Arch. Surgery* **120**: 315.

Griffith, A.A. (1920). The phenomena of rupture and flow in solids. *Phil. Trans. Roy. Soc.* (London), Ser A, **221**: 163–198.

Gross, S.B., Spindler, K.P., Brighton, C.T. and Wassell, R.P. (1988). The proliferative and synthetic response of isolated bone cells to cyclical biaxial mechanical stress. *Trans. Orth. Res. Soc.* **13**: 262–263.

Gross, D. and Williams, W.S. (1982). Streaming potential and the electro-mechanical response of physiologically moist bone. *J. Biomech.* **15**: 277–295.

Guillemin, R. (1985). The language of polypeptides and the wisdom of the body. *The Physiologist* **28**: 391–396.

Guillemin, R., Brazeau, P., Bohlen, P., Esch, F., Ling, N., and Wehrenberg, W.B. (1982). Growth hormone releasing factor from a human pancreatic tumor that caused acromegaly. *Science* **218**: 585–587.

Ham, R.G. (1981). Survival and growth requirements of nontransformed cells. In: *The Handbook of Experimental Pharmacology* **57**: 13–88.

Harrigan, T. and Mann, R.W. (1984). Characterization of Microstructural Anisotropy in Orthotropic Materials Using a Second Rank Tensor, *J. Material Sci.* **19**: 761–767.

Harris, A.K., Wild, P. and Stopak, D. (1980). Silicone rubber substrata: A new wrinkle in the study of cell locomotion. *Science*, **208**: 177–179.

Harris, A.K., Stopak, D. and Wild, P. (1981). Fibroblast traction as a mechanism for collagen morphogenesis. *Nature* **290**: 249–251.

Hart, R.T. (1983). "Quantitative Response of Bone to Mechanical Stress", (Doctoral Dissertation) Case Western Reserve University, Cleveland, Ohio.

Hayes, W.C. and Snyder, B. (1981). Toward a quantitative formulation of Wolff's law of trabecular bone. In Mechanical Properties of Bone, (S.C. Cowin, ed.), Am. Soc. Mech. Engineers, New York, Pub. No. AMD-Vol. 45, pp. 43–68.

Hefton, J.M., Madden, M.R., Finkelstein, J.L., Oefelein, M.G., LaBruna, A.N. and Staianao-Coico, L. (1987). The grafting of cultured human epiderman cells onto full-thickness wounds on pigs. *J. Burn Care* **19**: 29.

Hempstead, B., and Chao, M.V. (1989). The nerve growth factor receptor: biochemical and structural analysis. *Recent Prog. Hormone Res.* **45**: 441–466.

Herbison, G.J. and Talbot, J.M. (1985). Muscle atrophy during space flight: Research needs and opportunities. *The Physiologist* **28**: 520–527.

Hort, W. (1960). Makroskopische und mikrometrische untersuchungen am Myokard verschieden stark gefullter linker kammern. *Virchows Arch. Path. Anat.* **333**: 523–564.

Hudlicka, O. (1982). Growth of capillaries in skeletal and cardiac muscle. *Circ. Res.* **50**: 451–461.

Ingber, D.E. and Jamieson, J.D. (1985). Cells as tensegrity structures: architectural regulation of histodifferentiation by physical forces transduced over basement membrane. In *Gene Expression during Normal and Malignant Differentiation* (L.C. Anderson, C.G. Gahmberg, P. Ekblom, eds.) Academic Press, New York. pp. 13–32.

Ingber, D.E. and Folkman, J. (1989). Mechanochemical switching between growth and differentiation during fibroblast growth factor-stimulated angiogenesis in vitro: Role of extracellular matrix. *J. Cell Biol.* In press.

Ingvar, S. (1920). Reaction of cells to the galvanic current in tissue cultures. *Proc. Soc. Exper. Biol. Medicine* **17**: 198 ff.

Jarrell, B.E., Williams, S.K., Stokes, G., et al., (1986). Use of freshly isolated capillary endothelial cells for the immediate establishment of a monolayer on a vascular graft at surgery. *Surgery* **100**: 392–399.

Jaye, M., Howk, R., Burgess, W., Ricca, G.A., Chiu, I.-M., Ravera, M.W., O'Brien, S.J., Modi, W.S., Maciag, T. and Drohan, W.N. (1986). Human endothelial cell growth factor: cloning, nucleotide sequence, and chromosome localization. *Science* **233**: 542–545.

Kamiya, A., Bukhari, R. and Togawa, T. (1984). Adaptive regulation of wall shear stress optimizing vascular tree function. *Bull. Math. Biology.* **46**: 127–137.

Kamiya, A. and Togawa, T. (1980). Adaptive regulation of wall shear stress to flow change in the canine carotid artery. *Am. J. Physiol.* **239** (*Heart Circ. Physiol.*) **8**: H14–H21.

Karpovich, P.V. (1968). Exercise in medicine: a review. *Arch Phys. Rehabil. Medicine* **49**: 66–76.

Kawamura, K. (1982). Cardiac hypertrophy—scanned architecture, ultrastructure, and cytochemistry of myocardial cells. *Japanese Circ. J.* **46**(9): 1012–1030.

Kawamura, K., Imamura, K., Uehara, H., Nakayama, Y., Sawada, K. and Yamamoto, S. (1984). Architecture of hypertrophied myocardium: Scanning and transmission electron microscopy. In *Cardiac Function* (H. Abe et al., eds.), Japan Sci. Soc. Press, Tokyo/VNU Sci. Press, Utrecht, pp. 81–105.

Kennedy, D.F. and Cliff, W.J. (1979). A Systematic Study of Wound Contraction in Mammalian Skin. *Pathology* **11**(2): 207–222.

Knobil, E., and Greep, R.O. (1959). The physiology and growth hormone with particular reference to its action in the rhesus monkey and the "species specificity" problem. *Recent Prog. Hormone Res.* **15**: 1–69.

Krompecher, St. (1937). *Die Knochenbildung.* Verlag Fischer, Jena.

Lanyon, L.E. (1984). Functional strain as a determinant for bone remodeling. *Cal. Tiss. Int.* **36**: S56.

Lavoie, P.A., Collier, B. and Tennenhouse, A. (1976). Comparison of a α-bungarotoxin binding to skeletal muscle after inactivity or denervation. *Nature,* London, 349–350.

Leach, C.S. and Rambaut, P.C. (1977). Biochemical responses of skylab crew men: an overview. In *Biomedical Results from Skylab.* (Johnston, R.S. and Dietlein, L.F., eds.), Washington, D.C. NASA, pp. 204–216.

Lee, S., Fung, Y.C., Matsuda, M., Xue, H., Schneider, D. and Han, K. (1985). The development of mechanical strength of surgically anastomosed arteries sutured with dexon. *J. Biomech.* **18**: 81–89.

Leuret, F. and Lassaigne, J.-L. (1825). *Recherches Physiologiques et Chimiques pour Servir a l'Histoire de la Digestion.* Madame Huzard, Paris.

Levi-Montalcini, R. and Angeletti, R.U. (1966). Nerve growth factor. *Physiol. Rev.* **48**: 534–569.

Li, C.H., Evans, H.M. (1948). The biochemistry of pituitary growth hormone. *Recent Prog. Hormone Res.* **3**: 3–44.

Li, C.H., and Chung, D. (1971). Human pituitary growth hormone. *Int. J. Prot. Res.* **III**: 73–80. See also Li et al., ibid, 81–92, 93–98, 99–108, 185–189.

Li, C.H., and Ramasharma, K. (1987). Inhibin. *Ann Rev. Pharmacol. Toxicol.* **27**: 1–21.

Liebow, A.A. (1963). Situations which lead to changes in vascular patterns. In *Handbook of Physiology,* Sec. 2: *Circulation,* Vol. 2, Chap. 37, Washington, D.C., Am. Physiological Soc. pp. 1251–1276.

Limas, C.J. (1983). Biochemical aspects of cardiac hypertrophy. *Fed. Proc.* **42**: 2716–2721.

Linzbach, A.J. (1960). Heart failure from the point of view of quantitative anatomy. *Am. J. Cardiol.* **5**: 370–382.

Litwak, P., Ward, R.S., Robinson, A.J., Yilgor, I. and Spatz, C.A. (1988). Development of a small diameter, compliant vascular prostheses. In: *Tissue Engineering* (R. Skalak and C.F. Fox, eds.), Alan Liss, New York, pp. 25–30.

Lund, D.D. and Tomanek, R.J. (1978). Myocardial morphology in spontaneously hypertensive and aortic-constricted rats. *Am. J. Anatomy.* **152**: 141–152.

Lund, E.J. (1921). Experimental control of organic polarity by the electric current I. Effects of the electric current of regenerating internodes of obelia commisuralis. *J. Exper. Zool.* **34**: 471.

Lyman, D.J., Fazzio, F.J., Voorhees, H., Robinson, G. and Albo, D. Jr. (1978). Compliance as a factor effecting the patency of a copolyurethane vascular graft. *J. Biomed. Mat. Res.* **12**: 337.

Maciag, T., Cerundolo, I.S., Kelley, P.R. and Forand, R. (1979). An endothelial cell growth factor from bovine hypothelamus: identification and partial characterization. *Proc. Natl. Acad. Sci.* U.S.A. **76**: 5674–5678.

Mack, P.B., LaChange, P.A., Vose, G.P. and Vogt, F.B. (1967). Bone demineralization of foot and hand of Gemini-Titan IV, V, and VII astronauts during orbital flight. *Am. J. Roentgenol.*

100: 503–511.

Martin, R.B. (1972). The effects of geometric feedback in the development of osteoporosis. *J. Biomech.* **5**: 447.

Matthews, A.P. (1903). Electrical polarity in the hydroids. *Am. J. Physiol.* **8**: 394.

McGrath, M.H. and Simon, R.H. (1983). Wound geometry and the kinetics of wound contraction. *Plast. Reconstr. Surgery* **72**: 66–72, 1983.

Menton, D.N., Simmons, D.J., Chang, S.L., Orr, B.Y. (1984). From bone lining cell to osteocyte— An SEM study. *Anat. Rec.* **209**: 29–39.

Meyer, G.H. (1867). Die Architektur der spongiosa. *Archiv. fur Anatomie, Physiologie, und wissenschaftliche Medizin* (Reichert und wissenschafliche Medizin, Reichert und Du Bois-Reymonds Archiv) **34**: 615–625.

Meyrick, B. and Reid, L. (1979). Hypoxia and incorporation of 3H-thymidine by cells of the rat pulmonary arteries and alveolar wall. *Am. J. Path.* **96**: 51–69.

Meyrick, B. and Reid, L. (1982). Crotalaria-induced pulmonary hypertension: Uptake of 3H-thymidine by the cells of the pulmonary circulation and alveolar walls. *Am. J. Path.* **106**: 84–94.

Nakamura, T., Nakajima, T., Takishima, T., Suzuki, N., Arai, S. and Suwa, N. (1975). Analysis of diastolic pressure-volume relation of the canine left ventricle: Half-inflation pressure as an index of left ventricular compliance. *Tohoku J. Exp. Med.* **117**: 311–321.

Nakamura, T., Nakajima, T., Suzuki, N., Arai, S. and Suwa, N. (1976). Left ventricular stiffness and chamber geometry in the pressure-overload hypertrophied heart. *Tohoku J. Exp. Medicine* **119**: 245–256.

Nakamura, T., Abe, H., Arai, S., Kimura, I., Kushibiki, H., Motomiya, M., Konno, K. and Suzuki, N. (1982). The stress-strain relationship of the diastolic cardiac muscle and left ventricular compliance in the pressure-overloaded canine heart. *Japanese Circ. J.* **46**: 76–83.

Nattie, E.E., Wiley, C.W. and Bartlett, D., Jr. (1974). Adaptive growth of the lung following pneumonectomy in rats. *J. Appl. Physiol.* **37**: 491–495.

Nulend, J.D. (1987). Cellular Responses of Skeletal Tissues to Mechanical Stimuli", (Doctoral Dissertation) Amsterdam, Free University Press.

Palkovits, M. (1984). Distribution of neuropeptides in the central nervous system: a review of biochemical mapping studies. *Prog. Neurobiol.* **23**: 151–189.

Pauwels, F. (1980). *Biomechanics of the Locomotor Apparatus*, German edn. 1965. English trans. by P. Maqnet and R. Furlong, Springer-Verlag, Berlin, New York.

Pestronk, A., Drachman, D.B., Griffin, J.W. (1976). Effects of muscle disuse on acetylcholine receptors. *Nature*, London. **260**: 352–353.

Pittelkow, M.R. and Scott, R.E. (1986). New techniques for the in vitro culture of human skin keratinocytes and perspectives on their use for grafting of patients with extensive burns. *Mayo Clin. Proc.* **61**: 771–777.

Raben, M.S. (1959). Human growth hormone. *Recent Prog. Hormone Res.* **15**: 71–114.

Rakusan, K. and Turek, Z. (1986). A new look into the microscope: proliferation and regression of myocardial capillaries. *Can. J. Cardiol.* **2**: 94.

Rivier, J., Spiess, J., Thorner, M., and Vale, W. (1982). Characterization of a growth hormone-releasing factor from a human pancreatic islet tumour. *Nature* **300**: 276–278.

Rodbard, S. (1975). Vascular caliber. *Cardiology*, **60**: 4–49.

Roesler, H. (1981). Some historical remarks on the theory of cancellous bone structure (Wolff's law). In *Mechanical Properties of Bone*, (S.C. Cowin, ed.), Am. Soc. Mech. Engineers, New York, Pub. No. AMD-Vol. 45. p. 27–42.

Romanova, L.K., Leikina, E.M., Antipova, K.K. and Sokolova, T.N. (1970). The role of function in the restoration of damaged viscera. *Sov. J. Dev. Biol.* **1**, 384.

Roux, W. (1880–1895). *Gesammelte Abhandlungen über die entwicklungs mechanik der Organismen.* W. Engelmann, Leipzig, 1895. No. 3: Über die Leistungsfähigkeit der Deszendenzlehre zur Erklärung der Zweckmässigkeiten des fierischen Organismus", Bd. 1, pp. 102–134, (1st pub. 1880. No. 4: Der züchtende Kampf der Teile (Theorie der funktionellen Anpassung), Bd 1, pp. 137–422, (1st pub. 1881). Nr. 5: "Der zuchtende Kampf der Teile im Organismus"

(Autoreferat), Bd. 1, pp. 423–437 (1st pub. 1881).

Salzstein, R.A. and Pollack, S.R. (1987). Electromechanical potentials in cortical bone. II. Experimental analysis. *J. Biomech.* **20**: 271.

Schreiber, A.B., Winkler, M.E. and Derynck, R. (1986). Transforming growth factor 2: a more potent angiogenic mediator than epidermal growth factor. *Science* **232**: 1250–1253.

Sih, G.C.M. and Chen, E.P. (1981). *Cracks in Composite Materials.* M. Nijhoff, Hague/Boston.

Simard, C.P., Spector, S.A. and Edgerton, V.R. (1982). Contractile properties of rat hind limb muscles immobilized at different lengths. *Exp. Neurol.* **77**: 467–482.

Sisken, B.F. and Smith, S.D. (1975). The effects of minute direct electrical currents on cultured chick embryo trigeminal ganglia, *J. Embroyl. Exper. Morph.* **33**: 29.

Skalak, R. and Fox, C.F. (eds). (1988). *Tissue Engineering.* Alan Liss, New York.

Smith, S.D. (1974). Effects of electrode placement on stimulation of adult frog limb regeneration. *Ann. N. Y. Acad. Sci.* **238**: 500.

Snowden, J.M. and Cliff, W.J. (1985). Wound Contraction. Correlations Between the Tension Generated by Granulation Tissue, Cellular Content and Rate of Contraction. *Quart. J. Exp. Physio.* **70**(4): 539–548.

Sobin, S.S., Tremer, H.M., Hardy, J.D. and Chiody, H. (1983). Changes in arterioles in acute and chronic pulmonary hypertension and recovery in the rat. *J. Appl. Physiol.* **55**: 1445–1455.

Spector, S.A., Simrad, C.P., Fournier, M., Sternlicht, E. and Edgerton, V.R. (1982). Architectural alterations of rat hind-limb skeletal muscles immobilized at different lengths. *Exp. Neurol.* **76**: 94–110.

Stanley, J.C., Burkel, W.E., Graham, L.M. and Lindbald, B. (1985). Endothelial cell seeding of synthetic vascular prostheses. *Acta Cir. Scand. Suppl.* **529**: 17.

Stone, J.L., Snyder, B.D., Hayes, W.C. and Strang, G.L. (1984). Three Dimensional Stress Morphology Analysis of Trabecular Bone, *Trans. Orthop. Res. Soc.* **30**: 199.

Suwa, N. (1982). Myocardial structure of hypertrophied hearts. *Japan. Circ. J.* **46**(9): 995–1000.

Suwa, N. and Takahashi, T. (1971). *Morphological and morphometrical analysis of circulation in hypertension and ischemic kidney.* Urban & Schwarzenberg, München.

Takamizawa, K. and Hayashi, K. (1987). Strain energy density function and uniform strain hypothesis for arterial mechanics. *J. Biomech.* **20**: 7–17.

Tezuka, F. (1975). Muscle fiber orientation in normal and hypertrophied hearts. *Tohoku J. Exp. Medicine* **117**: 289–297.

Thoma, R. (1893). *Untersuchungen über die Histogenese und Histomechanik des Gefässystems.* Enke, Stuttgart.

Thomas, K.A., Rios-Candelore, M., Gimenez-Gallego, G., DiSalvo, J., Bennett, C., Rodkey, J. and Fitzpatrick, S. (1985). Pure brain derived acidic growth factor is a potent angiogenic vascular endothelial cell mitogen with sequence homology to interleukin 1. *Proc. Natl. Acad. Sci. U.S.A.* **82**: 6409–6413.

Vaishnav, R.N., and Vossoughi, J. (1987). Residual stress and strain in aortic segments. *J. Biomech.* **20**: 235–239.

Weiss, P.A. (1955). Nervous system (neurogenesis). In *Analysis of Development,* (B.H. Willier, P.A. Weiss and V. Hamburger, eds.), Saunders, Philadelphia, pp. 346–401.

Whitehouse, W.J. (1974). The Quantitative Morphology of Anisotropic Trabecular Bone, *J. Microscopy,* **101**: 153–168.

Whitehouse, W.J. and Dyson, E.D. (1974). "Scanning Electron Microscope Studies of Trabecular Bone in the Proximal End of the Human Femur," *J. Ana.* **118**: 417–444.

Wolff, J. (1869). Über die bedeutung der Architektur der spongiösen Substanz. *Zentralblatt für die medizinische Wissenschaft.* **VI. Jahrgang**: 223–234.

Wolff, J. (1870). Über die innere Architektur der Knochen und ihre Bedeutung für die Frage vom Knochenwachstum. *Archiv für pathologische Anatomie und Physiolgie und für klinische Medizin* (Virchovs Archiv) **50**: 389–453.

Wolff, J. (1892). *Das Gesetz der Transformation der Knochen.* Hirschwald, Berlin.

Yannas, I.V. and Burke, J.F. (1980a). Design of an artificial skin. I. Basic design principles. *J. Biomed. Mater. Res.* **14**: 65–81.

Yannas, I.V., Burke, J.F., Gordon, P.L., Huang, C. and Rubenstein, R.H. (1980b). Design of an artificial skin. II. Control of chemical composition. *J. Biomed. Mater. Res.* **14**: 107–131.

Yannas, I.V., Burke, J.F., Orgill, D.P. and Skrabut, E.M. (1982). Wound tissue can utilize a polymeric template to synthesize a functional extension of skin. *Science* **215**: 174–176.

Yasuda, I. (1974). Mechanical and electrical callus. *Ann. N.Y. Acad. Sci.* **238**, 457–465.

Yeh, C.K. and Rodan, G.A. (1984). Tensile forces enhance prostaglandin *E* synthesis in osteoblastic cells grown on collagen ribbons. *Cal. Tiss. Int.* **36**: S67.

Yue, X., van der Lei, B., Schakenraad, T.M., van Oene, G.H., Kuit, J.H., Feijen, T. and Wildevuur, C.R.H. (1988). Smooth muscle seeding in biodegradable grafts in rats: A new method to enhance the process of arterial wall regeneration. *Surgery* **103**: 206.

Zamora, J.L., Navarro, L.T., Ives, C.L., Weilbaecher, D.G., Gao, Z.R. and Noon, G.P. (1986). Seeding of arteriovenous prostheses with homologous endothelium. *J. Vasc. Surg.* **3**: 860.

We discussed animals. But biomechanics applies equally well to plants. BAMBOO by Ku An, 顧安. Ink on silk, 123 × 53 cm wall scroll, in National Palace Museum, Taipei. Ku was a native of East Whei, Kiangsu. He lived in 14th century, Yüan Dynasty.

Author Index

Subject Index

CPSIA information can be obtained at www.ICGtesting.com
Printed in the USA
BVOW04*1224060714

358274BV00004B/22/A